CONCISE ENCYCLOPEDIA
OF CONSTRUCTION
TERMS AND PHRASES

CONCISE ENCYCLOPEDIA OF CONSTRUCTION TERMS AND PHRASES

EDITED BY

KARL F. SCHMID, PE, LEED AP

MP MOMENTUM PRESS

MOMENTUM PRESS, LLC, NEW YORK

Concise Encyclopedia of Construction Terms and Phrases
Copyright © Momentum Press®, LLC, 2013.

First published by Momentum Press®, LLC
222 East 46th Street, New York, NY 10017
www.momentumpress.net

ISBN-13: 978-1-60650-618-9 (hardback, case bound)
ISBN-10: 1-60650-618-8 (hardback, case bound)
ISBN-13: 978-1-60650-619-6 (e-book)
ISBN-10: 1-60650-619-6 (e-book)

DOI: 10.5643/9781606506196

Cover design by Jonathan Pennell
Interior design by Exeter Premedia Services Private Ltd. Chennai, India

10 9 8 7 6 5 4 3 2 1

Printed in the United States of America

Contents

PREFACE

This lexicon, so named because it comes with numerous illustrations, was assembled to serve primarily contractors, but also designers, people who work on their own houses, students and apprentices, and others who are interested in the building trades. It was assembled by one person and so, undoubtedly, shows some preferences; there may be some topics with more information that you want and others that have been omitted. I hope not, but forgive me if this is so.

Nothing in this lexicon is original; it's a compilation of existing words and phrases that were obtained by searching the web's many glossaries, dictionaries, text books, trade magazines, and the like. The same sources were used where I went beyond providing just the simple meaning of a word, as was Wikipedia, the free encyclopedia. All the illustrations were found on either Google Images or Bing Images (listed as free). So you may ask, "Why should I purchase this book?" to which I give the following reasons: 1) Going through the pages may stimulate your thinking and cause you to think out of the proverbial box; 2) Many of the definitions have been expanded upon as a result of considerable research that you may not wish to do, and in some cases I have thrown in my own experience; and, 3) Whenever I thought words were not enough illustrations were added; again available on line but my time is cheaper than yours. You may also find the list of organizations that affect construction useful, perhaps not now, but one day when a unique project comes along, one that requires special knowledge.

The words are listed alphabetically, A – Z. There are a few places where I varied this if it seemed to make sense to keep topics together, but these cases are very few and should be obvious to you. Entries known to be trademarks or service marks are so labeled and, to the best of my knowledge properly treated, e.g. Thermoply™. No entry in the lexicon, however, should be regarded as affecting the validity of any trademark or service mark.

As already noted all pictorial illustrations were found on either Google Images or Bing Images, the majority from Google. They were selected chiefly for their ability to enhance and supplement the written word, a picture is worth 1,000 words has proven true numerous times. The abbreviations were gathered from hither and you, your own experience may have them in lower case whereas I have them in upper case, and vice-versa, they are, however, used as shown, but your way is also correct, it's an abbreviation.

Some words have various spellings, e.g., weather head and weatherhead and nonbearing and non-bearing. Where I found such situations I tried to point them out and do my best at discerning the most common usage. And some words are used to describe different things, in which case I attempted to show all meanings.

Lastly, to reiterate, nothing in this lexicon is original, it's a compilation of existing words and phrases that were obtained by searching the web's many glossaries, dictionaries, text books, trade magazines, and the like. Indeed so many references were used that space does not logically allow for me to list all of them, so I shall list none of them; please know that I am thankful to all.

Karl F. Schmid

KEYWORDS

Construction terminology, construction abbreviations, architectural terminology, construction math formulas, construction dictionary, construction practices, construction trade associations, construction government agencies, construction industry societies, building services, building inspection

ABBREVIATIONS

The abbreviations listed here are ones that you may see on drawings or other contract documents.

2/10 Rule: This rule states that the top of a flue must be at least two feet higher than any portion of the roof part within ten feet. Building codes usually have this rule or one similar to it in them.

4/1 Rule: A rule for safe placement of a ladder. The 4/1 rule states that for every four feet of working ladder length, the base of the ladder should be one foot out from the top support point.

4-inch Center: With regard to faucets, when the center of one handle to the center of the other handle is four inches with the spout between them. It is also referred to as 4" spread.

A 4-inch lavatory sink faucet.

30 lb Felt: 30 lb felt was the standard underlayment for quality metal roofing systems and is still used but has been replaced in many areas by the newer synthetic underlayments. It is typically made of a polyester fleece and infused with tar.

80/20 Rule (Pareto's Rule): The 80/20 Rule means that in anything a few (20%) are vital and the many (80%) are trivial. In Pareto's case, it meant 20% of the people owned 80% of the wealth. In Juran's (Joseph Moses Juran (December 24, 1904 to February 28, 2008) was a management consultant and engineer-remembered principally for his work on quality and quality management) initial work, he found that 20% of the defects caused 80% of the problems. You can apply the 80/20 Rule to many situations, from the science of management to the physical world.

ABS: Acrylonitrile butadiene styrene. A type of plastic used to make plumbing pipe.

AC or ac: Alternating current. The type of electrical current provided to most homes.

AC or A/C: Air conditioner or air conditioning.

ACM: Additional construction management.

ADA: The Americans with Disabilities Act.

A–E or A/E: An architect and engineering firm. Typically used to identify the principal designer(s) on a project.

AFF: Above finished floor.

AND: Air Force–Navy Aeronautical Design Standards.

APP: Atactic polypropylene. A group of high-molecular-weight polymers formed by the polymerization of propylene. It is used in the torch-down type of modified bitumen roof systems.

AQL: Acceptable quality level. This is usually defined as the worst case quality level, in percentage or ratio, that is still considered acceptable.

ASD: Allowable stress design. A structural design method by which a structural element is designed so that the unit stresses computed under the action of working or service loads do not exceed specified allowable values. See Working stress design and Elastic design.

BBC: Basic building code. A minimum model regulatory code for the protection of public health, safety, welfare, and property by regulating and controlling the design, construction, quality of materials, use, occupancy, location, and maintenance of all buildings and structures within a jurisdiction.

BCTMP: Bleached chemi-thermomechanical pulp.

BHP or bhp: Brake horsepower. The actual amount of horsepower consumed by a pump as measured on a dynamometer.

BIM: Building information modeling. A 3D, object-oriented approach to computer-aided architectural design. BIM enables data for manufacturer's details to be imported right into project design and presents 3D models of products in place in building. BIM also provides access and ability to add to detailed imagery and information to everyone involved in the building process and building operations after project completion.

BIVP: Building-integrated photovoltaics. A term used for products, such as commercial glazing, with solar-power collection cells built in.

BSP: British standard pipe.

BTU or btu: British thermal unit. The amount of heat required to raise the temperature of one pound of water one degree Fahrenheit.

BUR: Built-up roofing. See Built-up roof.

C: Centigrade. Relating to, conforming to, or having a thermometric scale on which the interval between the freezing point of water and the boiling point of water is divided into 100° with 0° representing the freezing point and 100° the boiling point.

CAD: Computer-aided drafting for which there are numerous commertially available programs on the market, e.g., Autocad®. Also, **Computer-aided design**.

CCA: Chromated copper arsenate. A pesticide that is forced into wood under high pressure to protect it from termites, other wood-boring insects, and decay caused by fungus.

CCF or ccf: 100 cubic feet.

CG&E: Chopped glass and emulsion. A roof coating that consists of asphalt or clay emulsion and glass fiber reinforcement. The glass fiber comes in rope form and is mechanically chopped into small pieces and then mixed with the emulsion at the end of the spray gun so that the mixture is complete by the time the surfacing hits the top of the roof. Standard mixture is 9 gallons of emulsion and 3 pounds of glass fiber for every 100 square feet (36.5 liters of emulsion and 1.5 kg of chopped glass for every 10 square meters). The CG&E coating is then usually surfaced with a fibered aluminum roof coating at a rate of 1.5 gallons per 100 square feet (6 liters per 10 square meters).

CFM or cfm: Cubic feet per minute. A rating that expresses the amount of air a blower or fan can move. The volume of air (measured in cubic feet) that can pass through an opening in 1 min. Also, Certified Facility Manager (CFM) by the International Facilities Management Association (IFMA).

CI: Cast iron.

CMU: Concrete masonry unit.

CPA: Copolymer alloy. CPA is made by mixing at least one more monomer type than is needed to create a polymer and then polymerizing to create a polymer that is composed of all the monomers present. This cumbersome definition is given because some polymers are made with just one monomer type and some are made from two polymer types. Also, Certified Public Accountant.

CPE: Chlorinated polyethylene resins and elastomers. CPE exhibit excellent physical and mechanical properties, including resistance to chemicals, oils, heat and weather, low-temperature performance, compression-set resistance, flame retardancy, high filler acceptance, tensile strength, and resistance to abrasion.

CPM: Critical Path Method. A project modeling technique. CPM is commonly used with all forms of projects, including construction, aerospace and defense, software development, research projects, product development, engineering, and plant maintenance, among others. Any project with interdependent activities can apply this method of mathematical analysis.

cps: Cycles per second.

CPVC: Chlorinated polyvinyl chloride. A heat-resistant and low-combustibility plastic that is typically used in water supply. This is also abbreviated as CPvCL.

CRF: Condensation resistance factor. A rating of a window's ability to resist condensation. The higher the CRF, the less likely condensation is to occur.

CRM: Customer relationship management. A computerized system for tracking all contacts with customers and prospects.

CSM: ASTM designation for chlorosulfonated polyethylene. See CSPE.

CSPE: Chlorosulfonatedpolyethylene. A product of the chemical modification of polyethylene by chlorine and sulfur dioxide. It is resistant to fire, oil, and the action of microorganisms and exhibits good adhesion to various surfaces.

Chlorosulfonated polyethylene is superior to other rubbers in its resistance to the effects of ozone and inorganic acids, such as chromic, nitric, sulfuric, and phosphoric acids, as well as to the effects of concentrated alkalies, chlorine dioxide, and hydrogen peroxide. It is resistant to light, is impermeable to gas, and has good dielectric properties.

Chlorosulfonated polyethylene is used in the production of industrial and household goods and of anticorrosion coatings to be applied by the rubberizing method. It is also used as a film-forming agent in varnishes and paints for the preservation of wood, metal, and reinforced concrete and as a base for adhesives and hermetic sealants.

CSST: Corrugated stainless steel tubing. Also known as "TracPipe." It has a polyethylene jacket.

cu. ft.: Cubic feet.

dB or db: Decibel. A unit for expressing the relative intensity of sounds. A zero-decibel reference is stipulated to be the lowest level audible to the human ear; the speaking voice of most people ranges from 45 to 75 decibels. At 130 decibels most people feel pain.

DBH: Breast height of a tree. DBH is considered to be four and one-half feet above the ground level.

DC: Direct current. A unidirectional electrical current that is usually provided by sources such as batteries, thermocouples, solar cells, and commutator-type electric machines of the dynamo type.

Dia: Diameter.

DOB: Department of Buildings. Often DOB is the local agency that issues building permits. Another common name for this agency is **Department of Building[s] and Safety (DBS).**

DW: Distilled water.

DWV: Drainage, waste, and vent. Pipes in a plumbing system that remove waste water; the section of a plumbing system that carries water and sewer gases out of a home. Not for drinking water.

ECO: Energy cut off. A safety device that is designed to shut power off to the water heater and prevent high temperature.

EF: Energy factor. A measure of the overall efficiency rating of the water heater based on the model's recovery, efficiency, stand-by loss and energy input.

EIBLL: Environmental Intervention Blood Lead Level. A confirmed concentration of lead in whole blood equal to or greater than 20 µg/dL for a single venous test or of 15–19 µg/dL in two consecutive tests taken at least 3 months apart.

EIFS: Exterior Insulation and Finishing System. Also known as synthetic stucco or Dryvit. An exterior finish for a building or home composed of polystyrene foam covered with a synthetic stucco. This type of stucco (in contrast to traditional, porous cement-based stucco) is waterproof and is sprayed on.

EJ: Expansion joint.

EMT: Electrical metal tubing. A thin wall galvanized steel pipe that is used to carry electrical or other types of conductors.

EOD: Edge of deck.

EOJ: Abbreviation for edge of joist.

EOS: Abbreviation for edge of slab.

EPL: Elevated Blood Level. An excessive absorption of lead that is a confirmed concentration of lead in whole blood of 20 µ/dL (micrograms of lead per deciliter of whole blood) for a single venous test or of 15–19 µ/dL in two consecutive tests taken 3–4 months apart.

ERP: Enterprise resource planning. A computerized system that is used to manage all aspects of a company's operations.

ERW: Electric resistance welded.

EVT: Equiviscous temperature. A measure of viscosity used in the tar industry, equal to the temperature in degrees Celsius at which the viscosity of tar is 50 s as measured in a standard tar efflux viscometer.

EVT is the temperature at which a bitumen attains the proper viscosity for use in built-up roofing. There is usually a 25°F variance permitted above and below the recommended EVT. The EVT is measured in application equipment just prior to application using a standard thermometer or it can be measured just after application using a laser thermometer.

Here are some sample EVT temperatures:

Asphalt—ASTM D 312

Asphalt Type	Mop Application	Mechanical Spreader
Type I		
Dead Level	350	375
Type II		
Flat	400	425
Type III		
Steep	425	450
Type IV		
Spec. Steep	450	475

Temperatures in °F
Temperatures may vary ± 25°F

E-Modulus: Modulus of elasticity.

F: Fahrenheit. Noting, pertaining to, or measured according to a temperature scale (Fahrenheit scale) in which 32° represents the ice point and 212° the steam point.

FAI: Fresh air intake.

FBM or fbm: Feet, board, measure.
1 Fbm (foot board measure) is equal to a board 12" wide × 12" long × 1" thick.
Fbm is calculated as thickness (inches) × width (inches) × length (feet)/12.
Thickness and width used in the fbm calculation are nominal (2 × 4, 2 × 6) although the finished size is 1.500 × 3.500, 1.500 × 5.500, etc. A 2 × 4 × 8" would equal 5.33 fbm (2 × 4 × 8/12 = 5.33).

FC: Field cut.

FF & E: Furniture, furnishings, and equipment.

FHT: Female hose thread, for example, the end of a garden hose.

FIP: Female iron pipe connection. Standard internal threads on pipe fittings, usually a

braided hose that connects a faucet or toilet to the water supply stop valve. The same function as a riser, but much more flexible and easier to install. They are most often Stainless Steel Braided hose or PVC/polyester re-enforced hose.

FMPX: Female pipe swivel connection.

Female pipe swivel connection.

FNPT: The female pipe thread, in which the threads are on the inner side of the connection fitting.

FOB or f.o.b: Free on board. A commercial term that signifies a contractual agreement between a buyer and a seller to have the subject of a sale delivered to a designated place, usually either the **place of shipment** or the **place of destination**. Therefore, a shipment **f.o.b. shipping point** requires the seller to bear the expense and the risk of putting the subject of the sale into the possession of the carrier, but the duty to pay the transportation charges from f.o.b. point is on the buyer. Where the shipment is **f.o.b. destination point**, the seller is responsible to bear the transportation charges and the risk of transport until the buyer point of destination.

F.o.b. is not just a pricing agreement as to who shall bear the cost of transportation but is also a delivery term designating where title and risk of loss pass.

When in addition to designating a delivery point, the agreement specificies a vessel, truck, or other vehicle, the seller must also load the goods aboard the vessel or vehicle at the seller's expense.

The shipper must assume the expense of loading the goods onto the truck as well the expenses and risk for shipping the goods to the FOB destination.

F.O.I.A.: Freedom of Information Act. A federal law that, with specified exceptions, requires that documents and materials generated or held by federal agencies be made available to the public and establishes guidelines for their disclosure.

fps: Feet per second.

fsp: Fire standpipe.

ft: Foot.

FVI: Flammable vapor ignition.

G or g: Acceleration due to gravity. g is the acceleration of gravity 9.8 (m/s^2) or the strength of the gravitational field (N/kg) (which it turns out is equivalent). G is the proportionality constant 6.67×10^{-11} (Nm^2/kg^2) in Newton's law of gravity. On the other hand, the force of gravity, or $F = mg$, at the surface of the earth, or $F = GMm/r^2$ at a distance r from the center of the earth (where r is greater than the radius of the earth).

When there is an earthquake, the forces caused by the shaking can be measured as a percentage of gravity, or percent g.

G-90: A coating weight for galvanized sheet metal, 0.90 ounces of zinc per sq. ft., measured on both sides of the sheet.

G-type joist girder: A type of joist girder where joists are located at panel points where diagonal webs intersect the top chord only.

ga: Gauge, the wall thickness of tubing.

gal: Gallon.

GFCI/GFI: Ground-Fault Circuit Interrupter. See **Ground-Fault Circuit Interrupter.**

GFE: Government furnished equipment.

GPF: Gallons per flush. A measure that is used in toilets and urinals. Current law requires a maximum of 1.6 gpf. Older styles were usually 3.5 gpf.

GPG: Grains per gallon. A measurement of the amount of dissolved material in water. One grain per gallon equals 17.1 ppm.

GPH or gph: Gallons per hour. A measure of flow rates.

GPM or gpm: Gallons per minute. A unit of measurement by which flow rates of faucets and showerheads are measured and regulated.

GPS or gps: Gallons per second.

HAP: Hazardous air pollutant. Pollutants that are known or suspected to cause cancer or other serious health effects, such as reproductive effects or birth defects, or adverse environmental effects.

HDPE: High-density polyethylene.

HEPA filter: High-efficiency particulate air. A high-efficiency particulate air-filter designed to remove lead-contaminated dust. Filters meeting the HEPA standard have many applications, including use in medical facilities, automobiles, aircraft, and homes. The filter must satisfy certain standards of efficiency set by the United States Department of Energy (DOE). To qualify as HEPA by US government standards, an air filter must remove 99.97% of all particles greater than 0.3 micrometre from the air that passes through. A filter that is qualified as HEPA is also subject to interior classifications.

A high-efficiency particulate air-filtered vacuum designed to remove lead-contaminated dust is a **HEPA vacuum**.

h.p.: Horsepower. A unit of power equal in the United States to 746 watts and nearly equivalent to the English gravitational unit of the same name that equals 550 foot-pounds of work per second.

HPA: High-powered amplifier.

HPF: Highest possible frequency. Also, high-power field.

HR: Hot runner. A premium grade of PVC that flows easily and smoothly into a mold. It also increases the range from the melting point to the burning point while in the molding process. Also, Human Resources, the department that deals with personnel matters in companies large enough for such a department.

hr.: Hour.

HVAC: Heat, ventilation, and air conditioning. Heating, ventilating, and air-conditioning systems that are used to provide thermal comfort and ventilation are commonly referred to as HVAC systems.

HW: High water. Also, highway. Also, hot water.

ID: Inside diameter. The diameter measurement from the inside of a pipe. Commonly used for a sizing pipe.

INR: Impact noise rating. A rating that is expressed by a single number. INR is a rough measure of the effectiveness of a floor's construction in providing isolation against the noise of impacts; in general, the higher the number, the greater the effectiveness.

IPS or I.P.S.: Iron pipe size (outside diameter or OD). Same as National Pipe Straight threads standard. Also, an internal pipe swivel connection (female).

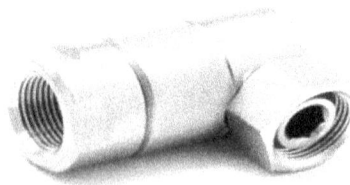

Internal pipe swivel (female).

JBE: Joist bearing elevation.

K: Kelvin, kilogram, knot, kip, potassium plus others. Also, in the legal sector an abbreviation for contract or contracts.

ka: cathode.

KB: Kilobyte.

Kc/sec: Kilocycles per second.

KD: Wood dried in a kiln with the use of artificial heat to a specified moisture content. Also, knocked down, e.g., an unassembled window or door.

kg: Kilogram.

KGPS: Kilograms per second.

kHz: Kilohertz.

kPa: Kilopascal. A metric unit for pressure. 100 kPa = one atmosphere.

kph: Kilometers per hour.

KSI or ksi: Kips per linear foot. 1000 pounds per square inch.

KSF or ksf: Kips per square foot. 1000 kips per square foot.

KW or kw: Kilowatt. A measure of the rate of supply of energy or power, equal to 1000 watts or 3,412 BTU per hour. See Kilowatt.

Kwhr or kwh: kilowatt-hour.

LB or lb: Pound. A number of different definitions have been used, the most common today being the international avoirdupois pound, which is legally defined as exactly 0.45359237 kilograms: 16 avoirdupois ounces.

LDO: Lint, dust, and oil.

LEED: Leadership in Energy and Environmental Design. A system established by the United States Green Building Council (USGBC) to define and measure "green" buildings.

LH: Left hand.

LPG: Liquified petroleum gas.

LPP: Liquid propane.

LRF: Lumber recovery factor. A measurement of lumber recovery or yield from a quantity of log volume. LRF is most commonly expressed as thousand board feet per cubic meter or board feet per cubic feet. A true measure of LRF is the finished shippable lumber per unit of logs delivered.

LRFD: Load and resistance factor design. A method of proportioning structural members such that no limit state is exceeded when all appropriate load combinations have been applied.

LRRP: The lead paint renovation, repair and painting program established by the Environmental Protection Agency for pre-1978 homes to address health and safety issues associated with lead paint. Firms involved in such projects must be trained and certified in lead-safe work procedures. Also referred to as RRP. Information is available on the EPA's web site.

LVL: Laminated veneer lumber. See Laminated veneer lumber, beam, and Laminated veneer lumber, panel.

LVLC: Laminated veneer lumber core. A door manufactured by laminating veneer with all grain laid-up parallel. It can be manufactured by using various species of wood fiber in various thicknesses.

LW: Low water.

Ma: One million years ago (Megannum).

MBF: Thousand board feet.

MCE: Maximum Credible Earthquake. An earthquake that is about 50% higher than the Design Base Earthquake (DBE).

MCM: Micro circular mills. American Standard Gauge unit of measure for wire sizes that are larger than four aught. See American Standard Gauge.

MDF: Medium density fiberboard. A type of pressed fiberboard often used in cabinet building.

MEC: Model Energy Code, established by Energy Policy Act of 1992 to serve as a baseline for state energy codes. Although referenced in some state codes, it has been succeeded by the International Energy Conservation Code (IECC).

MERV: Minimum efficiency reporting value, a value that is assigned to air filters installed in the ductwork of HVAC systems. The American Society of Heating, Refrigerating and Air-Conditioning Engineers (ASHRAE) developed this measurement method. MERV ratings (ranging from a low of 1 to a high of 20) also allow the comparison of air filters made by different companies.

The higher the MERV, the more efficient the air filter is at removing particles. At the lower end of the efficiency spectrum, a fiberglass panel filter may have a MERV of 4 or 5. Alteration projects frequently require that the HVAC system of the building be protected from construction dust with a MERV 8 filter. At the higher end, a MERV 14 filter is typically the filter of choice for critical areas of a hospital (to prevent transfer of bacteria and infectious diseases). Higher MERV filters are also capable of removing higher quantities of extremely small contaminant.

A higher MERV filter creates more resistance to airflow because the filter material is more dense; the filter media becomes denser as filter efficiency increases. For the cleanest air select the highest MERV filter that the HVAC unit's fan can drive air through it.

Care should be taken when considering filters that incorporate an electrostatic charge. While they may offer a reasonable MERV value, these filters will actually drop in efficiency as the filter loads with contaminant.

Mfbm: Thousand board feet.

MLW: Mean low water.

MMBF: Million board feet.

MMSF: A unit of measure for medium density fiberboard (MDF) equal to one million square feet on a 3/4 inch basis. See MSF.

MO: Masonry opening.

MSF: A unit of measure for medium density fiberboard (MDF) and plywood equal to 1,000 square feet, on a 3/4 inch basis for MDF and on a 3/8 inch basis for plywood.

MSR: Machine stress rated. Machine stress rated lumber is dimension lumber that has been evaluated by mechanical stress-rating equipment. The stress-rating equipment measures the stiffness of the material and sorts it into various modulus of elasticity (E) classes.

NC: National Coarse. A measurement of threads per inch on a tap.

NDS: National Design Specification for Wood Construction. A primary code for wood construction. Local codes that are backed by law may differ from the NDS.

NEC: National Electrical Code. A set of rules governing safe wiring methods. Local codes that are backed by law may differ from the NEC in some ways.

NEF: National Extra Fine. A measurement of threads per inch on a tap.

NF: National Fine. A measurement of threads per inch on a tap.

NH: No hub.

NIC: Not in contract.

NIMBY: Not in my backyard.

NOM: Nominal. Nominal usually refers to the inside diameter of trade sizes of copper pipe and some CPVC pipes. For example, a 1/2" NOM Comp. Fitting is actually 5/8" because 1/2" pipe has an OD of 5/8". But this is not so for iron pipe connections. See IPS.

NPS: National pipe straight threads standard. NPS threads are the same as IPS.

NPSP: Non-point source pollution, e.g., nutrients, sediments, toxic substances, and pathogens that degrade waterways. NPSP occurs mainly through storm water runoff.

NPSHA: Net Positive Suction Head Available. The suction head that is available to prevent cavitation of the pump.

It is defined as atmospheric pressure + gage pressure + static pressure − vapor pressure − friction loss in the suction piping.

NPT: National Pipe Taper. A measurement of threads per inch on the threaded pipe.

NTS: Not to scale.

OC or o/c or O.C.: A term that is used whenever measurements are taken from the center of one member to the center of the adjacent member. For example, it is the term used to define the spacing between studs, joists, rafters, etc.

OD: Outside diameter. The diameter of a pipe measured from the outside edge.

OEM: Original equipment manufacturer.

OSB: Oriented Strand Board. A decking made from wood chips and lamination glues.

OSM: Outside measurement.

PB: Polybutlene. A type of plastic plumbing pipe made from polybutylene or PB.

PDMS: Polydimethylsiloxane. PDMS belongs to a group of polymeric organosilicon compounds, which are commonly referred to as silicones. PDMS is the most widely used silicon-based organic polymer and is particularly known for its unusual rheological (or flow) properties.

PE: Professional engineer. Also, polyethylene plastic material. Also, plain end pipe.

PEX: Cross-linked polyethylene. PEX tubing is commonly used for hydronic radiant floor heat, but increasingly also used for water supply lines. PEX is stronger than PE.

pH: A measure of the acidity or alkalinity of a solution, numerically equal to 7 for neutral solutions, increasing with increasing alkalinity and decreasing with increasing acidity. The pH scale commonly in use ranges from 0 to 14.

PITI: Principal, interest, taxes, and insurance. The four major components of monthly housing payments.

PMR: Protective membrane roofing. A roofing membrane installed under the roof insulation as a protective measure against temperature differentials and UV light.

Plc: Programmable logic controller or programmable controller. A digital computer used for to automate electromechanical processes, e.g., machinery on assembly lines, amusement rides, and light fixtures.

PP: Polypropylene pipe.

ppm: Parts per million. A way of expressing the amount of a substance in a liquid in the terms milligrams (of substance) per liter of liquid.

ppt: Parts per thousand. A way of expressing the amount of a substance in a liquid in the terms grams (of substance) per liter of liquid.

Pr: Pressure regulator. A pressure regulator is usually required when water pressure ever exceeds 80 PSI on potable water supplies inside a structure.

psi: Pounds per square inch, or more accurately pound-force per square inch, is a unit of stress or pressure based on avoirdupois units. It is the pressure resulting from a force of one pound-force applied to an area of one square inch.

psia: Pounds per square inch absolute is used to make it clear that the pressure is relative to a vacuum rather than the ambient atmospheric pressure. To obtain the absolute pressure (in psia), you must add the gage pressure (psig) and the atmospheric pressure (14.7 psi at sea level). For example, a tire that is pumped up to 60 psi (psig) above atmospheric pressure will have an absolute pressure of 60 + 14.7 (atmospheric pressure at sea level) = 74.7 psia.

psig: Pounds per square inch gauge is the pressure that you read on the gauge; it is relative to atmospheric pressure.

psf: Pounds per square foot.

PVC or CPVC: Polyvinyl chloride. A type of plastic used to make elastomeric roof membranes as well as plumbing pipes, fittings, conduit, and fences.

R & D: Reamed and drifted. A pipe that is commonly used in water wells. It has a special, heavy-duty coupling and a guaranteed I.D. clearance. Also, research and development.

RCRA: Resource Conservation and Recovery Act. The Resource Conservation and Recovery Act (RCRA) was enacted in 1976. It is the principal federal law in the United States governing the disposal of solid waste and hazardous waste.

Reman: Remanufactured. Reman is used to describe wood boards that require further manufacturing.

RESFEN: A computer program designed to calculate energy use based on window selection in residential buildings. Created under sponsorship of the U.S. Department of Energy by Lawrence Berkeley Laboratory.

RF: Radio frequency. Technology used with bar code scanners and other input devices in plant and warehouse tracking systems.

RFI: Request for Information. A written request, usually a form, from the contractor to the architect that requests additional information on an issue in the contract specification or drawings. The answer to an RFI can result in a COP if either the architect or contractor believes there is a change in the contract terms.

RFID: Radio frequency identification. A technology that uses electronic tags and labels on products, pallets or carts along with wireless scanners and other devices to automatically track the location of components and products throughout the manufacturing and/or distribution process.

RFP: Request for Proposal. A request for uniform detailed information from prospective professionals who are being screened for a project. An RFP is also a document that is generated by the architect to request a cost proposal for a potential change to the contract. Generally an RFP is used when no determination has been made whether or not the change will be executed.

RO: The rough opening dimension for a door, window, or piece of equipment.

RPM: Revolutions per minute.

RRP: The renovation, repair and painting program established by the Environmental Protection Agency for pre-1978 homes to address health and safety issues associated with lead paint. Firms involved in such projects must be trained and certified in lead-safe work procedures. Also referred to as LRRP. Information is available on EPA's web site.

RTD: A temperature measuring device that measures the change in electrical resistance to determine temperature (resistive thermal device).

S1S2E: Surfaced one-side and two-edges.

S4S: Surfaced four sides.

SBC: Standard Building Code. A minimum model regulatory code for the protection of public health, safety, welfare, and property by regulating and controlling the design, construction, quality of materials, use, occupancy, location and maintenance of all buildings and structures within a jurisdiction.

SBR: Styrene butadiene rubber. The most widely used elastomer for pipe and fitting gaskets worldwide.

High-molecular weight polymers having rubber-like properties, formed by the random copolymerization of styrene and butadiene monomers.

SBS: Styrene butadine styrene copolymer. A plasticiser used in the hot-mop type of modified bitumen roof systems.

High-molecular-weight polymers that have both thermoset and thermoplastic properties, formed by the block copolymerization of styrene and butadiene monomers. These polymers are used as the modifying compound in SBS-polymer-modified asphalt roofing membranes to impart rubber-like qualities to the asphalt.

SBS-modified: Asphalt that has been combined with SBS polymers to increase its elasticity.

SDR or DR: Standard Dimension Ratio or Dimension Ratio. A sewer pipe term that is used to determine the minimum wall thickness of pipe and fittings. SDR = Average Outside Diameter divided by the Minimum Wall Thickness. The lower the SDR, the thicker the wall and thus the stronger the fitting.

Sfpm: Surface feet per minute of a saw.

SFRM: Spray-Applied Fire Resistive Materials.

SG: Specific gravity. Also called relative density. The ratio of the density of a substance to the density of a standard, usually water for a liquid or solid, and air for a gas.

As applied to wood, the ratio of the oven-dry weight of a sample to the weight of a volume of water equal to the volume of the sample at a specified moisture content (green, air-dry, or oven-dry).

SHGC: Solar Heat Gain Coefficient. The fraction of incident solar radiation admitted through a window, both directly transmitted and absorbed and subsequently released inward. SHGC is expressed as a number between 0 and 1. The lower a window's solar heat gain coefficient, the less solar heat it transmits.

The nationally recognized rating method by the National Fenestration Rating Council (NFRC) is for the whole window, including the effects of the frame. Alternately, the center-of-glass SHGC is sometimes referenced, which describes the effect of the glazing alone. Whole window SHGC is lower than glass-only SHGC and is generally below 0.8.

Solar heat gain can provide free heat in the winter but can also lead to overheating in the summer. How to best balance solar heat gain with an appropriate SHGC depends upon the climate, orientation, shading conditions, and other factors.

ENERGY STAR provides simplified guidance on recommended SHGC values.

SI: Système International [d'Unités]. The international system of weights and measures (metric system).

SJ: Slip joint.

Slip joint disconnected   Slip joint complete

Individual pieces sold separately

Slip joint.

SMLS: Seamless pipe.

SMV: Slow moving vehicle.

SPF: Spruce-Pine-Fir. Four species: White Spruce (Piceaglauca), Engelmann Spruce (Piceaengelmanni), Lodgepole Pine (Pinuscontorta), and Alpine Fir (Abieslaciocarpa) comprise the spruce-pine-fir species group.

SPF lumber is a distinctly white wood, with very little color variation between springwood and summerwood. The wood has a bright, clean appearance, ranging in color from white to pale yellow, with a fine straight grain and smooth texture.

SS: Stainless steel.

SSR: Standing Seam Roof. A type of roof system where the deck is attached to clips, which are then attached to the beam or joist. Usually this type of roof system cannot be counted on to provide lateral stability or support to the joist top chord.

ST: A hot rolled structural tee shape with symbol ST that is cut or split from S shapes.

STC: Sound transmission class. The measure of sound stopping of ordinary noise.

STD or std: Standard.

SV: Service victory. A designation for service weight cast iron drainage pipe.

SYP: Southern Yellow Pine. A species group, composed primarily of Loblolly, Longleaf, Shortleaf, and Slash Pines. Various subspecies also are included in the group.

T1-11: Texture 1-11. Sheets of wood siding, textured with a series of evenly spaced vertical grooves.

T & C: Threaded and Coupled. Some cast iron pipe is sold threaded with a coupling attached.

T & G: Tongue and groove. A joint made by a tongue (a rib on one edge of a board) that fits into a corresponding groove in the edge of another board to make a tight flush joint. Typically, the sub-floor plywood is T & G.

T & M: Time and Materials. It is an agreement between the owner and the contractor that provides for payment based on the contractor's actual cost for labor, equipment, materials, and services plus a fixed add-on amount to cover the contractor's overhead and profit.

T & P valve: Temperature and pressure valve. A valve that releases water pressure when temperature and pressure exceed a preset limit.

T & S: Tub and shower.

TBE: Threaded both ends.

TCG: Triple chip grind. A type of saw blade that provides versatility for very clean cuts whether ripping or crosscutting, and for precision smoothness, certification program.

TDS: Total dissolved solids. A measure of the amount of substances dissolved in a liquid. Often quoted in PPM or PPT.

THHN: A thermoplastic-insulated, nylon-jacketed conductor designed for use in dry locations and an operating temperature of up to 90°C.

TJI or TJ: A manufactured structural building component resembling the letter "I," used as floor joists and rafters. I-joists include two key parts: flanges and webs. The flange of the I joist may be made of laminated veneer lumber or dimensional lumber, usually formed into a 1 1/2-inches width. The web or center of the I-joist is commonly made of plywood or oriented strand board (OSB). Large holes can be cut in the web to accommodate ductwork and plumbing waste lines. I-joists are available in lengths up to 60-inches long.

TO: Threads only.

TOS: Top of steel elevation.

TPA: A Thermoplastic Tri-Polymer Alloy Single Ply Roof Membrane System.

TPD: Tons per day

TPE: Thermal Plastic Elastomer. Also called santoprene. It is a replacement for rubber.

TPI: Threads per inch.

TPO: Thermoplastic Olefin. Single-ply roofing membranes.

UBC: Uniform Building Code. The UBC was first published in 1927 by the International Council of Building Officials. Updated editions of the code were published approximately every 3 years until 1997. The UBC was then replaced in 2000 by the new International Building Code (IBC) that was published by the International Code Council (ICC). The ICC was a merger of three predecessor organizations that published three different building codes: International Council of Building Officials (ICBO) Uniform Building Code; Building Officials and Code Administrators International (BOCA), the BOCA National Building Code; and, Southern Building Code Congress International (SBCCI) Standard Building Code.

The new ICC is intended to provide consistent standards for safe construction and eliminate differences between the three different predecessor codes. It is primarily used in the United States.

ULF: Ultra low flush. A widely used description of 1.6 gpf or less toilets.

UL label: Underwriters Laboratory seal of approval based on accepted testing methods and results.

UNC: Unified national coarse thread. A standard kind of coarse straight thread that is used on fittings, nuts, and bolts. It is not used for pipes.

UNF: Unified national fine thread. A standard kind of fine straight thread that is used on fittings, bolts, and bolts. It is not used for pipes.

Uni-flex: One piece stop and riser combination.

UNO: Unless noted otherwise.

UPC: Uniform Plumbing Code. The handbook for plumbing installation used by IAPMO. Also, Uniform Product Code. The governing body for bar code implementation. No relation to the Uniform Plumbing Code organization.

UV: Ultraviolet and also ultraviolet light Insolation (incoming solar radiation) in the UV wavelength that over time damages many materials exposed to the sun.

VCP: Verified Clay Pipe.

VOC: Volatile Organic Compound. Organic chemicals and petrochemicals that emit vapors while evaporating. In paints, VOC generally refers to the solvent portion of the paint that, when it evaporates, results in the formation of paint film on the substrate to which it was applied. See **Volatile organic compounds (VOC).**

WC: Water closet (toilet).

WH: White finish.

WWF: Welded wire fabric.

XCM: Extended Services CM. A form of Construction Management (CM) where other services such as design, construction, and contracting are included with Additional Construction Management (ACM) services provided by the Construction Manager.

XH: Extra heavy.

XL: Extra large.

XLG: Extra long.

XS: Extra small.

YBP: Years before present.

A Concise (Oft Times Not So Concise), Illustrated, Lexicon, of Construction Terms and Phrases

A la carte real estate service: Transactions that are rendered one at a time instead of a commission-based, full service relationship.

A-frame design: A style of house incorporating a high, peaked roof. It has an open and airy interior, featuring open ceiling rafters.

A-valve: A manual gas shut-off valve.

Abacus: A thick square or rectangular plate of any size forming the top of a column.

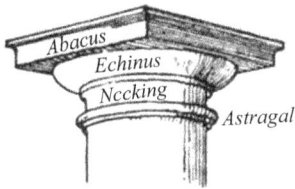

The top of a column.

Also, an oblong frame with rows of wires or grooves along which beads are slid to do calculating.

An Abacus.

Abandon: Leaving something in a place, without the intention of returning.

Abandonment: The voluntary surrendering of property rights but not transferring title to someone else.

In accounting, it is a donation or voluntary disposal of a business asset where it is cheaper to abandon the asset than to restore or salvage it. The book value of an abandoned asset is generally written off as a loss.

Abate: To eliminate, lessen, or reduce. To end or suppress. Also, to hammer metal, to remove or shape material to form a relief design.

Abatement: The ending, reduction, or lessening of something, e.g., noise abatement or an abatement in the purchase price.

Other examples of abatement: the decreasing of strength of a timber as it is shaped to the proper size for use, the wastage of wood as lumber is planed to size, and the shaping of metal by hammering, and the removal of a painted surface, covering a painted surface with an impermeable surface, or covering a surface with a heavy-duty coating (encapsulation).

Abatjour: A movable screen or movable slat for cutting off the view between an interior or porch and a lower area in front of a building.

Also, a skylight or different means of admitting light into a building and deflecting the light downward.

Abeyance: A lapse in succession during which title to a piece of property is not clearly established. Abeyance describes the legal status of real estate titles when lawful ownership of the property is in question and is being determined.

Ablation: The removal of material by melting and/or burning away. Ablation is the opposite of accumulation and it refers to all processes that remove snow, ice, or water from a glacier or snowfield. Ablation refers to the melting of snow or ice that runs off the glacier, evaporation, sublimation, calving, or erosive removal of snow by wind.

Ablative surface: A surface designed to melt or burn away at a controlled rate. An example would be a spacecraft where ablation is used to both cool and protect mechanical parts and/or payloads that would otherwise be damaged by extremely high temperatures.

Abney level: A hand-held level used in surveying to determine elevations and slope angles.

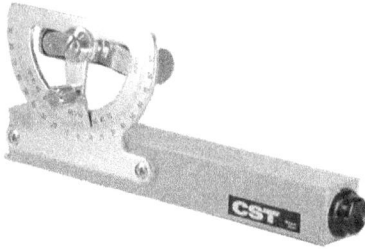

Abney level.

Abode: Any home or residence.

Abrade: To wear away or erode such as with sandpaper or emery board.

Abrasion: Wearing away.

Abrasion soldering: An intentional mechanical abrasion of the base metal during soldering.

Abrasive: Material used for grinding, sanding, polishing, or the wearing away of another material. Aluminum oxide, garnet, and silicon-based compounds are commonly used as abrasives for sanding and smoothing wood.

- **Abrasive coatings:** Two types of abrasive coatings are used in the manufacturing of coated abrasives: open coat and closed coat. With an open coat, 50–75% of the surface is covered by abrasive grain. There are evenly spaced voids between particles of grain that helps to reduce the effect of loading caused by wood dust or metal particles. With a closed coat, the entire surface is covered with abrasive grain, with no voids between the particles. This is the most typical coating; it permits the greatest degree of stock removal and longest product life.
- **Abrasive flapper:** Strips of material impregnated with an abrasive that are attached to a hub with a shank. When the shank is inserted into the chuck of a power

drill, the spinning strips can be used to sand irregular surfaces.

Abrasive flapper.

- **Abrasive paper:** Paper that has an abrasive surface, e.g., sandpaper, emery paper, garnet paper.
- **Abrasive stones:** Grinding stones used to sharpen metal blades. The blades are either rubbed along the stone or the stone is spun and the metal held against it, wearing down the metal and creating a sharp edge.
- **Abrasive surface:** A surface that has been roughened for safety or for warning.
- **Abrasive surface tile:** Floor tile that has been roughened to be slip-resistant.
- **Abrasive wheel:** A non-metallic disc that is impregnated with an abrasive, such as Carborundum and used in a power saw to cut masonry and metal.

Abreuvoir: The joint or interstice between stones in a stone structure that is filled with mortar.

Abreuvoir.

Absentee owner: A landlord who lives elsewhere.

Absolute: A measure having as its zero point or base the complete absence of the entity being measured. For example, the zero point of any thermodynamic temperature scale, such as Kelvin or Rankine, is set at absolute zero.

Absolute humidity: The mass of water in a given volume of air.

Absolute pressure: The pressure above zero absolute, i.e., the sum of atmospheric and gauge pressure. In vacuum related work it is usually expressed in millimeters or inches of mercury.

Absolute scale: A temperature scale that uses absolute zero as its zero point.

Absolute zero: The lowest temperature on the absolute or Kelvin scale, equal to $-459.7°F$.

Absorb: To fill or soak up.

Absorption: The process of drawing a fluid or gas into a porous material such as a sponge soaking up water. Absorption may be expressed as a percentage of the original weight of the material.

In commercial real estate, the amount of inventory or units of a specific commercial property type that become occupied during a specified time period (usually a year) in a given market, is typically reported as the absorption rate.

With regard to concrete, absorption is the increase in mass of the concrete that results from the penetration of water into the pores. Absorption is usually measured by submerging a concrete specimen in water. Absorption is considered to be a predictor of the durability of concrete: the higher the absorption the lower the durability.

With regard to masonry, the amount of water that a masonry unit (solid, hollow clay, concrete, or natural building stone) absorbs when immersed in either cold or boiling water for a stated length of time. It may be expressed as percentage of mass (clay masonry) or mass per unit volume (concrete masonry).

Terms beginning with absorption.

- **Absorption chiller:** A system that does not use a compressor, but uses thermal energy (low pressure steam, hot water, or other hot liquids) to produce the cooling effect.

- **Absorption coefficient:** The ratio of the sound absorbed to the sound incident on the material or device; the sound absorbed by a material or device is usually taken as the sound energy incident on the surface minus the sound energy reflected.

- **Absorption field:** A field engineered to receive septic tank effluent. An absorption field consists of a series of shallow trenches, parallel, round, or whatever the land allows, usually 18" to 24" wide, in which drain tiles (pieces of perforated pipe) are placed. If the permeability of the soil, as established by a percolation test is exceptional, then the pipes can be laid directly on it. In most cases the trenches must be lined with a 6" layer of gravel to help the effluent absorb properly into the soil. Also called a **leeching** or **seeping field**.

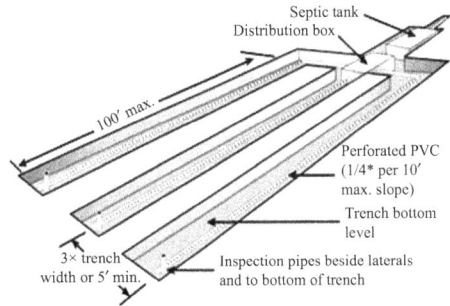

Typical absorption field.

- **Absorption rate:** With regard to real estate, the speed at which the real estate market can absorb new offerings of land or buildings during a specified period of time.

  With regard to masonry, the amount of water absorbed when a brick is partially immersed for 1 min. The absorption rate is usually expressed in either grams or ounces per minute per 30 square inches. Also called suction or initial rate of absorption.

The total amount of water that a masonry unit absorbs when immersed in water is referred to as **total absorption**.

* **Absorption refrigerator:** A refrigerator that creates low temperatures by using the cooling effect that is formed when a refrigerant is absorbed by a chemical substance.

Abstract: Design elements showing general forms instead of a detailed and realistic representation.

Abstract of judgment: A court judgment that is usually filed with the county. An abstract of judgement creates a lien against a piece of property.

Abstract or title search: A review of public records to determine whether liens or defects of title exist on a piece of property that could interfere with clear ownership transfer. Searches are done prior to closing of title on a sale.

The Abstract of title should provide a short history of the title of the land in question. It should be a summary of the facts relied on as evidence of title and must contain a note of all conveyances, transfers, and other facts relied on as evidences of a claimant's title, together with all such facts appearing of record as may impair title.

Abut: Boundaries of contiguous properties with no intervening land.

Abutment: A supporting structure at the end of a bridge or arch. Abutments carry the load from the deck into the foundation.

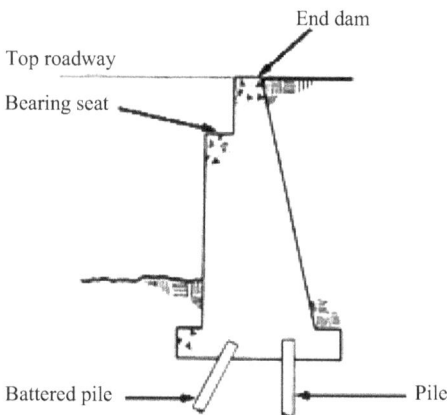

An abutment.

Abutment piece: The bottom plate of a wall.

Abutting: Property adjoining or bordering another property.

Acanthus: An ornamental design motif representing leaves of the acanthus plant, native to the Mediterranean.

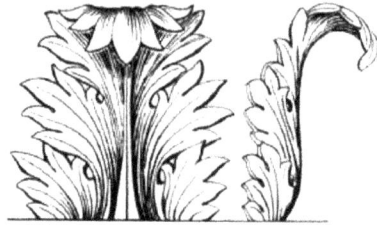

Acanthus leaf.

Accelerated cost recovery system: A tax calculation that provides greater depreciation in the early years of ownership of real estate or personal property.

Accelerated depreciation: A bookkeeping method that depreciates property faster in the early years of ownership. Speak with your accountant about both the accelerated cost recovery system and accelerated depreciation to see if they can help you.

Acceleration: A situation that forces a contractor to increase the work effort to meet the contract completion date and avoid possible liquated damages. Also, with regard to earthquakes, the time rate of velocity change, commonly measured in "g" (an acceleration of 32 ft/sec/sec or 980 cm/sec/sec = gravity constant on earth). Also, a vector quantity equal to the rate that velocity changes with time.

Acceleration clause: A provision that gives a lender the right to collect the balance of a loan if a borrower misses a payment.

Accelerating admixtures: Accelerating admixtures are added to concrete either to increase the rate of early strength development or to shorten the time of setting, or both. Chemical compositions of accelerators include some of inorganic compounds such as soluble chlorides, carbonates, silicates, fluosilicates, and some organic compounds such as triethanolamine.

Among all these accelerating materials, calcium chloride is the most common accelerator used in concrete. However, there is growing interest in using "chloride-free" accelerators as replacement for calcium chloride because calcium chloride may promote corrosion of steel reinforcement, especially in moist environments or when used carelessly.

Accelerogram: The record from an accelerograph showing acceleration as a function of time.

Accelerograph: A strong motion earthquake instrument that records ground (or base) acceleration. Also commonly called an **accelerometer**.

Kinemetrics FBA-23 accelerograph
(USGS photo).

Acceptable risk: The probability of social or economic consequences due to a hazard that is a realistic basis for determining design requirements or taking certain social or economic actions.

Acceptance: The seller's written approval of a buyer's offer.

Also, when a contract is substantially completed, the contractor can request final inspection of all or part of the contract. When work has been completed and accepted, the contractor is fully relieved of all liability for the maintenance, reconstruction, or restoration of that work. In practice a contract may be inspected and accepted as a whole or in parts.

Acceptance sampling: Statistical sampling to determine whether to accept or reject a production lot. Acceptance sampling is commonly used as a contract quality control technique in construction, manufacturing, and the military. Most often a producer/contractor supplies a customer/client with a number of randomly drawn samples and the decision to accept or reject the lot is made by determining the number of defective items in the lot. The lot is accepted if the number of defects falls below the previously agreed upon acceptance number. Otherwise the lot is rejected.

Access: Entrance, entry, way in, means of entry, a means of approach to a structure or a part thereof such as a road, street, pathway, or corridor.

Terms beginning with access, accessible, and accessibility:

- **Access chamber:** An underground chamber enabling access to drains or other underground services.
- **Access door:** Any door through an assembly structure that usually provides access to a mechanical, plumbing, or electrical circuit within the assembly structure. Access doors are usually flush with screw type openers. Also referred to as an **access panel**.

Access door/panel.

- **Access floor:** A floor structure normally constructed over the floor slab that allows access for cabling and related parts. Also referred to as raised flooring.
- **Access stair:** Usually considered to be a stair between two floors that does not serve as a required exit.
- **Accessibility:** Generally refers to a facility's capability to permit disabled people to enter and use the room or building.

Chapter 11 of the International Building Code is dedicated to accessibility.

- **Accessible equipment:** Equipment that is not guarded by locked doors, elevation, or other effective means.

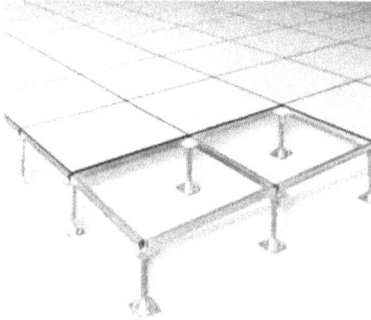

Access floor.

- **Accessible surface:** A term used in lead inspection reports to describe an interior or exterior surface accessible for a young child to mouth or chew, such as a window sill.
- **Accessible wiring methods:** Wiring that is installed so that it may be removed or exposed without damaging the building structure or finish or is not permanently closed in by the structure or finish of the building.
- **Accessible, readily:** Something is deemed **readily accessible** when it is capable of being reached quickly for operation, renewal, or inspections without requiring a person to climb over or remove obstacles or to resort to portable ladders, etc.

  An **accessible route** is a continuous unobstructed path that connects all accessible spaces and rooms in a building that can be negotiated by all categories of people having physical disabilities.

Accident frequency: The ratio of accidents and hours worked. While the ratio in most industries is usually calculated per million hours worked, smaller firms can use fewer hours worked to make comparisons between projects and project managers.

Accretion: The process of growth or enlargement by a gradual buildup, e.g., the increase

of a delta by the accumulation of silt and sand, rocks formed by the slow accretion of limestone, and ice buildup on a windshield.

Human Resource people often refer to increased job responsibilities as an **accretion of duties**.

Accumulator, hydraulic: A pressure storage reservoir in which a non-compressible hydraulic fluid (generally nitrogen) is held under pressure by an external source. The external source can be a spring, a raised weight, or a compressed gas.

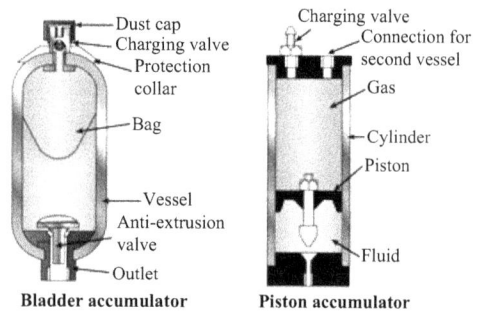

Hydraulic accumulators.

Accuracy: Accuracy is commonly defined as the difference between a measured value and the true value. Accuracy gets referred to in specifications in the form of "accurate to $\pm$ X".

Acetal plastic: A type of plastic used in the PB pipe.

Acid: A sour substance that liberates hydrogen ions in water and is sour and corrosive. Acids turn litmus red and have a pH of less than 7. They are generally divided into two classes: strong mineral or inorganic acids such as sulfamic, sulfuric, phosphoric, hydrochloric, or nitric, and weak organic or natural acids such as acetic (vinegar), citric (citrus fruit juices), oxalic, and fatty acids (oleic) such as palmitic and stearic.

Other terms beginning with acid.

- **Acid etch:** To clean or alter a surface using acid.

  Also, with regard to concrete, acid etching involves using muriatic acid to remove stains from concrete, preparing the surface for staining or creating designs for aesthetics. This is not to be confused with acid staining that can be done after preparing the concrete floor with muriatic acid. Also known as acid washing.

- **Acid number:** A designation of the amount of free acid in oils, flats, waxes, and resins. Acid number is expressed as the number of milligrams of potassium hydroxide required to neutralize one gram of the material being tested.
- **Acid rain:** Sulfur dioxide emissions that combine with water in the atmosphere and fall to the earth.
- **Acid-resistant grout:** An epoxy grout that resists the effect of prolonged contact with acids.
- **Acid resisting brick:** Brick that is suitable for use in contact with chemicals, usually in conjunction with acid-resistant mortars.
- **Acid-proof counter:** A horizontal work surface resistant to acid spills.

Acidity: The level of acid in substances such as water and soil. Acidity is important when it comes in touch with concrete because concrete is chemically basic; it has a pH of about 13. Concrete is, therefore, attacked by acids (pH less than 7). The American Concrete Institute has prepared a list of over 250 chemicals documenting their effect on concrete.

Acknowledgments: The paperwork that suppliers sends to designers to indicate how the supplier interpreted the designer's order.

Acme thread: The original trapezoidal thread form and probably the most commonly used worldwide. The Acme thread was developed sometime in the second half of the nineteenth century as a profile well suited to power screws. It had numerous advantages, over the then used square thread: it was easier to cut; it wore better; it was stronger; and, it made for smoother engagement of the half nuts on a lathe lead screw.

The trapezoidal metric thread form is similar to the Acme thread form, except the thread angle is 30°.

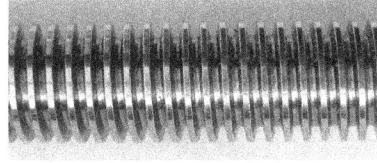

Acme thread.

Acoustical: Relating to sound or to the sense of hearing. Acoustics is the science of sound including its production, transmission, and effects.

Other terms beginning with acoustical.

- **Acoustical block:** A masonry block that is used for its sound-absorbing qualities.
- **Acoustical materials:** Materials that are capable of absorbing sound waves.
- **Acoustical panel:** Ceiling and wall mounted modular units composed of sound absorbing materials.

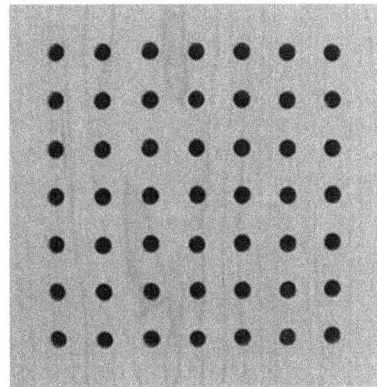

Perforated wooden acoustical panel.

- **Acoustical tile:** Ceiling panels in board form used for its sound absorbing properties, sometimes used on walls.

Acoustical ceiling tiles.

- **Acoustical treatment:** The act or process of applying acoustical materials to walls and ceilings.

Acre: A unit of measure equal to 4,840 square yards (43,560 square feet; 0.405 hectares).

Acre foot: The volume of material needed to cover an acre of land one foot deep. Your water bill may use acre feet as the basis of your rate.

Across the grain: The direction at right angles to the length of the fibres and other longitudinal elements of the wood.

Acrow: A telescopic prop used as a temporary support in construction. Named after the American manufacturer who introduced them.

Acrow props.

Acrylic: A synthetic resin from acrylic acid or a derivative thereof. The resin is known for its clarity.

Acrylic coating: A coating system with an acrylic resin base.

Acrylic resin: Polymers of acrylic or methacrylic monomers often used as a latex base for coating systems.

Act of God: A natural catastrophe that no one can prevent such as an earthquake, a tidal wave, a volcanic eruption, or a tornado. Acts of God are significant because of the havoc and damage they wreak, and also because construction contracts often state that "acts of God" are an excuse for delay or failure to fulfill a commitment or to complete a construction project. Many insurance policies exempt coverage for damage caused by acts of God. This, in turn leads to disputes as to whether a violent storm or other disaster was an act of God, and therefore exempt from a claim, or a foreseeable natural event; who said insurance companies have no religion?

Courts have recognized various events as acts of God, e.g., tornadoes, earthquakes, death, extraordinarily high tides, violent winds, and floods.

Action level: The level at which an employer must begin certain compliance activities outlined in the OSHA lead standard. The action level, regardless of respirator use, for the lead in construction standard is an airborne concentration of 30 $\mu g/m^3$ calculated as an 8-hour time weighted average.

Active: Something that will corrode in the presence of moisture or a "noble" metal.

The **noble metals** are metals that are resistant to corrosion and oxidation in moist air, unlike most base metals. They are considered to be (in order of increasing atomic number) ruthenium, rhodium, palladium, silver, osmium, iridium, platinum, and gold.

Terms beginning with active:
- **Active door:** The door in a double-door set that is opened first and to which a functionable lockset is applied.
- **Active falling area:** An area within two tree-length radius of where a faller (a person who fells trees for a living) or a mechanized falling machine is operating.
- **Active fault:** See fault, active.
- **Active panel:** The primary operating door panel.

Active solar system: A system that utilizes electric pumps or fans to transfer solar energy for storage or direct use.

Actual age: The number of years a structure has been standing.

Actuator: A device (motor or cylinder) for converting pneumatic or hydraulic energy into mechanical energy. It's energy source usually comes in the form of an electric current, hydraulic fluid pressure, or pneumatic pressure. It converts that energy into some kind of motion.

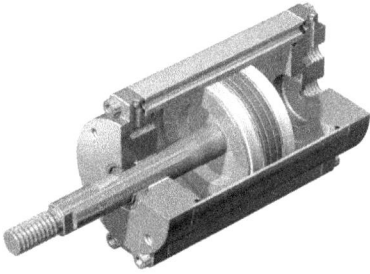

Pneumatic actuator.

Adapted vegetation and native vegetation: Plants that are indigenous to a locality (native) or plants that are adapted to the local climate and are not considered invasive species or noxious weeds (adapted); they require limited irrigation following planting, do not require active maintenance such as mowing, and provide habitat value.

Adapter: Any device for connecting two parts, especially ones that are of different sizes or have different mating fitments. For example, in plumbing, fittings that adapt from one system type to another, connecting different ODs (outside diameters) together and in electronics a plug that is used to connect an electrical device to a mains supply when they have different types of terminals. A device that is used to connect several electrical appliances to a single mains socket is also called an adapter. **Adaptor** is a variant of adapter.

Adaptive re-use: A use for a structure or landscape other than its historic use, normally entailing some modification of the structure or landscape.

Addenda: Corrections or changes that are made to the contract documents; addenda are published. Addenda are written by the person or firm responsible for the original set of contract documents.

Addition: An extension or increase in floor area or height of a building that increases its exterior dimensions; new construction added to an existing building.

Add on interest: A method of calculating interest by which the interest payable is determined at the beginning of a loan and is added onto the principal. The sum of the interest and principal is the amount repayable upon maturity.

For example, let's say Contractor A borrows $5,000 for two years from Bank B and that annual interest rates are 8%. Furthermore, Contractor A will repay the loan in two equal repayments at the end of each year. The interest charge on the loan is $800 ($5,000 × 8% × 2 years). Adding this to the principal gives a total of $5,800. The annual payment will be $2,900 ($5,800/2).

From this you can see that when multiple repayments are used in add-on interest, the effective lending rate becomes higher than the nominal rate because the borrower returns a portion of the principal with each payment, but is still being charged interest on the amount of the original loan. For example, if you took a straight loan of $5,000 for 2 years at 8% interest, with payments at the end of each year, your payments would be $2,803.85.

Add-on factor: The ratio of rentable to useable square feet. Also known as the load factor and the rentable-to-useable ratio. Also see efficiency percentage.

Add-on factor = (rentable square feet/useable square feet)

If you are renting/leasing space do not assume that you know the exact meaning of add-on factor, as what is and is not included changes from place to place.

Additive: A substance added to something in small quantities, typically to improve or preserve it. For example, with regard to cement, chemicals that are added to cement-based products (concrete, mortar, etc.) to impart various desirable properties, e.g., change curing time, increase strength, and enhance workability. The amount of additives added needs to be carefully monitored as too much or in combination they may result in undesirable effects.

Addendum: An addition or change to a contract.

Additional principal payment: Extra money that is included in the monthly payment to help reduce the principal and shorten the term of the loan.

Adhere: To cause two surfaces to be held together by adhesion, typically with asphalt or roofing cements in built-up roofing and with contact cements in some single-ply membranes.

Adherence: The extent to which a coating bonds to a substrate.

Adhesion: A property of prefaced masonry units. For example, a specification may require that when viewed without magnification, no visible failure of the adhesion of the facing material to the masonry unit shall occur after the unit has been subjected to a standard compressive strength test. Also, the state in which two surfaces are held together by interfacial forces that may consist of molecular forces or interlocking action, or both.

Adhesion is also the ability of dry paint to attach to and remain fixed on the surface without blistering, flaking, cracking, or being removed by tape.

Adhesive, edge joining: An adhesive used to bond together strips of veneer by their edges to make larger sheets.

Adhesive, heat activated: A dry adhesive that becomes tacky or fluid when heated or placed under pressure.

Adhesives: In addition to the adhesives mentioned above there are numerous other types of adhesives the definitions of which are found in their names, e.g., cold-setting, contact, hot melt, hot-setting, intermediate temperature setting, and multiple layer (for bonding dissimilar materials).

Adiabatic: Impassable to heat. Also, occurring without gain or loss of heat.

For example, compressing refrigerant gas without removing or adding heat is **adiabatic compression**, and the maintenance of ambient conditions during the setting and hardening of concrete so that heat is neither lost nor gained is **adiabatic curing**.

Adit: A sloping tunnel or shaft driven through a hill or mountainside to reach beds of rock. A crossword puzzle favorite.

Adit.

Adjustable: Alterable, adaptable, modifiable, convertible, changeable, variable, etc.

Many things are adjustable:
• **Adjustable bar hanger:** A metal hanger that can be made to fit the varying distances between floor and ceiling joists or rafters to securely hold electrical outlet boxes and devices.

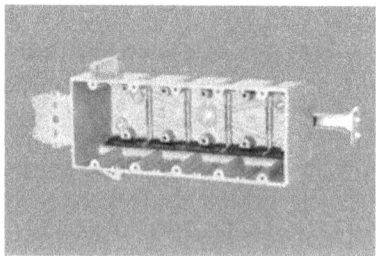

Adjustable bar hanger.

• **Adjustable bevel square:** A tool used to transfer any angle accurately. See Angle meter.

Adjustable bevel square.

- **Adjustable Countersink:** A device that cuts the recess for screw heads and threads. It has a drill bit or knife blade for cutting the recess for the screw thread whose length adjusts to the length of the screw head.

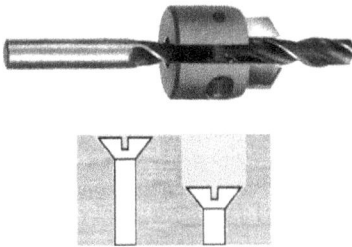

Adjustable counter sink boring bit.

- **Adjustable-rate mortgage (ARM):** A loan with an interest rate that is periodically adjusted to reflect changes in a specified financial index.
- **Adjustable suspension scaffold:** A suspension scaffold equipped with a hoist that can be operated by an employee on the scaffold.

Adjustable suspension scaffold.

- **Adjustable wrench:** A wrench that has an adjustable head to fit various sizes of nuts and bolts. Some adjustable wrenches feature a locking mechanism to prevent slippage.
- **Adjusted cost basis:** The cost of any improvements the seller makes to the property. Deducting the cost from the original sales price provides the profit or loss of a home when it is sold.
- **Adjusting link:** An adjustable strap or bar forming a connection between the lift rod of a faucet and the ball lever assembly of the drain.
- **Adjustment period:** The amount of time between interest rate adjustments in an adjustable-rate mortgage.

Administrative approval: The processing and proper approval of a work request, change order, and the like to ensure that funding and level of effort increases (decreases) will be met.

Administrative services: Generally, services that are not directly construction related, e.g., transportation, food services, security, office equipment, etc.

Administrator: A person appointed to handle the affairs of a person who has died intestate, i.e., without a valid will, one who manages the estate of a deceased person who left no executor. Also referred to as **Administratrix**.

Administrator's deed: A legal document that an administrator of an estate uses to transfer property.

Admixture: A material, other than water, cement, aggregate, or fiber reinforcement, that is used as an ingredient in a batch of concrete or mortar. Among the most common admixtures are those that improve plasticity, retard or advance hydration, and add color.

The most common admixtures that are blended into concretes or their ready-mix equivalents contain calcined kaolin, calcined diatomaceous earth, fiberous material, and reactive vinyl acetate co-polymers, with or without talc and/or bentonite or their functional equivalents.

Adobe: Unburned or unfired brick, dried in the sun. Construction that utilizes unburned (unfired) clay masonry units is referred to as **Adobe masonry**.

Adult wood: Wood that characteristically has relatively constant cell size, well-developed structural patterns, and stable physical behavior. Also referred to as **mature wood**.

Advanced bill: A preliminary bill of materials (BOM) that is prepared using the engineer-of-record's contract drawings. From the advanced bill, a purchase order is usually prepared and provided to the steel mill or manufacturer to reserve a time slot during which the steel order will be produced or to reserve the needed shapes that are produced by the mill.

Adverse grade: The acquisition of title to property through possession without the owner's consent for a certain period of time.

Adverse use: The access and use of property without the owner's consent.

Aeolian soil: Soil that is composed of materials deposited by the wind.

Aeration: A liquid may be aerated by passing it through air, e.g., by means of a fountain, or by passing air through the liquid, e.g., by means of bubble diffusers.

Aerated liquids are commonly used to smooth the flow of tap water at the faucet, for the production of aerated water or soda, to add oxygen for secondary treatment of sewage or industrial wastewater, and to dispel other dissolved gases such as carbon dioxide or chlorine.

Aeration also refers to the process of using mechanized equipment to either puncture soil with spikes (spike aeration) or remove ~1"X2" cores of soil from the ground (core aeration).

Aerator: The round screened screw-on tip of a sink spout. It mixes water and air for a smooth flow.

Aerial logging: A logging system that uses helicopters or balloons to fully suspend the logs. Not to be confused with cable systems that use cables and supports.

Aerobic: When referring to organisms, requiring the presence of air or free oxygen for life. See **anaerobic**.

Aerogel: A microporous, transparent silicate foam that is currently under development for potential use as a glazing cavity-fill material. It offers very high thermal performance.

Aerosol: Solid or liquid particles dispersed in gaseous media. (FYI: Aerosol paint products have not contained chlorofluorocarbons (CFCs) since 1978.)

Aesthetic: Something having a sense of beauty or is pleasing to the eye.

Affiant: A person who makes a sworn statement.

Affirmation: A substitution for an oath granted to people based on religious reasons.

Aftershock: One of a series of smaller quakes following the main shock of the earthquake.

Aga: A type of large heavy oven that is permanently lit. Generally, agas are used in farmhouses and other types of large houses. They are fueled by coke, oil, or gas.

An aga.

Age class: With regard to wood, any interval into which the age range of trees, forests, stands, or forest types is divided. Forest inventories commonly group trees into 20 year age class groups.

Agency: A relationship in which one person, the agent, acts on behalf of another with the authority of the latter, the principal. This is a fiduciary relationship that results from concent of the principal that the agent shall act on his/her behalf.

Agency CM (ACM): A contractual form of the construction management system that is exclusively performed in an agency relationship between the construction manager (CM) and owner. ACM is the form from which other CM forms and variations are derived.

Agency closing: The process by which a lender uses a title company or other firm as an agent to complete a loan.

Agent: One who, by mutual concent, acts for the benefit of another, one who is authorized by a person to act in that person's behalf.

As pertains to real estate, a person licensed by the state to conduct real estate transactions.

As pertains to construction management, an agent is authorized by a client (principal) to act in his/her stead or behalf and, in turn, the CM owes the client a "fiduciary duty" (Trust). While simply stated, there are mutations and permutations to CM agreements, so be careful. For example, a "construction manager for-fee" may be classified as an independent contractor for tax purposes. A "construction manager for-fee" does not have any financial responsibility, whereas a "construction manager at-risk" does have financial risk similar to that of a general contractor.

Agglomeration economies: Cost reductions or savings that come about from efficiency gains that are associated with the concentration or clustering of firms/producers or economic activities and the formation of a localized production network.

Aggregate: Sand, gravel, or a combination of both. The use and size of gravel varies depending on whether the product is concrete (sand plus gravel larger than 1/4 inch), grout (sand plus gravel 1/4 inch or smaller), mortar, or plaster (sand only).

Aggregates are generally classified as follows:
- **Aggregate, coarse:** One of the four ingredients of concrete, usually gravel, that is retained on a #4 sieve.
- **Aggregate, fine:** One of the four ingredients of concrete, usually sand, that will pass the #4 sieve and will be retained on the #200 sieve.
- **Aggregate, heavyweight:** Aggregate of high specific gravity such as barite, magnetite, limonite, limenite, iron, or steel used to produce heavy concrete.
- **Aggregate, lightweight:** Aggregate of low specific gravity, such as expanded or sintered clay, shale, slate, diatomaceous shale, perlite, vermiculite, or slag; natural pumice, scoria, volcanic cinders, tuff, and diatomite, sintered fly ash, or industrial cinders; used to produce lightweight concrete; aggregate with a dry, loose weight of 70 pounds per cubic foot or less.

Aggregate coated panel: Sheet material, usually plywood, with one side having a decorative face that has been applied with epoxy.

Aggregate coated (stone) panel trash receptacle.

Aggregate storage bins: In a concrete batching plant, the bins that store the necessary aggregate sizes and feed them to the dryer.

Agreed boundary: A compromise boundary to which property owners agree in order to resolve a dispute.

Agreement of sale: A document that the buyer initiates and the seller approves that details the price and terms of the transaction.

Agrillaceous: A fine-grained sedimentary rock with grains less than 1/16 mm, e.g., clay.

Terms beginning with air:

- **Air break:** A piping arrangement in that a drain from a fixture discharges through the atmosphere into a receptacle. This arrangement prevents the creation of a syphon.
- **Air brush:** A small spray gun with a fine spray. Also the British term for spray gun.
- **Air conditioning (AC):** The process that simultaneously controls temperature, humidity, movement, cleanliness, and odor of air circulated through a space.
- **Air conditioning coil:** A coil configured to remove heat from refrigerant gas.
- **Air content:** With regard to concrete, the volume of air voids in concrete, usually expressed as a percentage of total volume of the paste, mortar or concrete. See Air-entrained concrete.
- **Air cooled flue:** A flue that may consist of double or triple pipes in which heated gases rising out of the inner flue create a draft that pulls cool air through the outer pipe(s).

Air cooled flue.

- **Air cure:** One method by which liquid coatings cure to a dry film. Oxygen from the air enters the film and cross-links the resin molecules. Also called **Air Dry** and **Oxidizing**.
- **Air dried wood:** Lumber that was dried by exposure to open air, usually in a yard or shed, without artificial heat.
- **Air ducts:** Pipes that carry warm or cold air.
- **Air gap:** With regard to a drainage system, an unobstructed vertical distance through air between a waste water pipe outlet and the flood-level rim of the receptor into which it is discharging. An air gap prevents wastewater from backing up into the originating waste water pipe.

With regard to a water-distribution system, the unobstructed vertical distance through air between the lowest opening from a water supply discharge to the flood-level rim of a plumbing fixture.

- **Air handler:** A piece of mechanical equipment containing a blower and coils for the heating and or cooling of air in a HVAC circuit.
- **Air infiltration:** The amount of air leaking in and out of a building through cracks in walls, windows, and doors.
- **Air lock:** A blockage in the flow of liquid, especially on the suction side of a pump caused by an air bubble in the line.
- **Air quality control regions:** Geographical units of the country that reflect common air pollution problems. They are designated by the national government for purposes of reaching uniform standards. Local governments may also have a say in what may and may not be constructed because of air pollution.
- **Air, saturated:** Moist air in a state of equilibrium with a plane surface of pure water or ice at the same temperature and pressure.

Said differently, air whose vapor pressure is the saturation vapor pressure and whose relative humidity is 100.

Yet another try, a mixture of dry air (air containing no water vapor) and water vapor at its maximum concentration for the prevailing temperature and pressure.

- **Air-blown asphalt:** Asphalt produced by blowing air through molten asphalt held at an elevated temperature. This procedure is used to modify the properties of the asphalt.
- **Air-entrainment agent:** A type of admixture. This is actually a detergent that produces small, evenly spaced bubbles in concrete. It makes the concrete more plastic or workable and more frost resistant.
- **Air-entrained concrete:** Air-entrained technology was development in the late

1930s. Nowadays it is recommended for nearly all concretes, principally to improve resistance to freezing and thawing and deicing chemicals; because air-entrained concrete contains billions of microscopic air cells internal pressure on the concrete is relieved by providing tiny chambers for the expansion of water when it freezes.

Air-entrained concrete is produced through the use of air-entraining portland cement, or by introducing air-entraining admixtures as the concrete is mixed on the job. The amount of entrained air is usually between 5% and 8% of the volume of the concrete, but this may vary to meet special conditions.

The use of air-entraining agents results in concrete that is highly resistant to severe frost action and cycles of wetting and drying or freezing and thawing and has a high degree of workability and durability.

- **Aircrete:** A lightweight aerated cement-based material from which easily handled high insulating building-blocks are made.
- **Airless spraying:** An airless sprayer uses an electrically run hydraulic pump to move paint (often with thinners) from a bucket or container, through a tube, into a high-pressure hose, to a spray gun, and, finally, to the surface. The process has also been used in the reinforced plastics field for the spray-up technique.
- **Airspace:** The area between insulation facing and interior of exterior wall coverings. This is normally a 1" air gap.

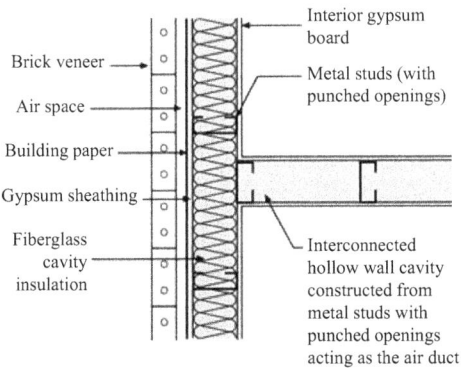

Brick veneer
Air space
Building paper
Gypsum sheathing
Fiberglass cavity insulation

Interior gypsum board
Metal studs (with punched openings)
Interconnected hollow wall cavity constructed from metal studs with punched openings acting as the air duct

Typical wall construction.

In windows, the space that is between the panes in a double-paned window.

The component that is placed at the perimeter of the insulating glass unit to separate the two lites of glass is called an **airspacer** or **spacer**.

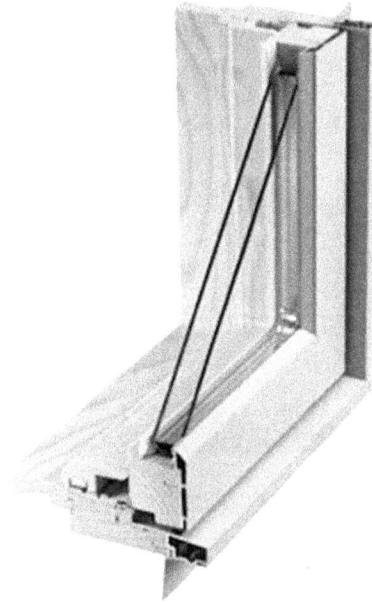

Typical double paned window construction.

AISC weight: The weight of structural steel as defined by the American Institute of Steel Construction(AISC).

Alabaster: A fine-textured, regularly white, gypsum that is easily carved and translucent when thin.

Albedo: Albedo is synonymous with solar reflectance.

Alclad: A trade name for an aluminum alloy coated with pure aluminum to give it high corrosion resistance.

Alcohol: A colorless volatile inflammable liquid that is miscible (capable of mixing in any ratio without separation of two phases) with water. Ethyl alcohol is commonly used as a shellac solvent and methyl alcohol and wood alcohol are used in paint removers.

Alcove: A recessed section of a room, such as a breakfast nook.

Algae: A simple nonflowering plant of a large group that includes the seaweeds and many single-celled forms. Algae contain chlorophyll but lack true stems, roots, leaves, and vascular tissue. Divisions of algae are Chlorophyta (**green algae**), Heterokontophyta (**brown algae**), and Rhodophyta (**red algae**).

With regard to roofs, rooftop fungus that can leave dark stains on roofing. Also known as algae discoloration and fungus growth.

Rooftop algae.

Alkali: Mineral salt found in soil. The adjective alkaline is commonly used in English as a synonym for base, i.e., a substance capable of reacting with an acid to form a salt and water, or (more broadly) of accepting or neutralizing hydrogen ions. A solution of a soluble base has a pH greater than 7.

Alkali-aggregate reaction: Alkali-aggregate reactivity is a type of concrete deterioration that occurs when the active mineral constituents of some aggregates react with the alkali hydroxides in the concrete. Alkali-aggregate reactivity occurs in two forms alkali-carbonate reaction (ACR) and alkali-silica reaction (ASR).

Indications of the presence of alkali-aggregate reactivity may be a network of cracks, closed or spalling joints, or displacement of different portions of a structure.

Alkali-carbonate reaction: The reaction between alkalies (sodium and potassium) in Portland cement and certain carboniferous rocks and minerals, e.g., calcitic dolomites and dolomitic limestones. The reaction may cause abnormal expansion and cracking of the concrete. These reactions may occur internally between the cement and the aggregates in the concrete, or externally with rocks in the surrounding environment.

Alkali-silica reaction: The reaction between the alkalies (sodium and potassium), in Portland cement and certain siliceous rocks or minerals, e.g., opaline chert and acidic volcanic glass present in some aggregates. These reactions may occur internally between the cement and the aggregates in the concrete. The reaction may cause abnormal expansion and cracking of the concrete, or, rarely, externally with rocks in the surrounding environment.

Alkali metals: A group in the periodic table consisting of the chemical elements lithium (Li), sodium (Na), potassium (K), rubidium (Rb), caesium (Cs), and francium (Fr). Because of their high reactivity, they must be stored under oil to prevent reaction with air. It is conceivable that extremists could attempt to use alkali metals as an alternative for explosives, given the wide availability of public information about the reactivity of alkali metals on the Internet. If you suspect such activity report it to the FBI.

Alkaline soil: Soil that contains a higher concentration of mineral salt than natural acid.

Alkalinity: pH values above 7.

Alkyd: Synthetic resin modified with oil. Coating that contains alkyd resins in the binder.

Alley: A lane behind a row of buildings or between two rows of buildings.

Alligatoring: A network of fine cracks on a surface with a definite pattern as indicated by the name. For example, paved highways, roofs, and painted surfaces frequently alligator. The effect is often caused by weather aging.

Allocate: To set aside funds for a project.

Allowable cut: The amount of wood that can be removed from a landowner's property during a certain time span. The allowable cut should not exceed the net growth during that same time on the property. **Allowable annual cut (AAC)** is the volume of timber that may be harvested annually from a specific timber tenure.

All-in ballast: Ballast suitable for making into concrete without the addition of any other aggregate.

All-inclusive Deed of Trust or Mortgage: A trust deed or mortgage that includes underlying financing; one payment is made to the all-inclusive mortgagee or beneficiary who then makes the payments on the underlying loans. Also called a **Wraparound Deed of Trust or Mortgage**.

All-purpose mud: A type of drywall mud containing adhesive chemicals that holds the drywall tape in place and helps the mud adhere to the drywall face.

Allowable soil pressure: The maximum stress permitted in soil of a given type under given conditions.

Allowable stress: The maximum stress permitted at a given point in a structural member under given conditions.

Allowance(s): A sum of money set aside in the construction contract for items that have not been selected and specified in the construction contract. For example, selection of tile for flooring may require an allowance for an underlayment material, or an electrical allowance that sets aside an amount of money to be spent on electrical fixtures.

Alloy: A metal containing other elements for property enhancement, e.g., brass is an alloy of copper and zinc. In plastics, a blend of polymers with other polymers or copolymers.

Alluvium: Loosely compacted sand, gravel, and silt deposited by streams in relatively recent geologic time.

Along the grain: The direction parallel with the length of the fibres and other longitudinal elements of the wood.

Alternate top bevel (ATB): A design for a circular saw blade where teeth are alternately beveled. ATB blades give the cleanest cross-cuts in hard and soft wood or plywood.

Alternate top bevel.

Alternate-fuel vehicles: Vehicles that use low-polluting, non-gasoline fuels such as electricity, hydrogen, propane, or compressed natural gas, methanol, and ethanol. Efficient gas-electric hybrid vehicles are included in that group.

Alternative mortgage: Any home loan that does not conform to a standard fixed-rate mortgage.

Alterations: Any construction or renovations to an existing structure other than repair or addition. Also, the moving of a building is frequently deemed an alteration.

Aluminized steel: Sheet steel with a thin aluminum coating on the surface to enhance the steel's ability to withstand weathering.

Aluminum (Al): Aluminum is the third most abundant element (next to oxygen and silicon) in the earth's surface rocks. It has been estimated that 8% of the earth's crust consists of aluminum as against 5% of iron. Aluminum never occurs as a native metal; its ultimate extraction is by means of electrolysis. Bauxite is the most important aluminum ore.

Aluminum is a non-rusting metal that is often used in roofing for metal roofing and the fabrication of gutter and flashings.

As a warning, aluminum should not be directly embedded in concrete, because the metal reacts with the alkalis in cement. **But aluminum does have many uses:**

- **Aluminum bench level:** A mechanic's bench level that is extremely stable and lightweight.
- **Aluminum file:** A tool that is used to deal with the problem that ordinary files pose when used on soft aluminum: clogging. The scalloped-tooth pattern cuts cleanly and leaves a fine finish.

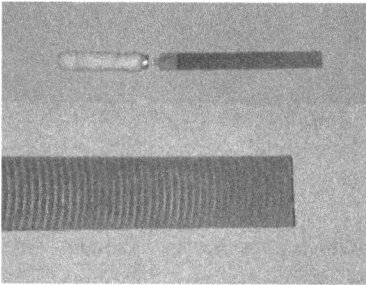

Aluminum file.

- **Aluminum oxide:** A chemical that is created when aluminum is exposed to the atmosphere. Aluminum oxide is an insulator, and when it builds up on aluminum power wire it can create heat that may loosen the connection, cause arcing, and even fire.
- **Aluminum siding:** A metal covering that provides an alternative to paint for owners of wood homes.
- **Aluminum surround:** The aluminum frame around a screen or energy panel.
- **Aluminum-clad windows:** Wooden windows with aluminum covering the exterior.

Ambient heat/pressure: With regard to operating equipment, the heat or pressure in the area where the equipment is located.

Ambient lighting: The light reaching an object in a room from all light sources, surrounding light.

Ambient temperature (ABM): The average temperature of the atmosphere in an area, the temperature of the surrounding environment. Technically, ambient temperature is the temperature of the air surrounding a power supply or cooling medium.

Amenities: Parks, swimming pools, health-club facilities, party rooms, bike paths, community centers, and other enticements offered by builders of planned developments.

Americans with Disabilities Act: A law passed in 1990 that outlaws discrimination against a person with a disability in housing, public accommodations, employment, government services, transportation, and telecommunications.

American Lumber Standards: These standards guide lumber manufacturers' associations in preparing or revising their grading rules that cover classification, nomenclature, basic grades, sizes, descriptions, measurements, shipping, grade marking, and inspection of lumber. Purchasers must, however, make use of association rules because the basic standards are not in themselves commercial rules.

American method: The application of giant individual shingles with the long dimension parallel to the rake. Shingles are applied with a 3/4-inch space between adjacent shingles in a course.

American Standard Gauge: Units used to measure the size of wire. The smaller the wire the larger will be its gauge number. For example, an 18 gauge wire is a very thin wire used for a doorbell and a 2 gauge is a large feeder wire. Wires larger than 2 gauge are measured in aughts. Wire larger than 4 aught are measured in micro-circular-mills (MCM).

American wallpaper roll: A roll of wallpaper that contains about 36 square feet.

Amide: A functional group that can act as an epoxy resin curing agent.

Amorphous: Something that is without a clearly defined shape or form.

Amortization: For accountants amortization is the preferred term for the charging (writing off or apportioning) of the cost of an intangible asset as an operational cost over the asset's estimated useful life. It is identical to depreciation, the preferred term for reducing the value of tangible assets. The purpose of both terms is to reflect a reduction in the book value of the asset due to usage and/or obsolescence, and to spread a large expenditure proportionately over affixed period, and thereby reduce the taxable income (not the actual or cashincome) of a firm; it is essentially a process by which invested capital of a firm is recovered by gradual sale of the firm's asset(s) to its customers over the years.

For bankers amortization is a payment plan by which a loan is reduced through regularly scheduled payments of principal and interest.

Ampacity: The current, in amperes, that a conductor can carry continuously under the conditions of use without exceeding its temperature rating.

Ampere: The unit by which the flow of current through a conductor is measured.

Amplification: With regard to earthquakes, the period (or frequency) of the ground motion coinciding with the period of the building causing significant increase of acceleration and damage.

Amplification factor: A multiplier of the value of moment or deflection in the unbraced length of an axially loaded member to reflect secondary values generated by the eccentricity of the load.

Amplitude: With regard to earthquakes, the maximum deviation from mean of centerline of a wave.

Also, a measure of floor vibration; the magnitude or total distance traveled by each oscillation of the vibration.

Anaerobic: Without oxygen. An anaerobic organism or anaerobe is any organism that does not require oxygen for growth. In fact, it could possibly react negatively and may even die if oxygen is present. For example, anaerobic digestion is a series of processes in which microorganisms break down biodegradable material in the absence of oxygen. A pond in which fish cannot live is anaerobic. See **aerobic** and **anoxic**.

Analog: A continuous range of numbers or values.

Analogous color: Hues that are next to one another on the color wheel. Examples would be red and orange or blue and green.

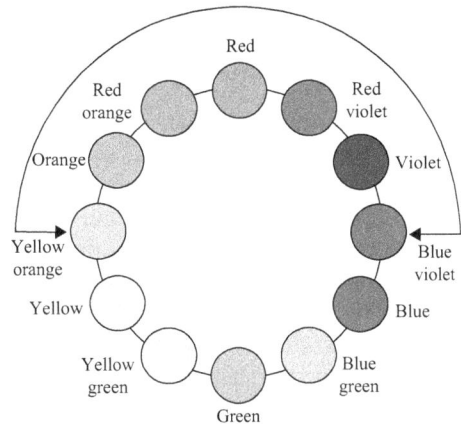

Analogous color scheme.

Anchor bolt: A bolt used to connect the foundation to the inside of the framing above it. Anchor bolts are placed in the foundation while the concrete is still wet, then after the concrete cures, the foundation sill plate is attached to the anchor bolts.

Anchor bolt plan: A plan view showing the size, location, and projection of all anchor bolts.

Anchorage: The process of fastening a joist or joist girder to a masonry, concrete, or steel support by either bolting or welding.

Also, a secure fixing, usually made of reinforced concrete to which the cables are fastened.

Also, an extension of reinforcing steel, either straight or in a bent shape, used to transfer the force in the reinforcement to a support. In pre-tensioned and post-tensioned concrete, a separate device used for the same purpose.

Also, a secure point of attachment for lifelines, lanyards, or deceleration devices.

Angiosperm: A plant that has flowers and produces seeds enclosed within a carpel. The angiosperms are a large group and include herbaceous plants, shrubs, grasses, and most trees. Compare with **gymnosperm**.

Angle: An amount of rotation. The measurement of angles using 360° in a whole circle, with each degree divided into 60 min of 60 s each, is of very great antiquity, going back to the Babylonians who used a number system based on 60 s rather than tens.

Terms beginning with angle:
- **Angle bar:** A steel structural member in the shape of an L. Angle bars are classified by the thickness of the stock and the length of their legs.
- **Angle bay window:** See **Bay window**.
- **Angle blasting:** Blast cleaning at angles less than 90°.
- **Angle block:** A square of tile specially made for changing direction of the trim.

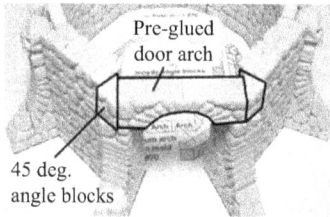

Pre-glued door arch

45 deg. angle blocks

- **Angle brick:** Any brick shaped to an oblique angle to fit a salient corner.
- **Angle controlled tightening:** A bolt tightening procedure in which a fastener is first tightened by a pre-selected torque (called the snug torque) so that the clamped surfaces are pulled together, and then is further tightened by giving the nut an additional measured rotation. Frequently bolts are tightened beyond their yield point by this method in order to ensure that a precise preload is achieved. Bolts of short length can be elongated too much by this method and the bolt material must be sufficiently ductile to provide for the plastic deformation involved. Because of the bolt being tightened beyond yield, its re-use is limited.
- **Angle driver:** A unique tool that simplifies work in corners and at awkward angles.

Angle driver.

- **Angle grinder:** A tool found mainly in metal and auto-body shops. It is useful to woodworkers and do-it-yourselfers, as well. An angle grinder is used for cleaning up pitted or rusted metal surfaces, to smooth out welded seam, and, when properly equipped, to cut metal. It is ideal at metal finishing.
- **Angle iron:** A steel section whose cross-section is L-shaped. If the vertical and horizontal legs of the "L" are the same length it is called an equal angle, if different, an unequal or odd leg angle. Angles are also available in other metals. Also referred to simply as an **angle**.
- **Angle meter:** A versatile tool used to measure any angle easily, including level and plumb. It features an easy-to-read dial. See **Adjustable bevel square**.

Angle meter.

- **Angle of incidence:** The angle between the axis of a light beam impinging on a surface and a perpendicular to the surface at the point of impact.

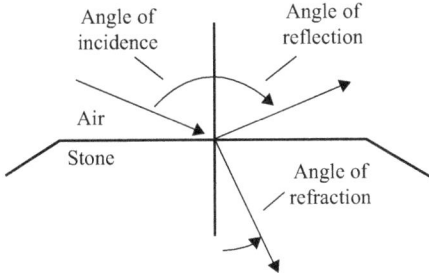

Angles of incidence, reflection, and refraction.

- **Angle of reflection:** Angle between the axis of a light beam reflected from a surface and the perpendicular to the surface.
- **Angle of refraction:** The angle formed by the path of refracted light or other radiation and a line drawn perpendicular to the refracting surface at the point where the refraction occurred.
- **Angle of repose:** The internal friction and cohesion of soil allowing it to assume a natural angle at rest.
- **Angle seat wrench:** A valve seat wrench with a handle that includes a 90° bend.

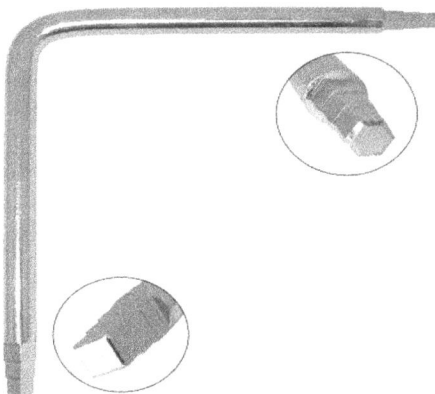

6-step down angle seat wrench.

- **Angle stop:** A shutoff valve between water pipes and a faucet. The inlet connects to the water-supply pipe in a wall, the outlet angles up 90° to the faucet.

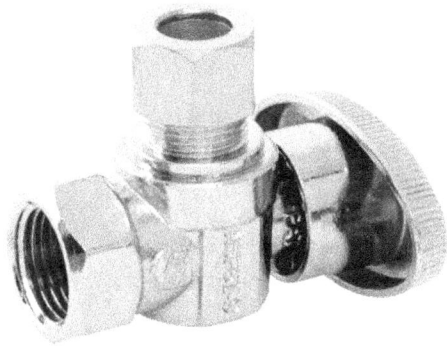

Angle stop.

- **Angle unit:** A member used as a joist substitute that is intended for use at very short spans (10 feet or less) where open web steel joists are impractical. They are usually used for short spans in skewed bays, over corridors, or for outriggers. It can be made up of two or four angles to form channel sections or box sections. Tube and channel sections are also used. See **Joist substitute**.

Angled fasteners: Roofing nails and staples driven into decks at angles not parallel to the deck.

Anisotropic: Materials that exhibit different properties along axes in different directions.

Annealing: Thermal treatment that changes the properties or grain structure of a metallic material.

Also, the process of relieving stresses in molded plastic articles by heating below deformation temperature, then maintaining this temperature for a predetermined length of time, and then using a controlled cooling process.

Also, a controlled cooling process to reduce thermal residue stress in glass; **annealed glass** is ordinary window glass that is cooled slowly to eliminate stresses in the material.

Annual: Something that happens once a year.

Annual growth: The layer of wood developed by a tree during a given year. **Annual growth rings** are the layer of growth that a tree puts on in one year. The annual growth rings can be seen in the end grain of lumber. Also referred to as **annual or seasonal increment**.

Annual percentage rate (APR): Annual cost of credit over the life of a loan, including interest, service charges, points, loan fees, mortgage insurance, and other items. APR allows a borrower to evaluate the cost of a loan in terms of a percentage. For example, if your loan has a 7% rate, you'll pay $7 annually per $100 borrowed. All other things being equal, you simply want a loan with the lowest APR.

Sounds easy, but loans can be confusing, even though we have a US Government passed Truth in Lending Act. This is because often not all others things are equal. APR may include Private Mortgage Insurance, processing fees, discount points, and other fees. You need to read the fine print, as difficult as that may be.

Here's an example to show how APR may vary: Assume that you borrow $100,000 for a warehouse/garage. The bank offers you a loan at 7% interest rate, but with $1,000 in closing costs. The APR in this case would be 7.10%. You can do the calculations using Excel or other readily available calculators; I'll use my handy hp 12C: number of payments = 30 × 12; interest = .07/12; present value = 100,000. The payment comes to $665.30; this is the payment the bank will expect.

But in fact you received only $99,000. Again using my handy hp 12C, I was forced into an iterative process of raising the interest rate step by step until I came up with the same monthly payment of $665.30 for a $99,000 loan. The result was a monthly interest rate of 0.592%, or 7.0999% annually, the APR. This is not much of a difference, but it will add up over 30 years.

Annual percentage yield (APY): Annual percentage yield is a tool for evaluating how much a deposit earns; it is a standardized way of comparing investments. APY is notable because it takes compounding into account.

In general, the more compounding periods the higher the APY. So if two CDs pay the same interest rate, choose the one that pays

out interest most often. If you're using APY instead of interest rate, pick the one with the highest APY.

Let's take a look: A CD of $1,000 that pays 6% at the end of a year will yield you $1,060, or an APY of 6%. But a $1,000 CD that pays 6% interest compounded daily (365 days) will yield you an APY of 6.183%. And, if interest is compounded four times per year your APY would be 6.136%. Compounding periods do matter.

Annual mortgagor statement: A yearly statement to borrowers that details the remaining principal and amounts paid for taxes and interest.

Annuity: The payment of a fixed sum to an investor at regular intervals.

Annular area: The part of a large circle that remains after a small circle with the same center has been removed.

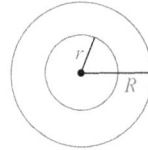

Area of annulus = $\pi R^2 - \pi r^2$

This ring shaped area often refers to the net effective area of the rod side of a cylinder piston, i.e., the piston area minus the cross-sectional area of the rod.

Anodic: When two metals are connected in an electrolyte, they will form a galvanic cell, with the higher metal in the galvanic series being the anode. The anodic will oxidize and produce an electrical current that protects the cathode from corrosion.

Anoxic: An environment that is depleted of oxygen. Anoxic waters are areas of sea water or fresh water that are depleted of dissolved oxygen. See **Anaerobic** and **Aerobic**.

Anticipatory breach (of contract): A communication that informs a party that the obligations of the original contract will not be fulfilled. It occurs when one party by declaration repudiates his/her contractual obligation before it is due. The repudiation required is "a positive statement indicating that the promisor will not or cannot substantially perform his/her contractual duties." Where the anticipatory repudiation is by the party's conduct rather than by declaration it is called "voluntary disablement." As an example, in a contract for sale of a used truck the seller transfers ownership of the truck to a third party during the executor interval before delivery is due on the first contract.

Anti: A prefix meaning against and opposite of. **Anti is used freely in combination with elements of any origin, for example:**

- **Antifoam:** A substance added to boiler water to control foaming in impure water that experiences rapid circulation currents.
- **Anti-fouling paints:** Paints formulated especially for boat decks and hulls, docks, and other below-water-line surfaces and structures to prevent the growth of barnacles and other organisms on ships' bottoms.
- **Anti-friction coatings:** Coatings that are dry lubricants consisting of suspensions of solid lubricants, such as graphite, (or others) of small particle size in a binder. Such coatings can be applied to fastener threads to replace metallic coatings such as zinc and cadmium and offer maintenance free permanent lubrication. The coatings are permanently bonded to the metal surface and provide a lubricating film and preventing direct metal to metal contact.
- **Anti-oxidant compound:** A compound applied onto aluminum wires to prevent aluminum oxide from forming.
- **Anti-sap stain:** A wood treatment used to prevent fungus from staining the wood.
- **Anti-seize compound:** An anti-seize compound is used on the threads of fasteners in some applications. The purpose of the compound depends upon the application. It can prevent wear caused by adhesion between sliding surfaces of mating surfaces; such compounds are frequently used with stainless steel fasteners to prevent this effect from occurring.

In some applications it is used to improve corrosion resistance so that parts can be subsequently disassembled. It can also provide a barrier to water penetration since the threads are sealed by use of the compound.

- **Anti-stick coating:** As used on saw blades, decreases friction and heat buildup and helps provide cleaner, smoother, and quieter cutting action. Also resists resin and pitch buildup and improves safety conditions.
- **Anti-stratification device:** A device which stirs water in a hot water heater to prevent the stratification of hotter water at the top and cooler water at the bottom.
- **Anti-vibration slots:** Slots cut in the body of a saw blade, usually in a star burst pattern. Anti-vibration slots reduce vibration so the blade runs more smoothly and produce a cleaner cut.

Apex: The highest point, e.g., the highest point on a joist or joist girder where the sloped chords meet. See also **Peak**.

Appearance grades (wood): High-line regular board and dimension grades that include tighter restrictions on certain appearance characteristics, particularly wane (the absence of square wood on the edge of a board from any source.)

Appliance: Utilization equipment, generally other than industrial, that is normally built in standardized sizes or types and is installed or connected as a unit to perform one or more functions such as clothes washing, air conditioning, food mixing, deep frying, etc.

Application: A document that details a potential borrower's income, debt, and other obligations to determine credit worthiness. An **application fee** is the fee that a lender charges to process a loan application. And, an **application for payment** is a contractor's written request for payment for completed portions of the work and for materials delivered or stored and properly labeled for the respective project.

Application rate: The rate at which a material is applied per unit area.

Applied force: See **external force**.

Appliqué: A technique whereby pieces of fabric are layered on top of one another and joined with decorative stitches.

Appraisal: An expert valuation of property. An **appraisal fee is** the fee that an appraiser charges to estimate the market value of the property. An **appraisal report** is a detailed written report on the value of a property based on recent sales of comparable sites in the area. And, an **appraised value** is an opinion of the current market value of a property.

Appreciation: An increase in the value of a home or other property.

Approved: Sanctioned, endorsed, accredited, certified, or accepted as satisfactory by a duly constituted and nationally recognized authority or agency.

Approved agency: An established and recognized agency that is regularly engaged in conducting tests or furnishing inspection services and that has been approved by the Department of Buildings and Safety or similar organization. See **Authority having jurisdiction**.

Approved plans: Plans sent by a manufacturer/fabricator, engineer, architect, contractor, or other person for approval by an owner and an authority having jurisdiction. Generally, plans include a framing plan, elevations, sections, material list, or whatever is appropriate to make the final installation clear.

Approved vendor list: A current list of vendors providing the owner with goods and services. Many government agencies award contracts only to firms on their approved vendor list.

Apron: A trim board that is installed beneath a window sill.

Also, the skirting (apron), that is the decorative portion of a bathtub that covers the rough-in area from the floor to the top rim of the tub. It is often sold separately from the tub.

Apron flashing: Apron flashings are typically used for roof to wall junctions such as along front of a dormer or front of a chimney.

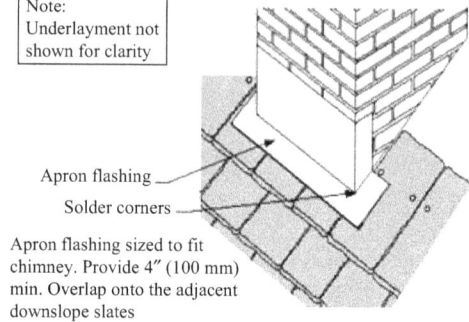

Note: Underlayment not shown for clarity

Apron flashing

Solder corners

Apron flashing sized to fit chimney. Provide 4″ (100 mm) min. Overlap onto the adjacent downslope slates

Note - All weights and dimensions are approximate

Apron flashing.

Apron sink: An exposed front to the sink as opposed to sitting inside the cabinet.

An apron sink.

Aquatint: Print making process used to create areas of solid color, as well as gradations of white through black tones. Usually has the appearance of transparent watercolor.

Aqueduct: A bridge or channel for conveying water, usually over long distances.

An aqueduct.

Another aqueduct.

Also an arbor.

Aquifer: A layer or zone below the surface of the earth which is capable of yielding a significant volume of water. The upper level of the aquifer is called the water table.

Arbitration: The settlement of a contract dispute by selecting an impartial third party to hear both sides and reach a decision. Many contracts include a clause that requires all disputes be resolved through arbitration. Before signing such a contract understand that by so doing you are giving up your right to sue in court.

Arbor: An axle or spindle on which something revolves. A device that holds a tool in a lathe. The shaft on which a circular saw blade or chip head is mounted. An arbor hole is the central hole in a saw blade through which the saw arbor fits.

Arc: An arc is a chain of volcanoes (volcanic arc) that sometimes forms on the land when an oceanic plate collides with a continental plate and then slides down underneath it.

Also, a discharge of electricity through a gas, such as air.

Arcade: A series of arches supporting a wall, or set along it. Also, a covered passageway with arches along one or both sides.

Arcade.

Also, short for video arcade.

Arbors.

Also, a shady garden alcove with sides and a roof formed by trees or climbing plants trained over a wooden framework.

Another kind of arcade.

Arch: A curved structure that supports weight over an area, such as a doorway.

The arch is a basic form of masonry construction dating back millenia. Brick arches are found spanning over window and door openings in Victorian and older buildings; their disadvantage is that they exert horizontal thrust at their bearings, which sometimes leads to distortion in poorly designed or maintained arches.

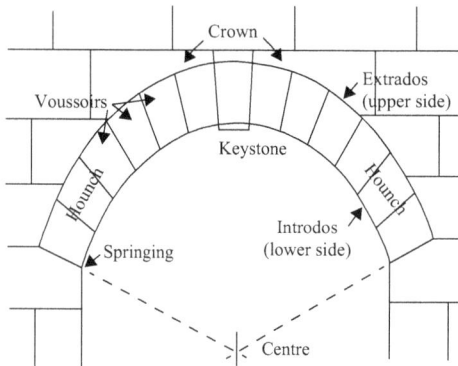

Masonry arch.

Architect: A person licensed to practice the profession of architecture under the education law of the state in which the person practices.

Architectural: Of or pertaining to the art and science of architecture.

Terms beginning with architectural:

- **Architectural conservation:** The science of preserving a historic structure's materials by observing and analyzing their deterioration, determining causes of and solutions to problems, and directing remedial interventions.
- **Architectural conservator:** A specialist in the scientific analysis of historic materials.
- **Architectural fee:** The fee an architect charges for services. In general, architects charge for their services by the hour, by the square foot, or by a percentage of the project budget.
- **Architectural history:** The study of architecture through written records and the examination of structures in order to determine their relationship to preceding, contemporary, and subsequent architecture and events. An architectural historian is a historian with advanced training in this specialty.
- **Architectural panel:** A metal roof panel that usually requires solid decking underneath. A metal siding panel may also be referred to as an architectural panel.
- **Architectural roofing:** Roofing systems that are non-weight load bearing and must be installed over solid decking rather than battens or purlins. Also referred to as **non-structural systems**. See **Architectural panel**.
- **Architectural shingles:** A new generation of high-quality asphalt shingles with distinctive appearances. Also known as laminated or dimensional shingles, architectural roofing shingles are among the highest quality roofing products made.

  Traditionally, they are composed of a heavy fiber glass mat base and ceramic-coated mineral granules that are tightly embedded in carefully refined, water-resistant asphalt. Recently, however, a new product has surfaced on the market, synthetic slate.

  Synthetic slate that simulates the appearance, texture, and contours of traditional natural slate also fits under the general category of architectural shingles. It is an engineered polymer composite roofing product that is formulated for beauty, durability, and handling.

  Also known as **laminated or dimensional shingles**.

Architectural shingles

Architect's supplemental instructions (ASI): A document written by the architect that contains additional instructions regarding the contract specifications or drawings. Generally, the architect uses an ASI when the information is not anticipated to cause a change to the contract. If the contractor believes these instructions are a change to the contract, he/she can write a change order proposal (COP).

Architrave: The molded frame or ornament surrounding a window, door or other rectangular opening.

Architrave.

Archival collection: An accumulation of manuscripts, archival documents, or papers having a shared origin or provenance, or having been assembled around a common topic, format of record, or association (e.g., presidential autographs). The term also refers to the total archival and manuscript holdings of an organization or institution.

Archives: The repository where archives and other historic documents are maintained. See also **historic document**.

Archivist: A professional who is responsible for managing and providing access to archival and manuscript collections.

Arc-fault circuit interrupter (AFCI): A circuit breaker designed to prevent fires by detecting an unintended electrical arc and disconnecting the power before the arc starts a fire. AFCIs are designed to distinguish between a harmless arc that occurs incidental to normal operation of, say, switches and plugs and an undesirable arc that may occur from, say, example, a cut electrical cord.

Conventional circuit breakers respond only to overloads and short circuits and do not protect against arcing conditions that produce erratic, and often reduced current. The AFCI circuitry continuously monitors the current and discriminates between normal and unwanted arcing conditions. When an unwanted arcing condition is detected, the AFCI opens its internal contacts, thereby de-energizing the circuit.

While an AFCI resembles a GFCI/RCD (Ground-Fault Circuit Interrupt/Residual-Current Device) in that it has a test button, it is important to distinguish between the two. A GFCI and RCD are designed to protect against electrical shock of a person, while an AFCIs is primarily designed to protect against arcing and fire.

Arc-over voltage: The minimum voltage required to create an arc between electrodes under specified conditions. Generally something to be avoided.

Arch-top: One of several terms used for a variety of window units with one or more curved frame members, often used over another window or door opening. Also referred to as circle-heads, circle-tops, and round-tops.

Antique vintage arch-top window.

Area: A unit of measure of length times width. May be expressed in square inches, square feet, acres, etc.

Area divider: A flashed assembly usually extending above the surface of the roof that is anchored to the roof deck. It is used to relieve thermal stresses in a roof system where an expansion joint is not required, or to separate large roof areas.

Area well: Used around basement windows to hold back the soil. Usually constructed of galvanized, ribbed steel, concrete, or masonry.

Argon: An inert, nontoxic gas used in insulating glass to reduce heat transfer.

Armored/BX cable: Cable with a flexible metal covering that is often used with appliances. Almost universally referred to as BX (only) cable.

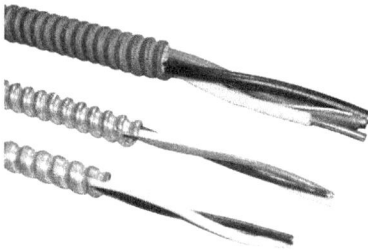

Armored BX cable.

Arpent: A French measurement of land equal to .84625 acres.

Arris: A sharp corner or junction of two planes or similar.

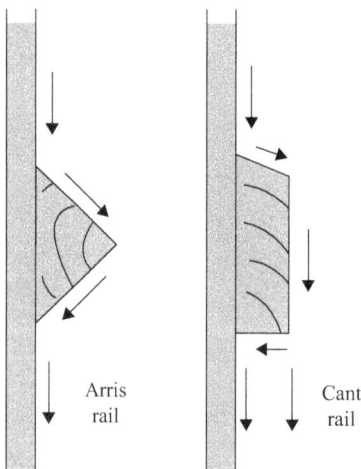

Arris and cant rails.

Arsenic (As): A widely distributed element that occurs in metallic form in the ores of lead, silver, and nickel. Arsenic is a strong poison. It is used in the making of weed killers and insecticides, and in some alloys, wood preservatives, paint, and medicine.

Articulation class: A single-number summation of how effective a ceiling is in absorbing sound reaching it from over low partitions.

Articulation index (AI): The articulation index measures the performance of all the elements of a particular configuration working together, including ceiling absorption, space dividers, furniture, light fixtures, partitions, background masking systems, and HVAC system sound.

Also, in audiology, the AI is a tool that is used to predict the amount of speech that is audible to a patient with a specific hearing loss.

Asbestos: A group of natural, fibrous, impure silicate materials used to reinforce some roofing products. Asbestos have been outlawed and are the responsibility of the building owner to abate.

As-built drawings: Construction drawings that have been revised to show changes made to the original plans during construction.

Aseismic: A term that describes a fault on which no earthquakes have been observed.

As-is condition: The purchase or sale of a property in its existing condition.

Ashlar: Smooth sawn stonework used in a wall.

Ashlar stonework.

Ashpit: The area below the hearth of a fireplace that collects the ashes.

Askarel: A generic term for a group of non-flammable synthetic chlorinated hydrocarbons used as electrical insulating media. Askarels of various compositional types are used. Under arcing conditions, the gases produced, while consisting predominantly of noncombustible hydrogen chloride, can include varying amounts of combustible gases, depending on the askarel type.

Asking price: A seller's initial price for a property.

Aspect ratio: The ratio of height to width of a shear wall. While an easy definition, you must read the appropriate building code to learn how such things as window and door openings affect your designs.

Also, for any rectangular configuration, the ratio of the lengths of the sides.

Asperity: An asperity is an area on a fault that is stuck. The earthquake rupture usually begins at an asperity.

Asphalt: A dark colored viscous to solid hydrocarbon residue, also referred to as bitumen. Asphalt is widely used in pavements, to bind and waterproof aggregate mixtures, and to seal surfaces. Asphalt is used in hot mixes and also warm mixes. Asphalt emulsions and, rubberized asphalt that mixes crumb rubber from used tires with asphalt are also used.

Asphalt is categorized by the temperature at which it is applied: hot mix asphalt (HMA), warm mix asphalt, or cold mix asphalt. Hot mix asphalt is applied at temperatures over 300°F (150°C). Warm mix asphalt is applied at temperatures of 200–250°F (95–120°C), resulting in reduced energy usage and emissions of volatile organic compounds. Cold mix asphalt is often used on lower volume rural roads.

With specific regard to roofing, asphalt can be refined to conform to various roofing grade specifications:
- **Dead-Level Asphalt:** A roofing asphalt conforming to the requirements of ASTM Specification D 312, Type I. This asphalt is for use in roofs that do not exceed a 1/4 in 12 slope (2%).
- **Flat Asphalt:** A roofing asphalt conforming to the requirements of ASTM Specification D 312, Type II. This asphalt is for use in roofs that do not exceed a 1/2 in 12 slope (4%).
- **Steep Asphalt:** A roofing asphalt conforming to the requirements of ASTM Specification D 312, Type III. This asphalt is for use in roofs that do not exceed a 3 in 12 slope (25%).
- **Special Steep Asphalt:** A roofing asphalt conforming to the requirements of ASTM Specification D 312, Type IV. This asphalt is for use in roofs which do not exceed a 6 in 12 slope (50%).

Other asphaltic materials include:
- **Asphalt concrete primer:** A solution of asphalt in petroleum solvent, used to prepare concrete roof decks for the application of hot asphalt. The primer lays dust and improves the adhesion of the molten asphalt to the roof deck.
- **Asphalt emulsion:** Emulsified asphalt consists of asphalt droplets suspended in water. Under normal circumstances oil and water do not mix, but when an emulsifying agent is added to the water the asphalt will remain dispersed. Most emulsion are used for surface treatments. Asphalt emulsions are complicated and use a great deal of chemistry to get the emulsion properties desired.
- **Asphalt plastic roofing cement:** Asphalt-based sealant material, meeting ASTM D4586 Type I or II. Used to seal and adhere roofing materials.

  Asphalt plastic roofing cement is a trowelable mixture of solvent-based bitumen, mineral stabilizers, other fibers, and/or fillers. Also referred to as **mastic, blackjack, roof tar**, and **bull**.

Aspiration: The process of mixing air with discharging water to enhance the hydromassage effect of a whirlpool.

Assessed value: A tax assessor's determination of the value of a home in order to calculate a tax base.

Assessment: A tax levied on a property, or a value placed on the worth of a property.

Assessment rolls: A list of taxable property compiled by the assessor.

Asset: An item of value, that includes cash, real estate, securities, and investments. In the technical terminology of accounting, assets are the resources with which a business operates, e.g., buildings, equipment, tools, and cash.

Asset management: Managing assets so as to maximize value for the owner.

Also, the management of real property, installed equipment, and furniture, furnishings, and equipment using computers for scheduling work assignments, monitoring equipment usage, controlling inventory and the like.

Assignor: A person who transfers rights and interests of a property.

Assumable mortgage: A mortgage that can be transferred to another borrower.

Assumption: Allows a buyer to assume responsibility for an existing loan instead of getting a new loan.

Assumption clause: A provision that allows a buyer to take responsibility for the mortgage from a seller.

Assumption fee: A fee that is charged by the lender to process new records for a buyer who assumes an existing loan.

Asthenosphere: The soft and probably partly molten layer of the earth below the lithosphere. Distinguished by low seismic-wave velocities and high seismic wave attenuation.

Astragal: A molding that is attached to one of a pair of swinging double doors, against which the other door strikes.

The vertical strip that is between the doors is an astragal.

Also, a small convex molding.

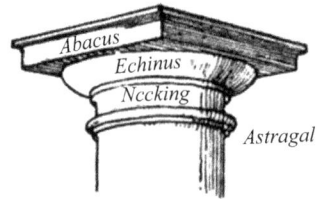

Another form of astragal.

As-built drawings: Drawings that are marked up to reflect all changes that were made during the construction process or after construction. As-built drawings amend the contract drawings to show the exact location, geometry, and dimensions of the constructed project. It is good practice to make as-built drawings by marking the changes as the project unfolds. As-built drawings are not the same as **record drawings**. See **Record drawings**.

Atmospheric pressure: Pressure exerted by the atmosphere at any specific location. (Sea level pressure is ~14.7 pounds per square inch absolute.)

Atomic absorption: A laboratory method of measuring elements such as lead. The lead is vaporized at high temperature and light of a specific wavelength is shined through the vapor. If lead is present the light is partially extinguished. The instrument converts this change into a number that describes how concentrated the lead is in the test material.

Atrium: An opening or skylighted lobby through two or more floor levels other than an enclosed stairway, elevator, etc. Atria are commonly found in large modern hotels.

Attached drawer: A type of drawer in which the drawer face is attached to the front of a self-contained drawer box. See **Integral drawer**.

Attachment plug: A device that, by insertion in a receptacle, establishes a connection between the conductors of the attached flexible cord and the conductors connected permanently to the receptacle. Also referred to as attachment cap, plug cap, or plug.

Attenuation: As an example, when you throw a pebble in a pond, it makes waves on the surface that move out from the place where the pebble entered the water. The waves are largest where they are formed and gradually get smaller as they move away. This decrease in amplitude of the waves is called attenuation. Seismic waves also become attenuated as they move away from the earthquake source.

Attic: The accessible space between the ceiling framing of the topmost story and the underside of the roof framing. Inaccessible areas are considered are called structural cavities.

Attic access: An opening that is placed in the ceiling of a home to provide access to the attic.

Attic ventilators: In houses, screened openings provided to ventilate an attic space.

Auger: A tool that includes a screw shaped shaft that digs a hole when turned.

Auger bit: A long, 7-inch to 10-inch bit typically used with a brace for drilling holes in wood. An auger bit bores a faster, cleaner hole because of its screw point and spur design.

Aught: American Standard Gauge unit of measure for wire sizes that are larger than two gauge. See **American Standard Gauge**.

Authentic divided lites (ADL): Permanent, stationary muntins and bars that separate the glass in a window or door sash to give the sash two or more lites of glass. Also referred as **true divided lites**.

Authentic divided lite/true divided lite doors.

Authority having jurisdiction: The organization, office, or individual responsible for approving equipment, materials, and installation, or a procedure.

Note: The phrase "authority having jurisdiction" is used in National Fire Protection Association (NFPA) documents in a broad manner, since jurisdictions and approval agencies vary, as do their responsibilities. Where public safety is primary, the authority having jurisdiction may be a federal, state, local, or other regional department or individual such as a fire chief, fire marshal, chief of fire prevention bureau, labor department, or health department, building official, electrical inspector, or others having statutory authority. For insurance purposes, an insurance inspection department, rating bureau, or other insurance company representative may be the authority having jurisdiction. In many circumstances, the property owner or his or her designated agent assumes the role of the authority having jurisdiction; at government installations, the commanding officer or departmental official may be the authority having jurisdiction.

Authorized person: A person approved or assigned by the employer to perform a specific type of duty or duties or to be at a specific location or locations at the jobsite. See **Designated person**.

AutoCAD: Probably the world's most popular computer-aided drafting software product for the personal computer.

Autoclave: An instrument used to sterilize equipment and supplies by subjecting them to high-pressure saturated steam. They are widely used in hospitals and laboratories and for the pre-disposal treatment and sterilization of waste material, such as pathogenic hospital waste.

Also, a cylindrical vessel in which wood is impregnated with a preservative under high pressure. Used to treat lumber, roundwood posts, and poles or any other wood based product that requires rot or insect protection.

Autoclave crazing: A failure of the ceramic glazing on structural clay facing tile, facing brick, and solid masonry units when subjected to a standard steam pressure in an autoclave for a specified time.

Automatic: Self-acting, operating by its own mechanism when actuated by some impersonal influence, as, for example, a change in current, pressure, temperature, or mechanical configuration.

Automatic center punch: A tool that allows for one-handed operation by pressing down on the tool and a spring-loaded mechanism strikes a blow. More precise than using a hammer and punch.

Auxiliary load: Any dynamic live loads such as cranes, monorails, and material handling systems.

Average annual effective rate: In commercial real estate, the average annual effective rent divided by the square footage.

Average annual effective rent: In commercial real estate, the tenant's total effective rent divided by the lease term.

It is the true rent, after taking into account rent concessions that are usually expressed as a dollar amount per square foot. Some parts of the country quote rents as dollars per foot per month; others use dollars per foot per year.

Example: If 6,000 square feet of office space normally rents for $22 per foot per year, but the landlord gives 6 months' free rent on a 5-year lease, then the effective rental rate is calculated as follows:

6,000 feet × $22 per foot = $132,000 per year.

6 months free rent $0 + 4.5 years paid rent @ $132,000 per year = $594,000.

$594,000 paid rent spread over a 5-year term = $118,800 per year effective rent.

$118,800 effective rent of 6,000 feet (118,800 ÷ 6,000) = $19.80 per foot effective rental rate.

Average price: The price of a home determined by totaling the sales prices of all houses sold in an area and dividing that number by the number of homes.

Average value: A group of calculations that test properties such as adhesion, fabric, and paper tear resistance, and surface friction. Calculations available are average peaks, average troughs, average peaks and troughs, first peak, and interval peaks.

Aviation snips: Snips that are used to cut straight or curved lines. They are generally used for heating, air conditioning, gutter work, and general industrial use. Aviation snips are available in right- or left-handed styles.

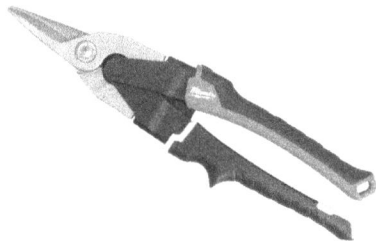

Aviation snips.

Avigation easement: An easement that grants the right to fly airplanes over property, even if the practice causes damage, inconvenience, or loss of property value. Such an agreement usually restricts the property owner from building or growing anything over a specified height.

Avonite: A solid surface material resembling granite. Avonite is used for counter tops. It can be worked and polished using woodworking tools.

Awl, scratch awl: An tool that is used for all types of precision layout work and general scribing.

Awning window: A window unit that opens by moving the bottom of the window sash outward. The top of the window sash is attached with hinges.

Axed arch: A brick arch in which the bricks are cut (traditionally with an axe) to a wedge shape. The mortar joints are of even thickness.

Axed arch.

Axial: Relating to, characterized by, or forming an axis. Also, located on, around, or in the direction of an axis:

- **Axial force:** A push (compression) or pull (tension) acting along the primary axis of a member.
- **Axial load:** A load whose line of action passes through the centroid of the member's cross-sectional area and is perpendicular to the plane of the section.
- **Axial strain:** The strain in the direction that the load is applied, or on the same axis as the applied load.
- **Axial stress:** The axial force acting at a point along the length of a member divided by the cross-sectional area of a member.
- **Axial strut load:** A structural member designed to transfer a axial tension or compression load only.

- **Axial tension:** An axial force that causes tension in a member.

B tank: An acetylene tank holding 40 cu.ft. of gas and used for plumbing. B tanks were once used to fuel boats, thus the name.

Acetylene B tank refill.

b value: A measure of the relative frequency of the occurrence of earthquakes of different sizes.

Back: An often used word:

- **Back addition:** Originally, a traditional terraced house was comprised of rooms between the front and rear external walls. When "indoor plumbing" came into being an extension was built at the back of the house to contain the bathroom, wc, kitchen, and scullery. The rear wing of a house is still called the back addition, even if it was built at the same time as the rest of the house.

House with a back addition.

- **Back charge:** Billings for work performed or costs incurred by one party that, in accordance with the agreement, should have been performed or incurred by the party to whom billed. Owners bill back charges to general contractors, and general contractors bill back charges to subcontractors. Examples of back charges include charges for cleanup work or to repair something damaged by another subcontractor.
- **Back check valve:** A valve inserted into a water line circuit to prevent the back flow of water during flow operations. Also known as a **backflow preventer**.
- **Back clearance:** As pertains to saws, the angle between a tangent to the cutting circle of a tooth and a line along the top of the tooth intersecting this tangent. Also referred to as **clearance angle**.
- **Back flow:** See Backflow and Backflow preventer.
- **Back gouging:** The forming of a bevel or groove on the other side of a partially welded joint to assure complete penetration upon subsequent welding from that side.
- **Back pressure:** The resistance to a moving fluid due to obstacles in the confinement vessel (pipe, air duct, etc.) in which it is moving. The term back pressure is misleading. Resistance would probably be a better term as the flow continues in the same direction but is reduced due to resistance. For example, an automotive exhaust muffler with many twists, bends, and angles could be described as having particularly high back pressure.
- **Back priming:** A coat of paint that is applied to the back of woodwork and exterior siding to prevent moisture from entering the wood and causing the grain to swell.
- **Back saw:** A tool that is perfect for the fine, accurate cut. It includes the dovetail and tenon saw. A backsaw is distinguished (from other hand saws) by a stiff length of brass or steel set over the top edge of the saw blade, providing support for the blade and preventing the blade to twist or flex.

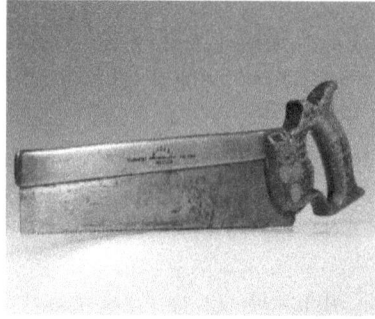

Back saw.

- **Back surfacing:** Granular material added to the backs of shingles to assist in keeping them separate during delivery and storage.
- **Back title letter:** A letter that a title insurance company gives to an attorney who then examines the title for insurance purposes.
- **Back vent:** A branch vent that is installed for the purpose of protecting fixture traps from siphoning. Back vents include most of the vents that are not installed specially to permit circulation between vent stacks and soil or waste stacks.
- **Back-nailing:** The method of fastening the back or upper side of a ply of roofing felt or other component in a roof system so that the fasteners are covered by the following ply.
- **Back wall:** The wall facing an observer who is standing at the entrance to a room, shower, or tub shower.
- **Back-plastering:** Plaster that is applied to one face of a lath system following the application and subsequent hardening of plaster being applied to the opposite face; used primarily in construction of solid plaster partitions and certain exterior wall systems.
- **Back-to-back escrow:** Arrangements that an owner makes to oversee the sale of one property and the purchase of another at the same time.
- **Back vent:** A plumbing fixture's separate vent in a building drainage system.

- **Backarc:** The region landward of the volcanic chain on the other side from the **subduction zone** (the subduction zone is the place where two lithospheric plates come together, one riding over the other. Most volcanoes on land occur parallel to and inland from the boundary between the two plates).
- **Backband:** A rabbeted moulding used to surround the outside edge of casing. Also spelled back band.

Baseboard and backband.

- **Backbedding:** Material or compound used to seal the glass to a window sash.
- **Backcut:** The final cut in felling a tree by hand. The back cut is made on the side opposite the intended direction of fall, after the undercut. The backcut disconnects almost all of the tree from the stump leaving a hinge that helps to control the tree's fall.
- **Backer board:** A board is not part of the original framing. It is placed in a wall or joist system to provide backing for the attachment of drywall, sheathing, or an intersecting wall.
- **Backer rod:** A flexible plastic foam material that is inserted into a joint to provide a limit to building caulking.
- **Backfill:** The replacement of excavated earth into the excavation.
- **Backflow, drainage:** Reversals of drainage flow.
- **Backflow preventer:** A backflow preventer is used to protect potable water supplies from contamination due to polluted water backing up. Also known as a **back check valve**, and a **backflow prevention assembly**.

Most water supply systems are maintained at a high enough pressure to enable water to flow from the tap. When pressure fails or is reduced as may happen if a water main bursts, pipes freeze, or there is unexpectedly high demand on the water system due to, say, a major fire, the pressure in the pipe may be reduced. This may allow contaminated water from the ground, from storage or from other sources to be drawn into the system.

Pumps in the water distribution system, boilers, heat exchanging equipment, or power washing equipment may also cause back pressure. So too may commercial/industrial descaling (boilers) or when bleaches are used for residential power washing.

To prevent such contamination some jurisdictions require an air gap or mechanical backflow prevention assembly between the delivery point of mains water and local storage or use. When mandated, most laws require that a **double check (DC)**, **reduced pressure principle device (RP)**, or an **air gap** be installed.

- **Backhaul:** A delivery originates from where the truck or trailer is loaded. Then the load is delivered to a destination. And then the trucker returns home. If the return is also a paying load to be delivered to the vicinity of origin, that load is called a backhaul. If the trucker returns home empty, that run is called a **deadhead**.
- **Backhearth:** The part of the hearth inside the fireplace.
- **Backhoe:** Excavation equipment that is used to dig with by drawing a steel bucket towards the operator of the equipment. Buckets are designed in one cubic yard increments.
- **Backing:** A framing member that is installed at a non-layout position so that other framing members can be securely attached later in the construction process. Also, carpet backing holds the pile fabric in place.

- **Backing bar:** A welding aid used to prevent melting through of a joint when performing, e.g., a complete-joint penetration groove weld.

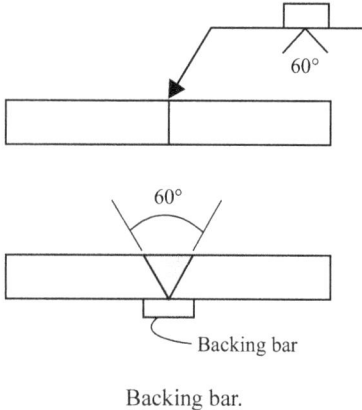

Backing bar.

- **Backing layer, vinyl flooring:** One of three layers of material typically found in vinyl floor covering. The backing layer is the bottom layer. See the **Pattern layer and Wear layer**.

  Some vinyl tile consists of four primary: a urethane **wear layer** to resist scratches and scuffs, a protective **clear film layer** to protect against rips, tears, and gouging, a **pattern layer** that carries the realistic colors and patterns, and a structural vinyl **backing layer** that adds strength and durability.

- **Backing veneer:** The layer of veneer used on the reverse side of a piece of plywood from the face or decorative side.

- **Backloaded insulation:** Thermal/acoustical insulation that is placed above the ceiling suspension system and laid across the horizontal grid members above the acoustical panels or tile. Also known as **backloading**.

- **Backout:** Work that the framing contractor does after the mechanical subcontractors (heating, plumbing, and electrical) finish therough phase of their work (before insulation). Backout is done to prepare for a building department frame inspection. Generally, the framing contractor repairs anything that was disturbed by others and completes all framing necessary to pass a **rough frame inspection**.

- **Backpressure:** Pressure that is greater than the water supply pressure and, therefore, potentially causes a backflow. See **Backflow preventer**.

- **Backset:** A dimension used in hardware applications indicating the distance from the edge of a door to the center of the door-opening device. (doorknob, lever, etc.)

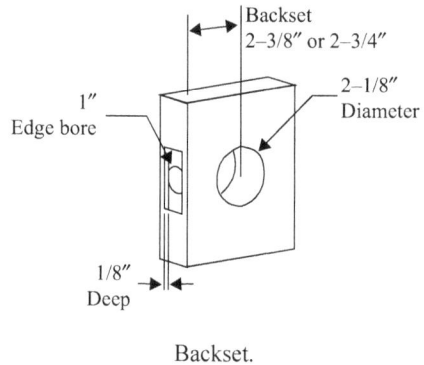

Backset.

- **Backsiphonage:** The flowing back of contaminated water into potable water due to negative pressure. See **Backflow preventer**.

- **Backsplash:** A part that fits on the wall behind the countertop and is designed to protect the wall from countertop and sink splashes. The backsplash is often made from the same materials as the countertop. See **Full backsplash** and **Block backsplash**.

- **Backup:** The overflow of a plumbing fixture due to drain stoppage.

- **Backup offer:** A secondary bid for a property that the seller will accept if the first offer fails.

- **Backwater valve:** A valve in a sewer line that prevents sewage from flowing back into a house.

Bacterial corrosion: Corrosion that results from substances (e.g., ammonia or sulfuric acid) produced by the activity of certain bacteria.

Bacterial decay: Bacteria and actinomycetes are probably the most common wood-inhabiting microorganisms; they are the most adaptable in terms of environmental influences. Regarding wood decay, they may be roughly classified into four groups for convenience and ease of comparison:

- Those bacteria that affect the permeability to liquids of wood but have no significant effect on strength properties.
- Those types that attack the wood structure.
- Bacteria that only function as integral members of the total microflora and are associated in the ultimate breakdown of the wood.
- The "passive" colonizers that have no effect upon the wood but have a marked influence on the remainder of the population by their antagonistic activities.

Bactericide: An agent capable of destroying bacteria.

Baffle: A device (deflector) used to stop the transmission of a material such as sound, light, or a liquid. A non-combustible baffle should be placed around a recessed light fixture to prevent heat produced by the light from being trapped by insulation where it can build up and cause a fire.

A baffle is also a barrier in a dry kiln. It is used to deflect and control airflow through the lumber.

And, a baffle is a device, usually a plate, that is installed in a reservoir to separate the pump inlet from return lines.

Baghouse dust collector: An air pollution control device that captures particulate in filter bags.

Baked-enamel finish: The application of special enamel paint onto a non-porous surface by baking it onto a surface, usually metal.

Baking: The process of drying or curing a coating by the application of heat in excess of 65°C. Below this temperature, the process is referred to as **forced drying**. Not to be confused with baking food in an oven.

Balance: A mechanical device (normally spring loaded) that is used in single- and double-hung windows as a means of counterbalancing the weight of the sash during opening and closing. Also know as **sash balance**.

Balance sheet: The balance sheet, or statement of financial position, presents the financial condition of a business as of a moment of time. It is akin to a photograph in that it portrays in monetory terms the economic resources controlled by the business and the claims on and interests in those resources as of a specific date. See also **Income statement** and **Statement of changes in financial position**.

Balanced match: See **Veneer**.

Balancing valve: A water heater valve that controls water flow and balances heat distribution to different locations.

Balcony: An elevated platform or seating space of an assembly room projecting from a wall of a building.

Ball: When rubbing down coatings or films with abrasive paper, the material removed by this action may form into soft or sticky lumps or balls. A coating or film that exhibits this latter phenomenon is said to ball.

Ball passage: The size of a ball that can pass through the trapway of a toilet. This also relates to trapway size. In general, the trap size will be 1/8' larger than the maximum size ball that can pass through it.

Ball peen hammer: A hammer that has a hardened head. It is used for striking cold chisels and punches for general metalwork. The ball peen was originally used to mushroom rivet heads. Sizes 8 to 32 oz.

Ball peen hammer.

Ball valve: A valve that uses a ball to seal against the seat.

Ball valve.

Ballast: Smooth aggregate placed on the surface of a roof to weigh down the roofing. Also protects the roof materials from ultraviolet light. A roof that has ballast placed on it is called a **ballast roof**.

Also, a transformer that steps up the voltage in a florescent lamp.

Ballcock: The fill valve that controls the flow of water from the water supply line into a gravity-operated toilet tank. It is controlled by a float mechanism that floats in the tank water. When the toilet is flushed, the float drops and opens the ballcock, releasing water into the tank and/or bowl. As the water in the tank is restored, the float rises and shuts off the ballcock when the tank is full.

Balloon framed wall: Framed walls (generally over 10' tall) that run the entire vertical length from the floor sill plate to the roof. This is done to eliminate the need for a gable end truss.

Balloon loan: A loan that has a series of monthly payments with the remaining amount due in a large lump sum payment at the end.

Balloon payment: The final lump sum payment due at the end of a balloon mortgage.

Balloon-frame construction: A type of framing used in two-story homes in which studs extend from the ground to the ceiling of the second floor.

Balloon-frame construction.

Balsa: A tree that is native to tropical South America. Balsa is the lightest commercial wood. Balsa is used for life-belts, as insulation, in aircraft construction, as shock absorbers, and by hobbists for model aircraft and boats.

Balsam: The term is applied to certain resins obtained from a number of different trees. Balsam is a thick brown and black oily substance that is used medicinally or in the making of perfume.

Baluster: One of a series of vertical posts that are placed at regular intervals along the length a balustrade. Balusters are typically similar in pattern to the newel post but are much smaller in diameter or thickness. Balusters are attached to a top rail and to a bottom rail or directly to a stair tread or floor. See **Balustrade**.

Balustrade: A railing system that is found on stairs or along open areas between floors. Most balustrades contain newel posts, balusters, and top rails. Some balustrades contain bottom rails. See **Newel post, Baluster, Top rail, Bottom rail, Bread loaf top, Filler, Volute, Goose neck, One-quarter turn, Rosette, Skirt, Stair bracket, False tread,** and **Riser**.

Bamboo: A type of grass consisting of many different species, of which some grow to over

100 feet in height. Bamboo is put to many uses in light construction.

Bamboo construction.

Band joist: A piece of lumber to which the ends of the joists are nailed or screwed. A band joist is critical to the strength of the floor system because it holds the regular joist ends in their vertical position.

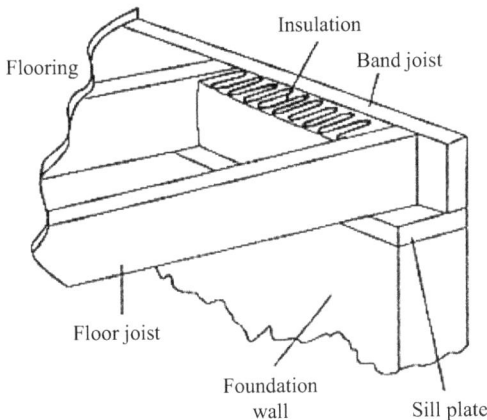

Band joist with floor joist and sill plate.

Band saw: A saw that uses a blade that is butt welded into an endless belt with teeth on one or both edges arranged to cut sequentially. The band usually rides on two wheels rotating in the same plane, although some bandsaws may have three or four wheels. Bandsawing produces uniform cutting action as a result of an evenly distributed tooth load. Bandsaws are used for woodworking, metalworking, and for cutting a variety of other materials. They are especially useful for cutting irregular or curved shapes, but can also be used to produce straight cuts.

Band sawmill: A sawmill that uses a thinner band saw blade (less kerf therefore less sawdust waste) than a circular saw. A bandsaw can also have teeth on both sides that allows cuts to be made in two directions instead of just one, improving efficiency and productivity. Band sawmills are generally portable.

Banister: A handrail for a staircase.

Bank inspector: An independent consultant engaged by a construction lender to provide periodic site inspections, to report on the progress of construction and general compliance, and to make recommendations for contractor payment applications. Inspections are made during construction and are performed for the benefit of the bank.

Bankruptcy: A proceeding in which an insolvent debtor can obtain relief from payment of certain obligations. Bankruptcies remain on a credit record for 7 years and can severely limit a person's ability to borrow.

Bar: The part of a chainsaw on which the cutting chain moves. Muntins are also referred to as bars. Also, a square or round piece of solid steel that is usually 6' or less in width.

Bar clamps: Bar clamps provide strong clamping pressure, permit two-handed tightening, and are ideal for joining boards to make wider panels. They are available in ranging capacities.

Bar clamps.

Bar joist: See **Steel joist**.

Barbed nail: A nail with a barbed shank that resists withdrawal.

Barbed nail.

Barber pole effect in book match: See **Veneer**.

Barberchair: A vertical split in a tree, that is often caused by an insufficient undercut or by neglecting to cut the sapwood on both sides of a heavy leaning tree before felling. This results in a stump that looks like a high-backed chair. The situation that creates a barberchair can be very dangerous to the faller.

Barge: A beam rafter that supports shorter rafters.

Barge board: A decorative board covering the projecting rafter (fly rafter) of the gable end. At the cornice, this member is a fascia board. Also spelled **bargeboard**.

Barge board.

Bark: The outermost, protective layer, of a tree.

Bark pocket: A small area of bark surrounded by normal wood. Also an opening between annual growth rings that contains bark. Bark pockets appear as dark streaks on radial surfaces and as rounded areas on tangential surfaces.

Bark pocket maple. Often sold as an exotic wood.

Barn poles: Roundwood used for fence posts, highway signs, etc. They can be purchased as treated or untreated lumber.

Barometric damper: An automatic adjustable device for regulating the draft through a fuel-burning appliance, thereby making operation of the appliance nearly independent of the chimney draft over its normal range of operation.

Barometric damper.

Barratry: The persistent incitement of litigation.

Barrel roof: A roof configuration with a partial cylindrical shape to it.

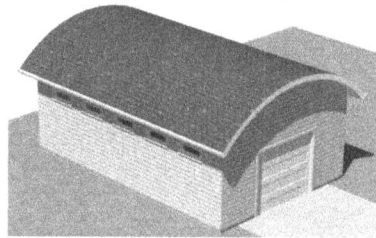

Barrel roof.

Barrier tape: Non-adhesive tape that is used to mark dangerous areas. The tapes have imprints such as "POLICE LINE-DO NOT CROSS" or "DANGER-DO NOT ENTER."

Barrier wrap: Treated or untreated roundwood encased in a wrap for protection of the wood and to keep wood treatments from leaching into the surroundings.

Basal area: The cross-sectional area of a tree, in square feet, measured at breast height. Used as a method of measuring the volume of timber in a given stand.

Base: In lease terminology a face, quoted, dollar amount that represents the rate or rent in dollars per square foot per year and typically referred to as the **base rate**. See **Average annual effective rent and Base rent**.

Also, as regards to trim and moldings, the base is applied where floor and walls meet to form a visual foundation. The base protects walls from kicks, bumps, furniture, and cleaning tools. A base may one, two, or three members. The base shoe and base cap are used to conceal uneven floor and wall junctions. Also known as **baseboard** and **base molding**. The shoe molding is sometimes called a **carpet strip**.

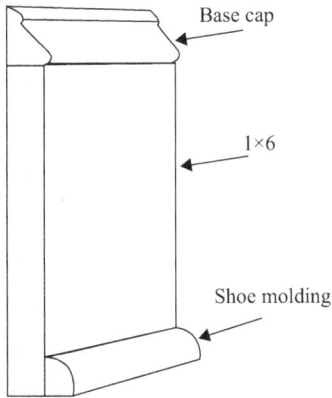

Three member baseboard.

Base block: A square or rectangular piece that is placed between the bottom of the casing and the floor. Also called **plinth block**. See **Corner block**.

Base cap: A component of a two part wood base molding. Base cap moldings tend to be tall and substantial. They are appropriate for applications that require exceptional depth of wood work design.

Base cap molding.

Base coat: The first coat of any finish substance on a surface, such as the first coat of synthetic stucco applied on the stucco sheathing over a wire or glass fiber mesh.

Base flashings: A continuation of a built-up roofing membrane, at the upturned edges of the watertight tray. They are normally made of bitumen-impregnated, plastic, or other non-metallic materials and applied in an operation separate from the application of the membrane itself. See **Counter flashings**.

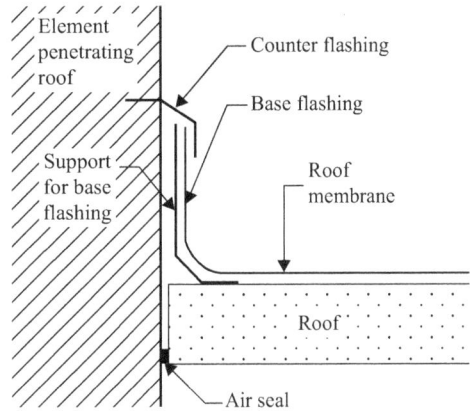

Base and counter flashings.

Base isolation: A method that uses flexible bearings to detach a building superstructure from its foundation in order to reduce earthquake forces.

Base line: A virtual line that establishes the point of departure for all building layouts in the field.

Base metal: The metal to be welded or cut.

Base molding: See **Base**.

Base of tree: That portion of a natural tree that is three feet and less above ground level.

Base plate: A steel plate that is welded to the base of a column to distribute the column loads over an area of foundation large enough to prevent crushing of the concrete and usually secured by anchor bolts. Sometimes spelled **baseplate**.

Base ply: The lowermost ply of roofing in a roof membrane or roof system.

Base rent: In commercial real estate, the minimum rent due to the landlord. Typically, it is a fixed amount. This is a face, quoted, contract amount of periodic rent. The annual base rate is the amount upon which escalations are calculated. See **Base**.

Base shear or Equivalent lateral force (ELF): Total shear force acting at the base of a structure. An important consideration in earthquake design.

Base sheet: The first layer of a multiple ply membrane roof system. Base sheets are installed by rolling strips of special base sheet style roll roofing over the roof deck and nailing the base sheet to the roof deck. Once the base sheet is in place, hot tar is mopped over it.

Base sheets that have been mechanically fastened.

Base shoe: See **Base**.

Baseboard: See **Base**.

Baseboard electric heat: Heating units that are installed in the floor. Usually they can be controlled by a central thermostat.

Baseline irrigation water use: The amount of water that would be used by a typical method of irrigation for the region.

Basement: A story partly underground, but having less than one-half its clear height (measured from finished floor to finished ceiling) below the curb level (NYC Department of Buildings definition). See **Cellar**.

In many areas a basement is simply any floor below the first story in a building. Also, igneous and metamorphic rocks underlying the sedimentary rocks of a region and extending to the base of the crust.

Basement window: A wood or metal in-swinging sash that is hinged at either the top or the bottom. Also referred to as **basement sash** and **cellar sash**.

Basement window inserts: The window frame and glass unit that is installed in the window buck.

Baseplate: See **Base plate**.

Basic building code (BBC): A minimum model regulatory code for the protection of public health, safety, welfare, and property by regulating and controlling the design, construction, quality of materials, use, occupancy, location and maintenance of all buildings and structures within a jurisdiction.

Basin: A bowl or sink.

Basin cock: A separate single hot or a single cold faucet.

A basin cock.

Basin wrench: A wrench with a long handle with jaws mounted on a swivel that allows the jaws to reach and handle nuts to fasten faucets to a previously installed sink.

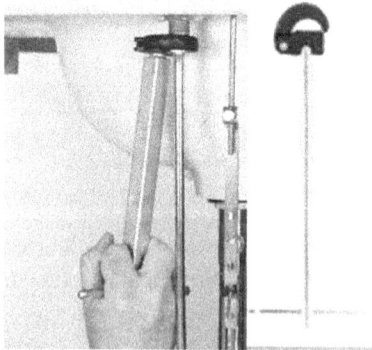

Basin wrench.

Basis: The total amount paid for a property, including equity capital and the amount of debt incurred.

Basis point: A basis point is used in finance to describe the percentage change in the value or rate of a financial instrument. One basis point is equivalent to 0.01% (1/100th of a percent) or 0.0001 in decimal form. In most cases, it refers to changes in interest rates and bondyields.

For example, if the Federal Reserve Board raises interest rates by 25 basis points, it means that rates have risen by 0.25% percentage points.

In the bond market, a basis point is used to refer to the yield that a bond pays to the investor. For example, if a bond yield moves from 5.50% to 5.70%, it is said to have risen 20 basis points.

The usage of the basis point measure primarily refers to yields and interest rates, but it may also refer to the percentage change in the value of an asset such as a stock. It may be heard that a stock index moved up 110 basis points in the day's trading. This represents a 1.10% increase in the value of the index.

Bas-relief: Raised or indented sculptural patterns that remain close to the surface plane.

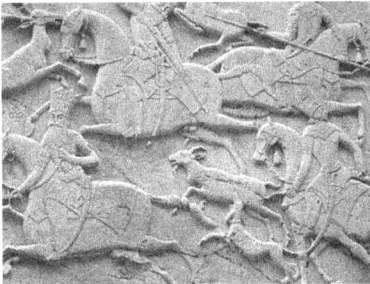

A sculpture in bas-relief.

Bastard: As used here, a scoundrel, villain, rogue, rascal, weasel, snake, snake in the grass, miscreant, good-for-nothing, reprobate; informal lowlife, creep, nogoodnik, scamp, scalawag, jerk, beast, rat, ratfink, louse, swine, dog, skunk, heel; slimeball, son of a bitch, SOB, scumbucket, scuzzball, scuzzbag, dirtbag, sleazeball, sleazebag; dated hound, cad; archaic blackguard, knave, varlet, whoreson, or in other words the guy who cheated you.

Other terms that begin with bastard:
- **Bastard file:** A course file for rough shaping of metal or wood.
- **Bastard granite:** A quarriers term for nearly any stone which may not be considered a true granite, particularly applied to gneiss.
- **Bastard sawn:** Lumber, primarily hardwoods, in which the annual rings make angles of 30–60° with the surface of the piece.

Bat: A half-brick.

Batch plant: A concrete mix batching plant that establishes all the proper proportions of ingredients that comprise a concrete design mix.

Bathroom: An area including a basin with one or more of the following: a toilet, a tub, or a shower.

Batt: A section of fiber-glass or rockwool insulation measuring 15 or 23 inches wide by 4–8 feet long and various thickness. Sometimes batts are faced, meaning they have a paper or foil covering on one side or they may be unfaced, i.e., without paper. It is possible to purchase batts that are faced on both sides.

Batten: A long, flat strip of squared wood or metal used to hold something in place or as a fastening against a wall. Also, a symmetrical pattern used to conceal the line where two parallel boards or panels meet.

Also, with regard to roofing: a cap or cover; in a metal roof, a metal closure set over, or covering the joint between, adjacent metal panels; a strip of wood usually set in or over the structural deck used to elevate and/or attach a primary roof covering such as tile; in a membrane roof system, a narrow plastic, wood, or metal bar that is used to fasten or hold the roof membrane and/or base flashing in place.

Batten type standing seam: A standing seam roof in which the panels are raised up and fastened together and then a batten strip is placed over the seam to form a watertight seal.

Batterboard: One of a pair of horizontal boards nailed to posts set at the corners of an excavation. They are used to indicate the desired level and also as a fastening for stretched strings to indicate outlines of foundation walls.

Bay: The opening between two columns or walls that forms a space.

Bay window: A window unit installed in an area that projects out from the wall. The exterior wall typically forms a 30° or 45° angle on each end of the bay window area. An **Angle Bay Window** refers to the angle departure from the plane of the wall. See also **Bow window**.

Bay window.

Beach marks: The progression marks appearing on a fatigue fracture surface indicating successive positions of an advancing crack front. The term came from an ocean beach as the tide goes out. Debris is left at the last previous high level creating a line of debris, or **beach mark**. The beach mark is always oriented perpendicular to the direction of the stress, so a failure analyst can tell how the failed part was loaded and where the crack started.

Bench marks are also surveyors' marks that are cut into a wall, pillar, or building and used as a reference point in measuring altitudes.

Beam: A structural member transversely supporting a load. A structural member carrying building loads (weight) from one support to another. Sometimes called a girder.

Other terms beginning with beam:

- **Beam chair:** A wire seat or support for reinforcing bars designed to maintain their location while concrete for a beam is poured around them.
- **Beam clamp:** Beam clamps allow workers to hang boxes, pipes, and strut from thick structural beams.

Beam clamp.

- **Beam column:** A structural member whose main function is to carry loads both parallel and transverse to its longitudinal axis.
- **Beam fireproofing:** Fire-resistant materials that cover a horizontal structural member, to insure structural integrity in the event of a fire
- **Beam formwork:** The system of support for freshly placed concrete for a horizontal structural member.
- **Beam furring:** Strips of wood or metal fastened to a horizontal structural member to form an airspace. Beam furring is used to give the appearance of greater thickness, or for the application of an interior finish such as plaster.
- **Beam pocket:** Notch or opening at the top of a bearing wall or supporting column that secures and bears the weight of a beam.

Bearing: A surface or point at which a load is transferred from one building element to another. Also, the distance that the bearing shoe or seat of a joist or joist girder extends over its masonry, concrete, or steel support. Also, a structural support, usually a beam or wall, that is designed by the specifying professional to carry reactions to the foundation.

Bearing area: The part of a structural element (beam, column, etc.) that rests on a support.

Bearing capacity: In geotechnical engineering, bearing capacity is the capacity of soil to support the loads applied to the ground, i.e., the maximum average contact pressure between the foundation and the soil that should not produce failure in the soil; there are three modes of failure that limit bearing capacity: general shear failure, local shear failure, and punching shear failure.

Ultimate bearing capacity is the theoretical maximum pressure that can be supported without failure. Allowable bearing capacity is the ultimate bearing capacity divided by a factor of safety.

Sometimes, on soft soil sites, large settlements may occur under loaded foundations without actual shear failure occurring. In such cases, the allowable bearing capacity is based on the maximum allowable settlement.

Bearing header: A beam placed perpendicular to joists and to which joists are nailed in framing for a chimney, stairway, or other opening.

Also, the horizontal structural member over an opening (for example over a door or window). For example, a wood lintel is a bearing header.

Bearing partition: A partition that supports any vertical load in addition to its own weight. See **Bearing wall**.

Bearing points: The location or point[s] on a member where another member supports it. Typically, a truss has two bearing points, one at each exterior wall.

Bearing wall: A wall that supports any vertical load in addition to its own weight such as a roof system or a floor system. A bearing wall is a structural part. See **Bearing partition**.

Bed: A coating of sedimentary rock.

Also, the top or bottom horizontal surface of a piece of, say, granite, that is covered when the piece is set in place.

Also, a filled or open space extending horizontally between adjacent pieces set in place.

Bed joint: The horizontal mortar joint in masonry construction.

Bedrock: A subsurface layer of earth that is suitable to support a structure, a solid, uniform, layer of rock of known density.

Beehive burner: A wood waste incinerator.

Before-tax income: Total income before taxes are deducted.

Before-tax investment value: The sum of the present values of the mortgagor and mortgagee of property.

Bell: In piping, bell typically refers to a gasketed joint that will fit over a pipe or the spigot of another fitting.

Bell-bottom pier hole: A type of shaft or footing excavation, the bottom of which is made larger than the cross section above to form a belled shape.

Bell-bottom pier hole.

Bell and spigot joint: A type of joint used in the cast iron pipe where the male pipe end (the spigot) fits into the female end (the bell) and is then caulked with oakum and sealed with lead.

Belt sander: This tool saves on elbow grease compared to a hand plane. Belt sanders are powerful and able to remove stock aggressively when fitted with a coarse abrasive belt. They come in a range of sizes.

Bench grinder: Bench grinders can buff, clean, polish, sharpen, and remove rust, among many uses.

Bench level: This tool is designed primarily for machinists who do precision work.

Benched excavation: An excavation procedure that utilizes platforms to stabilize the slope of a deep excavation. Also referred to as a benching system.

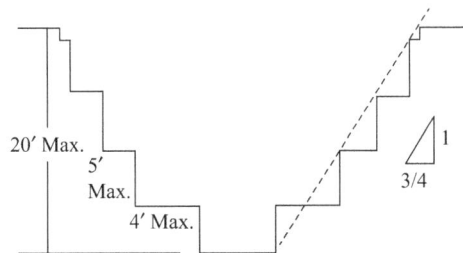

Multiple benched excavation.

Benchmark: A surveyor's mark that is cut into a wall, pillar, or building and used as a reference point in measuring altitudes.

Bend: To curve. When bending occurs one side squeezes together in compression, and the other side stretches apart in tension. See **Bending stress**.

Bend test: A requirement for steel wires and bars that are used as pre-tensioning steel and post-tensioning steel for concrete reinforcement to establish their ability to be bent around a pin without cracking.

Bendable concrete: A new product, spray-on "bendable" concrete may sound oxymoronic, but it is being used to shore up vulnerable older buildings in Christchurch, New Zealand. The product has polyvinyl alcohol synthetic fibers bonded to the concrete to give it tensile strength.

Benderboard: Pliable, lightweight board that is used for making concrete patios, in gardens, and as woven fencing: often 4–6 inches (10–15 cm) wide and 1/4–1/2" (0.6–1.2 cm) thick. Benderboard is usually made from California redwood.

Bending moment: The internal load generated within a bending element whenever a pure moment is reacted, or a shear load is transferred by beam action from the point of application to distant points of reaction.

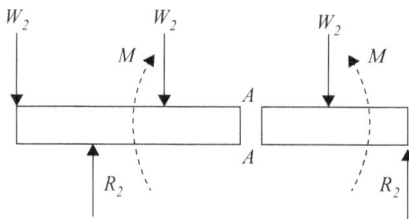

If the bending moments of the forces to the left of A-A are clockwise, then the bending moment of the forces to the right of A-A must be counter-clockwise.

Bending moment at A-A is defined as the algebraic sum of the moments about the section of all forces acting on either side of the section.

Bending moments are considered positive when the moment on the left portion is clockwise and on the right counter-clockwise. This is referred to as a sagging bending moment as it tends to make the beam concave upward at A-A. A negative bending moment is termed hogging.

Bending strength: An alternate term for **flexural strength**. It is most commonly used to describe flexure properties of cast iron and wood products.

Bending stress: Bending stress is zero at the neutral axis and assumed to increase linearly to a maximum at the outer fibers of the section.

The formula for bending stress in a beam in the elastic range:

Bending stress (in psi) = $(M \times c)/I$, where "M" is the calculated bending moment at the section in in-lbs, "I" is the moment of inertia of the section in inches around the neutral axis, and "c" is the verticle distance from the neutral axis to the point at which the stress is desired in inches.

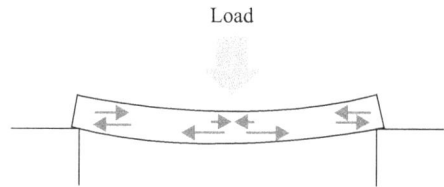

Bending stress.

Beneficial occupancy: The point at which the project is complete in nature so as to allow the owner to utilize the project for its intended use. The mechanical systems, life safety systems, telecommunications systems, and any other systems that are required to properly utilize the project are complete and in good working order. The remaining items to be completed are of a nature that the correction does not cause inconvenience to the owner or disruption to the owner's normal operations.

Beneficiary: The lender who makes a loan, also called a mortgagee. The person borrowing money is the mortgagor.

Benioff zone: A dipping planar (flat) zone of earthquakes that is produced by the interaction of a downgoing oceanic crustal plate with a continental plate. These earthquakes can be produced by slip along the subduction thrust fault or by slip on faults within the downgoing plate as a result of bending and extension as the plate is pulled into the mantle. Also known as the Wadati-Benioff zone.

Bent: Part of a bridge substructure. A rigid frame commonly made of reinforced concrete or steel that supports a vertical load and is placed transverse to the length of the bridge. Bents are commonly used to support beams and girders. An end bent is the supporting frame forming part of an abutment. The vertical members of a bent are called columns, piers, or piles. The horizontal member resting on top of the columns is a bent cap. The columns stand on top of some type of foundation or footer that is usually hidden below grade. A bent commonly has at least two or more vertical supports.

Bent is also used to describe a capped pile pier. A support having a single column with bent cap is sometimes called a **hammerhead pier.**

Bentonite: An absorbent aluminium phyllosilicate, essentially impure clay that consists mostly of montmorillonite. Bentonite usually forms from the weathering of volcanic ash in the presence of water. There are different types of bentonite, each named after the respective dominant element, such as potassium (K), sodium (Na), calcium (Ca), and aluminum (Al).

Bentonite slurry is used in construction, where the slurry wall is a trench filled with a thick colloidal mixture of bentonite and water. A trench that would collapse due to the hydraulic pressure in the surrounding soil does not collapse as the slurry balances the hydraulic pressure. Forms for concrete, and rebar, can be assembled in a slurry filled trench, and then have concrete fill the trench from the bottom up. The liquid concrete being heavier than the bentonite slurry displaces it. The bentonite slurry can be re-used.

Bentonite is also used to seal large ponds and for similar uses.

Berm: A flat strip of land, raised bank, or terrace. Berms often border a river or canal.

Berm.

Bermuda seam: A metal roof that has a step profile.

Bermuda seam (roof).

Bernoulli's principle: So named after the Dutch–Swiss mathematician Daniel Bernoulli who published this principle in his book "Hydrodynamica" in 1738. Bernoulli's principle states that in fluid dynamics an inviscid flow (having no or negligible viscosity), an increase in the speed of the fluid occurs simultaneously with a decrease in pressure or a decrease in the fluid's potential energy.

There are actually different forms of the Bernoulli equation for different types of flow. The simplest form is valid for incompressible flows (e.g., most liquid flows) and also for compressible flows (e.g., gases) moving at low Mach numbers. Other more advanced forms may in some cases be applied to compressible flows at higher Mach numbers.

Bernoulli's principle can be derived from the principle of conservation of energy that states in a steady flow, the sum of all forms of mechanical energy in a fluid along a streamline is the same at all points on that streamline, i.e., the sum of kinetic energy and potential energy must remain constant throughout. Therefore an increase in the speed of the fluid occurs proportionately with an increase in both its dynamic pressure and kinetic energy, and a decrease in its static pressure and potential energy. If the fluid is flowing out of a reservoir, the sum of all forms of energy is the same on all streamlines because in a reservoir the energy per unit volume the sum of pressure and gravitational potential is the same everywhere.

Bernoulli's principle can also be derived directly from Newton's second law. If a small volume of fluid is flowing horizontally from a region of high pressure to a region of low pressure, then there is more pressure behind than in front. This gives a net force on the volume, accelerating it along the streamline.

Fluid particles are subject only to pressure and their own weight. Therefore, when a fluid is flowing horizontally and along a section of a streamline, if the speed increases, this increase can only be because the fluid on that section has moved from a region of higher pressure to a region of lower pressure. If its speed decreases, it can only be because it has moved from a region of lower pressure to a region of higher pressure. As a result, within a fluid flowing horizontally, the highest speed occurs where the pressure is lowest, and the lowest speed occurs where the pressure is highest.

As you look at the above picture intuition may tell you that the pressure will be greatest in the narrow portion of the tube. But in fact, if we conducted an experiment by putting pressure gauges at various points along the tube, we would have found that the air pressure is lowest where the air is moving fastest.

Bequest: Personal property given to a person through a will.

Berber carpet: A style of carpet with a distinctive, short looped pile. See **Wool carpet**, **Nylon carpet**, **Indoor–outdoor carpet**, **Sculptured carpet**, and **Shag carpet**.

Betterment: An improvement that increases a property's value as opposed to repairs that maintain the value.

Bevel: A machine angle other than a right angle, i.e., a 3° bevel that is equivalent to a 1/8 inch drop in a 2 inch span (1 mm in 16 mm).

Bevel cut: A cut at a non-right angle to the main surface, forming a sloping surface. Such a cut is beveled. See **Miter cut**.

Bevel cut.

Beveled edge: An edge of the door that forms an angle of less than 90° with the face of the door, such as a 3° beveled edge.

Beveled glass: Plate glass that has its perimeter ground and polished at an angle.

Bevel-edge chisel: Used in fine cabinetry work. These chisels often feature a chrome-vanadium steel blades and double-hooped boxwood handles; boxwood is considered to be the best wood for tool handles because of strength and shock-absorbing qualities.

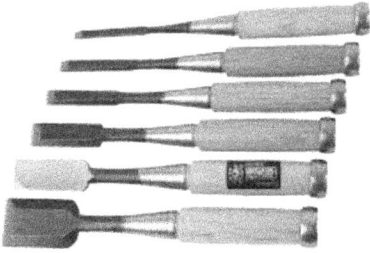

A set of bevel-edge chisels.

Bevel protractor: Used to read, transfer and mark angles quickly and accurately.

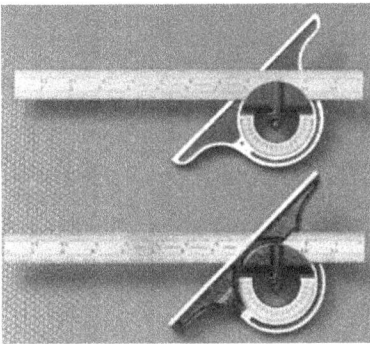

Bevel protractor.

Beveled edge square: Useful when extreme accuracy is required to determine 90° angles.

Beveled edge square.

Bicycle rack: Outdoor bicycle racks, bicycle lockers, and indoor bicycle storage rooms.

Bid: A formal offer by a contractor, in accordance with plans, specifications, terms, and conditions, to do a project in all or in part at a certain price. A bid is a complete proposal that is submitted in competition with other bidders to execute a specified job(s) within a prescribed time, and not exceeding a proposed amount. A bid usually includes labor, equipment, and materials, a guarantee to complete a project on time and within budget. The bid-receiving party may reject the bid, make a counter offer, or turn it into a binding contract by accepting it. See also **Offer** and **Proposal**.

In financial markets a bid is the price at which prospective buyers are willing to buy commodities, foreign exchange, or securities.

Bid bond: A bond issued by a surety on behalf of a contractor that provides assurance to the recipient of the contractor's bid that, if the bid is accepted, the contractor will execute a contract and provide a performance bond. Under the bond, the surety is obligated to pay the recipient of the bid the difference between the contractor's bid and the bid of the next lowest responsible bidder if the bid is accepted and the contractor fails to execute a contract or to provide a performance bond.

Bid documents: The documents that are distributed to contractors by the owner for bidding purposes. They include drawings, specifications, form of contract, general and supplementary conditions, proposal forms, and other information including addenda.

Bid security: Funds or a bid bond submitted with a bid as a guarantee to the recipient of the bid that the contractor, if awarded the contract, will execute the contract in accordance with the bidding requirements of the contract documents.

Bid shopping: A practice by which contractors, both before and after their bids are submitted, attempt to obtain prices from potential subcontractors and material suppliers that are lower than the contractors' original estimates on which their bids are based, or after a contract is awarded, seek to induce subcontractors to reduce the subcontract price included in the bid. Bid shopping is considered to be unethical.

Bidding requirements: The procedures and conditions for the submission of bids. The requirements are included in documents, such as the notice to bidders, advertisements for bids, instructions to bidders, invitations to bid, and sample bid forms.

Bidding war: Offers from multiple buyers for a piece of property. Agents also sometimes compete to list a house for sale.

Bidet: A bathroom fixture used for washing one's genital and anal area.

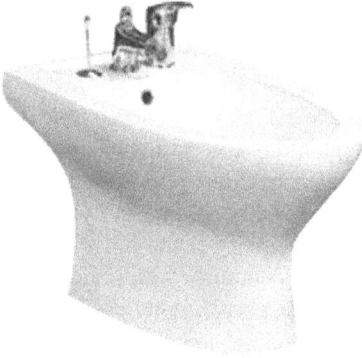

A bidet.

Bifurcate, Bifurcation: A division into two or more branches so as to form a fork, e.g., a fork in the road or river.

Also, the phenomenon whereby a perfectly straight member may either assume a deflected position, deflect then twist out of plane, or may remain in an undeflected configuration; the bending of a structural member about two perpendicular axes at the same time.

Big box stores: Large, warehouse-type stores. Costco, Walmart, Home Depot, and Lowes are examples.

Bight: A work area made hazardous by a line or equipment under tension.

Bilateral contract: A contract in which the parties involved give mutual promises. Also called **reciprocal contract**.

Bill of lading: In commercial law, the receipt a carrier gives to a shipper for goods given to the carrier for transportation. The bill evidences the contract between the shipper and the carrier, and can also serve as a document of title creating in the person possessing the bill ownership of the goods shipped.

A bill of lading usually provides a list that gives each part or mark number, quantity, length of material, total weight, or other description of each piece of material that is shipped to a jobsite. See **Shipping list**, Bill of materials, and Cut-list.

Bill of materials: A list of items or components used for fabrication and accounting purposes. See **Cut-list**, **Shipping list**, and **Bill of lading**.

Bill of sale: A document that transfers ownership of personal property.

Binder: A receipt for a deposit to secure the right to purchase a home at an agreed term by a buyer and seller.

Also, a report issued by a title insurance company that details the condition of a home's title and provides guidelines for a title insurance policy.

Also, asphalt binder that functions as an inexpensive waterproof, thermoplastic adhesive; it is the glue that holds the road together. In its most common form, asphalt binder is simply the residue from petroleum refining. To achieve the necessary properties for paving purposes, binder must be produced from a carefully chosen crude oil blend, and processed to an appropriate grade. For a few applications, additives (usually polymers) are blended or reacted with the binder to enhance its properties.

Also, solid ingredients in a coating that hold the pigment particles in suspension and attach them to the substrate. Regarding paint, binders consist of resins (e.g., oils, alkyd, and latex). The nature and amount of binder determine many of the paint's performance properties, e.g., washability, toughness, adhesion, color retention, etc.

Bio-based materials: Engineering materials that are made from substances derived from living matter. Typically the term refers to modern materials that have undergone more or less extensive processing and not to unprocessed materials such as wood and leather. Bio-based materials fall under the broader category of bioproducts or bio-based products that include materials, chemicals, and energy derived from renewable biological resources. Bio-based materials are often, but not always, biodegradable. They are also referred to as **biomaterials**.

Biodegradation: Bio-degradation is the natural process of breaking down organic matter into nutrients that can be used by other organisms. Degradation means decay, and the "bio-" prefix means that the decay is carried out by a huge assortment of bacteria, fungi, insects, worms, and other organisms that eat dead material and recycle it into new forms.

Biofuel-based electrical systems: Electrical power systems that run on renewable fuels derived from organic materials, such as wood by-products and agricultural waste.

Biological control: The use of chemical or physical water treatments to inhibit bacterial growth in cooling towers.

Biological monitoring: The analysis of a person's blood and/or urine, to determine the level of a contaminant, such as lead, in the body.

Biomass: Plant material from trees, grasses, or crops that can be converted to heat energy to produce electricity.

Biomass boiler: Boilers that burn bark, sander dust, and other wood-related scrap that is not usable in product production. These boilers are also known as **hogged fuel boilers**. Biomass boilers are usually used to make steam and heat for on-site use.

Bird bath: Small, inconsequential amounts of water on a roof that quickly evaporate.

Bird peck: A mark or wound in a tree or piece of wood caused by birds pecking on the growing tree in search of insects.

Bird screen: Material, usually galvanized or stainless steel wire mesh, to prevent birds from entering HVAC and other intakes and openings located at the closure assembly of a building. See **Bird stop**.

Bird stop: Material, usually galvanized or stainless steel wire mesh, used to fill the space under the first course of tile at the eave line to prevent birds from nesting in the roofing. See **Bird screen**.

Biscuit: A joint between two boards made by using a biscuit saw to notch out the ends of the joined boards. A premanufactured biscuit fits into the slots made by the biscuit saw. The glued biscuit swells as the glue soaks in, forming a very tight fit when the joint dries.

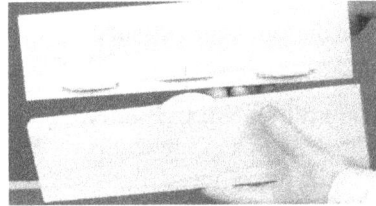

Biscuit joint.

Biscuit saw: A special saw used to cut a notch in boards that will be joined with a biscuit joint.

Bisque: The rough, unpolished finish found in unglazed areas of vitreous china fixtures, such as inside the tank or the bottom of the bowl.

Bit: A drill point that has a variety of uses with braces and drills. Each bit is designed for a specific application such as masonry, wood, steel, or other materials.

Bite: The thickness of chip material that each cutting edge of a saw removes with one pass.

In general, the formula is: bite = distance between teeth (in inches) times feed speed (in feet per minute) divided by saw velocity (in feet per minute). Stated more simply, it is the distance the log advances into the saw between successive teeth.

Bitumen: A class of amorphous, black or dark colored, (solid, semi-solid, or viscous) cementitious substances, natural, or manufactured that is composed principally of high-molecular-weight hydrocarbons that are soluble in carbon-di-sulfide. Bitumen is found in petroleum asphalts, coal tars and pitches, wood tars and asphalts. Also, a generic term used to denote any material composed principally of bitumen, typically asphalt or coal tar.

Bituminous emulsion: Bituminous particles suspended in water or other solution. See also **Asphalt emulsion**.

Biweekly mortgage: A mortgage that requires payments every 2 weeks.

Bi-fold door: A door with two slabs that are connected to each other with hinges. When closed the slab ends butt against each other. When opened the slabs fold onto each other. A track at the top of the by fold door holds the slabs in position. Sometimes spelled **Bifold**.

Bi-metal utility blades: Bi-metal blades are produced through a patented process of combining two types of metal, each with its own attributes and benefits. The first is spring steel that provides flexibility to the blade, resulting in a blade that will not break under normal work conditions. The second metal, high-speed steel that delivers a hardened cutting edge that stays sharp longer than traditional carbon blades.

Black label shingles: Utility grade shingles with sapwood, flat grain, and large knots. Commonly used on garages and barns.

Black oiled: A pipe surface that is protected with a varnish-type oil on the O.D. It is for temporary corrosion protection during transit and in short-term storage.

Blackberry: A small bubble found in the flood coat of an aggregate-surfaced built-up roof. They are usually the result of trapped moisture vapor. Blackberries are also referred to as **tar boils** and **blueberries**.

Blackwater: There is no single definition for blackwater. Wastewater from toilets and urinals is always deemed blackwater. Wastewater from kitchen sinks, showers, and bathtubs may also be considered blackwater. You should check your local codes for the definition used in your area.

Bladder pressed panel: See **Membrane pressed panel**.

Blade diameter: The measurement of a saw blade measured on the extreme outside edge of two opposite tips.

Blanket insulation: Insulation in rolled-sheet form, often backed by treated paper or foil that forms a vapor barrier.

Blanket insurance policy: A policy that covers more than one person or piece of property.

Blanket mortgage: A mortgage that covers more than one property owned by the same borrower.

Blankets: Fiberglass or rock-wool insulation that comes in long rolls 15 or 23 inches wide.

Blanks: An assembly of identical or nearly identical multiple structural elements or built-ups that are fabricated prior to engineering or detailing being complete. Blanks are fabricated to take advantage of economics of scale or to gain time.

With regard to lead testing, non-exposed samples of the medium that is being used for lead testing (i.e., wipe or filter) that are analyzed to determine if the medium has been contaminated with lead (e.g., at the factory or during transport). Also, empty spaces on a document that are to be filled in.

Also, manufactured articles of a standard shape or form that are ready for final processing, as by stamping or cutting such as a key blank.

Blasting: A method of cleaning or of roughening a surface by a forceable stream of sharp angular abrasive.

Also, an explosive charge, such as dynamite or a bomb that sets off a violent explosion. The violent effect of such an explosion consists of a wave of increased atmospheric pressure followed immediately by a wave of decreased pressure.

Bleaching: Using chemicals, such as bleach, to lighten or to remove color in a wood.

Bleed: Paint is said to bleed when it penetrates through another coat of paint.

Bleed-off: To draw off (liquid or gaseous matter) from a container. Also, to exude a fluid such as sap.

Also, to divert a controllable portion of hydraulic pump delivery directly to reservoir.

Also, the release of built-up solids in a cooling tower, accomplished by removing a portion of the concentrated recirculating water that carries dissolved solids. Such bleed-off is also referred to as **blowdown**.

Bleed-off rate: The frequency with which the dissolved minerals and dirt are removed from a cooling tower. This rate varies depending on the mineral content and scaling tendency of the entering water.

Bleeder strip: A starter-strip that is used along rake edges in conjunction with asphalt shingle roofing. Also known as a **rake-starter**.

Bleeder strip/rake-starter.

Bleeding channels: Passageways left by escaping excess water during curing of the concrete. These channels weaken the concrete and allow water, salts, and other chemicals to attack the concrete interior.

Blended repair tapering: Referring to end splits of wood, a repair using wood or filler similar in color so as to blend well with adjacent wood.

Blending: A color change that is detectable at a distance of 6–8 ft. (1.8–2.4 m) but that does not detract from the overall appearance of the woodwork.

With regard to repairs, wood, or other filler insertions similar in color to adjacent material so as to blend well.

Blind: Lacking perception or discernment; not able to notice; something; concealed or closed.

Terms beginning with blind:
- **Blind hole:** A hole that is not completely drilled through.
- **Blind miter:** A joint made by butting the first piece of material into a corner and then shaping the second piece so that it conforms to the outline of the first piece. Also called a **cope joint**.
- **Blind mortise and tenon:** A method of construction of stile and rail wood doors where openings are machined into, but not through the stiles and where the ends of the rails are so machined as to fit these openings.
- **Blind nailing:** Driving a nail into a part of the board that will not be visible on the finished product. See also **Face nailing**.
- **Blind sample:** A subsample submitted for lead analysis with a composition and identity known to the submitter but not to the analyst. Blind samples are used to test the analyst's laboratory proficiency in conducting lead measurements.
- **Blind stop:** A piece of rectangular molding that is utilized in the construction of window frames. A blind stop is nailed between the outside trim piece and the outside sashes. It is designed to serve as a stop for the sash or a screen. Also spelled **blindstop**.

A blind stop has the additional benefit of keeping air from getting into the building through the edges of the window. The blindstop serves the same purpose as the inside stop, except it is positioned for function on the exterior of the window.

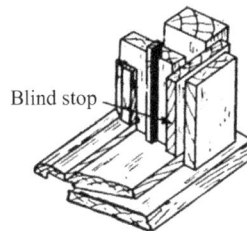

Blind stop.

- **Blind thrust fault:** A thrust fault that does not rupture all the way up to the surface so there is no evidence of it on the ground.

It is buried under the uppermost layers of rock in the crust.

Blinding: A layer of concrete covering the ground so that reinforcement can be laid out without becoming contaminated.

Blistering: Formation of dome-shaped projections in paints or varnish films resulting from local loss of adhesion and lifting of the film from the underlying surface.

Also, a spot or area where veneer does not adhere.

Block: A masonry product that is used in the assembly of footings, foundation walls, along with both interior and exterior walls. Blocks are precast to specific dimensions and are available in many shapes and styles.

Also, a short term used to describe any concrete masonry unit.

Also, a piece of wood built into a roof assembly to stiffen the deck around an opening, support a curb, or for use as a nailer for attachment of membranes or flashing.

Block backsplash: A thick, relatively short backsplash with a square top. See **Backsplash** and **Full backsplash**.

Block backsplash. This one has an integrated knife holder.

Block building system: A building system that involves using block to construct the perimeter foundation system and the exterior bearing walls. The footings in a block building system may also be built from block but are often made from concrete.

Block cell: The open cavities often found in blocks. These cells may be filled with either grout or insulation.

Block out: To install a box or barrier within a foundation wall to prevent the concrete from entering an area. For example, foundation walls are sometimes "blocked" in order for mechanical pipes to pass through the wall, to install a crawl space door, and to depress the concrete at a garage door location.

Block plane: A tool that is used for trimming end grain, smoothing small wood pieces and for edge-planing plywood, particleboard, and plastic laminate.

Block plane.

Block reinforcing: The insertion of steel reinforcing rods and or steel wire truss reinforcing in a block assembly wall.

Blocking: Short pieces of material used to provide solid bridging over bearing points and to block fire from quickly spreading into other parts of the framing. Usually found in joists over every bearing wall or beam and in studs at every connection with stair stringers or dropped ceilings.

Also, sections of, often preservative treated, wood that are built into a roof assembly. They are usually attached above the deck and below the membrane or flashing to stiffen the deck around an opening, act as a stop for insulation, support a curb, or to serve as a nailer for attachment of the membrane and/or flashing.

Also, with regard to doors, wood shims used between the door frame and the vertical structural wall framing members; **door blocking**.

Also, rafters are said to be **blocked, rafters** when short "2 by 4's" are used to keep the rafters from twisting. They are usually installed at the ends and at mid-span.

Blowdown: To open a valve in operating equipment to eject any sediment that has collected. Dissolved solids and particles entering a boiler and other operating systems through the make-up water remain behind during operation. Chemical treatments and blowdowns are used to eliminate these solids at the same rate as they are added from make up water.

Also, an accident in a nuclear reactor in which a cooling pipe bursts causing the loss of essential coolant.

Also, tree or trees felled by wind. Also known as **windfall**.

Blown asphalt: See **Air blown asphalt**.

Blow-offs: When shingles are subjected to high winds, and are blown off the roof deck.

Blow-out water closet: A water closet bowl that has a non-siphoning trapway at the rear of the bowl, and an integral flushing rim and jet.

Blown-in insulation: Insulation that is broken down by a machine that blows it into place. Blown-in insulation is often installed in inaccessible areas above ceilings and in wall cavities.

Blue print(s): A type of copying method often used for architectural drawings. Usually blue prints describe the drawing of a structure that is prepared by an architect or designer for the purpose of design and planning, estimating, securing permits, and actual construction. Also referred to a **blue line**.

Blue label shingles: Shingles that are made from the highest quality all heartwood and all clear cedar with 100% edge grain. Most residential structures use red or blue label shingles.

Blue sky laws: Regulations on the sale of securities to prevent consumers from investing in fraudulent or high-risk companies without being informed of the risks.

Blue stake: Another phrase for **utility notification**. This is when a utility company (telephone, gas, electric, cable TV, sewer, and water, etc.,) comes to the job site and locates and spray paints the ground and/or installs little flags to show where their service is located underground.

Blue stain: A bluish or dim-grayish discoloration of the sapwood that is caused by certain dark-colored fungi, on both the surface and in the interior of the wood. Also known as **sap stain** or **sapwood stain**.

Blue-ribbon condition: A house that has been maintained close to its original condition. Also called **mint condition**.

Board: Lumber that is nominally less than two inches thick and two inches or more wide.

Board and batten siding: Vertical siding in which boards are installed first with small spaces between them. Narrower boards called battens are then installed over the small spaces. Also spelled **board-and-batten**.

Board and batten siding.

Board foot: Lumber quantities are expressed in feet, board measure (BM or ft. b.m.) or in board feet (bd. ft.), or in thousand board feet (M bd. ft.). One board foot is the amount of lumber in a rough-sawed board 1-foot long, 1-foot wide and 1-inch thick (144 cubic inches) or the equivalent volume in any other shape, e.g., 1-foot long by 6-inches wide by 2-inches thick. The originals or "nominal" dimensions and volumes determine the number of board feet in a quantity of dressed lumber. The process of surfacing or other machining does not lessen the board feet that are sold. Examples: $1" \times 12" \times 16' = 16$ board feet, $2" \times 12" \times 16' = 32$ board feet.

Importantly, thickness and width used in fbm calculations are nominal and not the finished sizes.

Board on board siding: Vertical siding installed with gaps between boards. Boards of the same size are then installed over the gaps. See **Board on batten siding**.

Board of equalization: A state board charged with ensuring that local property taxes are assessed in a uniform manner.

Board on board siding.

Board siding: Siding made from wood, hardboard, or pressed wood byproducts; usually installed horizontally, one board at a time.

Body: Regarding coatings, the thickness or viscosity of the fluid.

Body belt: A strap with means both for securing it about the waist and for attaching it to a lanyard, lifeline, or deceleration device. Body belts are not acceptable as part of a personal fall arrest system. See **Body harness**.

Body force: An external force acting throughout the mass of a body. Gravity is a body force. An inertial force is a body force.

Body harness: Straps that may be secured about the employee in a manner that will distribute the fall arrest forces over at least the thighs, pelvis, waist, chest, and shoulders, with means for attaching it to other components of a personal fall arrest system.

Full body harness.

Body jets: Wall-mounted shower sprays that are usually installed in multiples at various heights.

Body wave: A seismic wave propagated in the interior of the earth. P and S waves are examples.

Body-wave magnitude: See **Magnitude** and **Body wave**.

Boiled Oil: Linseed (sometimes soya) oil that was formerly heated for faster drying. Today, chemical agents are added to speed up the drying process.

Boiler: A closed vessel or arrangement of enclosed tubes in which water is heated to supply steam to drive an engine or turbine or provide heat.

Simplistically stated, steam generation equipment.

Boiler feed: A check valve that controls inlet water flow to a boiler.

Boilerplate: Form language that is used in deeds, mortgages, and other documents. Details can be added by individual parties.

Boiler plug: A threaded plug of solid construction fitted to boilers. Removal allows inspection and washing out.

Bollard: A short pipe length that is placed vertically in the ground and filled with concrete to prevent vehicular access or to protect property from damage by vehicular encroachment. Used extensively where acts of terrorism are likely.

Also, a steel or cast iron post to which ships are tied.

Bolt: An externally threaded fastener with a preformed head on one end.

In reference to wood, a short section of a tree trunk. And in veneer production, a short log of a length suitable for peeling in a lathe.

Bolt circle: A circle scribed around the arbor hole of a circle saw that has one or more holes laid out to receive bolts.

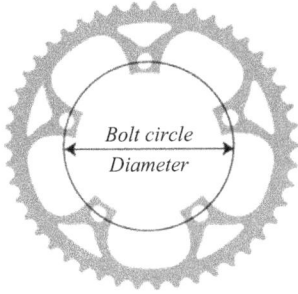

Bolt circle.

Bolt preload: The high tension stress that is intentionally developed in a bolt during installation and assembly.

Bolted splice: The connection between two structural members that are joined at their ends by bolting to form a single, longer member.

Bona fide: A legal term that refers to actions or persons that are honest and in good faith.

Bond: Siding made from wood, hardboard, or pressed wood byproducts; usually installed horizontally, one board at a time.

Also, the arrangement or pattern of bricks (or other masonry units) in a wall. Each unit should overlap the unit below by at least one quarter of a unit's length, and sufficient bonding bricks should be provided to prevent the wall splitting apart. Common bond patterns are **Flemish**, **Stretcher**, **English**, and **English Garden Wall**.

English bond.

Also, an amount of money (usually $5,000–$10,000) deposited with a governmental agency to secure a contractor's license. The bond may be used to pay for the unpaid bills or disputed work of the contractor.

You may be required to provide any or all of the following bonds:

- Performance bonds guarantee the faithful performance of all work required to complete the contract.
- Payment bonds guarantee the payment of all bills for labor and materials used in the work, including materials purchased for the project but not included in it.
- Bid bonds guarantee that the contractor, upon being declared the successful bidder, will enter into a contract with the owner for the amount of the submitted bid and will provide contract bonds as required.
- Maintenance bonds guarantee the performance of the contract. They cover the time after the building is completed. This bond stipulates the building stability and maintenance responsibility of the construction company after the building has been completed.
- Permit bonds are required by some local authorities to indemnify the authority against any costs arising from the contractor using public street, crossing curbs, connecting to public utility lines, etc.
- Supply bonds guarantee the quality and quantity supplied to the owner. These are rarely used on construction projects.

Bond beam: A horizontal beam poured inside the U block for reinforcement of block walls. A bond beam is made by filling the block cells with either grout or insulation up to the level of the bottom of the U block. Reinforcing steel is placed, and the U block is filled with grout.

Bond beam.

Bond strength: The unit load applied in tension, compression, flexure, cleavage, or shear, required to break an adhesive assembly, with failure occurring in or near the plane of the bond.

Bonded core: Stiles and rails (edge bands) are securely glued to the core prior to application of crossbanding, three ply skins, veneers, or laminate.

Bonded glass: Glass pieces that have been adhered together by glue, resin, or cement.

Bonding, bonded: The permanent joining of metallic parts to form an electrically conductive path that ensures electrical continuity and the capacity to conduct safely any current likely to be imposed.

Bonding agent: A chemical agent that is used to create a bond between two layers.

Bonding company: A properly licensed surety that executes surety bonds, or bonds payable to the owner. The bonding company secures the performance on a contract either in whole or in part, or for payment for labor and materials.

Bonding jumper: The permanent joining of metallic parts to form an electrically conductive path that ensures electrical continuity and the capacity to conduct safely any current likely to be imposed. (This is not the same as grounding, but bonding jumpers are essential components of the bonding system that is an essential component of the grounding system.

Note: The NEC does not authorize the use of the earth as a bonding jumper. That is because the resistance of the earth is more than 100,000 times greater than that of a bonding jumper.)

Bone-dry ton (BDT): The volume of wood chips (or other bulk material) that would weigh one ton (2000 pounds, or 0.9072 metric tons) at zero percent moisture content. Also known as an **ovendry ton** and **bone dry metric ton**.

Note: There is a difference between "ton" (also known as US short ton) and "metric ton" (also known as tonne). A ton weighs 2000 pounds, while a metric ton weighs 2204.623 pounds.

Bone-dry unit (BDU): A unit of volume that is used in the forest industry to measure bulk products such as wood chips. It is a unit of volume that would weigh 2,400 pounds at zero percent moisture content.

Book matched veneer: A veneer pattern produced by turning over every other veneer strip. On a surface the strips look much like mirror images of each other. See **Veneer**, **Whole piece veneer**, **Slip matched veneer**, and **Unmatched veneer**.

Book size: The height and width of a door prior to prefitting.

Book value: The value of a property as a capital asset based on its cost plus any additions, minus depreciation.

Booking: The process of folding the pasted side of wallpaper over onto itself. Booking allows the glue to cure without drying unevenly.

Boom: A truck used to hoist heavy material up and into place. A boom is used to put trusses on a home or to set a heavy beam into place.

Also, raft of logs or a string of logs chained together, end to end, used to hold floating logs. A means of log storage or transportation.

Boomboat: Any boat used to push or pull logs, boom, bundles, or bags, in booming ground operations.

Booster tile: Small roof tile placed under the cap tiles on the starter course only.

A tile roof that is showing booster tiles.

Boot: A piece of material preformed to protect roof penetrations from dirt, moisture, and other foreign and/or damaging substances.

Border cut: Referring to ceilings, a cut made on both ceiling panel and grid at the perimeter.

Bore: The large hole that is physically drilled through a door to allow for installation of hardware (such as a latchset like a door knob, lever, deadbolt, entry set, etc.). Bore holes are usually 2–1/8" in diameter.

Boring test: An analysis of soil in which holes are bored into the ground and samples are removed.

Borough: A section of a city that has authority over local matters, e.g., New York City has five boroughs: Manhattan, Brooklyn, Queens, Staten Island, and the Bronx (New Yorkers always refer to it as "The Bronx").

Borrow: Sand, gravel, or other material used for grading.

Borrow pit: The hole at a site that has been excavated.

Botched job (trivia): There is a myth associated with the Forth (railroad) Bridge in Scotland that is twinned with the Forth Road Bridge for vehicle traffic. The myth harks back to the bridge's design. The original plan by Thomas Bouch, in 1873, called for a suspension bridge similar to the one he had built across the Firth of Tay. Unfortunately, that structure collapsed in 1879, killing dozens and causing officials to disapprove his design for the Forth Bridge. Bouch was so disgraced that people associated his name with a poorly executed task; a "Bouch job." This eventually gave us the phrase **botch job** or **botched job**.

However, according to the Oxford English Dictionary, the word botch, or some version of it, has been around for centuries.

Bottom bearing: A bearing condition where the joist or joist girder bears on its bottom chord and not at an underslung condition.

Bottom chord: Lower or bottom member of a truss.

Bottom chord extension (BCX): The two angle extended part of a joist bottom chord from the first bottom chord panel point toward the end of the joist.

Bottom chord extension. (In this case the extension was added during construction.)

Bottom chord strut: A bottom chord of a joist or joist girder designed to transfer a axial tension or compression load.

Bottom plate: Bottom plates are essentially extension of the foundation upon which the roof, ceilings, and walls rest. In a wood-frame building, the bottom plates are usually of wood, though various other materials may be used, such as steel, concrete, or composite materials.

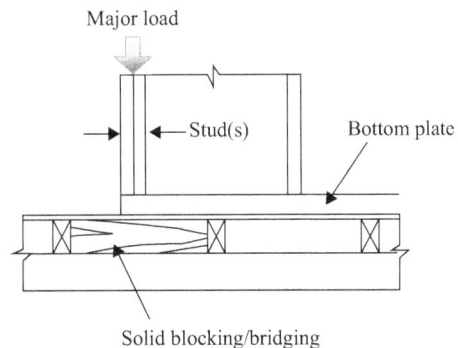

Bottom plate.

Bottom plate: The 2 × 4's or 6's that lay on the subfloor upon which the vertical studs are installed. The bottom plate sets on the subfloor, nails through the subfloor into floor joists. Also referred to as the **sole plate** and **sill plate**.

Bottom rail: The lower rail of a balustrade into which the bottom ends of balusters connect. See **Balustrade** and **Top rail**.

Also, the bottom horizontal member of a window sash or door panel.

Bound moisture: Moisture that is closely bound to the cell wall constituents of wood. See **Bound water**.

Bound water: Water that is bound within the cell wall of wood. Water held in wood below the fiber saturation point. Bound water moves by diffusion. See **Bound moisture**.

Boundary: The dividing line between two adjacent properties.

Boundary condition: An idealization to model how a structure is attached to its "external" points of support, for example, pin, fixed, roller, or shear release.

Bow: A deviation flatwise from a straight line drawn end to end. A bow is measured at the point of greatest distance from the line.

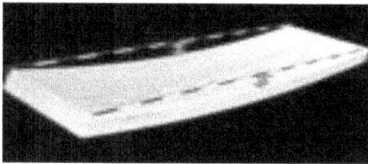

Bowed lumber.

Bow saw: A tool that is ideal for pruning, landscaping work and sawing firewood. It has a tubular steel frame with blade-tensioning lever that snaps closed to form a handle.

Bow string joist: A non-standard type of joist where the top chord is curved and the bottom chord is straight or level. Also spelled **bowstring joist**.

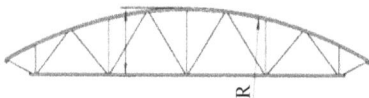

Bow string joist.

Bow window: A window unit that projects out from the wall in an arch. A bow window commonly consisting of five sashes. See **Bay window**.

Bow window.

Bow's notation: Used in a graphical analysis of a joist or joist girder. It is a notation for denoting truss joints, members, loads, and forces.

When several members are pinned together and the joint is in total equilibrium (not moving), the resultant force must be zero. This means that if we add up all the forces as vectors, they must form a closed polygon. If one or even two of these forces is unknown, then it must be the vector that closes the polygon.

Box, cabinet: The storage section of the cabinet. See **Face frame** and **Face**.

Box bay: A combination of window units that projects to the exterior. Usually features a large center unit with two flanking units at 90° angles to the wall.

Box beam: Box beams may be made from steel or wood. They are formed like a long box with four sides and are hollow in the center.

Decorative box beams.

Box gutter: A rain gutter on a roof. It is usually rectangular in shape and it may be lined with metal, asphalt, or roofing felt.

Box gutters are usually placed between parallel surfaces, as in a valley between parallel roofs or at the junction of a roof and a parapet wall. They should not be confused with valley gutters or valley flashings that are installed at the non-parallel intersection of roof surfaces, typically at right angled internal corners of pitched roofs. Also referred to as **parallel gutter** and **trough gutter**.

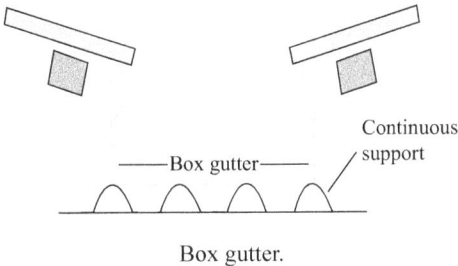

Box gutter.

Box heart: Wood that is sawn or milled so that the pith (the small, soft core in the center of a tree and its branches) is contained within the profile. Most lumber production excludes this part of the tree and is termed **free-of-heart center (FOHC)**.

Box nails: Framing fasteners with a slightly smaller shank or shaft than common nails, but with the same length and the same size heads as comparable common nails. Box nails are less likely to split the wood than common nails but they are not quite as strong. See **Common nails**, **Gun nails**, and **Sinkers**.

Boxed mullion: A hollow mullion between two double-hung windows to hold the sash weight.

Brace: An inclined piece of framing that is applied to wall or floor to strengthen the structure. Braces are often used on walls as temporary bracing until framing has been completed.

A brace.

Also, a tool in carpentry having a crank handle and a socket to hold a bit for boring.

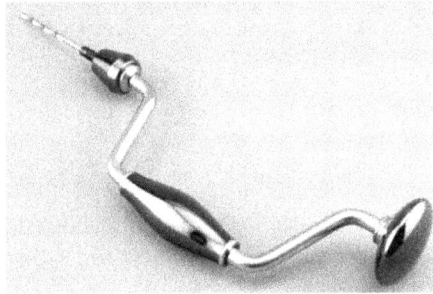

A brace.

Braced framing: A construction method in two-story homes in which the frame is reinforced with posts and braces.

A braced frame resists lateral loads by the use of diagonal bracing, K-braces, or other system of bracing.

Bracket: A structural support attached to a column or wall on which to fasten another structural member.

Brad: A small wire nail with a small, often asymmetrical head.

Brad point bit: A drill bit that is used for precision drilling in wood. Brad point bits are designed for boring exact size holes for a clean, finished look that is required in doweling, cabinetry, and other fine woodworking.

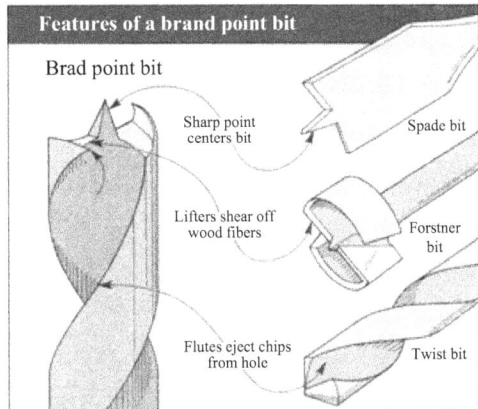

Bred point bit.

Brake: Hand- or power-activated machinery used to form metal.

Also, an electro-mechanical device that is used to prevent an elevator from moving when the car is at rest and no power is applied to the hoistway motor.

Brake drum: With regard to elevators, a round, machined surface on the motor shaft that the brake clamps for stopping.

Branch circuit: The circuit conductors between the final overcurrent device protecting the circuit and the outlet(s).

Also, a branch circuit that supplies energy to one or more outlets to which appliances are to be connected and that has no permanently connected luminaries (lighting fixtures) that are not part of an appliance.

Also, a branch circuit that supplies two or more receptacles or outlets for lighting and appliances.

An individual branch circuit, as defined in Article 100 of the National Electric Code (NEC), "is a branch circuit that supplies only one utilization equipment." Although a duplex receptacle is installed and mounted by one strap or yoke, it is considered two receptacles. A branch circuit supplying only a duplex receptacle and no other device is not an individual branch circuit.

A multiwire branch circuit consists of two or more ungrounded conductors that have a voltage between them, and a grounded conductor that has equal voltage between it and each ungrounded conductor of the circuit and that is connected to the neutral or grounded conductor of the system.

The bottom line: be sure you know what you are doing before installing electrical wiring and electrical appliances.

Brands: Airborne burning embers released from a fire.

Brashness: A condition of wood that is characterized by a low resistance to shock and by abrupt failure across the grain without splintering.

Brazing: A metal-joining process that is similar to soldering, except the temperatures used to melt the filler metal are higher. In brazing the filler metal is heated above its melting point and distributed between two or more close-fitting parts by capillary action. The filler metal is brought slightly above its melting temperature while protected by a suitable atmosphere, usually a flux. It then flows over the base metal (known as wetting) and then it is cooled to join the pieces together.

Breach of contract: The failure to perform provisions of a contract without a legal excuse.

Breach of covenant: A failure of the seller's guarantee of good title which occurs when the buyer is evicted by a person claiming under a paramount title. Since it is a future covenant it is not breached until that eviction occurs. In short, the failure to obey a legal agreement.

Breach of warranty: A seller's inability to pass clear title to a buyer. A warranty is a guarantee and is breached when a thing so guaranteed is deficient according to the terms of the warranty.

Bread loaf top rail: A common type of top rail that has a profile shaped like a loaf of bread. See **Balustrade** and **Top rail**.

Breakaway torque: The torque necessary to put into reverse rotation a bolt that has not been tightened.

Break loose torque: The torque required to effect reverse rotation when a prestressed threaded assembly is loosened.

Breaking stress: The force required to rupture a specimen in a tension test under specified conditions.

Breaker panel: The electrical box that distributes electric power entering the home to each branch circuit (each plug and switch) and composed of circuit breakers.

Breaking strength: A measure of the strength of steel strand fabricated from individual wires of varying tensile strength. Breaking strength is the equivalent tensile strength of the steel strand.

Break-even point: The point at which an investment brings in an amount that is sufficient to cover recurring expenditures. For investment real property, this is the point at which gross income is equal to normal operating expenses, including debt service, i.e., the point at which you start to earn money. Also known as the **default point**.

Breast height: The height at which the diameter of a tree is measured: four feet, six inches above the ground.

Breather tube: A tube placed through air spacer and seal of insulating glass that allows unit to accommodate changes in pressure between time and location of manufacture and time and location of installation, where it is sealed. Usually used to accommodate changes in altitude between plant and job site. Also referred to as **capillary tube**.

Breccia: A clastic (denoting rocks composed of broken pieces of older rocks) sedimentary rock with angular fragments.

Breezeway: A roofed passageway with open sides.

Bressumer beam: A timber lintel flush with the surface of the brickwork above.

A bressumer beam.

Brick: Building material made from clay molded into oblong blocks and fired in a kiln.

Brick bond: The pattern of brick in a masonry wall. See **Bond**.

Brick course: The horizontal layer of brick masonry.

Brick ledge: Part of the foundation wall where brick (veneer) will rest.

Brick lintel: The metal angle iron that brick rests on, especially above a window, door, or other opening.

Brick mason's hammer: A tool designed exclusively for setting and splitting bricks, masonry tile, and concrete block. It has a forged-steel head with a square striking face opposite a flat, sharp cutting edge. It should never strike metal, including a brick set or stone chisel.

Brick mason's hammer.

Brick mold: Trim that is used around an exterior doorjamb to which siding butts. Also known as **brickmould** and **brick mould casing (BMC)**.

Brick saw: A saw that is used to saw bricks. It features a coarse-cutting blade that cuts on both the push and pull strokes.

Brick shelf: A constructed or inserted bearing surface for brick masonry to sit upon.

Brick tie: A small, corrugated metal strip, approximately 1" × 6" – 8" long or similar nailed to wall sheeting or studs. They are inserted into the grout mortar joint of the veneer brick, and hold the veneer wall to the sheeted wall behind it. See **Cavity wall tie**.

Brick veneer: A vertical facing of brick laid against and fastened to sheathing of a framed wall or tile wall construction.

Bricklayer: A tradesmen or mechanic who lays brick and CMUs.

Bridge faucet: A faucet that has the hot and cold handles connected to the spout by an exposed pipe that spans above the sink.

Some of these faucets are designed to be wall-mounted, and some are designed to be countertop- or sink-mounted.

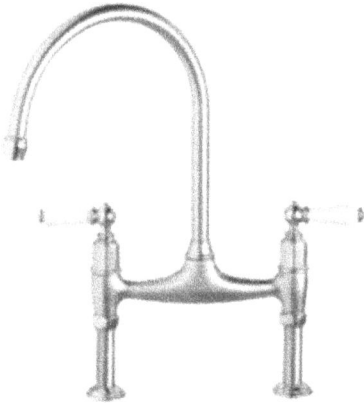

A bridge faucet.

Bridge loan: A short-term loan for borrowers who need more time to find permanent financing.

Bridging: With regard to general framing, small wood or metal members that are inserted in a diagonal position between the floor joists or rafters for the purpose of bracing the joists/rafters and spreading the load. See **Horizontal bridging** and **Diagonal bridging**.

Also, a method of reroofing with metric-sized shingles.

Bridging anchor: An angle or bent plate attached to a wall where the bridging will be attached or anchored, either by welding or bolting. The ends of all bridging lines terminating at walls or beams should be anchored.

Bridging clip: A small piece of angle or plate with a hole or slot that is welded to the top and bottom chord angles so that bridging may be attached.

Bridging clip.

Bridging diagram: A diagram of the profile of a joist used to show the number and location of the rows of bridging.

British thermal unit (BTU): The quantity of heat required to raise the temperature of one pound of water by 1°F.

Brittle: A brittle structure or material exhibits low ductility, meaning that it exhibits very little inelastic deformation before complete failure.

Failure, i.e., the sudden loss of strength at some critical stress either by breaking along a new fracture or, most commonly, by frictional sliding on an already existing fracture is referred to as **brittle behavior**.

And, material that fails suddenly and without warning signs is referred to as **brittle failure** or **brittle fracture**.

Brittle-ductile boundary: The depth in the crust where the crust changes from being brittle (tending to break) above, to being ductile (tending to bend) below. Most earthquakes occur in the brittle portion of the crust above the brittle-ductile boundary.

Bronze: An alloy of copper and tin that is generally used in casting. The term is often applied to brown-colored brasses.

Broker: A person licensed by the state to deal in real estate.

Brokerage: The act of bringing together two or more parties in exchange for a fee or commission.

Broom clean: The ideal condition of a building when it is turned over to an owner or tenant. Many construction contracts require broom cleaning prior to turnover.

Broom finish: The most common exterior flatwork finish; a slightly rough texture achieved by running a broom over freshly troweled concrete.

Brooming: An action carried out to facilitate embedment of a ply of roofing material into hot bitumen by using a broom, squeegee, or special implement to smooth out the ply and ensure contact with the bitumen or adhesive under the ply.

Brown coat: Second coat of stucco, applied over the scratch coat. The purpose of the brown coat is to provide a relatively smooth surface for the finish coat. The brown coat is trowled over the scratch coat and then smoothed with a long float. See **Scratch coat** and **Finish coat**.

Brownstone: A vintage row house constructed of red sandstone.

Bruised composition shingles: A composition shingle that has been permanently dented by a hailstone but has not fractured. See **Fractured composition shingles** and **Granular loss**.

Brush texture: A finish that is applied to drywall with a brush.

Buck: Often used in reference to rough frame opening members, e.g., door bucks are used in reference to metal doorframes. See **Window bucks**.

Bucket handle pointing: Pointing that is recessed in the half-round shape of an old-fashioned metal bucket. See **Pointing**.

Buckle: An upward, elongated tenting displacement of a roof membrane frequently occurring over insulation or deck joints. A buckle may be an indication of movement within the roof assembly.

Also, to bend under compression.

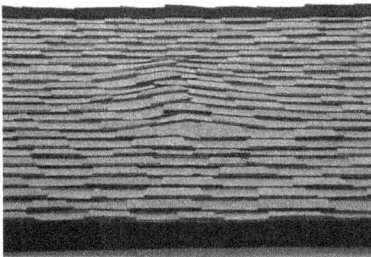

A column that buckled.

Buckling load: The load at which a straight member under compression transfers to a deflected position.

Budget analysis: An analysis as to sufficiency of funds. A budget analysis takes into consideration various factors especially a cost-to-complete estimate compared to the balance of funds. The analysis may also include: consideration for the contractor performance; percent of completion and schedule status; status of contract buyout and contingency; potential for shared savings; various other risk factors as deemed pertinent; etc. This estimate typically does not include projections for discretionary change orders, unless otherwise noted.

Budget summary table: A table summarizing the hard cost budget funds and committed hard cost contingency to date to determine the remaining (balance) hard cost contingency. The purpose of this table is to help provide the earliest possible notice of a budget overrun.

Buffer: A pertains to elevators, a device designed to stop a descending elevator car or counterweight beyond its normal limit and to soften the force with which the elevator runs into the pit during an emergency.

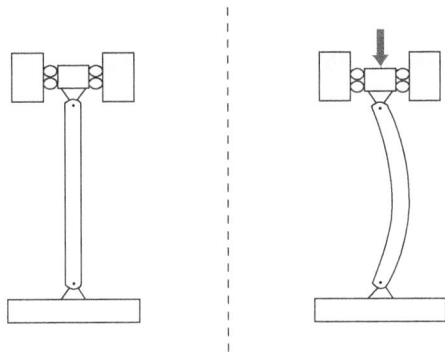

A buckled roof.

Elevator buffer.

The channel in the elevator pit floor of a traction elevator that supports buffers and guide rails is referred to as **a buffer channel**.

The large diameter springs that are permanently placed in a traction elevator pit for the purpose of stopping a descending car or counterweight beyond its normal limit of travel are **buffer springs**. The distance that the springs compress is called the **buffer stroke**.

Buffer strip: A parcel of land that separates two or more properties.

Builder upgrades: Extra house features or better finishing materials that a builder offers.

Builder's risk insurance: Builder's Risk Insurance is insurance coverage taken on a specific construction project for the period of construction. Some contracts require this insurance to protect the owner.

Building: A usually roofed and walled structure built for permanent use (as for a dwelling). Buildings stand alone or are cut off from adjoining structures by firewalls with all openings between structures protected by approved fire doors.

Terms beginning with building:
- **Building and loan association:** An organization that raises money to helps its members purchase real estate or construct a building.
- **Building automation system (BAS):** A computer-based monitoring and control system that coordinates, organizes, and optimizes building control subsystems, including lighting and equipment scheduling, and alarm reporting.

- **Building code:** A comprehensive set of laws that control the construction or remodeling of a home or other structure. Building codes vary from jurisdiction to jurisdiction. Do not presume you know the code. Read the local code prior to bidding on a project.
- **Building engineer:** A qualified engineering professional with relevant and sufficient expertise who oversees and is responsible for the operation and maintenance of the building exterior (roof, envelope, foundation, windows, and doors, etc.) interior finishes, mechanical, electrical and plumbing systems in the project building and its surrounding grounds. Also, the building engineer usually oversees renovation and capital repair projects. Also referred to as **building designer** and **specifying professional**.
- **Building envelope:** The elements of a building that enclose the conditioned spaces and keep water from entering; the outer structure of the building. Also known as **building shell**.
- **Building footprint:** The area of the site occupied by the building structure, not including parking lots, landscapes, and other non-building facilities.
- **Building frame terms:**
  - **Braced frame:** One having diagonal braces for stability and capacity to resist lateral forces.
  - **Concentric braced frame:** The centerlines of brace, supporting beam and column coincide.
  - **Eccentric bracing:** The centerlines of brace, beam, and of column and do not coincide allowing deformation, thereby utilizing ductility.
  - **Moment frame:** Frames in which structural members and joints resist lateral forces by bending. There are "ordinary," "intermediate," and "special" moment frames. The latter provide the most resistance.
- **Building house drain:** That part of the lowest piping in a drainage system that receives the discharge from the soil, waste, and other drainage pipes and conveys it

to the building house sewer by gravity. A building house drain may be combined, i.e., it may convey storm water in combination with sewage and other drainage, or it may be for sanitary or storm drainage only.

- **Building house sewer:** That part of the horizontal piping of a drainage system that extends from the end of the building house drain and that conveys it to a public sewer, private sewer, individual sewage-disposal system, or other point of disposal. As with a building house drain a building houde sewer may be combined, or it may be for sanitary or storm drainage only.
- **Building insurance:** Building insurance covers the structure.
- **Building inspector:** An individual trained and certified to inspect completed construction work for compliance with accepted building codes and ordinances.
- **Building line or setback:** Guidelines that limit how close an owner can build to the street or an adjacent property. These guidelines are usually found in zoning codes.
- **Building moratorium:** A halt on home construction to slow the rate of development.
- **Building official:** The officer or other authority that has the duty of administration and enforcement of a building code.
- **Building operating plan:** A document that summarizes the intended operation of each base building system that is described in the systems narrative. (The building operating plan may also be known as **Owner's Operating Requirements** or similar.) The operating plan includes the time-of-day schedules for each system for each of the eight day types (Monday to Sunday plus holidays), the mode of operation for each system when it is running (occupied vs. unoccupied; day vs. night, etc.), and the desired indoor conditions or setpoints for each schedule or mode. The operating plan accounts for any differences in needs or desired conditions for different portions of the project building, as well as any seasonal variations in operations patterns. The plan accounts for all the monitored space conditions used to control the base systems, i.e., air temperature, relative humidity, occupancy, light level, CO_2 levels, room pressurization, duct static pressure, etc.

- **Building paper:** A general term for papers, felts, and similar sheet materials used in buildings without reference to their properties or uses. Generally comes in long rolls.
- **Building permit:** A permit issued by the local government, usually the county or city, after a fee has been paid and plans have been reviewed and approved. Normally construction cannot begin until after the permit is issued nor will a **Certificate of Occupancy** be issued until a building inspector has approved all work.
- **Building restrictions:** Regulations that limit the manner in which property can be used.
- **Building services:** Plumbing, electrical wiring, HVAC, gas supply, and other support systems in a building.
- **Building sub-house drain:** That portion of a house drainage system that cannot drain by gravity into the building house sewer.

Built-in cabinets: Cabinets that are hand-built on site. See **Milled cabinets, Custom cabinets,** and **Mass produced cabinets**.

Built-ins: Appliances or other items that are framed into a home or permanently attached.

Built-up member: A structural element that is fabricated from any number of other structural elements. They may be connected by welding, bolting, or other means. Also known as **built-up section**.

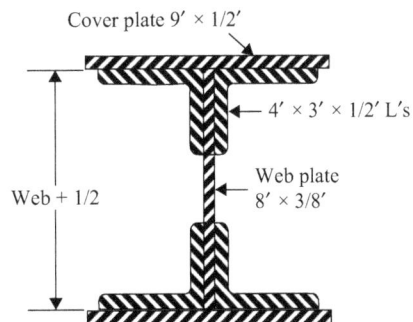

Built-up member.

Built-up roof (BUR): Usually BUR systems have alternating layers of bitumen and reinforcing fabrics (also called roofing felts or ply sheets) that create a finished membrane. The number of plies in a cross section is the number of plies on a roof: The term "four plies" denotes a four ply roof membrane construction. Sometimes, a base sheet, used as the bottommost ply, is mechanically fastened. Built up roofs are considered to be fully adhered if applied directly to roof decks or insulation. Also referred to as **tar and gravel roofs**.

The bitumen typically used in BUR roof systems is asphalt, coal tar, or cold-applied adhesive. Cold-applied adhesives typically are solvent-based asphalts that don't have to be heated in a kettle or tanker.

Surfacings for built up roof systems include aggregate (such as gravel, slag, or mineral granules), glass-fiber or mineral surfaced cap sheets, hot asphalt mopped over the entire surface, aluminum coatings, or elastomeric coatings. A roof system composed of a built up roof membrane with two or three plies and a polymer-modified bitumen membrane cap sheet is commonly referred to as a **hybrid system**.

See **Multiple ply membrane**.

Bulb-tee: A steel reinforcing member (a rolled steeltee with a formed bulb on the edge of the web)that is used when constructing pre-stressed, poured decks. When the deck is poured, it surrounds the bulb-tee.

Bulb-tee.

Bulkhead: A term used to describe a barrier in concrete formwork. A bulkhead is used when the pour will be continued at another time.

Also, a mechanism to discontinue formwork with a predictable designed procedure.

Also, a retaining wall designed to hold back water from the ocean or another body of water.

Also, an enclosed structure on or above the roof of any part of a building. A bulkhead encloses a shaft, stairway, tank, or service equipment, or other space that is not designed or used for human occupancy.

Bulk specific gravity: With regard to masonry, a property of natural building stone equal to the mass of an oven-dried specimen divided by the difference between the mass of the soaked and surface-dried specimen in air and the mass of the soaked specimen in water.

Bulking: One cubic of soil or rock at the in situ expands and does not translate into one cubic of fill in the dump truck, stockpiled or placed, and compacted on the site. Bulking or **swell factors** for some common materials:

Material	Density at the borrow (kg/m³)	Bulking (swell) factor (%)
Basalt	**2.4–3.1**	**75–80**
	1.8–2.6	20–40
Dolomite	2.8	50–60
Earth		20–30
Gneiss	2.69	75–80
Granite	2.6–2.8	75–80
Gravel, dry	1.80	20–30
Gravel, wet	2.00	20–30
Gravel, wet w/clay		50–60
Limestone	2.7–2.8	75–80
Loam		15–25
Quartz	2.65	75–80
Rock		40–80
Sand, dry	1.60	20–30
Sand, wet	1.95	20–30
Sandstone	2.1–2.4	75–80
Slate	2.6–3.3	85–90
Soil	1.2–1.6	20–30

Example:

After excavation 200 cubic yards with bulking factor 40% swells to

(200 cy) (100% + 40%)/100% = 280 cy

Something to be aware of when estimating excavation.

Bull nose: Rounded corners, e.g., drywall and tile corners.

Bull float: Tool with a long handle that slides on the surface of concrete to press the course aggregate down and to raise the cream.

Bulldozer: A dozer blade in the front of a bulldozer is used to push soil and other materials. Usually a ripper that is shaped like a claw is also connected to the back of the bulldozer to loosen the soil or other material.

Bullet resistant doors: Doors that resist penetration by shots of varying caliber. Resistance may be rated as resistant to medium power, high power, or high power small arms, and high power rifles.

Bundle: A package of shingles. Normally, there are 3 bundles per square and 27 shingles per bundle.

Bundle of rights: The various interests or rights an owner has in a property.

Bungalow: A small one-story house or cottage.

Bunk: As pertains to lumber, there are numerous definitions, among them: a cross support for a load of lumber; a unit of lumber consisting of several pieces or sticks that are banded together for convenience in shipping and delivery; a piece of wood placed on a lumberman's sled to sustain the end of heavy timbers; and a stacked bundle of lumber used as a specification for packaging wood according to species and size.

Burl: A swirl or twist in the grain of wood, usually occurring near a knot, but which itself does not contain a knot. Burl is valued as the source of veneers that are used for ornamental purposes.

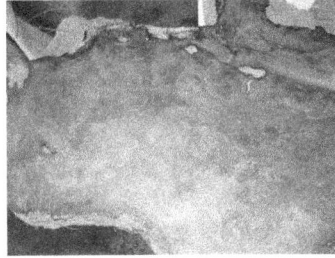

Redwood burl.

Burl, bending: A bending burl is detectable at 1.8–2.4 m (6–8ft) as a swirl. Intricately patterned burl wood is often used by wood turners and furniture makers.

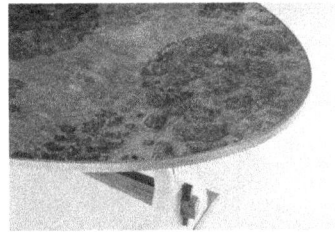

Burl, bending.

Burn pattern: The direction a fire burns. Fire burns up and away from the point of origin toward an oxygen source.

Burnish: A process that uses non-diamond tooling or rubbing to polish something, e.g., a floor or metal, to a high reflectivity.

Busbar: A solid bar of aluminum or copper conductor for primary distribution of electric service throughout a building.

Bushings: One end is spigot and the other end is hub or gasket. The spigot end is the larger size and the hub or gasket is the smaller size. Bushings always reduce in size. There are concentric and eccentric bushings. They are threaded inside and out.

Bushing fitting.

Business risk: The uncertainty associated with the possible profit outcomes of a business venture.

Busduct: The assembly structure surrounding, isolating, and containing the busbars.

Butadiene: A gas that is chemically combined with styrene to create a resin used in latex binders, styrene-butadiene.

Butcher block: A surface made from thick cubes or strips of hardwood used for cutting or chopping food items.

Butt: The bottom of a felled part of a tree. Also, the large end of a log.

Butt cut: The first cut above the stump of a tree. Also known as a **butt log**.

Butt edge: The lower edge of the shingle tabs.

Butt hinge: A hinge made up of two flat, rectangular plates with a pin connecting them.

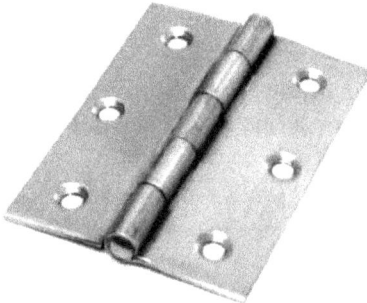

Butt hinge.

Butt joint: A joint made by placing two square-cut pieces of material end to end without any overlap. The seam that results when two members are joined together end to end without overlapping the members is a **butt seam**.

Butt plate: The end plate of a structural member usually used to rest or butt against a like plate of another member in forming a connection.

Butt rot: Decay or rot characteristically confined to the base or lower bole of a tree.

Butterfly roof: A roof style consisting of two planes that slope inward forming a V, lower in the center than at the outside edge.

Butterfly roof.

Butterfly valve: A valve made of a square, rectangular, or round disk attached to a shaft inside a body of the same shape. Rotating the shaft 90° opens or closes the valve.

Butterfly valve.

Button plug: Regarding boiler operations, a form of an easily melted plug containing an easily split (fissile) material and a solid core of none fissile material. On overheating the easily melted material melts releasing the solid core. The stated advantage is that failure of the plug is complete, i.e., the entire diameter of the plug's central hole is opened on melting of the easily melted material. A plug with a core of only easily melted material may partially, not completely, melt. This may lead to a very small discharge of steam and water that can lead to low water levels not being brought to the attention of the operators. Also known as a **drop plug**.

Buttress: A support that transmits a force from a roof or wall to another supporting structure. See **Counterfort**.

A (flying) buttress.

Also, the raised portion of a shower curb that is on more than one level. Also referred to as tile rise.

A shower stall showing a buttress.

Buttress dam: A gravity dam that is reinforced by structural supports.

A buttress dam.

Butyl: A hydrocarbon radical, C_4H_9. Butyl has a rubber-like consistency, that is formed from the copolymerization of isobutylene and isoprene. Butyl is used primarily in sealants and adhesives.

Butyl rubber: A butyl-based, synthetic elastomer.

Butyl tape: A sealant tape that is used in numerous sealant applications such as sealing sheet metal joints.

Buy down: A subsidy (usually paid by a builder or developer) to reduce monthly payments on a mortgage.

Buyer broker: A real estate broker who exclusively represents the buyer's interests in a transaction and whose commission is paid by the buyer rather than the seller.

Buyer's market: A slow real estate market in which buyers have the advantage.

Buyer's remorse: An emotion felt by first-time homebuyers after signing a sales contract or closing the purchase of a house.

BX cable: Armored electric cable wrapped in plastic and protected by a flexible steel covering.

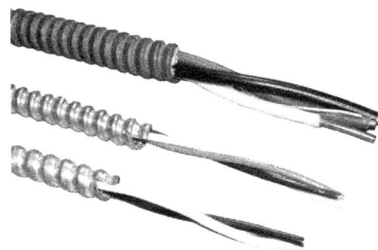

BX cable.

Bylaws: The rules and regulations that a homeowners association or corporation adopts to govern activities.

Bypass doors: Door with two flush slabs that are mounted on tracks. Each door part slides parallel to the other. Bypass doors are common on closets and patios. Also called **sliding doors**.

C Section: A structural member cold-formed from sheet steel in the shape of a block "C" that can be used by itself or back to back with another C Section.

C Shapes: A hot rolled shape called an American Standard Channel with symbol C.

C-channel: A structural framing member that, when viewed cross-sectionally, has the shape of a "C."

C-clamps: This tool is similar to the vise grip but offers a wider opening. It holds objects in place while sawing or joining. C-clamps feature clamp pads that protect surfaces and allow for gripping tapered pieces. C-clamps are ideal for laminating or veneering.

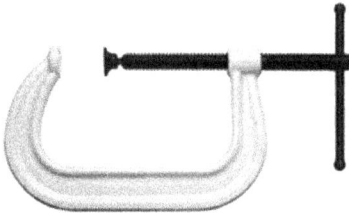

C-Clamp.

Cabinet: An enclosure that is designed for either surface mounting or flush mounting and is provided with a frame, mat, or trim in which a swinging door(s) is/are or can be hung.

Cabinet rasp: This rasp provides for a quick way to remove woodworking stock. It has a round and flat face.

Cabinet scraper: Cabinet scrapers are available in three basic shapes: rectangular, straight with concave and convex ends, and gooseneck. These tools works by cutting, not abrading, the wood. They are excellent at smoothing wood.

They are sharpened by burnishers.

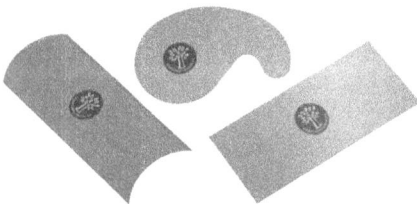

Cabinet scrapers.

Cabinetmaker's screwdriver: This tool usually features a turned oval hard wood handle and a blade designed for wood screws. Part of the blade is flattened to handle a wrench if more torque is needed.

Cable(s): Cable refers to the construction of a wire rope that includes multiple groups of strands. For example, seven groups of 1×7 strand would be a cable referred to as 7×7. 7×19 would be seven groups of 1×19 strand. See **Strand**.

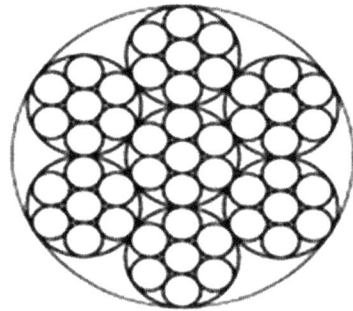

7×7 cable.

In mechanics, cables, also known as **wire ropes**, are used for lifting, hauling, and towing or conveying force through tension.

In electrical engineering, cables are used to carry electric currents.

In optics, an optical cable contains one or more optical fibers in a protective jacket that supports the fibers.

With regard to elevators, wire ropes are used to support an elevator car; there are usually 4–6 in number. Wire ropes pass over the drive sheave to the counterweight, either pulling up the car or lowering it. The amount of the drive sheave actually in contact with the cable is called cable wrap.

Cable railings: Temporary steel cables strung between steel or concrete columns for the protection of workers at the edge or around openings in buildings while under construction.

Cable tray: A steel tray for the containment of computer, telephone, and fiberoptic cable that is either hung from the floor above, or inserted into a floor assembly.

Cable-stayed bridge: A bridge in which the roadway deck is suspended from wire ropes anchored to one or more towers.

Cable-stayed bridge.

Cableway: A power operated system for moving loads in a generally horizontal direction in which the loads are conveyed on an overhead cable, track, or carriage.

Cage: A term used to describe a cage of reinforcing bars that have been prefabricated for columns and or beams and girders in concrete construction.

Caisson: A 10- or 12-inch diameter hole drilled into usually through unsatisfactory soil and embedded into bedrock 3–4 feet. A caisson serves as the structural support for a type of foundation wall, porch, patio, monopost, or other structure. Usually reinforcing bars (rebar) are inserted into and run the full length of the hole and concrete is poured into the caisson hole.

Calcium silicate bricks: Smooth bricks made by compressing and heating a mixture of sand, or ground flint, and lime. They have a tendency to shrink.

Caldera: A large, roughly circular volcanic depression whose diameter is many times greater than that of its vent or vents.

Calibration: The act or process of adjusting control or recording equipment so that it shows correct readings.

Also, a set of gradations that show positions or values. Often used in the plural, e.g., the calibrations on a temperature gauge.

Calipers (outside and inside): These two tools are designed for accurately transferring and measuring outside and inside dimensions. The bowlegged caliper is used to measure outside dimensions, with the straight-legged caliper measure inside. Most woodturners consider them invaluable.

Call option: A clause in a loan agreement that allows a lender to ask for the balance at any time.

Calorific value: The calories or thermal units contained in one unit of a substance and released when the substance is burned. For example, the heat-production value of a wood source.

Calrod burner: An electrical burner made from coiled steel. See **Ceran top** and **Hologen burner**.

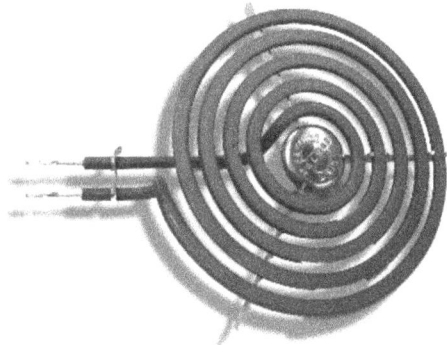

Calrod burner.

Cam lock: A lever operated lock that is used to prevent intrusion through the sash.

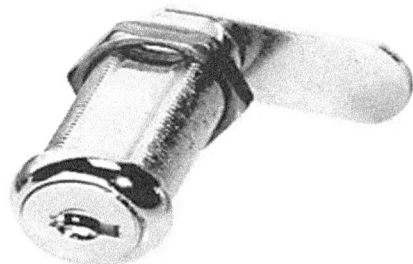

Cam lock.

Cam pivot: A zinc pivot pin attached to the top and bottom sash stiles of double hung units (bottom sash on single hung units). Cam pivots rest on the clutch system of the balance tube assembly which allow opening and closing of the sash.

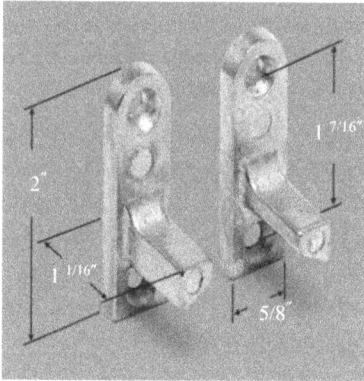

A pair of cam pivots.

Camber: Refers to the slight bow or arch that is found in many building materials. Camber is sometimes used as a synonym for **crown**. However, crown usually refers to the natural distortion that occurs in lumber, whereas camber usually refers to a built-in bow that was engineered by the manufacturer. See **Crown**.

Also, camber is an upward curvature of the chords of a joist or joist girder induced during shop fabrication to compensate for deflection due to loading conditions; this is in addition to the pitch of the top chord.

Cambium: The layer of tissue dividing the bark from the wood that forms new bark to the outside and new wood to the inside as the tree grows.

Caming: The metal used in the construction of decorative glass panels. Usually zinc or brass, it is also applied to single glass lites to create a decorative glass look.

Cam-action clamps: These tools come aslight, medium duty, and veneering clamps. They have cork faces that won't mar delicate work. The cam works as a lever to apply varying degrees of pressure.

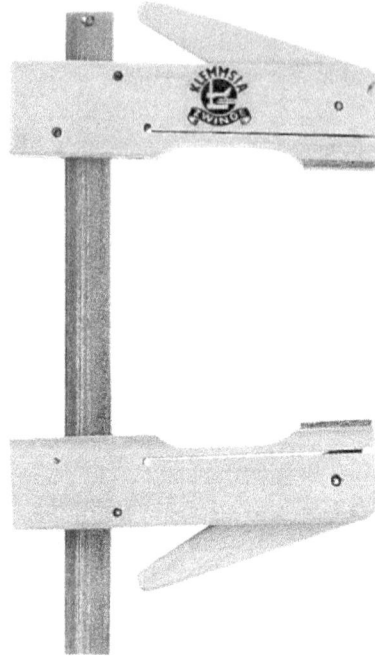

Cam-action clamp.

Can: A housing or container for a recessed light unit. The can is installed during electrical rough-in.

Can lights: Cylindrical chambers with bulbs recessed into the ceiling.

Cancellation clause: A clause that details the conditions under which each party may terminate the agreement.

Candela: The SI unit of luminous intensity. One candela is the luminous intensity, in a given direction, of a source that emits monochromatic radiation of frequency 540×10^{12} Hz and has a radiant intensity in that direction of 1/683 watt per steradian. See **Steradian**.

Candlepower: The unit of luminous intensity approximately equal to the horizontal light output from an ordinary wax candle.

Canopy: An overhang, usually over entrances or driveways.

Also, the physical exterior portion of a kitchen range hood (shell).

Also, refers to the mature tree crowns that form the upper layer of a tree community and includes other biological organisms.

Sometimes the term canopy is used to refer to the extent of the outer layer of leaves of an individual tree or group of trees. Shade trees normally have a dense canopy that blocks light from lower growing plants. For those of a certain age who served in Vietnam, the jungle had what was called a triple canopy that blocked out almost all light.

Cant: A piece of wood produced by a canter that requires further breakdown. See **Canter**.

Also, short for cant strip.

Cant strip: A beveled piece of material placed where the roofing material turns up such as at the intersection of a parapet wall and the roof deck. It is used to soften the angle that must be covered by the roofing membrane.

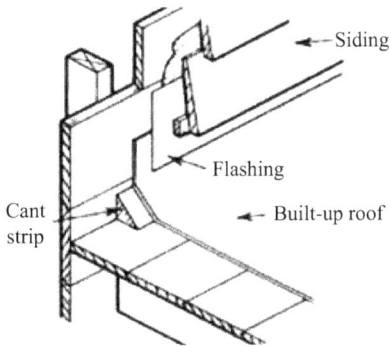

Cant strip.

Canter: A machine that converts logs into a square, rectangular, or two-sided cant for further processing. Canters are used to process logs or blocks into flitches, cants, and boards. A canter usually consists of a log turner, double or single length infeed, chipping section and a saw section.

Cantilever: A beam or joist with an end portion that hangs out past the structural part that supports it, e.g., a diving board is cantilevered.

Cap: The upper member of a column, pilaster, door cornice, molding, or fireplace.

Also, a limit on the amount the interest rate or monthly payment can increase on an adjustable-rate mortgage.

Cap coping: A term used to describe the course covering material used over a parapet wall or at the top of an assembly wall.

Cap flashing: The portion of the flashing attached to a vertical surface to prevent water from migrating behind the base flashing.

Cap plate: A steel plate welded to the top of a column that a joist, joist girder, or other structural member can bear on.

With regard to framing, the upper plate that sits on top of the top plate. Also referred to as a double top plate.

Double top plate or top plate and cap plate.

Cap row: Top course of shingles that does not have another course overlapping it. There is a cap row on the ridge of the roof.

Cap sheet: The top layer of a multiple ply membrane roof system that is usually covered with one of three finishes: (1) a smooth, flood coat, that is painted to prevent sun damage; (2) a flood coat with aggregate covering; or (3) a cap sheet with mineral granules embedded into the surface. See **Flood coat**.

Cap tile: A U-shaped roofing tile that forms the peaks in a barrel tile roof. Cap tiles are placed at the intersection of pan tiles with the U facing down. Water drains off of the peaks formed by the cap tiles and into the pan tiles. Specially designed cap tiles are also used on the rake, ridges, and hips.

Capacitance meter: A device for locating moisture within a roof system by measuring the ratio of the change to the potential difference between two conducting elements that are separated by a non-conductor.

Cape Cod style: A wood-frame or shingled house with a steep roof and several windows projecting from the second floor.

Early 20th century cape cod style house.

Capillary action: Capillary action is the rising of water above the free water table. It is the combination of solid–liquid adhesion and surface tension.

In the case of soils, the smaller the soil particles, the more susceptible to capillary action they are. Clays have the ability to wick water upward of 30 feet. Silts and fine sands can raise water one to ten feet. The combination of capillary action and surface freezing causes the water to expand and create pot holes in roads.

Capillary tube: A tube placed through air spacer and seal of insulating glass that allows unit to accommodate changes in pressure between time and location of manufacture and time and location of installation, where it is sealed. Usually used to accommodate changes in altitude between plant and job site. Also referred to as a **breather tube**.

Capital: The principal part of a loan, i.e., the original amount borrowed.

Also, money used to create income, such as funds invested in rental property.

Also, the top component of a column.

Capital and interest: A repayment loan and the most conventional form of home loan. The borrower pays an amount each month to cover the amount borrowed (or capital or principal) plus the interest charged on capital.

Capital expenditure: Property improvements that cannot be expensed as a current operating expense for tax purposes. Examples include a new roof, tenant improvements, or a parking lot. Such items are added to the basis of the property and then can be depreciated over the holding period. Capital expenditures are distinguished from current operating expenditures such as routine maintenance, re-painting, or plumbing repairs that can be expensed in the year they occur. See operating expenses.

Capital gains: Profits or losses an investor makes from the sale of real estate or investments. Capital gains or losses are calculated as the difference between the amount realized on the sale or exchange and the taxpayer's basis in the assets. Long-term capital gains or losses are capital gains or losses from the sale or exchange of capital assets held for the required holding period. Capital gains and long-term capital gains are taxed differently.

Capital gains tax: A tax placed on the profits from the sale of real estate or investments. See **Capital gains**.

Capital improvement: Any improvement that extends the life or increases the value of a piece of property. See **Capital expenditure**.

Capital tax: Any tax on a change in capital value including capital gains tax, estate tax, or inheritance tax. Capital taxes are distinguished from taxes on income.

Capitalization: A mathematical formula that investors use to compute the value of a property based on net income.

Capitalization rate: The percentage rate of return estimated from the net income of a piece of property.

Capped rate: The mortgage interest rate that will not exceed a specified value during a certain period of time, but it may fluctuate up and down below that level.

Capping head: Equipment used to form a cap (chamfer) on the end of roundwood posts.

Capstock: A material co-extruded with PVC formulated to offer a specific color, finish and/or function, such as heat resistance.

Car: As regards elevators, the load carrying unit including its platform, car frame, enclosure, and car door.

Car counterweight: A set of weights roped directly to the elevator car of a winding-drum type installation. In practice, this weight is equal to approximately 70% of the car weight.

Car doors: As applied to elevators, elevators use two different sets of doors: doors that are on the cars and doors that open into the elevator shaft. The elevator car doors have a clutch mechanism that unlocks the outer doors at each floor and pulls them open. In this way, the outer doors will only open if there is a car at that floor (or if they are forced open) thereby keeping the outer doors from opening up into an empty elevator shaft.

Car-switch operation: The operation of an elevator wherein the movement and direction of travel of the car are directly and solely under the control of the operator by means of a manually operated car switch or of continuous-pressure buttons in the car.

Car operating panel: A panel mounted in the elevator car that contains the car operating controls, such as call register (floor) buttons, door open and close, alarm, emergency stop, and any other buttons or key switches that may be required for operation. Also referred to as **car station**.

Carbide: A carbide alloy that is composed of cobalt and tungsten. This alloy gives saw blades and router bits longer lasting tips, sharper cutting edges, and greater impact resistance.

Carbide hole cutter: A drill bit that is designed specifically for the fast, easy production of holes in thin material. Carbide cutters are used primarily for use with mild steels like those found in electrical enclosures as well as stainless steel.

Carbide teeth: Specially treated tungsten carbide teeth on a saw blade.

Carbon 14 age: An absolute age obtained for geologic materials containing bits or pieces of carbon using measurements of the proportion of radioactive carbon (14 C) to daughter carbon (12 C). These dates are independently calibrated with calendar dates. This is used to determine when past earthquakes occurred on a fault.

Carbonate: An anion with a charge of −2. In the context of Porta Treatment (Porta Treatment is an internal boiler water treatment based on tannin and carbonate but also relies on the use of powerful antifoams to produce excellent results.) carbonate ions form the bulk of the alkalinity in the boiler water.

Carbonation: Reaction between calcium hydroxide or oxide in cement, mortar, or concrete with carbon dioxide to form calcium carbonate that can reduce the pH of the concrete.

Carbon content and chemical composition: An indicator of the property of steel bars to establish corrosion resistance and mechanical properties. Carbon content and carbon equivalent content are needed to establish weldability of the metal.

Carbon dioxide (CO_2) levels: An indicator of ventilation effectiveness inside buildings. CO_2 concentrations greater than 530 parts per million (ppm) above outdoor CO_2 conditions generally indicate inadequate ventilation. Absolute concentrations of CO_2 greater than 800–1,000 ppm generally indicate poor air quality for breathing.

Carbon equivalent content: The equivalent carbon content concept is used on ferrous materials, typically steel and cast iron, to determine various properties of the alloy when more than just carbon is used as an alloyant, which is typical. The percentage of alloying elements other than carbon is converted to the equivalent carbon percentage, because the iron-carbon phases are better understood than other iron-alloy phases. Most commonly this concept is used in welding, but it is also used when heat treating and casting cast iron.

Carpenter: A tradesperson who constructs things of wood.

Carpenter's pencil: A rectangular shaped pencil, about 1/4"× 1/2", with a 1/16"× 3/16" lead.

Carpet cove: Carpet that wraps a small distance up the wall.

Carpet grain: Direction in which the carpet fibers slant.

Carpool: An arrangement in which two or more people share a vehicle for transportation.

Carport: A roof that covers a driveway or other parking area.

Carryover: Water droplets and dissolved and suspended solids being carried out of a boiler by escaping steam.

Cartridge: The replaceable element of a fluid filter, e.g., a laser printer toner cartridge.

Also, a cartridge packages the bullet, propellant, and primer into a single unit. Also known as a **round**.

Also, the pumping unit from a **vane pump** that is composed of the rotor, ring, vanes, and one or both side plates.

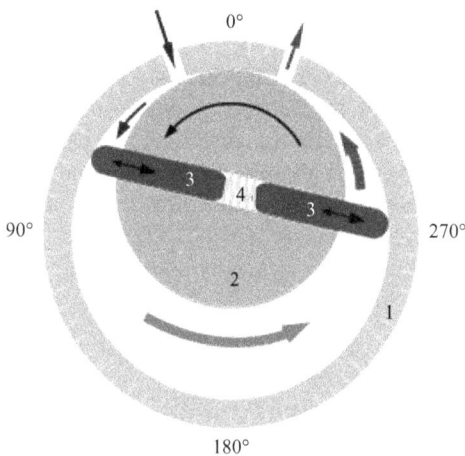

Rotary vane pump.

Casehardening: A condition of stress-and-set in dry wood whereby the outer fibers are under compressive stress and the inner fibers under tensile stress. The stresses persist after the lumber is dry and cause warp if the wood is remachined after drying. As opposed to **reverse casehardening** that is a final stress-and-set condition in dry lumber whereby the outer fibers are under tensile stress and the inner fibers are under compressive stress. This condition is not reversible.

Casement: A window unit that opens by swinging the side of the window sash outward like a door. The opposite side of the window sash is attached with hinges. Also known as a **casement window**.

Casement windows.

Casework: The term used to describe all cabinet, counter, and built in furniture work in a building.

Casing-off: The elimination of the frictional forces between a portion of a pile and the surrounding soil by the use of a sleeve between the pile and the soil.

Cash flow: Incomings and outgoings of cash. Cash flow represents the operating activities of an organization. In accounting, cash flow is the difference in amount of cash available at the beginning of a period (opening balance) and the amount at the end of that period (closing balance): cash flow is called positive if the closing balance is higher than the opening balance, and it is called negative when it is not. Cash flow is increased by such actions as winning contracts, selling an asset, reducing costs, increasing your margin on completed work, collecting faster, paying slower, bringing in more equity, and taking a loan. A positive cash flow is usually good, but, as you can see from the above list, it may not always be so.

Cash flow after taxes (CFAT): For real estate investments, it is the result of first calculating the net operating income, less mortgage and construction loan interest, less cost recovery for improvements and personal property, less amortization of loan points and leasing commissions. These sum to real estate taxable income. Then this real estate taxable income is multiplied by the applicable marginal tax rate resulting in the tax liability (savings). Then, from the net operating income, annual debt service is subtracted to equal **the cash flow before taxes (CFBT)**. Finally, the cash flow after taxes (CFAT) is calculated from the CFBT, less the tax liability (savings), plus investment tax credit.

The **Cash Flow Analysis Worksheet** can be used to calculate a property's gross operating income, net operating income, real estate taxable income and tax liability or (savings), CFBT, and CFAT.

Net operating income − (Interest + Cost recovery + Amortization of loan points) = Real Estate taxable income.

Real Estate taxable income × Investor's marginal tax rate = Tax liability (savings)

Then:

Net operating income − Annual debt service = Cash flow before taxes

Cash flow before taxes − Tax liability (savings) = Cash flow after taxes

There are rules that establish what can and cannot be included in the various categories such as operating expenses. To stay out of trouble with the tax people it is generally best to hire a CPA, whose fees are an expense.

Cash flow before tax (CFBT): For real estate, it is the result of calculating the effective rental income, plus other income not affected by vacancy, less total operating expenses, less annual debt service, funded reserves, leasing commissions, and capital additions.

Cash-out refinance: The refinancing of a mortgage in which the money received from the new loan is greater than the amount due on the old loan. The borrower can use the extra funds in any manner.

Cashier's check: A check the bank draws on itself rather than on a depositor's account.

Casing: Exposed molding or profile around a window or door, on either the inside or outside, to cover the space between the window frame or door jamb and the wall.

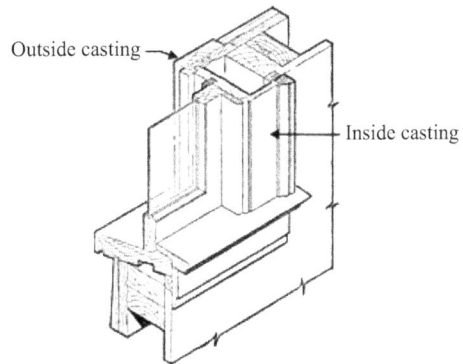

Window casing.

Also, a pipe or tube used to line a hole or shaft.

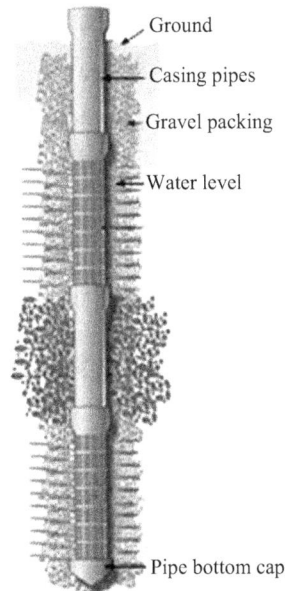

Well pipe casing.

Cast brass construction: Regarding faucets, sturdy, seamless, durable, and easy to clean one-piece faucet bodies.

Cast in place: Concrete that is formed and placed (poured) on site.

Cast iron: A hard, relatively brittle alloy of iron and carbon that can be readily cast in a mold and contains a higher proportion of carbon than steel (typically 2.0–4.3%). Cast iron is used in pipes, sinks, and bathtubs (surfaces are coated with porcelain-enamel to provide attractive, easy-to-clean finishes). Cast iron is resistant to destruction and weakening by oxidation (rust). It is not, however, used much in structural engineering.

Cast iron no hub pipe: Pipe that is used in certain locales and for commercial buildings for soil stacks. It cannot be soldered, threaded, or welded, and can only be connected by steel banded rubber sleeved adapters. Also referred to as the **soil pipe**.

Cast iron pipe joint set in concrete.

Cast-iron soil pipe: A pipe that is fabricated of an iron alloy containing carbon and silicon. It is usually lined with cement or coaltar enamel and coated externally with one of a variety of materials to reduce corrosion by soils; known technically as gray cast-iron pipe. Also called the **cast-iron pipe**.

Casting: The process of pouring molten metal or glass, clay slip, etc., into a hollow mold to harden. Some casting processes permit more than one reproduction.

Cat: Often short for Caterpillar tractor, or any other brand of bulldozer-type tractor. Generally refers to an earthmoving bulldozer but may also refer to a **skidding tractor**. (A tractor for work-in the logging industry. Skidding tractors are equipped with devices for collecting, loading, transporting, and unloading trees, trunks, and logs.)

Cat face: A deformed tree trunk surface usually caused by fire, disease, or rot.

Cat skinner: The person who operates a cat. See **Cat**.

Catalyst: A substance that increases the rate of a chemical reaction without itself undergoing any permanent chemical change, e.g., an acid catalyst added to an epoxy resin system to accelerate drying time.

Catch basin: An underground structure for drainage into which the water from a roof or floor drains. It is connected with a sewer drain or sump pump.

Catch platform: A platform or other construction projecting from the face of a building and used to intercept the fall of objects and to protect people and property from falling debris.

Cathedral ceiling: A high open ceiling formed by finishing exposed roof rafters.

Cathedral grain: A grain appearance characterized by a series of stacked and inverted "V"s, or cathedral type of springwood (earlywood) summerwood (latewood) patterns common in plain sliced (flat cut) veneer. See **Split heart**.

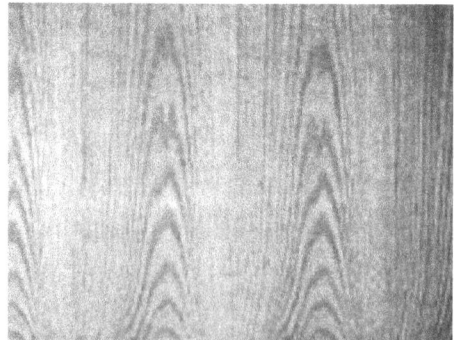

Typical "cathedral" grain pattern of oak veneer.

Cathodic: Metals low in the galvanic series. The galvanic series (or electropotential series) determines the nobility of metals and semi-metals. When two metals are submerged in an electrolyte, while electrically connected, the less noble (base) will experience galvanic corrosion as determined by the electrolyte and the difference in nobility. The difference can be measured as a difference in voltage potential.

Catwalk: A support member attached near the center of the bottom chord of a truss system to hold the trusses in a vertical position.

Also, a narrow, often elevated walkway, as on the sides of a bridge or in the flies above a theater stage.

Also, a temporary foot bridge that is used by bridge workers to spin the main cables (several feet above each catwalk) and to attach the suspender cables that connect the main cables to the deck.

Caulking: A flexible material used to seal a gap between two surfaces, e.g., between pieces of siding or the corners in tub walls.

Also, to fill a joint with mastic or asphalt plastic cement to prevent leaks.

Caustic: Able to burn or corrode organic tissue by chemical action.

Also, in optics, a caustic is the envelope of rays reflected or refracted by a curved surface or object, or the projection of that envelope of rays on another surface. The caustic is a curve or surface to which each of the light rays is tangent, defining a boundary of an envelope of rays as a curve of concentrated light.

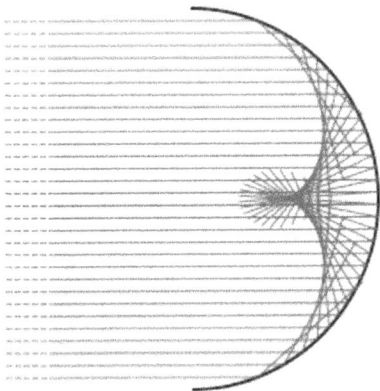

A caustic.

Caustic embrittlement: A type of embrittlement in the metal at joints and the ends of tubes in steam boilers. Caustic embrittlement is due to the chemical composition of the boiler water. It may lead to failure of the metal.

Caveat: A formal notice that asks a court to suspend action until the party which filed the challenge can be heard.

Caveat emptor: A legal principle derived from Latin than means "let the buyer beware." Caveat emptorexpresses the rule of law that the purchases buys at his/her own risk.

Cavedium: A courtyard or atrium.

Cave-in: The separation of a mass of soil or rock material from the side of an excavation, or the loss of soil from under a trench shield or support system, and its sudden movement into the excavation, either by falling or sliding, in sufficient quantity so that it could entrap, bury, or otherwise injure and immobilize a person.

Cavitation: Cavities or bubbles that are formed in a liquid that is being pumped. These cavities form at the low pressure or suction side of the pump, causing all or some of the following to occur: the cavities or bubbles collapse when they pass into the higher regions of pressure, causing noise, vibration, and damage to many of the components; there is a loss in capacity; the pump can no longer build the same head (pressure); and, the pump's efficiency drops.

Cavity wall: The term used to describe a masonry assembly wall where the outer wythe of masonry is separated from the inner masonry assembly by a continuous air space. The cavities may have insulation in them.

Cavity wall construction.

Cavity wall tie: See **Brick tie**.

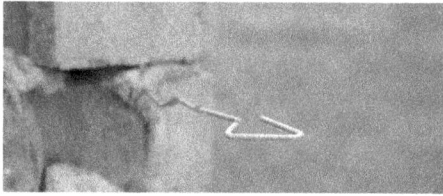

Celotex ™: Black fibrous board that is used as exterior sheathing.

Ceiling attenuation system: A system that rates a ceiling's efficiency as a barrier to airborne sound transmission between adjacent rooms. (For more information you are referred to ASTM E 413 and ASTM E 1414.)

Ceiling duty classification: The load carrying capability of grid main beams (per ASTM C635) pounds per lineal foot (Light: 5 lbs; Intermediate: 12 lbs; Heavy: 16 lbs).

Ceiling extension: A ceiling extension is similar to a bottom chord extension except that only one angle of the joist bottom chord is extended from the first bottom chord panel point toward the end of the joist.

Also, before most home construction became standardized for 8 foot tall ceilings, ceiling extensions were needed to lower a hanging lamp for lighting and refilling.

Ceiling extension for lowering a hanging lamp.

Ceiling height: The standard height of a ceiling is eight feet.

Ceiling joist: One of a series of parallel framing members used to support ceiling loads and supported in turn by larger beams, girders, or bearing walls. Also called **roof joists**.

Ceiling suspension system: A system of metal members that are designed to support a suspended ceiling. Ceiling suspension systems are usually designed to accommodate lighting fixtures and diffusers.

Celadon: The French name for a green, gray-green, blue-green, or gray glaze produced with a small percentage of iron as the colorant.

Cellar: A story partly or wholly underground, but having one-half or more of its clear height (measured from finished floor to finished ceiling) below the curb level. Cellars are usually not counted as stories in measuring the height of buildings. See **Basement**.

Cellular PVC: Extruded polyvinyl chloride material used in window and door components and trim. Unlike rigid (or hollow) vinyl, it features a foam or cell-structure inside. It can often be nailed, sawn, and fabricated like wood.

Cellulose: The carbohydrate that is the principal constituent of wood and forms the framework of the wood cells.

Cellulose insulation: Insulation that is made from shredded paper. Cellulose insulation should be treated with a fire retardant chemical or it may create a fire hazard.

Cellulosic composite: Generally, a material combining an organic material, such as wood fiber, extruded with a plastic; a composite whose ingredients include cellulosic elements is referred to as a **cellulosic material**. These cellulosic elements can appear in the form of, but are not limited to distinct fibers, fiber bundles, particles, wafers, flakes, strands, and veneers. These elements may be bonded together with naturally occurring or synthetic polymers. Additives such as wax or preservatives may also be added to enhance performance.

Celcon nut: Connectors that are used to hand-tighten faucets to lavatory from underside of lavatory.

Celsius (°C): The international temperature scale in which water freezes at 0 and boils at 100 under normal atmospheric conditions: $^{\circ}C = (^{\circ}F - 32) \div 1.8$.

Cement: The gray powder that is the "glue" in concrete (and should not be confused with concrete). Also, any adhesive.

Cement-aggregate reaction: A reaction between cement and constitutes of aggregates that may include hydration of anhydrous sulfates, rehydration of zeolites, wetting of clays and reactions involving solubility, oxidation, sulfates, and sulfites, in which case abnormal expansion and cracking may occur in the concrete.

Cement board underlayment: An underlayment made from cement board that has a smooth finish for use beneath vinyl floors. Cement board underlayment is almost impervious to water damage. See **Underlayment, Particleboard underlayment, Plywood underlayment, Lauan plywood underlayment, Gypsum-based underlayment**, and **Untempered hardboard underlayment**.

Cement content: The mass of cement in concrete. This is an important factor, especially for large pours, because it determines the heat generated during hydration and the susceptibility of the concrete to shrinkage. Durability and impereability of concrete are affected by cement content. Therefore, cement content should be specified.

Cement mixer: A mechanical device consisting of a rotating drum with fixed paddles inside, used for mixing cement with aggregate and water to produce concrete, mortar, or any other cement-based mixture.

Cementitious board: A type of cement board attached to a substrate to create an isolation membrane. The joints between the boards are then filled and leveled. See **thinset tile, Isolation membrane**, and **Mortar bed**.

Center bearing wall: A wall on the interior of a structure that is built to support the weight of the floor system above it. The center bearing wall is usually constructed along the center line of the structure. This is a structural part.

Center match: Regarding veneer, a symmetrical appearance. Each face has an even number of veneer pieces of uniform width before trimming. Thus, there is a veneer joint in the center of the panel, producing symmetry. This match reduces veneer yield and, therefore, is used in premium grade veneers only.

With reference to doors, an even number of veneer components or leaves of equal size (prior to edge trimming) matched with a joint in the center of the panel to achieve horizontal symmetry.

Also, known as **tongue and groove** and is commonly used as a colonial style flooring product.

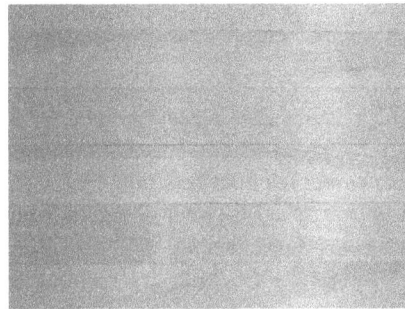

Center match tongue and groove flooring.

Center set: A style of bathroom faucet having combined spout and handles, with handles 4 inches apart, center-to-center.

Center of gravity: The location of the resultant of gravity forces on an object or objects; e.g., the point in the building plan at which the building would be exactly balanced. Also referred to as the **center of mass**.

Center of resistance: The resultant of resistance provided by walls and frames.

Center set faucet.

Also, a single-handle faucet installed on 4 inches center-to-center faucet holes.

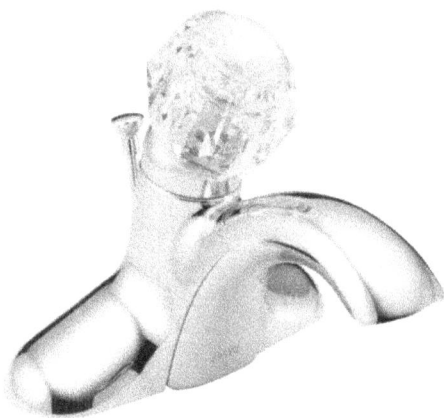

Another center set faucet.

Center-to-center hole spacing: The vertical distance measured between the center-points of two holes. For example, this measurement is often needed for tubular entry door handle sets, where a bore for the deadbolt is located above the bore for the latch.

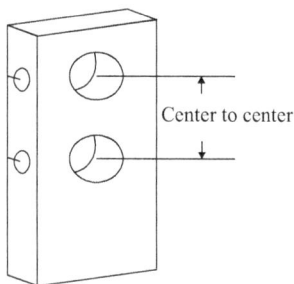

Center to center

Center-to-center hole spacing.

Centering: Framing that is used to support an arch or dome while it is under construction.

Centering for an arch.

Centerline span: A theoretical span definition which is the distance between the actual centerlines of a beam, column, joist, or joist girder. Also referred to as center-to-center.

Centimeter (centimeter) (CM): A metric unit of length equal to one hundredth of a meter (meter): 2.54 cm is equal to one inch.

Central air conditioning: A device that generates cold air through an outside unit that is connected to ductwork inside the house.

Central business district: The area of a city where most large businesses are located.

Centrifugal chiller: HVAC equipment that utilizes centrifugal force to assist the phase change cooling process of water that in turn is circulated throughout a chilled water circuit to individual air handlers.

Centrifugal force: The force that impels a thing, or parts of a thing, outward from a center of rotation.

An object traveling in a circle behaves as if it is experiencing an outward force. This centrifugal force depends on the mass of the object, the speed of rotation, and the distance from the center: the more massive the object, the greater the force; the greater the speed of the object, the greater the force; and the greater the distance from the center, the greater the force.

Centroid: Similar to the concept of center of gravity, except that it applies to a two-dimensional shape rather than an object. For a given shape, the centroid location corresponds to the center of gravity for a thin flat plate of that shape, made from a homogeneous material.

Ceramic disc valve: Part of a faucet's valving system. A ceramic disc valve creates a water-tight seal when the faucet is shut off. Unlike old rubber washers, ceramic disc valves are second in hardness only to diamonds. Ceramic discs are low maintenance and provide easy quarter turn operation.

Ceramic mosaic tile: A small tile, usually 1 × 1", 1 × 2", or 2 × 2", that are typically made from porcelain or natural clay. Generally, used in bathtub and shower enclosures and on counter tops. See **Glazed wall tile and Quarry tile.**

Ceramics: The art and science of forming objects from earth materials containing or combined with silica; the objects are then heated to at least 1300°F to harden.

Ceran top: A brand name for a type of specialty glass used in flat-surface cook-tops. The burners are placed under the glass and cooking utensils are placed on top of the glass.

Certificate of deposit (CD): An acknowledgment by a bank of receipt of money, with an obligation to repay it. The writing may or may not be a negotiable instrument depending on whether it meets the negotiability.

Also, a document that shows that the bearer has a specified amount of money on deposit with a bank, stock-brokerage firm, or other financial institution.

Certificate of deposit index: An index based on the interest rates on six-month CDs. It is used to determine the interest rate for some adjustable-rate mortgages.

Certificate of eligibility: A document issued by the Veterans Administration that verifies the eligibility of a veteran for a loan program.

Certificate of occupancy (CO and CofO): This certificate is issued by the local municipality and is required before anyone can occupy and live within the building. It is issued only after the local municipality has made all inspections and all monies and fees have been paid.

Also, a document that states that a home or other building has met all building codes and is suitable for habitation.

Certificate of sale: A document issued at a judicial sale, which entitles the buyer to receive a deed after court confirmation of the purchase of the property.

Certificate of title: A written opinion on the status of a piece of property based on an examination of the public record.

Certified: Equipment is "certified" if it meets the following criteria:(1) has been tested and found by a nationally recognized testing laboratory to meet nationally recognized standards or to be safe for use in a specified manner; or (2) is of a kind whose production is periodically inspected by a nationally recognized testing laboratory; and (3) it bears a label, tag, or other record of certification.

Certified lead abatement worker: An individual who has been trained by an accredited training program and certified by a state agency or by the EPA to perform lead hazard abatements.

Certified lead risk assessor: An individual who has been trained by an accredited training program and certified by a state agency or by the EPA to conduct lead risk assessments. A risk assessor also samples for the presence of lead in dust and lead in soil for the purposes of lead abatement clearance testing.

Certified welder: A welder who has been certified by a competent experienced welding inspector or a recognized testing facility in the field of welding. The welder must be certified to make certain welds under qualified procedures. The welder must be qualified for each position, type weld, electrode, and thickness of base metal that is to be welded in the shop or field.

Certified wood: Wood products that have been qualified by an independent third-party agency as satisfying their proprietary requirements for responsible environmental practices.

Chain: As pertains to surveying, a surveyors' unit of length in the Imperial system. **Gunter's chain**, named after its inventor, comprises 22 yards or 66 feet, ~20.117 meters. A Gunter's chain is useful for deriving areas in acres.

A lesser-known chain known as the engineer's chain or **Ramsden's chain** is 100 feet long and was used for measuring linear distances.

In Texas and elsewhere in the Southwestern United States, the **vara chain** of 20 varas (16.93 m, or ~55½ ft.) was used in surveying Spanish land grants.

Chain of title: The official record that details the ownership history of a piece of property.

Chain-of-custody: A tracking procedure for documenting the status of a product from the point of harvest or extraction to the ultimate consumer end use, including all successive stages of processing, transformation, manufacturing, and distribution. Establishing chains of custody is important when LEED certification is desired. Doing so can add to your costs.

Chainsaw firewood processor: Equipment used to produce firewood as a business.

Chair rail: Interior trim material that is installed about 3–4 feet up the wall, horizontally.

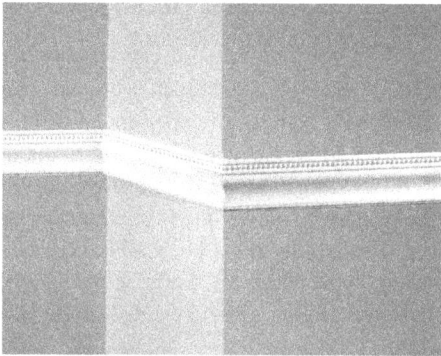

Chair rail.

Chalk: A powdery residue on the surface of a material.

Chalk line: A line made by snapping a taut string or cord dusted with chalk. Used for alignment purposes.

Chalking: Formation of a powder on the surface of a paint film caused by disintegration of the binder during weathering. Chalking can be affected by the choice of pigment or binder.

Chamfer: A beveled edge connecting two surfaces. If the surfaces are at right angles, the chamfer will typically be symmetrical at 45°.

   Also, the easing of an edge of a solid.

   With regards to metal roofs, material at 45° to each face.

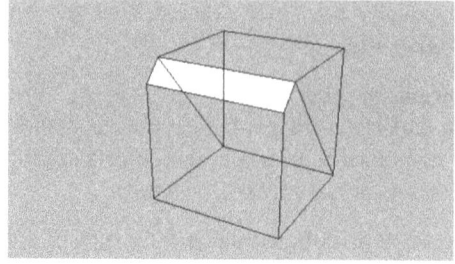

The highlighted yellow represents a chamfered edge.

Change frequency: The adjustment schedule on an adjustable-rate mortgage.

Change order: A written document that modifies the plans and specifications and/or the price of the construction contract.

Change order proposal (COP): A written document before it has been approved and acted on by the contractor, architect, and owner. A change order proposal can be issued by the contractor, architect, or the owner. The change order proposal becomes a change order only after it has been approved and affected by the contractor, architect, and owner.

Change order request (COR): A written document requesting an adjustment to the contract sum or an extension of the contract time. A COR may be issued by the architect, owner, or contractor.

Changed conditions: Conditions or circumstances, physical or otherwise, that surface after a contract has been signed and that alter the circumstances or conditions on which the contract is based, i.e., concealed conditions or latent conditions.

Channel: A fluid passage, the length of which is large when compared to its cross-sectional dimension, e.g., the bed of a stream or river. The deeper part of a river or harbor, especially a deep navigable passage. A broad strait, especially one that connects two seas. A **trench, furrow,** or **groove**.

   Also, a structural steel component which is C-shaped in cross section. Also known as a **channel section**.

Also, the marked location on a wall where another wall intersects it.

Also, in electronics, a specified frequency band for the transmission and reception of electromagnetic signals, as for television signals.

Channel flashing: Channel flashing is used for steep-slope roof construction and on tile roofs. It is a type of flashing used at roof-to-wall junctures and other roof-to-vertical plane intersections where an internal gutter is needed to handle runoff.

Channel flashing.

Character-defining feature: A prominent or distinctive aspect, quality, or characteristic of a historic property that contributes significantly to its physical character. Structures, objects, vegetation, spatial relationships, views, furnishings, decorative details, and materials may be such features.

Characteristic strength: The strength at which a member tested would fail, normally with 95% confidence.

Charge: To replenish a hydraulic system above atmospheric pressure. Also referred to as **supercharge**.

Also, to fill an accumulator with fluid under pressure. [The pressure at which replenishing fluid is forced into the hydraulic system (above atmospheric pressure) is called the **charge pressure**, and the pressure of compressed gas in an accumulator prior to the admission of liquid is called the **precharge pressure**.]

Charpy impact test: A standardized high strain-rate test that determines the amount of energy absorbed by a material during fracture. This absorbed energy is a measure of a given material's toughness and acts as a tool to study temperature-dependent brittle-ductile transition. Also known as the **Charpy v-notch test**.

Charring: As wood burns, the areas of the wood being consumed will begin to char. Charred portions of wood are not structurally stable. When the surface of structural framing members is charred more than 1/8" the framing member will generally have to be replaced.

Chart of accounts: An alpha/numeric accounting system that is used by the contractor to ensure that all debits/credits are properly assigned to the project.

Chase: A framed enclosed space around a flue pipe or a channel in a wall, or through a ceiling for something to lie in or pass through.

Chasing: A technique in which steel punches are used to decorate and/or texture of metal surface.

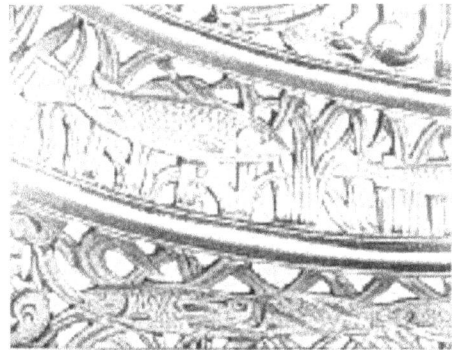

A metal surface that has been chased (or engraved).

Chattel: Personal property such as furniture, clothing, or a car.

Chattel mortgage: A lien on personal property used as collateral for a loan.

Chatter: Line[s] appearing across the face of wood at right angles to the grain giving the appearance of one or more corrugations resulting from a bad setting of sanding equipment.

Check: A splitting of the wood fibers within or on a log or lumber. Checks result from uneven wood shrinkage.

Check rail: The bottom rail on the upper sash and the upper rail of the lower sash of a double-hung window unit, where the lock is mounted. Also referred to as a **meeting rail**.

Check valve: A valve that permits fluid flow in one direction only.

Chemical composition: With regard to pre-tensioning steel and post tensioning steel, indicators of the steel's properties. This information is especially useful for predicting corrosion resistance and mechanical properties.

Chemical content: With regard to concrete, the presence or amount of foreign compounds in the concrete; most foreign compounds reduce compressive strength.

Chemical degradation: With regard to concrete, degredation of concrete due to chemical attack such as acid and sulfate attack, and alkali-aggregate reactions.

Chemical resistance: With regard to ceramic glazing on structural clay facing tiles, facing bricks, solid masonry units, and chemical resistant units, an indication of their resistance to changes in color and texture when subjected to immersion in a specified acid solution for a specified time period and at a specified temperature.

Chemical runoff: Water that transports chemicals from the building landscape, as well as surrounding streets, and parking lots, to rivers and lakes. Runoff chemicals may include gasoline, oil, antifreeze, and salts.

Chemical treatment: The use of biocidal, conditioning, dispersant, and scale-inhibiting chemicals to control biological growth, scale, and corrosion in cooling towers. Alternatives to conventional chemical treatment include ozonation, ionization, and UV light. Reducing or eliminating chemical treatment through effective alternatives reduces the environmental and human health risks associated with the chemicals used in conventional treatment protocols.

Wood also receives chemical treatment, but it is usually referred to as pressure treated wood/lumber or simply treated wood/lumber. Pressure treating wood forces a chemical preservative deep into the wood. The wood product is placed into a large cylindrical holding tank, and the tank is depressurized to remove all air. It is then filled with the preservative under high pressure, forcing it deeply into the wood. The tank is then drained and the remaining preservative reused.

Chem sponge: Also called a dry sponge or chemical sponge, a chem sponge is treated with special chemicals for use in removing soot from walls. Neither water nor other liquids are used with a chem sponge.

Chewable surface: An interior or exterior surface painted with lead-based paint that a young child can mouth or chew. Hard metal substrates and other materials that cannot be dented by the bite of a young child are not considered chewable.

Chicken tracks: An expression for scars that give the particular effect of a chicken's footprint. It is caused by air roots or vines.

Child-occupied facility: A lead inspection term that refers to a building, or portion of a building, constructed prior to 1978, visited regularly by the same child, six years of age or under, on at least two different days within any week, provided that each day's visit lasts at least 3 h and the combined weekly visit lasts at least 6 h, and the combined annual visits lasts at least 60 h. Child-occupied facilities may include, but are not limited to, day-care centers, preschools, and kindergarten classrooms.

Chimney: A vertical enclosure that contains one or more flues that are used to remove hot gasses from burning fuel, refuse, or from industrial processes.

Terms that begin with chimney:
- **Chimney back:** The back wall or lining of a fireplace or furnace chimney.
- **Chimney connector:** A pipe or metal breeching that connects combustion equipment to a chimney.

- **Chimney cricket:** A small roof built behind the chimney to move precipitation around the chimney and off the roof. Also called a chimney saddle.
- **Chimney flue:** The passage inside a chimney that channels smoke and heat to the outside.
- **Chimney pot:** A short pipe at the top of a chimney that increases ventilation to the fireplace and reduces smoke.

Chink: To install fiberglass insulation around all exterior door and window frames, wall corners, and small gaps in the exterior wall.

Also, the weatherproofing material placed between logs in a log home.

Chip: To mechanically reduce logs or whole trees to small pieces for fuel, pulp, or chipboard manufacture. Chippers are used to transform wood to chips. A chipper canter is a headrig machine that reduces debarked logs directly to chips and cants without producing sawdust.

Chipboard: A manufactured wood panel made out of 1–2" wood chips and glue. Often used as a substitute for plywood in the exterior wall and roof sheathing. Also called **OSB (Oriented Strand Board)**, **wafer board** and **chip board**.

Chisel plane: An excellent tool for removing glue and trimming work. It features a blade that extends in advance of the body, so the plane can be used to reach the farthest corner of a joint to get it clean.

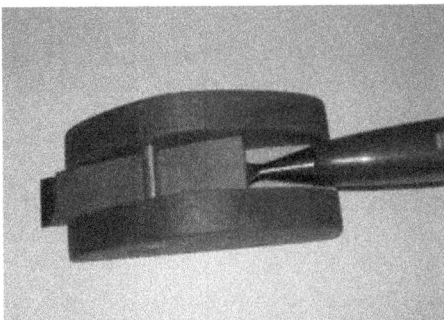

Chisel plane.

Chloride attack: Chlorides and nitrates of ammonium, magnesium, aluminum, and iron attack concrete, ammonium being the most harmful. Sodium chloride does not harm concrete but corrodes reinforcing steel.

Chloride content: With regard to concrete, the mass of free chloride in the concrete. Metal reinforcing corrodes rapidly when exposed to chloride ions. The diffusion of the chloride ion through concrete follows **Fick's law**. See **Fick's law**.

Chlorinated Polyethylene (CPE): CPE is a flexible material with high tear strength, good chemical resistance, and patency toward UV radiation. As a result of the high chlorine content (typically 30%) it is inherently difficult to ignite, but releases hydrogen chloride during combustion. It suffers from an extremely high permeability to gas. Resistance to most inorganic chemicals is generally good, while resistance to hydrocarbons increases with increasing chlorine content. The material is used mainly as an impact modifier for PVC and, to a lesser extent, LDPE and HDPE films.

Chlorinated polyethylene as defined by the National Roofing Contractors Association: A synthetic, rubber-like thermoset material, based on high-molecular-weight polyethylene with suphonyl chloride, usually formulated to produce a self-vulcanizing membrane. Chlorosulfonated Polyethylene or CSPE. Best known as **Hypalon™**, it was developed in 1951 by DuPont.

Chlorinated Polyethylene (CPE): Chlorinated polyethylene resins and elastomers (CPE) exhibit excellent physical and mechanical properties, including resistance to chemicals, oils, heat and weather, low-temperature performance, compression-set resistance, flame retardancy, high filler acceptance, tensile strength, and resistance to abrasion.

Chlorinated Polyvinyl Chloride (CPVC): A heat-resistant and low-combustibility plastic that is typically used in water supply.

Chlorofluorocarbons (CFCs): Hydrocarbons that are used as refrigerants and cause depletion of the stratospheric ozone layer.

Chlorosulfonated polyethylene (CSPE): A product of the chemical modification of polyethylene by chlorine and sulfur dioxide. It is resistant to fire, oil, and the action of microorganisms and exhibits good adhesion to various surfaces.

Chlorosulfonated polyethylene is superior to other rubbers in its resistance to the effects of ozone and inorganic acids, such as chromic, nitric, sulfuric, and phosphoric acids, as well as to the effects of concentrated alkalies, chlorine dioxide, and hydrogen peroxide. It is resistant to light, is impermeable to gas, and has good dielectric properties.

Chlorosulfonated polyethylene is used in the production of industrial and household goods and of anticorrosion coatings to be applied by the rubberizing method. It is also used as a film-forming agent in varnishes and paints for the preservation of wood, metal, and reinforced concrete and as a base for adhesives and hermetic sealants.

Chock: A wedge, block, or large stone placed against the tires of a vehicle to prevent its moving, especially on an incline.

Choke: To check or slow down a movement, growth, or action, e.g., a garden may be choked by weeds.

To obstruct by filling or clogging, e.g., a drainpipe may be choked with leaves or highways may be choked with stalled traffic. Usually the word refers to a restriction, the length of which is large with respect to the cross-sectional dimension.

Chopped Glass and Emulsion (CG & E): A roof coating that consists of asphalt or clay emulsion and glass fiber reinforcement. The glass fiber comes in rope form and is mechanically chopped into small pieces and then mixed with the emulsion at the end of the spray gun so that the mixture is complete by the time the surfacing hits the top of the roof. Standard mixture is 9 gallons of emulsion and 3 pounds of glass fiber for every 100 square feet (36.5 L of emulsion and 1.5 kg of chopped glass for every 10 square meters). The CG&E coating is then usually surfaced with a fibered aluminum roof coating at a rate of 1.5 gallons per 100 square feet (6 L per 10 square meters).

Chord: The two angle top or bottom member of a joist or joist girder, usually with a gap between the angles.

Chromium (Cr): A very hard, whitish metallic element derived mainly from the ore chromite. It is used in steel manufacture as a hardening agent, for producing stainless steels, to provide a hard, shiny, non-tarnishing coating, in dyeing and tanning, and in the manufacture of green, red, and yellow paints.

Chroma: A measurement of color. The degree of saturation of a hue. A color at its full intensity has maximum chroma.

Chuck: A clamping device at the end of a power drill for holding a drill bit.

Churn: The movement of workspaces and people within a space.

Cinder block: A masonry unit made from Portland cement and cinder. Cinder blocks are lighter and have better insulation qualities than concrete masonry units.

Cinder fill: Cinders used below a basement or around a foundation to promote drainage.

Circle: A shape in which all points on the perimeter are the same distance from the center. The formula for calculating the area of a circle is "pi × radius² = area." The formula for the perimeter is "pi × diameter = circumference."

Circle-top: One of several terms used for a variety of window units with one or more curved frame members, often used over another window or door opening. Also referred to as **arch-tops**, **circle-heads**, and **round-tops**.

Circle top window.

Circuit: The path of electrical flow from a power source through an outlet and back to ground.

Also, an arrangement of components interconnected to perform a specific function within a system.

Also, any complete movement of fluid, air, gas, or electricity through a pipe, duct, or conductor.

Circuit breaker: A device that looks like a switch and is usually located inside the electrical breaker panel or circuit breaker box. It is designed to shut off the power to portions or the entire house and to limit the amount of power flowing through a circuit (measured in amperes). 110-volt household circuits require a fuse or circuit breaker with a rating of 15 or a maximum of 20-amps. 220-volt circuits may be designed for higher amperage loads, e.g., a hot water heater may be designed for a 30-amp load and would, therefore, need a 30-amp fuse or breaker.

Circular headsaw: A circular plate having cutting teeth on the circumference and used to ripsaw logs.

Circular hollow: A structural steel component in the shape of a round tube.

Circular saw: A work-horse tool. A circular blade rotates on an arbor at a high speed so that the teeth on the blade cut the material, usually wood. The portable circular saw is critical to any building or framing project because it can cut lumber quickly, with power and with accuracy.

Circular stair: A stair system that winds in a curving pattern, usually, but not always around a common center.

Circumference: The length of the perimeter of a circle, calculated with the formula "pi × diameter = circumference." Can be used interchangeably with **perimeter**.

Cistern: A reservoir or tank for holding water or other fluids.

Civil engineer: A person who practices civil engineering: planning, designing, constructing, maintaining, and operating infrastructures while protecting the public and environmental health. In some places, a civil engineer may perform land surveying. On some US military bases, the personnel responsible for building and grounds maintenance, such as grass mowing, are called civil engineers.

Civil law: Roman law embodied in the Justinian Code and presently prevailing in most Western European States. It is also the foundation of the law of Louisiana. See **Common law**.

The term may also be used to distinguish that part of the law concerned with noncriminal matters, or may refer to the body of laws prescribed by the supreme authority of the state, as opposed to natural law.

Cladding: Material that is placed on the exterior of wood frame and sash components to provide ease of maintenance. Common cladding materials include vinyl and extruded or roll-formed aluminum.

Claim: A formal notice sent by a contractor to an owner asserting the fact that the terms of the contract have changed and compensation is being sought by the contractor from the owner.

Clamp: A tool, typically shaped like a capital G, for clamping two objects together for gluing or other work. Also referred to as a cramp.

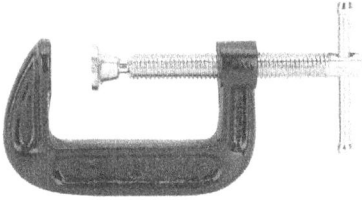

Clamp.

Class A, B, C/I-II-III flame spread; Class A-B-C roof coverings; and Hourly fire-resistance ratings: These terms get confusing (please follow along): Class A, B, C/I, II, III flame spread is based on ASTM E-84/UL 723 "Test for Surface Burning Characteristics of Building Materials." Class A-B-C roof coverings is based on ASTM E-108/UL 790, "Test for Fire Performance of Roofing Materials." And, hourly fire-resistance ratings is based on ASTM E-119, "Fire Tests of Building Materials."

Class A-B-C roof coverings: Class A-B-C roofing systems are sometimes confused with class A-B-C/I-II-III flame spread categories, but there is no correlation. The ASTM E-108/UL 790 roof coverings test is a pass-fail test under which a product either passes the criteria as a Class A (severe exposure), B (moderate exposure), or C (light exposure) roof or not; it does not produce a flame spread rating.

Prior to testing for fire exposure, wood shakes and shingles must go through ASTM D 2898, "Standard Rain Test." Although called a test 2898 actually prepares the wood shakes or shingles for the actual fire exposure test, i.e., ASTM E 108/UL 790, by "pre-weathering" them. Some wood shakes and shingles have achieved a Class A rating, but there is still a debate as to whether or not they're effective

over time. Some communities, especially in California, have prohibited their use in their jurisdictions.

Hourly fire resistance ratings: Hourly ratings are a function of the assembly being used, e.g., wall, floor, ceiling, and door, etc., and generally required the use of noncombustible membrane, e.g., gypsum and masonry. ASTM E-119, "Fire Tests of Building Construction Materials" is the test used to determine the hourly rating of an assembly. It exposes an assembly to heat and flame on one side and tests for heat transmission, burn-through, structural integrity, and ability to withstand a hose stream from a fire hose.

If you are still with me, the flame spread classification per ASTM E-84, 30-min duration, has no relation to a 30-min duration as determined by ASTM E-119. ASTM E-119 is not a required test for fire retardant treated wood and therefore under ASTM E-119 FRTW cannot be substituted for other noncombustible materials in a rated assembly. FRTW does have an advantage over untreated wood because it does not ignite or contribute to flame spread.

Class A, B, and C ratings for fire extinguishers
• **"A" Trash-wood-paper:** Fire extinguishers with a Class A rating are effective against fires involving paper, wood, textiles, and plastics. The primary chemical in them is monoammonium phosphate.
• **"B" Liquids:** Fire extinguishers with a Class B rating are effective against flammable liquid fires, e.g., cooking oils, oil, gasoline, kerosene, and paint. Monoammonium phosphate has proven effective at smothering these fires, while sodium bicarbonate induces a chemical reaction that extinguishes the fire.
• **"C" Electrical equipment:** Fire extinguishers with a Class C rating are suitable for fires in "Live electrical equipment." Both monoammonium phosphate and sodium bicarbonate are commonly used to fight these kinds of fires because of their nonconductive properties.

Class life: The useful economic life of an asset set by the Internal Revenue Service. See **(United States) Internal Revenue Service**.

Classical orders of architecture: Building styles originating from the construction of temples in ancient Greece and Rome. Orders are defined by their varying styles of column, although the orders also include information on the proportions of the building. The Greeks originally had three orders: the **Doric, Ionic, and Corinthian**. Doric is the simplest, Ionic more elaborate, and Corinthian more decorative still. The Romans added the **Tuscan and Composite** orders which are respectively plainer and more highly decorated than the Greek orders.

Classified property tax: A tax that varies in rate depending on the use of the property.

Clastic: Sediments formed from the breaking up of earlier rocks.

Claw hammer: A tool used for finish carpentry and light-duty nailing jobs. It features two sharp, beveled edges for gripping and drawing out stubborn nails.

Clay: A stiff, sticky fine-grained earth, typically yellow, red, or bluish-gray in color and often forming an impermeable layer in the soil. It can be molded when wet and is dried and baked to make bricks, pottery, and ceramics.

Also, solid particles contained within boiler feedwater. Most commonly seen at the lower points of the boiler where they form a sticky clay like substance. This clay is largely immobile and thus causes localized overheating and circulation problems.

Clay block: Masonry unit made from clay. Most often used on commercial structures. Also commonly called "Atlas Block."

Clay tile: Tile made from clay that has been forced through an extruder, cut to size, air-dried, and then fired in a kiln.

Cleanout: An opening that provides access to a drain line. Clean outs are closed with threaded plugs. Also referred to as cleanout plugs.

Clear all heart: High-grade redwood lumber that is free of knots, pitch, and blemishes. The grain of clear all heart is usually fairly straight.

Clear coating: A transparent protective and/or decorative film; generally the final coat of sealer applied to automotive finishes.

Clear lumber: Lumber or logs that are free or practically free of defects. First quality lumber or log.

Clear span: The actual clear distance or opening between supports for a structural member, i.e., the distance between walls or the distance between the edges of flanges of beams.

Clear title: A property that does not have liens, defects, or other legal encumbrances.

Clear-cut: An area in which all trees have been or will be felled, bucked, and skidded in one operation. When all trees in a given area are felled.

Clearance angle: As pertains to a saw, the angle between a tangent to the cutting circle of a tooth and a line along the top of the tooth intersecting this tangent.

Clearance examination: A visual examination and collection of environmental samples by a lead inspector or lead risk assessor and analysis by an accredited laboratory upon completion of a lead abatement project, interim control intervention, or maintenance job that disturbs lead-based paint.

Cleat: A metal strip, plate, or metal angle piece, either continuous or individual ("clip"), used to secure two or more components together.

Also, a structural block used at the end of a platform to prevent the platform from slipping off its supports. Cleats are also used to provide footing on sloped surfaces such as crawling boards.

Cleavage: The tendency of some rocks to split or break along smooth planes that are more or less parallel.

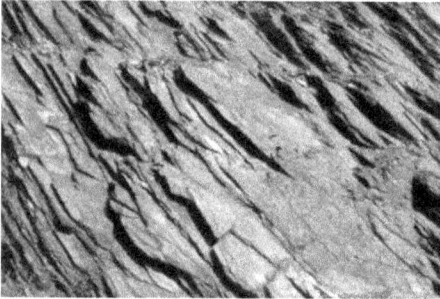

Rock cleavage.

Clerestory: A window in the upper part of a high-ceilinged room that admits light to the center of the room.

Clerestory roof: A roof style consisting of two sides that slope in opposite directions with a vertical wall section extending between the peaks. The vertical wall contains windows that provide light and/or ventilation into the building. The clerestory roof is common on condominium and passive solar homes.

Clerestory roof (allows light to enter building).

Clerk-of-the-works: An individual employed by an owner to represent him on a project at the site of the work. The clerk-of-the-work's abilities, credentials, and responsibilities vary at the discretion of the owner.

Clevis hanger: A steel "U" shaped hanger for the support of piping off a structural floor that allows for thermal expansion and pitch adjustment in piping installations.

Climate change: Refers to any significant change in measures of climate (such as temperature, precipitation, or wind) lasting for an extended period (decades or longer) (US Environmental Protection Agency, 2008).

Clip: An individual (discrete) cleat. See **Cleat**.

Also, with regard to metal roofing, a small metal component used to secure two pieces of metal to each other or to secure metal shingles or standing seam to solid decking.

Clip angle: A structural angle that attaches to the side of a wall, column, beam, etc., where a joist, joist girder, or other structural member bears.

Clip ties: Sharp, cut metal wires that at one time held the foundation form panels in place and protrude out of a concrete foundation wall.

Clipped gable: A gable cut back at the ridge in a small hip configuration.

Closed valley: The shingles on one roof face run past the valley and extend up the other face (for at least 12 inches). That face of the roof can be done all the way up, and then a chalk line snapped to mark the centerline of the valley, which acts as a guide for cutting the shingles for the adjacent face.

The shingles from the other face run up to the valley and are cut along the chalk line. A bead of roofing tar is generally applied under each cut shingle where it ended at the valley to prevent water from getting under the shingles. Also referred to as closed cut valley.

Close-coupled water closet: A two piece toilet, consisting of one tank and one bowl.

Closed-cut valley: A method of valley application in which shingles from one side of the valley extend across the valley while shingles from the other side are installed over the top of those and then trimmed back approximately 2 inches from the valley centerline.

Center line of valey

Cut 2′ diagonally off corner of trimmed shingle

Cut overlapping shingles 2′ back from valley center line

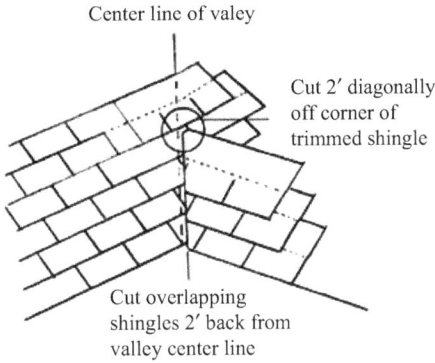

Closed-cut valley.

Closed-grain: A wood that exhibits narrow, inconspicuous, annual growth rings is considered closed-grain. Examples are cherry and maple.

Closer: A piece of door hardware that enables a door to close and latch by itself.

Closet bend: A curved fitting mounted immediately below the toilet that connects the closet flange to the toilet drain.

Closet flange: An anchoring ring that attaches to the closet bend and is secured to the floor. The heads of closet bolts that are used to secure the toilet in place are insert into slots in the closet flange.

Closet flange.

Closing: The final procedure in which documents are signed and recorded and the property is transferred. Expenses that are incidental to the sale of real estate, including loan, title, and appraisal fees are known as **closing costs**. And, a document that details the final financial settlement between a buyer and seller and the costs paid by each party is a **closing statement**.

Closure strip: A material used to close openings created by joining metal panels or sheets and flashings.

Cloud on title: Any matter appearing in the record of a title to real estate that on its face appears to reflect the existence or encumbrance that, if valid, would defeat or impair title; clouds on title may be removed by a court of equity by means of an action to quiet title (an equitable action to determine all adverse claims to the property in question).

Cluster development: A method of squeezing more homes into less space.

Clustered: When a defect described in the grading rule is sufficient in number and sufficiently close together to appear to be concentrated in one area.

Clustered wiring: A group of wires, each covered with insulation (except in some cases for the copper ground) and then the entire cluster of wires is also covered by plastic insulation.

Clutch: A device used in elevator power door operation to engage the car door to the landing door by a grasping and holding movement.

Clutch-head screwdriver: This screwdriver features a distinctive bow-tie shaped head that drives screws featured in mobile homes, cars, boats, appliances, and electric motors.

Coal ash: Sometimes coal ash is used as an additive in concrete. Recent research has devise a recipe for a concrete that has the potential to reuse larger amounts of coal ash thereby putting to use this toxic material.

Coal tar felt: A roofing membrane saturated with refined coal tar.

Coal tar pitch: Hydrocarbon substance created by processing coal, may be used to waterproof membrane roofing.

Coal tar roof cement: A trowelable mixture of processed coal tar base, solvents, mineral fillers, and/or fibers.

Coalescent aid: The small amount of solvent contained in latex coatings. Actually a coalescent aid is not a true solvent in that it does not dissolve the latex resins. The coalescent aid helps the latex resins flow together, aiding in film formation.

Coarse aggregate: Crushed stone or gravel used in concrete. Coarse aggregate will not, when dry, pass through a sieve with 1/4-inch-diameter (6-mm) holes.

Coarse orange peel surface texture: A surface showing a texture where nodules and valleys are approximately the same size and shape. This surface is generally acceptable for installing a protective coating.

Coated base sheet: An asphalt-saturated base sheet membrane later coated with harder, more viscous asphalt, thereby increasing its impermeability to moisture.

Coated felt: An asphalt-saturated ply sheet that has also been coated on both sides with harder, more viscous asphalt.

Coated glass: A window glass with an outside surface provided with a mirror reflective surface; the shading coefficient ranges from 20% to 45%.

Coating: A paint, varnish, lacquer, or other finish used to create a protective and/or decorative layer. Generally used to refer to paints and coatings applied in an industrial setting as part of the original equipment manufacturer's (OEM) process.

Coatings are often specified based on their property requirements, these are **coating properties:** For example, coating properties may be specified for epoxy-coated wires and bars; the amount of adhesion, continuity, and thickness of the epoxy would usually be included in the specification; these wires and bars are frequently used as pre-tensioning steel and post tensioning steel in concrete.

Cobalt HSS: In drill bits, a special high-speed steel blended with a significant percentage of cobalt. Cobalt adds hardness and abrasion qualities plus superior resistance to heat. Typically used for extra-tough or production drilling.

Code: A systematic compilation of laws. In the case of building codes they are generally enacted into law by the local governing body. Many building codes refer to one or more of the standard codes, e.g., the International Building Code.

Coefficient of friction: The coefficient of friction is a dimensionless number that represents the friction between two surfaces. Between two equal surfaces, the coefficient of friction will be the same. The symbol usually used for the coefficient of friction is μ.

The maximum frictional force (when a body is sliding or is in limiting equilibrium) is equal to the coefficient of friction × the normal reaction force:

$$F = \mu R,$$

where μ is the coefficient of friction and R is the normal reaction force.

This frictional force, F, will act parallel to the surfaces in contact and in a direction to oppose the motion that is taking/trying to take place.

There are several types of friction: dry friction that resists relative lateral motion of two solid surfaces in contact; fluid friction that describes the friction between layers within a viscous fluid that are moving relative to each other; lubricated friction that is a case of fluid friction where a fluid separates two solid surfaces; skin friction that is a component of drag, the force resisting the motion of a solid body through a fluid; and internal friction that is the force resisting motion between the elements making up a solid material while it undergoes deformation.

When surfaces in contact move relative to each other, the friction between the two surfaces converts kinetic energy into heat.

Coefficient of linear expansion: The change in length, per unit, for a change of one degree of temperature.

Coffer: A sunken panel in a ceiling.

Coffer.

Cofferdam: A temporary dam built to divert a river around a construction site so the dam can be built on dry ground.

Cofferdam.

Cogeneration: An example of this would be a central heating plant that produces steam at, say, 200 psi. This 200 psi would then drive a generator, thereby producing electricity. This process would reduce the steam pressure to, say, 100 psi, a manageable pressure that would then be shipped through the central heating system. An added efficiency would be to capture the remaining hot water and returning it to the central heating plant for reuse. Co-generation systems are typically 60–80% efficient, which is significantly more efficient than traditional power plants that have efficiencies of approximately 30%.

Cohesion: Mutual attraction by which the elements or particles of a body or substance are held together.

Also, a bonding together of a single substance to itself. Internal adhesion.

Cohesive soil: Soil consisting of silt and clay where the soil particles tend to stick together producing a sticky plastic soil characteristic.

Coign: The cornerstone of a building that differs in shape or color from the rest of the wall.

Coil coating: The application of a finish to a coil of metal or other material.

Terms beginning with cold:
- **Cold air return:** The ductwork (and related grills) that carries room air back to the furnace for re-heating.

- **Cold chisel:** A stone-cutting tool that has an integral handle and blade made of steel. The handle is struck by a hammer to cut material. It can cut sheet metal, remove rivets, bolts, nails, or cut away ceramic tiles adhered to a surface.
- **Cold deck:** A stack of logs left in the woods to be picked up, or a deck of logs at the mill for winter use.
- **Cold forming:** The process of shaping metal into desired configurations at ambient room temperature.
- **Cold joint:** As pertains to concrete, the joint that occurs when a batch of fresh concrete is placed next to concrete that is set or concrete that is less plastic with no vibration or rodding to cause the two batches to consolidate.
- **Cold process built-up roof:** A roof consisting of multiple plies of roof felts laminated together with adhesives that usually come right out of a can or barrel and require no heating.
- **Cold roof:** With regard to roofing, a cold roof incorporates "above sheathing ventilation" in order to help prevent hot spots on the roof and subsequent wintertime ice dams.

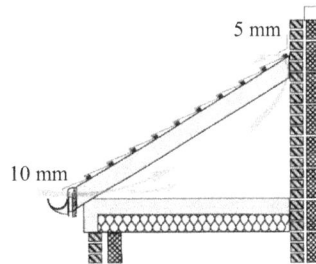

Air flow through a roof using soffit and ridge vents.

- **Cold-formed steel:** Sheet steel that is formed at room temperature into its final shape.

Collapse: Besides the usual meaning, i.e., to fall down or in, as the roof collapsed on top of me, in the lumber industry collapse means the flattening of single cell or rows of cells during the drying or pressure treating of wood.

Collapse is often characterized by a caved-in or washboarded appearance of the wood surface.

Collar: A pre-formed flange placed over a vent pipe to seal the roof around the vent pipe opening. Also called a vent sleeve and pipe boot.

Rain collar for no caulk roof flashing.

Also, a flange mounted on the saw arbor to support the blade on one or both sides. If the collar is fixed to the arbor, it is called a **fixed collar**. If not, it is a **loose collar**.

Also, a horizontal timber joining two opposing rafters together. Generally collars are nominally 1- or 2-inch-thick members that connect opposite roof rafters. They serve to stiffen the roof structure. They are also known as **collar ties** and **collar beams**.

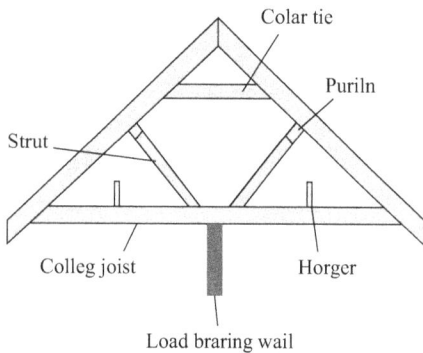

Collar with other parts of a roof truss.

Collarless saws: Collarless saws have the saw keyed directly to the arbor itself.

Collector head: A component used to direct water from a through-wall scupper to a downspout. Also known as a **conductor head** and **leader head**.

Collateral load: All additional dead loads other than the weight of the building, such as sprinklers, pipes, ceilings, and mechanical or electrical components.

Collateral security: Additional security that a borrower supplies to obtain a loan.

Collateral warrantee: A legal agreement between, for example, a developer and a building contractor or designer, that allows the contractor or designer to be made responsible to a third party, such as a finance provider or a purchaser, for the execution of their duties.

The need for collateral warranties may exist when the party that commissions a building will not carry the burden in the event of defects. For example, when an architect is appointed to design a group of dwellings for a developer. If the developer builds on "spec" and intends to sell the buildings quickly, the architect would normally be contractually liable only to the developer should defects arise. The collateral warranty establishes a contractual relationship between the new owners and the architect against defect.

Collection: The series of steps a lender takes to bring a delinquent mortgage up to date.

Colloid: A suspension of solid particles in a liquid.

Collusion: The action of two or more people to break the law.

Colonia: Unincorporated communities along the US-Mexico border.

Colonial: An architectural style associated with an early American period; Early American style c. 1730.

Colonial base and casing: A commonly used molding pattern.

Colonial windows: Windows with small rectangular panes, or divided lites, designated as 12-lite, 16-lite, etc.

Colonist door: Brand name of door that is made of pressed wood fiber and has numerous raised panels.

Colorant: Concentrated color (dyes or pigments) that can be added to paints to make specific colors.

Colorfast: Non-fading in prolonged exposure to light.

Color retention: The ability of paint to keep its original color. Major threats to color retention are exposure to ultraviolet radiation and abrasion by weather or repeated cleaning.

Column: A vertical structural member placed on a footing or foundation to support horizontal above-ground building components. A column is a vertical structural member that is designed to take primarily a compressive load.

Terms beginning with column:
- **Column base:** A (usually thick) plate at the bottom of a column through which the column's forces are transferred to the foundation.

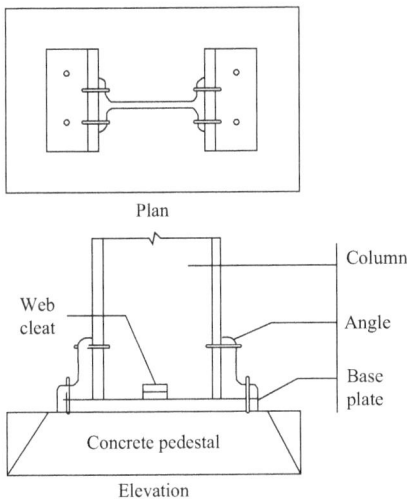

Plan

Column to base plate connection.

- **Column bay:** The total floor area exerting a load on a particular column and or the regular spacing of columns in a structural frame.
- **Column curve:** A curve that shows the relationship between axial column strength and slenderness ratio (the slenderness ratio is the ratio between the height or length of a structural element and the width or thickness of the element).
- **Column footing:** The support base for a load-bearing column. The footing is usually made of reinforced concrete.

- **Column splice:** The joining of two columns on the axis with the aid of steel plates, or a welded connection.

Columnar jointing: In igneous rocks, a regular six-sided form of jointing that produces regularly shaped pillars or columns.

Columnar jointing.

Comb grain: A quality of rift cut veneer with exceptionally straight grain and closely spaced growth increments resembling the appearance of long stands of combed hair.

Combination door: A door that is used in combination with a primary door. The door has interchangeable screen and glass panels.

Combination hand vise: A vise that holds all types of small parts securely. It can be fitted on a handle or on a clamp-like mechanism that is secured to a work surface.

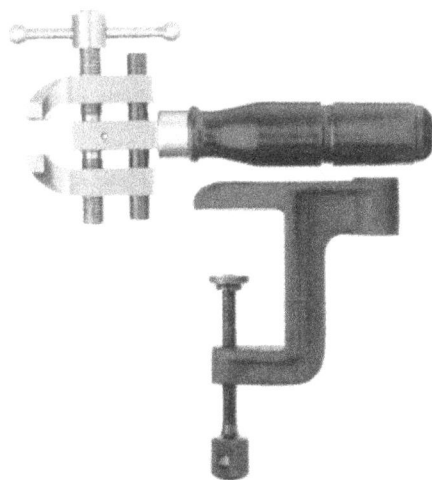

Combination hand vise.

Combination square: A tool, that has a 6 inch long steel rule for marking 90° and 45° angles.

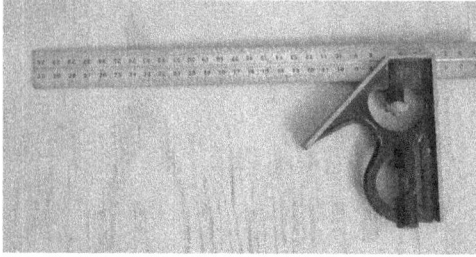

Combination square.

Combination window: A wood or clad wood frame storm sash with self-storing screen. Bottom glass panels such as those installed on a double hung unit operate by moving the plungers in and sliding the glass panel up to the desired position. Side glass panels such as those installed on gliders slide to the left or right to the desired position. All inserts are removable from the inside.

Combing ridge: An installation of finishing slate at the ridge of a roof whereby the slates on one side projects beyond to the apex of the ridge.

Combustible: Capable of igniting and burning.

Combustion air: The duct-work installed to bring fresh, outside air to the furnace and/or hot water heater. Normally two separate supplies of air are brought in: one high and one low.

Combustion chamber: The part of a boiler, furnace, or woodstove where the burn occurs; normally lined with firebrick or molded or sprayed insulation.

Comfort criteria: Specific original design conditions that include temperature (air, radiant, and surface), humidity, air speed, outdoor temperature design conditions, outdoor humidity design conditions, clothing (seasonal), and expected activity (ASHRAE 55-2004).

Comingling recycling: A process of recycling materials that allows consumers to dispose of various materials (such as paper, cardboard, plastic, and metal) in one container that is separate from waste. The recyclable materials are not sorted until they are collected and brought to a sorting facility.

Terms that begin with commercial:
- **Commercial bank:** A financial institution that provides a broad range of services, from checking and savings accounts to business loans and credit cards.
- **Commercial building inspection:** The process by which an inspector visually examines the readily accessible systems and components of a building used for commercial purposes and which describes those systems and components in a property condition assessment report.
- **Commercial property:** An area that is zoned for businesses.
- **Commercial real estate:** Any multifamily residential, office, industrial, or retail property that can be bought or sold in a real estate market.
- **Commercial standard:** A voluntary set of rules and regulations covering quality of product (or installation), method of testing, rating of the product, certification, and labeling of manufactured products.

Commingling: The mixing of money held in trust with other funds.

Comminutors: Comminutors are used to reduce the particle size of wastewater solids. They cut up and grind the coarse solids that enter a sewage treatment plant into smaller sizes thereby eliminating the problems caused downstream, especially clogging happening in pumps. Comminutors are particular important in treatment plants located in cold climates areas because collected waste does not normally get trapped on freezing screens. Also known as grinders and macerators.

Comminutor

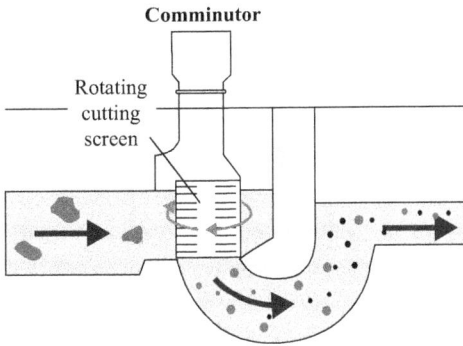

Comminutors.

Commission: The negotiable percentage of the sales price of a home that is paid to the agents of the buyer and the seller.

Commissioning: The process at or near construction completion when a facility is put into use to see if it functions as designed.

The schedule of activities relating to **existing building commissioning**, including the investigation and analysis phase, implementation phase, and ongoing commissioning is the **commissioning cycle**.

The commissioning cycle is a quality-oriented process for achieving, verifying, and documenting that the performance of facilities, systems, and assemblies meet defined objectives and criteria. It is an inclusive process for planning, delivering, verifying, and managing risks of and to the facilities' critical functions.

Commitment: A promise by a lender to make a loan with specific terms for a specified period. The fee a lender charges for promising to make a loan **is the commitment fee.**

Commode: Toilet

Common area: A portion of a building that is generally accessible to all occupants, such as hallways, lobbies, stairways, laundry and recreational rooms, playgrounds, garages/parking lots, and fences.

The actual, fine print definition of common area varies depending on location and owner; read the fine print.

Common area maintenance (CAM): Charges paid by the tenant for the upkeep of areas designated for use and benefit of all tenants. CAM charges are common in shopping centers. Tenants are charged for parking lot maintenance, snow removal, and utilities.

Common gable truss: A truss used to make a gable roof. It usually spans from outside wall to outside wall without relying on interior bearing walls for support. All trusses in a gable truss roof will be common gable trusses except the last truss on each end that are gable end trusses.

Common law: The system of jurisprudence that originated in England and was later applied in the United States. Common law is based on judicial precedent rather than statutory laws that are enacted by the legislature. It is to be contrasted with civil law, the descendent of Roman Law. Civil Law is prevalent in other western countries.

Common nails: The standard fasteners used by framers. See **Box nails** and **Sinkers**.

Common rafter: A rafter that is placed at 90° and extends from the cap plate to the ridge board.

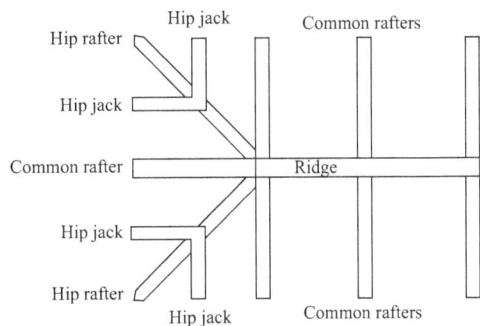

Various rafters including common rafters.

Common wall: A single wall that serves two dwelling units in a building. See **Party wall**.

Community Reinvestment Act: A federal law that encourages financial institutions to loan money in the neighborhoods where minority depositors live.

Commutator: A rotary electrical switch in certain types of electric motors or electrical generators. Commutators periodically reverse the current direction between the rotor and the external circuit. As a motor, it applies power to the best location on the rotor, and as a generator, picks off power similarly.

A commutator is a common feature of direct current rotating machines. By reversing the current direction in the moving coil of a motor's armature a torque is produced. Similarly, as a generator, reversing the coil's connection to the external circuit provides unidirectional (i.e., direct) current to the external circuit.

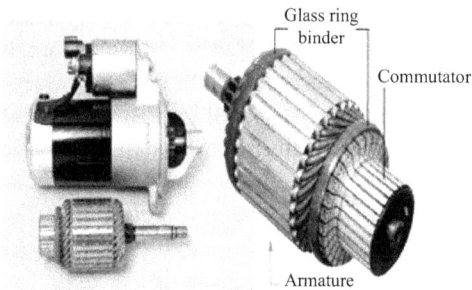

Commutator with armature.

Compact section: A steel section whose flanges must be continuously connected to the webs and the width-thickness ratios of its compression element cannot exceed the limiting width-thickness ratios designated in the AISC Manual.

Compactable fill: Soil that is capable of being compacted so as to provide a solid substance under the structural parts. Compactable fill is almost always placed in layers (also called "lifts") that are thin enough (usually 4–6 inches) to allow the compaction device to be effective throughout the layer. Each layer is thoroughly compacted before the next one is placed.

Compaction: The mechanical process of compacting soil, backfill and borrow fill. This may be done by hand tamping or with the aid of mechanical equipment.

Comparables: Properties used as comparisons to determine the value of a certain property.

Comparative market analysis: An estimate of the value of a property based on an analysis of sales of properties with similar characteristics. A comparative market analysis is generally essential if you expect to win an argument with the tax assessor's office.

Compass saw: This saw is similar to a coping saw, but more heavy duty. It has a thin blade set into a pistol-grip handle, allowing the user to quickly cut curves, circles and cutouts in wood, plywood, and wallboard. It is useful for cutting access holes when installing pipes and electrical boxes. A smaller version is the keyhole saw.

Compatible: In harmony with the location, context, setting, and historic character of a building or area. For example, when relating door edge to face appearance, the edge may not be the same species as the face; however, it may be similar in overall color, grain, character, and contrast as the face. See **Matching edge band**.

Compensator control: A displacement control for variable pumps and motors. (A variable displacement pump converts mechanical energy into hydraulic (fluid) energy. The displacement, or amount of fluid pumped per revolution of the pump's input shaft can be varied while the pump is running. Some variable displacement pumps are "reversible," i.e., they can act as a hydraulic motor and convert fluid energy into mechanical energy.) Displacement is altered in response to variances from pre-designated pressure setting.

Competent person (per OSH): One who is capable of identifying existing and predictable hazards in the surroundings, or working conditions which are unsanitary, hazardous, or dangerous to employees, and who has authorization to take prompt corrective measures to eliminate them.

Completed Contract Method: An accounting method that recognizes revenues and gross profit only when the contract is completed.

Component: As pertains to veneer, an individual piece of veneer or leaf that is joined to other pieces to achieve a full length and width face.

As pertains to a vector, any vector can be expressed as a collection of vectors whose sum is equal to the original vector. Each vector in this collection is a component of the original vector. It is common to express a vector in terms of components which are parallel to the x and y axes.

Composite: As used here, a solid material that is composed of two or more substances having different physical characteristics and in which each substance retains its identity while contributing desirable properties to the whole. For example, a structural material that is made of plastic within which a fibrous material (as silicon carbide) is embedded.

Terms beginning with composite:
- **Composite column:** The composite column was the result of combining both the Ionic and Corinthian columns.

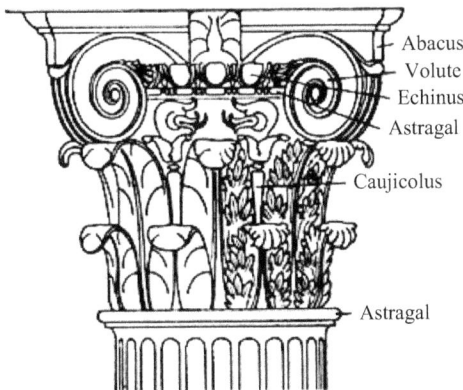

- **Composite beams:** Steel beams and a concrete slabs that are connected, usually by shear stud connectors so that they act together to resist the load on the beam.
- **Composite decks:** Corrugated and upset steel decking used as a lost form in the placement of concrete. Composite decks provide tensile strength to a concrete floor assembly.
- **Composite panels:** Door panels composed of a wood derivative such as medium density fiberboard (MDF). UComposite panels are generally used for opaque finishes.
- **Composite windows and doors:** Window or door components that consist of two or more materials, such as glass fibers or wood and plastic. The term is also used for windows and doors that combine two or more materials in the frame or sash construction, such as a product with a wood interior and a vinyl or aluminum exterior.
- **Composite wood:** These products are manufactured to designed specifications and are tested to ensure that they meet the national or international standards for which they were designed. Composite wood products are used in a variety of applications, from home construction to commercial buildings to industrial products.

  Typically, composite wood products are made from sawmill scraps and other wood and fiber (rye straw, wheat straw, rice straw, hemp stalks, kenaf stalks, or sugar cane residue) wastes. Also, referred to as engineered wood.

Composition shingles: Shingles that are used in steep-slope roofing. They are generally comprised of weathering-grade asphalt, a fiber glass reinforcing mat, an adhesive strip, and mineral granules.

Composting toilet system: Sometimes called biological toilets, dry toilets, and waterless toilets, they contain and control the composting of excrement, toilet paper, carbon additive, and, optionally, food wastes. Unlike a septic system a composting toilet system relies on unsaturated conditions (material cannot be fully immersed in water), where aerobic bacteria and fungi break down wastes, just as they do in a yard waste composter.

A composting toilet is self-contained and uses the process of aerobic decomposition (composting) to break down feces into humus and odorless gases.

Compound: Pertaining to the taping of gypsum board, a mixture of plaster and other materials that are used to cover joints and corners in gypsum board surfaces and assemblies.

Compound interest: The interest paid on the principal balance in a mortgage and on the accrued and unpaid interest of the loan.

In most capital budgeting decisions the decision makers will use annual cash flows and an annual rate of interest; in other types of financial analysis, particularly with loan transactions months may be used. There are tables available for all periods.

It is useful to distinguish between the effective and the nominal rate of interest, when the interest rate is stated in terms of one period of time but applied over a different period of time with compounding [reinvestment of the principal plus accrued interest]. For example, if the nominal annual rate of interest is 10% (0.10) per year and the nominal quarterly effective rate of interest is 2.5% (0.025), with interest being compounded quarterly, the annual rate is 10.38% (0.1038).

Let: r = the effective annual rate
j = the nominal rate for a year
j/m = the effective rate for a fraction of a year m
m = the number of compoundings in a year, for example, for quarterly compoundings m = 4

The nominal annual rate is equal to j. For this example, j = 0.10 and m = 4, and the effective rate for a quarter is 0.025. The effective annual rate r may be computed using

$$1 + r = (1 + j/m)^4$$

or

$$r = (1 + j/m)^4 - 1$$

for this example, we have

$$r = (1.025)^4 - 1 = 0.1038$$

As another example, suppose that a bank pays a nominal 7.75% annual interest rate on savings, but compounds quarterly. The nominal rate per quarter is 7.75%/4 = 1.9374% per quarter. The annual effective rate is $(1.019375)^4 - 1 = 08057$.

In this case going from quarterly to daily compounding increases the annual effective rate by about 0.0008 per year. This difference may seem negligible to your individual account of several thousand dollars, but it is important to a corporate treasurer responsible for investing temporary excess funds that may amount to millions of dollars.

As already stated the choice of time period varies from situation to situation. Most short-term instruments have interest compounded on a daily basis. A $1,000 90-day loan at nominal 8.5% annual interest assuming a 365-day year would require a payment of

$$\$1,000 \ (1 + 0.085/365)^{90} = \$1,021.18.$$

Stating this loan as a realized annual interest rate yields

$$(1 + 0.085/365)^{365} - 1 = 0.0887$$

or 8.87% per year.

See also **Annual percentage yield (APY)**.

Compound miter cut: An angled cut to both the edge and face of a board. Also referred to as **compound cut**.

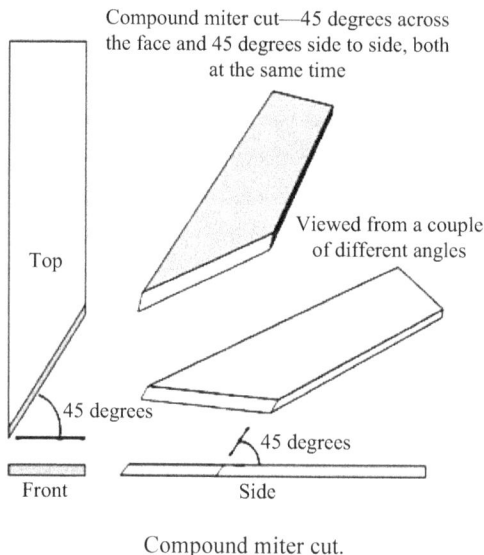

Compound miter cut—45 degrees across the face and 45 degrees side to side, both at the same time

Viewed from a couple of different angles

Top

Front

45 degrees

45 degrees

Side

Compound miter cut.

Compound miter saw: A saw that is comprised of amiter box and hand saw that are used in conjunction to make compound-angle (45 and 90°) miter cuts. It can also be used with a measuring accessory for cutting frames.

Basically a chop saw with a tilt mechanism added to the pivoting head. Miters are set by rotating the tool's turntable and the head is tilted for bevel cuts. A sliding version has the in-and-out capability of a radial-arm saw that enables it to make most any kind of cut.

A motorized compound miter saw.

Compounded thermoplastics: As defined by the National Roofing Contractors Association (NRCA): A category of roofing membranes made by blending thermoplastic resins with plasticizers, various modifiers, stabilizers, flame retardants, UV absorbers, fungicides, and other proprietary substances, alloyed with proprietary organic polymers.

Compounding sheave: With regard to elevators, a pulley that is located on the car, and on the counterweight, under which the hoist cables run to double the capacity and reduce the speed of an elevator.

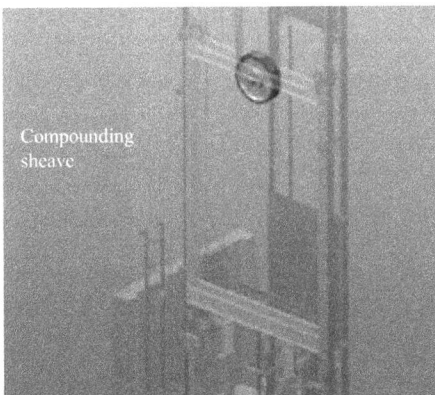

Compounding sheave.

Comprehensive historic preservation planning: The logical organization of preservation information pertaining to the identification, evaluation, registration, and treatment of historic properties and the setting of priorities for accomplishing preservation activities.

Compressed-air chamber: The space at the bottom of a caisson into which air is introduced. The air pressure excludes water so that excavation can take place.

Compressed workweek: The rearrangement of the standard five consecutive 8-h days workweek by increasing the daily hours and decreasing the number of days in the workweek, e.g., 10-h days for four days per week, or 9-h days for nine of 10 consecutive days. May save on overtime if, say, a project includes work that cannot be completed in an 8-h but can be completed in one 10-h shift.

Compression: A condition caused by the action of squeezing or shortening of a component. Compression is the opposite of tension.

Other terms that begin with compression:
- **Compression fitting:** A fitting that is used on all types of pipe, where a ferrule or a gasket is compressed against the fitting by tightening a threaded nut.
- **Compression member:** Any member in which the primary stress is longitudinal compression.
- **Compression parallel to grain:** A measure of wood's resistance to axial stresses along the access of the fibers.
- **Compression perpendicular to grain:** A measure of wood's resistance to load when applied perpendicular to the axis of the fibers.
- **Compression test:** A compression test determines behavior of materials under crushing loads. The specimen is compressed and deformation at various loads is recorded. Compressive stress and strain are calculated and plotted as a stress-strain diagram which is used to determine elastic limit, proportional limit, yield point, yield strength, and, for some materials, compressive strength.

- **Compression web:** A member of a truss system that connects the bottom and top chords and provides downward support.

Compression stress: As regards earthquakes, the stress that squeezes something. It is the stress component perpendicular to a given surface, such as a fault plane, that results from forces applied perpendicular to the surface or from remote forces transmitted through the surrounding rock.

Compression wave: See **P wave**.

Compressive strength: The measured maximum resistance to axial compressive loading as force per unit cross-sectional area. Standard strength tests are specified for each material.

Compressive strength is one of concrete's most important properties. It is how concrete in usually designated. Beyond this, compressive strength is directly and indirectly related to many other concrete properties: density, modulus of elasticity, modulus of rupture, permeability, and soundness. Concrete strength is determined using standardized ASTM testing procedures.

Compressor: A mechanical device that pressurizes a gas in order to turn it into a liquid, thereby allowing heat to be removed or added. A compressor is the main component of conventional heat pumps and air conditioners. In an air conditioning system, the compressor normally sits outside and has a large fan to remove heat.

Computer simulation: Computer software that models actions or occurrences in the real world.

Com-Ply: A type of wood product that is used for sheathing. It consists of two thin layers or veneers of wood on the outside and in the middle has a wood flake and resin center.

Concealed: A place rendered inaccessible by the structure or finish of the building. Wires in concealed raceways are considered concealed, even though they may become accessible by withdrawing them.

Concealed mounting system: A tile ceiling suspension system that uses T-Bars and splines which fit into kerfs cut into tile edges. Unlike exposed-grid systems, concealed mounting systems are not visible from below the ceiling. Inverted tee, "H and T," or "Z" profile grids are common for these applications with provisions for full plenum access usually incorporated into the grid design. Also known as **spline ceiling**.

A down side of a concealed mounting system is the difficulty of removing tiles to get above the ceiling. The tiles are usually damaged, which requires replacement with new tiles that inevitably have a different color due to the fading and accumulating of dirt of the existing tiles. Also, in a strong union situation the electrician or HVAC mechanic that needs to get above the ceiling may not remove the tiles.

Concealed nail method: Application of roll roofing in which all nails are driven into the underlying course of roofing and covered by a cemented, overlapping course. All nails are protected from the weather.

Concealed trap: A toilet bowl whose side is smooth or skirted. The trap way does not show.

A concealed trap toilet bowl.

Concentrate: A product that must be diluted by at least eight parts by volume water (1:8

dilution ratio) prior to its intended use (Green Seal GS-37).

Concentrated force: A force considered to act along a single line in space. Concentrated forces are useful mathematical idealizations, but cannot be found in the real world, where all forces are either body forces acting over a volume or surface forces acting over an area. Also referred to as **concentrated load**.

Concentration ratio or cycles of concentration: As regards HVAC systems, the ratio of the concentration of dissolved solids in the recirculating water to the concentration found in the entering makeup water. Higher cycles of concentration require a lower bleed-off rate, though increasing the cycles of concentration above a certain point leads to scaling, and water savings diminish after a certain level. The "cycle" refers to the number of times dissolved minerals in the water are concentrated compared with makeup water, not to water flow over the tower or to on-off cycles.

Concrete: The mixture of Portland cement, sand, gravel, and water. It is commonly reinforced with steel rods (rebar) or wire screening (mesh).

Other terms that begin with concrete:
- **Concrete board:** A panel made out of concrete and fiberglass usually used as a tile backing material.
- **Concrete duct bank:** A series of electrical conduit, cables, and or buss bars encased in concrete and usually buried under the ground.
- **Concrete masonry block (CMU), Concrete block:** A masonry unit made from Portland cement and aggregate. Blocks may or may not have pigment added. CMU is the most common type of block.
- **Concrete placement:** The act of placing concrete into formwork.
- **Concrete pump:** A piece of construction equipment (usually a truck designed for this purpose) which has the capacity to pump concrete through a series of joined pipes and flexible connections to a point of placement.

- **Concrete slab:** A shallow, reinforced-concrete structural member that is very wide compared with its depth. Spanning between beams, girders, or columns, slabs are used for floors, roofs, and bridge decks. If they are cast integrally with beams or girders, they may be considered the top flange of those members and act with them as a T beam.
- **Concrete stamp:** A form used to make patterns in concrete. After the concrete has been screed and rough finished, the stamp is pressed into the concrete. Colors may be added to make the concrete appear more natural by adding a dye admixture to the concrete or by spreading a dye on the surface.
- **Concrete surfaces, specifically, Portland cement concrete surfaces:** Concrete road surfaces are created using a concrete mix of Portland cement, coarse aggregate, sand, and water along various admixtures that are added to increase workability, reduce the required amount of water, mitigate harmful chemical reactions, lessen freeze-thaw disintegration, and for other beneficial purposes. There may also be Portland cement substitutes added, such as flyash, to reduce costs and improve surface properties.

  Concrete pavements are typically stronger and more durable than asphalt roadways. They can be grooved to provide a durable skid-resistant surface. But a notable disadvantage is that they typically have a higher initial cost and take longer to construct. This cost may be offset through the long life cycle of the pavement, especially if it is maintained.

  Concrete surfaces have been separated into three common types: **jointed plain (JPCP)**, **jointed reinforced (JRCP) and continuously reinforced (CRCP)**. Distinguishes each type is the jointing system that is used to control cracking.
- **Jointed Plain Concrete Pavements (JPCP):** Contain enough joints to control the location of all the expected shrinkage cracks. Jointed plain pavements do not

contain any steel reinforcement. However, there may be smooth steel bars at transverse joints and deformed steel bars at longitudinal joints.

- **Jointed Reinforced Concrete Pavements (JRCP):** Contain steel mesh reinforcement (sometimes called distributed steel). Today very few agencies employ this design and its use is generally not recommended.
- **Continuously Reinforced Concrete Pavements (CRCP):** Do not require any transverse contraction joints. Transverse cracks are expected in the slab, usually at intervals of 3–5 ft. CRCP pavements are designed with enough steel, 0.6–0.7% by cross-sectional area, so that cracks are held together tightly. Determining an appropriate spacing between the cracks is part of the design process for this type of pavement.
- **Concrete tile:** Tiles made from a stiff, low slump concrete. Concrete tiles are usually heavier and less expensive than clay tiles. Concrete tiles are heavier than clay tiles and add considerable extra dead load to roofs.
- **Concrete tilt-up:** The process of pouring concrete into forms on the ground, allowing the forms to cure and then raising the material to a vertical position to form walls.

Condemnation: The process the government uses to take private property for public use without the consent of the owner.

Condensate: Liquid formed by condensation. Also, water formed by removing heat from the steam that is used in HVAC systems.

Condensate gutter: A gutter located at the bottom of a window or skylight to collect condensate that forms on the inside of glass. The condensate is in turn weeped to the exterior of the assembly through the frame.

Condensation: Water vapor from the air deposited on any cold surface that has a temperature below the dew point.

Condensation is beads or drops of water (and frost in extremely cold places) that accumulates on the inside of the exterior covering of a building. Sometimes this becomes a problem on cold (and poorly insulated) window glass or framing that is exposed to humid indoor air. Use of louvers or attic ventilators will reduce moisture condensation in attics. A vapor barrier under the gypsum lath or dry wall on exposed walls will reduce condensation. See **Condensate**.

Condensing unit: The outdoor component of a cooling system. It includes a compressor and condensing coil designed to give off heat.

Conditional commitment: A promise by a lender to make a loan if the borrower meets certain conditions.

Conditions, covenants, and restrictions (CC and Rs): The standards that define how a property may be used and the protections the developer makes for the benefit of all owners in a subdivision.

Condominium/Cooperative: Individual units in a building or development in which owners hold title to the interior space while common areas, such as parking lots, community rooms, and recreational areas, are owned by all the residents.

A condominium is distinguished from a cooperative (CO-OP), which consists of a corporation or business trust entity that holds title to the premises and granting rights of occupancy to particular apartments by means of proprietary leases or similar arrangements.

Condominium conversion: The change in title from a single owner of an entire project or building to multiple owners of individual units.

Conduction: The direct transfer of heat energy through a material.

Conductivity: The rate at which heat is transmitted through a material.

Conductivity meter, electrical conductivity (EC) meter: A device that measures the amount of nutrients and salt in water.

Conductor: A substance or body that allows a current of electricity to pass continuously along it. Metals, such as copper or aluminum, are good conductors. In a circuit, current-carrying wires are termed "conductors," as in a flexible cord.

A conductor that has no covering or electrical insulation is a **bare conductor**. A conductor encased within material of composition and thickness that is not recognized by the NEC as electrical insulation is a **covered conductor**. A conductor encased within material of composition and thickness that is recognized by the NEC as electrical insulation is an **insulated conductor**.

Conduit: As regards electrical wiring, a tube or pipe through which electrical wires run.

A separate portion of a conduit or tubing system that provides access through a removable cover(s) to the interior of the system at a junction of two or more sections of the system or at a terminal point of the system is a **conduit body**.

Conduit body.

Configuration terms that pertain primarily to earthquake design:
- **Building configuration:** Size, shape, and proportions of the building; size, shape, and location of structural elements; and the type, size, and location of nonstructural elements.
- **Regular configuration:** Building configurations that resist lateral forces with shear walls, moment resistant frames or braced frames, all in simple and near symmetrical layout.

- **Irregular configuration:** A deviation from the regular, symmetrical building configurations. Irregular configurations do not have a repetitive plan.
- **Structural configuration:** The size, shape, and arrangement of the vertical load carrying the lateral force resistance components of a building.

Configurator: Software that allows users to input window and door sizes, options, and other information and create a quote and/or order. Typically, it can also be used to provide information to manufacturing software.

Conglomerate: A rock composed of rounded fragments, anything from a few millimeters to several centimeters in diameter.

Conglomerate.

Conifer: A softwood tree type, with needles and cones rather than flat, broad leaves.

Connections: A connection consists of two or more members joined with one or more mechanical fasteners. Connections provide continuity to the members and strength and stability to the system. They are usually made of wood or metal. Most wood structure failures are attributed to improper connection designs, construction (fabrication) details, and serviceability.

Mechanical connections generally fall under the following categories (adhesive connections are not included): **nails, spikes, and staples; lag screws and wood screws; bolts, drift bolts, and pins; metal connector plates;** and **timber connectors**.

Mechanical connections are constructed using two general fastener types: **dowel and bearing**. Dowel type fasteners, such as nails, screws, and bolts, transmit either lateral or withdrawal loads. Metal connector plates are a special case of dowel-type fasteners. They combine the lateral load actions of dowel fasteners and the strength properties of the metal plates. Bearing-type connections transmit lateral loads only. Bearing-type fasteners, such as shear plates and split ring connectors, transmit shear forces through bearing on the connected materials.

Hanger-type connections are a combination of dowel and bearing-type fasteners. They generally support one structural member and are connected to another member by a combination of dowel and bearing action.

Connector(s): A device that is used to couple (connect) parts of the personal fall arrest system and positioning device system together. It may be an independent component of the system, such as a carabiner (a coupling link with a safety closure, used by rock climbers.), or it may be an integral component of part of the system, such as a buckle or D-ring sewn into a body belt or body harness, or a snaphook spliced or sewn to a lanyard or self-retracting lanyard.

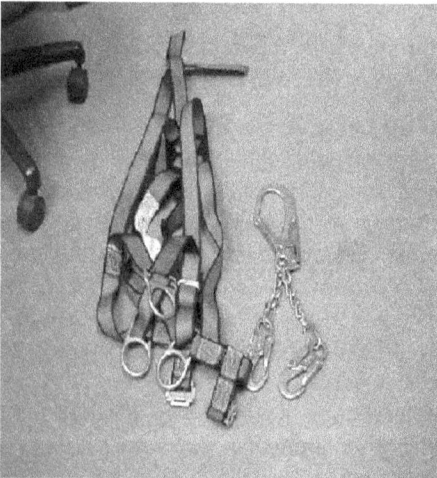

Connectors.

Also, a device for keeping two parts of an electric circuit in contact.

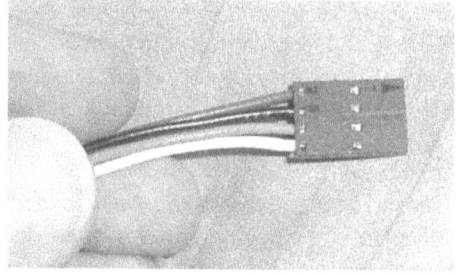

Connector.

Also, a short road or highway that connects two longer roads or highways.

Also, a pressure (solderless) device that establishes a connection between two or more conductors or between one or more conductors and a terminal by mean of mechanical pressure and without the use of solder.

Also, with regard to wood, connectors are used to make connections and to fasten wood members together. Bolts, drift bolts, nails, screws, lag screws, spikes, split ring connectors, toothed connectors, metal plate connectors, shear plate connectors, staples, etc. are all connectors. See **Connections**.

Consent judgment: A binding written agreement between two parties entered upon the record with the approval of a court of competent jurisdiction. This contract cannot be nullified or set aside without the consent of the parties thereto except for fraud or mistake. Also known as **consent order**.

Conservation district: A locally designated area, in which regulations for alteration or removal apply only to specific historic buildings within the district's boundary.

Conservator: A court-appointed guardian.

Consideration: Anything that is legal, has value, and induces a person to enter into a contract.

Constructability: The optimizing of cost, time, and quality factors with the material, equipment, construction means, methods, and techniques used on a project; accomplished by matching owner values with available construction industry practices.

Construction: While local entities may have their own definition, the term **building construction** will usually be more or less as follows: any and all work or operations necessary or incidental to the erection, demolition, assembling, installing, or equipping of buildings, or any alterations and operations incidental thereto. The term "construction" shall include land clearing, grading, excavating, and filling. It shall also mean the finished product of any such work or operations.

Heavy construction: Mainly involves the preparation and excavation of commercial and institutional sites. It involves the clearing of land with heavy equipment to prepare the site for roads, industrial plants, gravel pits, dams, etc. It also involves the construction of dams and minor bridges. Heavy construction is done by operating engineers, laborers, and teamsters. **Road building** may be classified separately from heavy construction.

Other terms that begin with construction:
- **Construction and demolition (C&D) debris:** Waste and recycles generated from construction, renovation, and demolition or deconstruction of preexisting structures.
- **Construction budget:** The best estimate cost figure covering the construction phase of a project, the target budget. It includes the cost of contracts with trade contractors, construction support items other purchased labor, material and equipment, and the construction manager's cost but not the cost of land, A/E fees, or consultant fees.
- **Construction change directive:** A written order signed by the owner and architect directing the contractor to make a change in the work and stating a proposed basis for any appropriate adjustment in the contract sum or the contract time. This may result in a change order if an agreement is reached between the contractor and the owner. But even in the absence of total agreement, the construction change directive by itself constitutes a change in the contract documents, with determination of any appropriate adjustment in contract sum or contract

time being handled in accordance with specified detailed procedures.
- **Construction chisel:** This tool is designed for rough carpentry, framing, and construction. It is made of a single piece of hand-forged alloy steel.
- **Construction contract:** A legal document that specifies the what, when, where, how, how much, and by whom in a construction project. A construction contract usually includes:
  1. The contractor's registration number.
  2. A statement of work quality such as "Standard Practices of the Trades" or "according to Manufacturers Specifications."
  3. A set of blue prints or plans.
  4. A construction timetable including starting and completion dates.
  5. A set of specifications.
  6. A basis for payment, e.g., fixed price or time and materials, and payment schedule.
  7. Any allowances.
  8. A clause that outlines how any disputes will be resolved.
  9. A written warrantee.
- **Construction, demolition, and land clearing (CDL) debris:** CDL includes all construction and demolition debris plus soil, vegetation, and rock from land clearing.
- **Construction document review:** An independent (third-party) party's review of another party's construction documents. Such a review is for the purpose of confirming that these documents and estimates are feasible and are in accordance with the proposed loan or project appraisal.
- **Construction documents:** All drawings, specifications, and addenda associated with a specific construction project. These documents delineate and graphically represent the physical construction requirements established by the A/E.
- **Construction documents phase:** The third phase of the architect's basic services wherein the architect prepares

working drawings, specifications, and bidding information.

- **Construction drywall:** A type of construction in which the interior wall finish is applied in a dry condition, generally in the form of sheet materials or wood paneling as contrasted to plaster.
- **Construction, frame:** A type of construction in which the structural components are wood or depend upon a wood frame for support.
- **Construction indoor air quality (IAQ) management plan:** Measures to minimize contamination in a specific project building during construction and describes procedures to flush the building of contaminants prior to occupancy.
- **Construction loan:** Short-term loans a lender makes for the construction of homes and buildings. The lender disburses the funds in stages.
- **Construction manager (CM), construction project management (CPM):** The overall planning, coordination, and control of a construction project from beginning to completion. CPM is aimed at meeting a client's requirement in order to produce a functionally and financially viable project.

  There was a time when it was usual for an owner to enter into a contract with a general contractor (GC), giving the GC plans that were developed by an architect. (The contractor would have been selected by using one of the three selection methods: low-bid selection, best-value selection, or qualifications-based selection.) The GC would then manage the entire project including sub-contractors. This form of construction management continues, but nowadays for large projects the hiring of Construction Management firms is more common.

  This form of construction management occurs when an owner hires a CM firm to manage the contract. The CM, in turn, develops the schedule, breaks down the work and hires contractors to do the various pieces of construction to complete

the project; in short completely manage the construction project. The argument for such an arrangement is normally that a CM firm can bring to bear a more competent staff in all areas of construction than most GCs can field, and it relieves the owner, whose business is normally other than construction, from diverting staff from their usual work.

You will seldom hear the words construction manager and construction management; CM (only) is universally used.

- **Construction management contract:** A written agreement that delineates responsibilities for coordination and accomplishment of overall project planning, design, and construction that are given to a construction management firm. The building team generally consists of the owner, contractor and designer or architect.
- **Construction management fee plus reimbursables:** A form of payment for CM services where the construction manager is paid a fixed or percentage fee for CM expertise, plus pre-established hourly, daily, weekly, or monthly costs for field personnel and equipment.
- **Construction management partnering:** A contractual commitment by the owner, A/E, and CM to achieve a common goal, and doing so without a stake holder's exposure to a potential for conflict of interest in pursuit of that goal.
- **Construction management services:** The scope of services provided by a construction manager and available to owners in whole or in part. CM services are not consistent in scope or performance from one CM firm to another.
- **Construction monitor:** An independent construction consultant engaged by a member(s) of the financial team to observe and report the current status of a construction project schedule, project budget, hard cost payment approvals, and contract administration compliance, and tracks critical action items. The Construction Monitor acts as an independent facilitator and

reports all findings in an objective manner. See **Budget analysis**.

- **Construction schedule:** A graphic, tabular, or critical path method representation or depiction of the construction portion of the project-delivery process, showing activities and durations.
- **Construction to permanent loan:** The conversion of a construction loan to a longer-term traditional mortgage after construction has been completed.
- **Construction work:** Work for construction, alteration, and/or repair, including painting and decorating.

Consumptive water use: The total amount of water used by vegetation, man's activities, and evaporation of surface water.

Contact adhesive: A resin-based type of adhesive. The surfaces of two materials that are to be glued together are first coated with the adhesive; then the adhesive is allowed to dry (10–20 min). When the two surfaces touch each other, they adhere. See **Plastic laminate countertop**.

Containment: A process in lead hazard reduction to protect workers and the environment by controlling exposures to the lead-contaminated dust and debris created by a lead abatement.

Contaminated aggregate: The presence of impurities that may interfere with the chemical reactions of hydration, or induce chemical degradation of concrete.

Contaminated mixing water: Impurities in the mixing water that may interfere with chemical reactions of hydration, adversely affect concrete strength, cause staining of the surface, and also lead to corrosion of reinforcing steel.

Contemporary: Reflecting the characteristics of the current period. For example, with regard to a building, contemporary means that it was constructed in the present or recent past and is not imitative or reflective of a historic design.

Contiguous lots: Pieces of property that are adjoined.

Contingency: Line-item amounts in the project budget, or contractor's schedule of values intended to cover costs of unknown, unforeseeable, or missed cost items. Contingency may be owner controlled or contractor controlled.

Contingency listing: A property listing with a special condition attached.

Contingent fee: A fee that must be paid if a certain event occurs.

Continuity: The term given to a structural system denoting the transfer of loads and stresses from member to member as if there were no connections.

Continuity tester: A device that tells whether a circuit is capable of carrying electricity.

Continuous: Uninterrupted, unbroken, constant, etc., as in the following:

- **Continuous footing:** A footing design where all parts of the footing are connected. Concrete runs continuously from one section of the footing to the next with no breaks or gaps. This helps the footing resist movement during earthquakes or other types of earth movement.
- **Continuous load:** A load where the maximum current is expected to continue for three hours or more.
- **Continuous span:** A span that extends over several supports and having more than two points.

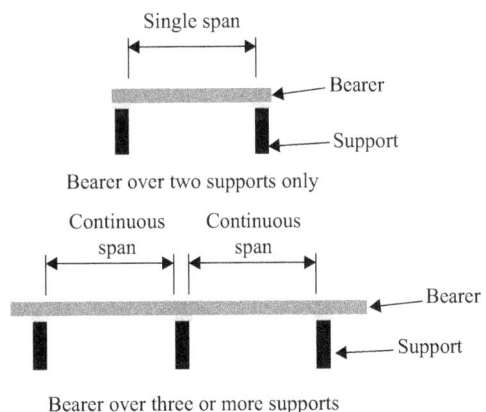

Top picture shows a single span. Bottom picture shows a continuous span.

- **Continuous span beam bridge:** A bridge made by linking one beam bridge to another. Some of the longest bridges in the world are continuous span beam bridges.

A continuous span bridge.

- **Continuous weld:** A weld that extends continuously from one end of a joint to the other.

Contract: A legal document or agreement, enforceable by law, between two or more parties for the doing of something specified, such as the construction of a building or furnishing materials.

Types of contracts include:
- **Aleatory contract:** An agreement whose performance by one party depends upon the occurrence of a contingent event.
- **Bilateral contract:** A contract in which there are mutual promises between two parties to the contract. Each party to a bilateral contract is both a promisor and a promisee.
- **Conditional contract:** A contract, the performance of which depends on an event that is not certain to occur but must occur unless its non-occurrence is excused before performance under the contract becomes due.

Other terms associated with the word contract:
- **Contract administration:** Overseeing the contractual duties and responsibilities of the A/E, contractor and CM during the construction phase of a project and ensuring that the interactive provisions in the contract for construction are occurring.

- **Contract document phase:** The final phase of design on an architectural project when construction documents are completed and bidding documents are formulated. See also "Construction Documents Phase."
- **Contract documents:** All documents (any general, supplementary, or other contract conditions, the drawings and specifications, all bidding documents, addenda issued prior to execution of the contract and post-award Change Orders, and any other items specifically stipulated as being included in the contract documents) that collectively form the contract between the contractor and the owner.
- **Contract for deed:** A contract in which the seller agrees to defer all or part of the purchase price for a specified period of time.
- **Contract drawings:** All the architectural, structural, mechanical, electrical, etc., plans that make up a legal set of contract documents to build a building by.
- **Contract performance bond:** A pledge of satisfactory performance from a surety company to the owner, on behalf of a contractor or subcontractor. A contract performance bond guarantees the completion of the work in accordance with the terms of the contract.
- **Contract substantial completion:** Usually the project contract documents define substantial completion; typically the definition states that substantial completion is achieved when the facility has received a certificate of occupancy and can be used for its intended use. However, contract substantial completion may have additional requirements to be achieved before a project is deemed substantially complete by the owner.
- **Contract sum:** The total dollar amount payable by the owner to the contractor for the performance of the work under the contract documents.
- **Contract time:** The time period set forth and established in the contract documents for completing a specific

- **Contract to purchase:** A contract the buyer initiates. A contract to purchase details the purchase price and conditions of the transaction and is accepted by the seller. Also known as an **agreement of sale**.

Contractor: A person or company licensed to perform certain types of construction activities. In many states contractor and sub-contractor licenses require extensive training, testing, and insurance.

Contractual lien: A voluntary obligation such as a mortgage or trust deed.

Controls: Devices that are used to regulate the function of a unit, e.g., **hydraulic controls**, **manual controls**, **mechanical controls** and **compensator controls**. Controls are frequently housed in a **control console that is a** fabricated, usually metal, cabinet that houses buttons and switches that control a machine or system. **Control valves** are used to control the flow of liquids and gases.

Control joint: Tooled, straight grooves made on concrete floors to "control" where the concrete should crack. Also, joints between adjacent parts of masonry to accommodate contraction of adjacent materials, thereby avoiding the development of high stresses and cracks.

Controlled access zone (CAZ): An area in which certain work (e.g., overhand bricklaying) may take place without guardrail systems, personal fall arrest systems, or safety net systems, and access to the zone is controlled. See **Lockout**.

Controlled growth: Any restrictions imposed on the amount or type of new development in an area.

Controller: With regard to electrical, a device (or group of devices) that serves to govern, in some predetermined manner, the electrical power delivered to the apparatus to which it is connected. A device that serves to control the equipment to which it is connected (machine, door operator, etc.). See **Controls**.

Convection: Currents created by heating air, which then rises and pulls cooler air behind it. See radiation.

Conventional irrigation: The most common irrigation system used in the region where a building is located. A conventional irrigation system commonly uses pressure to deliver water and distributes it through sprinkler heads above the ground.

Conventional jack: The type of hydraulic elevator mechanism whose cylinder must be installed in the ground. Installation must be perfectly vertical.

Conventional loan: A mortgage loan not insured by a government agency (such as FHA or VA).

Convertible adjustable-rate mortgage: A mortgage which starts as an adjustable-rate loan, but allows the borrower to convert the loan to a fixed-rate mortgage during a specified period of time.

Convertibility: The ability to change a loan from an adjustable rate schedule to a fixed rate schedule.

Conveyance: The transfer of title of property. A **conveyance tax** is frequently imposed on the transfer of real property.

Cooking unit, counter mounted: A cooking appliance designed for mounting in or on a counter and consisting of one or more heating elements, internal wiring, and build-in or mountable controls.

Cooling load: The amount of cooling required to keep a building at a specified temperature during the summer.

Cooling tower: A piece of equipment, usually a large tower, that uses water to regulate air temperature in a facility by absorbing heat from air-conditioning systems and equipment and transferring the heat to the atmosphere.

Cooperating broker: A real estate broker who finds a buyer for a property that another broker has listed.

Cooperative corporation: A business trust that holds the title to a cooperative residential building and grants occupancy rights to shareholders in the corporation.

Cooperative mortgages: Any loans related to a cooperative residential project.

Cooperative project: A project in which a corporation holds title and sells shares representing individual units to buyers who then receive a proprietary lease as their title.

Coped: When the top and bottom flanges of the end(s) of a metal I-beam have been removed. This is done to permit it to fit within, and bolted to, the web of another I-beam in a "T" arrangement. Should not be done without the approval of a structural engineer.

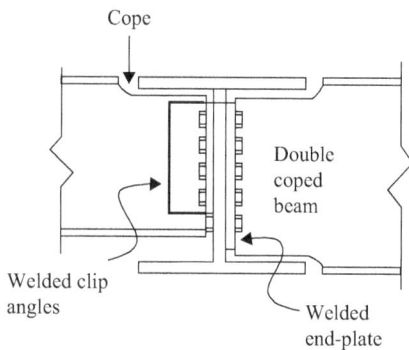

Double coped beam.

Coped joint: Cutting and fitting woodwork to an irregular surface.

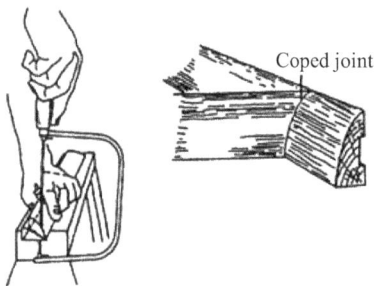

Coped joint.

Coping: A protective capping on the top of a parapet or free standing wall.

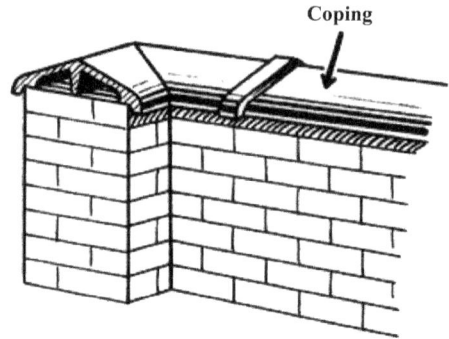

Coping.

Coping saw: This tool has a narrow metal frame which supports a thin blade that is held in place with a hook, loop, or pin on each end of the blade. It can be rotated in the frame to make intricate curved cuts. A coping saw makes a finer cut than a compass saw.

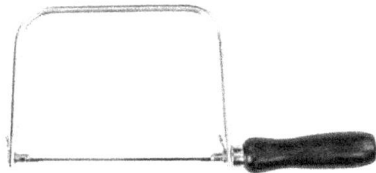

Coping Saw.

Copper: A reddish-brown element that conducts heat and electricity very well. It is also used as a primary roof material as well as a flashing component. Copper turns a greenish color after being exposed to the weather for a length of time and appears in the middle of the Galvanic Series.

Copper-clad aluminum conductors: Conductors drawn from a copper-clad aluminum rod with the copper metallurgically bonded to an aluminum core. The copper forms a minimum of 10% percent of the cross-sectional area of a solid conductor or each strand of a stranded conductor.

Copper oxide: A chemical that is created when copper is exposed to the atmosphere. Copper oxide is not an insulator and creates no hazard.

Corbel: A triangular, decorative, and supporting member that holds a mantel or horizontal shelf. Also referred to as a corbel block.

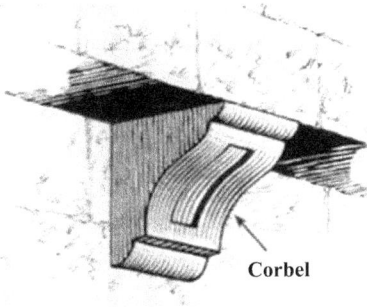

Corbel

Corbel

Also, projecting brick or masonry courses; from Norman-French meaning "crow" after carved stone projections used in medieval times to support roof trusses.

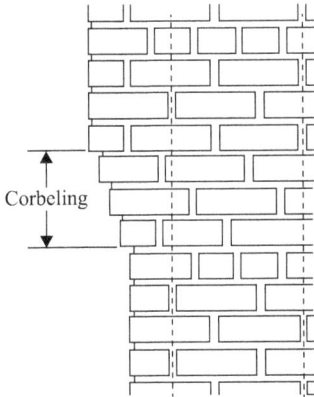

Corbeling.

Cord: A 4'× 4'× 8' stack of wood (128 cubic feet; 3.62 cubic meters), including air space and bark. The weight of a cord varies depending on whether it is green (freshly cut), seasoned (partially air dried), or dry (KD or kiln-dried). Fire wood is often sold by the cord.

Corded drill: Among the most popular power tool, the corded drill is a jack-of-all-trades. Among its many uses, it can bore holes in a range of materials, drive screws and nuts, brush away paint and rust, sand edges and stir paint.

Cordless drill: With recent developments in batteries, a cordless drill has all the benefits of a corded drill, but without a connecting cord. The tradeoff is that a cordless may not be able to handle more rugged work or have quite the power of a corded drill.

Core: The innermost layer or section in component construction.

Also, the central region of a skyscraper that usually houses elevators and stairwells.

Also, with regard to earth science, the central part of the earth, beginning at a depth of about 2900 km, probably consisting of iron-nickel alloy; it is divisible into an outer core that may be liquid and an inner core about 1300 km in radius that may be solid.

A fire resistant **core material** is generally used in wood doors requiring fire ratings of 3/4 h or more: Engineered composite products meeting the minimum requirements of Window and Door Manufacturers Association (WDMA).

With regard to earthquakes, the **central part of the earth** below a depth of 2,900 kilometers is the core. It is thought to be composed of iron and nickel and to be molten on the outside with a central solid inner core.

The **innermost portion of plywood**, fiberboard, particleboard, and veneer is called the core. This may be a solid or discontinuous center ply. Also known as the **center** of the panels and **core stock**.

Test samples of concrete, usually 12 inches high and 6 inches in diameter are called **core samples**. When properly cured, core samples may be used to determine the strength of the concrete.

Corinthian order: The most elaborate and decorated of the three ancient Greek orders of architecture, its capital is carved in imitation of the growth of acanthus leaves.

Silver Corinthian order columns.

Cork flooring: Cork flooring comes from the bark of the cork tree. Therefore, it is considered eco-friendly, flooring from a sustainable and renewable resource. Cork is sound absorbent, a good insulator, quite durable, and very easy to install.

Corner: Bend, curve, crook, dog-leg; turn, turning, jog, junction, fork, intersection, hairpin turn.

Other terms that begin with corner:
- **Corner bead:** A strip of formed sheet metal placed on outside corners of drywall before applying drywall "mud."
- **Corner block:** A decorative piece that is placed between the vertical side casing and the horizontal top casing. Also see base block.
- **Corner boards:** Used as trim for the external corners of a house or other frame structure against which the ends of the siding are finished.
- **Corner braces:** Diagonal braces at the corners of the framed structure designed to stiffen and strengthen the wall.
- **Corner chisel:** This tool is used for cutting clean, sharp inside corners. A perfect tool for the serious woodworker and tool collector.

Spring loaded corner chisel

- **Corner influence:** The effect on the value of a property because it is situated on a corner or near a corner.

Cornice: The overhang of a pitched roof, usually consisting of a fascia board, a soffit, and appropriate trim moldings.

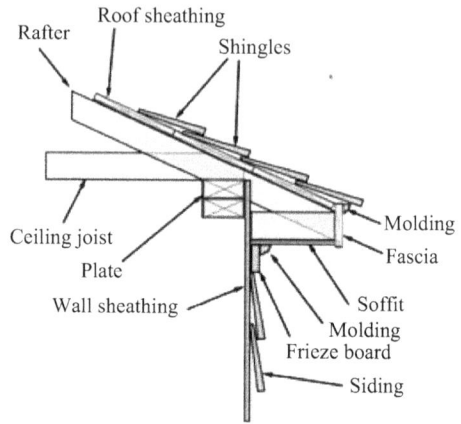

Roof sheathing
Rafter
Shingles
Ceiling joist
Plate
Wall sheathing
Molding
Fascia
Soffit
Molding
Frieze board
Siding

The cornice consisting of a fascia board, a soffit and trim molding.

Also, the top horizontal band of an entablature, found above the frieze.

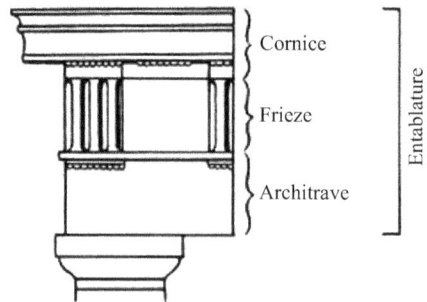

Cornice
Frieze
Architrave
Entablature

Cornice + Frieze + Architrave = Entablature.

Corrective work: Necessary or desired repairs to remedy problems uncovered by an insection inspection.

Corridor: Usually defined as an enclosed public passage that provides a means of access from rooms or spaces to an exit.

Corrosion: The disintegration of an engineered material into its constituent atoms due to chemical reactions with its surroundings. In the most common use of the word, this means electrochemical oxidation of metals in reaction with an oxidant such as oxygen.

Corrosion can also occur in materials other than metals, such as ceramics or polymers, although in this context, the term degradation is more common.

Corrosion inhibitive: A type of metal paint or primer that prevents rust by preventing moisture from reaching the metal. Zinc phosphate, barium metaborate and strontium chromate (all pigments) are common ingredients in corrosion-inhibitive coatings. These pigments absorb any moisture that enters the paint film.

Corrugated roof panel: A roofing sheet often made out of galvanized steel or fiberglass. Shaped in alternating ridges and valleys.

Corundum: A mineral consisting of the crystalline form of alumina. Corundum is second in hardness to the diamond and is used for polishing, especially glass.

Cosmetic damage: Damage to an item or surface that affects the way an item looks but not the way it functions.

Cost: Price, asking price, market price, selling price, unit price, fee, tariff, fare, toll, levy, charge, rental. Also, value, valuation, quotation, rate, worth.

Other terms that begin with cost:
- **Cost breakdown:** A financial statement furnished by the contractor that delineates the portions of the contract sum allotted for the various parts of the work and used as the basis for reviewing the contractor's applications for progress payments. See also **Schedule of values**.
- **Cost codes:** A numbering system given to specific kinds of work for the purpose of organizing the cost control process of a specific project.
- **Cost of construction:** The target cost figure covering the construction phase of a project. It includes the cost of subcontracts, construction support items, other purchased labor, material and equipment, and the construction manager's cost but not the cost of land, A/E fees, or consultant fees.

- **Cost-of-work:** All costs incurred by the contractor in the performance of a project as required by the plans and specifications for a specific project. Usually the term is more specifically defined in all guaranteed maximum price contracts.
- **Cost plus fee agreement:** A written agreement under which services are provided by a contractor, architect, engineer, or CM for the cost of work plus a specified fee.
- **Cost-plus contract:** A construction contract that determines the builder's profit based on a percentage of the cost of labor and materials. Also referred to as cost plus fee agreement.
- **Cost-to-complete:** All costs remaining to complete construction of the project to obtain a certificate of occupancy and occupy the project for its intended use. These costs may be further defined in the contract.

Cottage: A small, one-story house.

Cottage cheese texture: See **Popcorn texture**.

Cottage double-hung window: A double-hung window in which the top sash is shorter than the bottom sash. Also referred to as a **cottage window**.

Cotter pin: A metal pin used to fasten two parts of a mechanism together. It is a split pin that is opened out after being passed through a hole.

Counter batten: Wood strips installed vertically on sloped roofs over which horizontal battens are secured.

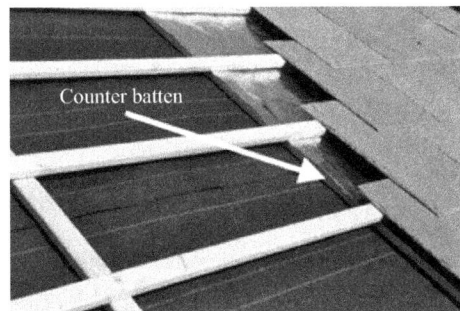

Slate counter batten.

Counter balance: The effect created by a force that is acting in opposition to another force; for example, the weight of an overhead door is counterbalanced by springs making it feel lighter when opening.

Counter flashings: Counter flashings are normally made of sheet metal to shield the exposed joints of base flashings. See **Base flashings**. Also written as **counterflashing**. Also known as **cap flashings**.

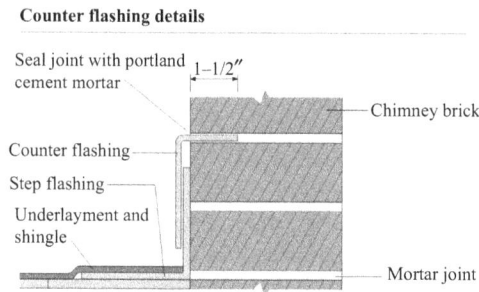

Counter flashing details.

Counterfort: A foundation wall section (buttress) that strengthens (and is generally perpendicular to) a long section of foundation wall. See **Buttress**.

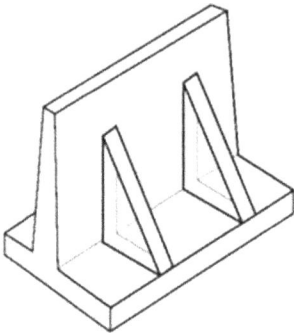

Counteroffer: A response to aeroffer.

Counterweight: A weight that is used as a counterbalance. Regarding elevators, a weight that counterbalances the weight of an elevator car plus ~40% of the capacity load.

Counter-balance valve: A regulating mechanism on valve-powered cylinders, such as those on bucket trucks. The counter-balance valves serve to limit the amount of fluid change within the flange. This prevents sudden changes in pressure and resultant instability.

The valves work by reacting to the weight of the load as that weight is translated into pressure on the pump cylinder. For example, in a bucket truck the weight of the bucket (bucket plus repairperson, tools, and equipment) supplies the downward pressure to the cylinder. When the counter-balance valves are directly triggered by the force of the weight that occurs when the load is being lowered or if the cylinder malfunctions, the valve restricts the amount of air or fluid that it releases. Because the fluid is released gradually the load is stabilized and prevented from falling uncontrolled.

Couple: A system of forces that is composed of two equal forces of opposite direction, offset by a distance. A couple is statically equivalent to a moment whose magnitude equals the magnitude of the force times the offset distance.

Coupler, **coupling:** A device for mechanically joining two linear components like pipes, scaffold tubes, or a drill bit with an extension.

Course: A row of shingles or roll roofing that runs the length of the roof. Also, a parallel layer of a building material such as bricks or siding that is laid-up horizontally; the number of courses in a vertical dimension of a wall is referred to as **coursing**.

Court: An open space that is surrounded partly or entirely by a building.

Cove: In roofing, a heavy bead of sealant material installed at the point where vertical and horizontal planes meet. It is used to eliminate the 90° angle. See also **Fillet**.

Cove molding: A molding with a concave face that is used as trim or to finish interior corners. Cove moldings are often used as a ceiling cornice. Small coves may be used as an inside corner guard. A specialty tile trim piece that is used to trim corners in a tile surface is also referred to as cove molding or **cove piece**. See **Tile base** and **Double bullnose**.

Covenant: A legal assurance or promise in a deed or other document, or implied by the law.

Covenants, conditions, and restrictions (CC&Rs): Rules and regulations for a development, such as acceptable landscaping or improvements that can be made to individual units.

Cover: With regard to reinforced concrete, the least distance between the surface of the reinforcement and the surface of the concrete.

Cover plate: A metal strip that is sometimes installed over the joint between formed metal pieces. A cover plate is usually welded to the top or bottom flange of a rolled steel beam or to the bottom chord of a joist or joist girder to increase the load carrying capacity of that member.

Coverage: The surface area that is covered by a material.

Also, the total amount and type of insurance that is carried.

Co-generation operation: Co-generation is the production of electricity and heat from a single fuel source. Co-generation captures heat lost during the production of electricity and converts it into useful thermal energy, usually in the form of steam or hot water.

Co-housing: Individual housing units that are clustered around a common building where residents share cooking and other activities.

Co-insurance: Coverage that involves the use of two or more insurers.

Co-maker: A person who signs a promissory note with the borrower and assumes responsibility for the loan/debt. Also known as **co-signer**.

Cracking: A separation of concrete into parts that are characterized by length, width, and depth, and whether the cracks are active or passive. Passive cracks are usually caused by construction errors, shrinkage, variations in internal temperature, and shock waves. Active cracks are usually caused by variations in atmospheric or internal temperature, absorption of moisture, reinforcement corrosion, chemical reactions, settlement, and loading conditions.

Crack perimeter: The total length of the crack around a sash through which outdoor air could leak into the room. In a double-hung window, the total crackage is three times the width plus two times the height of the sash.

Cracking pressure: The pressure at which a pressure actuated valve starts to pass fluid.

Cracks: Discontinuities in, for example, the weld filler material or in adjacent base metal in welded joints.

Craftsman style: An architectural style that evolved as part of the Arts and Craft movement near the turn of the century. Craftsman style homes are common in older neighborhoods of many American cities.

Craftsman style house. Especially common in Southern California.

Crane: A machine that is used to move material by means of a hoist. A crane can usually (but not always) move and is used to lift heavy materials or to lift members that are to be erected in a structure.

Cranked-neck rasp: A tool that is ideal for carefully shaping flat or slightly concave surfaces and is especially useful in woodcarving.

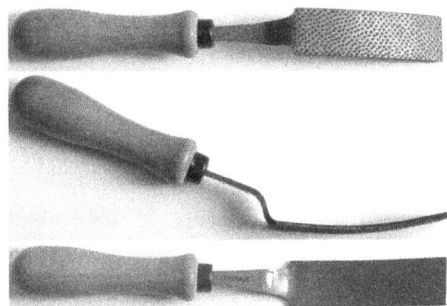

Cranked-neck rasp.

Crawler: Any vehicle that has tracks in place of wheels.

Crawl space: A shallow space below the living quarters of a house, normally enclosed by the foundation wall and having a dirt floor.

Cream: In reference to concrete, cream is the thin layer of fine mixture that comes to the surface of concrete when the course aggregate is pushed down with a concrete finishing tool such as a gandy or bull float.

Cream time: With regard to roofing, the time in seconds at a given temperature when the A and B (isocyanate and resin) components of spray polyurethane foam (SPF) will begin to expand after being mixed.

Creative financing: Innovative home-financing arrangements that help sell a property, but may not help the buyer in the long run; beware.

Credit: The money a lender extends to a buyer for a commitment to repay the loan within a certain time frame.

Other terms that begin with credit:
- **Credit history:** A record of an individual's current and past debt payments.
- **Credit life insurance:** Insurance that pays off a mortgage in the event of the borrower's death.
- **Credit rating:** A report ordered by a lender from a credit agency to determine a borrower's credit habits.
- **Credit report:** A credit bureau report that shows a loan applicant's history of payments made on previous debts. Several companies issue credit reports, but the three largest are Trans Union Corp., Equifax, and Experian (formerly TRW).
- **Credit repository:** Large companies that gather financial and credit information from various sources about individuals who have applied for credit. See **Credit report**.
- **Credit union:** A nonprofit cooperative organization that provide banking and financial services, including mortgages, home improvement loans, and home equity loans, to their members.

Creditor: An individual or institution to which a debt is owed.

Creep: Besides the guy who is dating your sister; materials tend to deflect when loaded statically below the yield stress point for a long period of time. Materials differ so you must be knowledgeable of the materials you are using. For example, in the case of concrete it can generally be assumed that 90% of creep occurs in the first 2 years of life and that all creep deflection has occurred in older structures. However, creep may begin again if additional loading is applied.

Creep resistance is an important property of gasket materials. Gasket materials are designed to flow under stress to fill any irregularities in the flange surface. The amount of creep sustained tends to increase with temperature. However once the tightening of a bolt is completed it is important that no further flow occurs since such deformation will lead to a reduction in bolt extension. Subsequently the stress acting on the gasket may reduce below a level and high rate of leakage could occur.

Also, with regard to earthquakes, creep along a fault is very slow periodic or episodic movement along a fault trace without earthquakes.

Creosote: A liquid coating made from coal tar. Creosote was once used as a wood preservative, especially on telephone poles. It has been banned for consumer use because of potential health risks.

Cricket: A second roof built on top of the primary roof to increase the slope of the roof or valley.

Cricket.

Also, a saddle-shaped, peaked construction connecting a sloping roof with a chimney, designed to encourage water drainage away from the chimney joint.

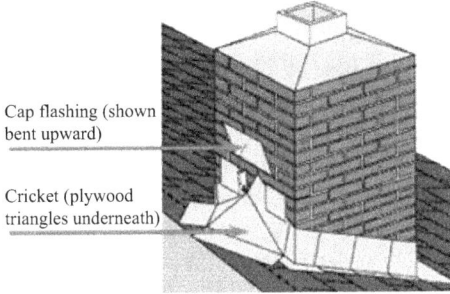

Cricket.

Crimp ring: Used with fitting in the poly-butylene (PB or Poly-B) pipe. The PB pipe is pressed firmly around the fitting with the crimp ring. See **Polybutylene (PB).**

Crimped angle web: A regular angle whose ends have been "crimped." Usually the actual crimped portion of the angle is only a few inches on each end and the end is inserted between top or bottom chord members to be welded.

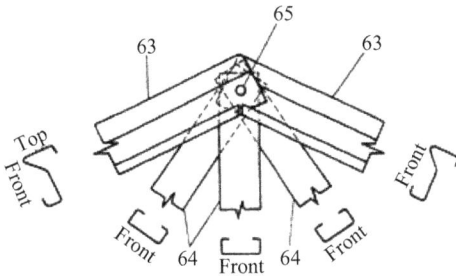

Crimped angle webs.

Cripple(s): A cut in an unseasoned joist, bearer, or stud designed to reduce movement in a floor or wall as the structural timber seasons.

Short studs that are used to fill the gap under the windowsill and between the header and the top plate if there is a gap are called **cripples**.

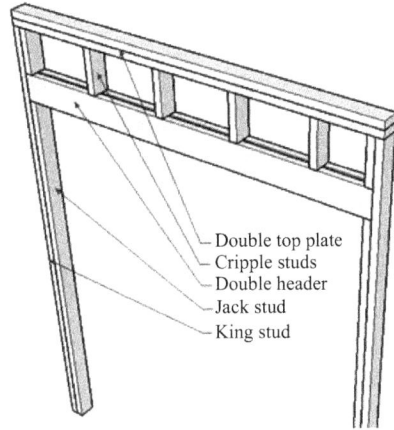

Door framing showing cripple studs.

Cripple rafter: A cripple rafter runs from a hip rafter to a valley rafter. A cripple rafter never reaches the wall top plate or the ridge board.

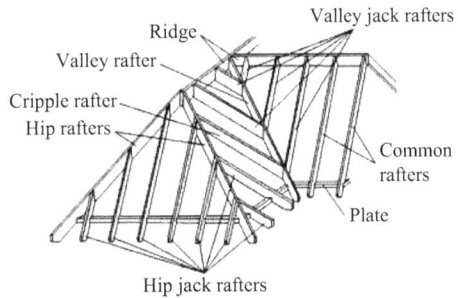

Rafters, including cripple rafters.

Wooden floors and stud walls that are built on top of an exterior foundation to support a house and create a crawl space are called rafters. They carry the weight of the house. During an earthquake, these walls are subject to collapse if not properly braced to resist horizontal movement.

Cripple wall failure due to an earthquake.

Critical facilities: Structures whose functioning during an emergency is essential or whose failure would endanger many lives. May include: structures such as nuclear power plants or large dams whose failure could be catastrophic; major communication, utility, and transportation systems; involuntary or high occupancy buildings such as prisons or schools; emergency facilities such as hospitals, police and fire stations, and emergency-response facilities.

Critical load: The load at which deflection of a member or structure occurs as determined by stability analysis.

Critical Path Method (CPM): A project modeling technique, CPM is commonly used with all forms of projects, including construction, aerospace and defense, software development, research projects, product development, engineering, and plant maintenance, among others. Any project with interdependent activities can apply this method of mathematical analysis.

CPM provides the continuous chain of activities from project-start to project-finish, whose durations cannot be exceeded if the project is to be completed on time.

Critical visual tasks: Visual tasks completed by building occupants, including reading and computer monitor use.

Crocus cloth: Finer than sandpaper grit on a cloth backing.

Crook: A deviation edgewise, measured at the point of greatest distance from a straight line that is drawn end to end of a piece of lumber. See also **Warp**.

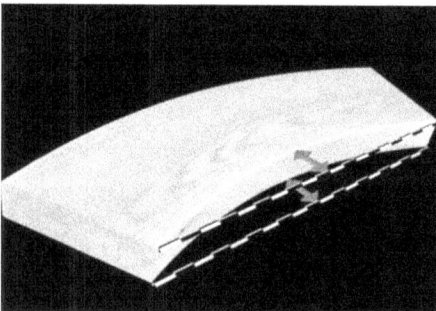

Crook in lumber.

Terms that begin with cross:

- **Cross banding:** A ply placed between the core and face veneer in 5-ply construction or a ply placed between the back and face of a 3-ply skin in 7-ply construction, typically of hardwood veneer or engineered wood product.

- **Cross bar:** With regard to veneer, an irregularity of grain resembling a dip in the grain running at right angles, or nearly so, to the length of the veneer.

- **Cross braces:** Two braces which cross each other in the form of an X.

  Also, the horizontal members of a shoring system that is installed perpendicular to the sides of the excavation, the ends of which bear against either uprights or wales.

- **Cross break:** A separation (break) of the wood cells across the grain. Such breaks may be due to internal strains resulting from unequal longitudinal shrinkage, or to external forces.

- **Cross bridging:** Diagonal bracing between adjacent floor joists, placed near the center of the joist span to prevent joists from twisting.

- **Cross connection:** Any plumbing arrangement that allows flow between potable water and a contaminant such as drain water, a gas, a chemical, or even steam. Cross connections are found in many building locations, including: boilers, lawn irrigation systems, between public and private water systems, at bidets, in toilet tanks, and at bathtubs.

- **Cross figures:** A series of naturally occurring figure effects characterized by mild or dominant patterns across the grain in some faces. For example, a washboard effect occurs in fiddle-back cross figure; and cross wrinkles occur in the mottle figure.

- **Cross laminated timber:** An engineered building product made by gluing layers of edge glued lumber panels in a cross pattern. This produces a panel that is further processed into specific sizes of cross laminated timber to be generally used for external and internal walls, ceilings and roofs.

- **Cross tee:** A short metal "T" beam used in suspended ceiling systems to bridge the spaces between the main beams.
- **Cross ventilation:** The effect of air moving through a roof cavity between the vents.
- **Crosscut:** To cut (wood or stone) across its main grain or axis. Also short for crosscut saw.
- **Crosscut saw:** A saw that is designed for making crosscuts. A crosscut is a cut made horizontally through the trunk of a standing tree, but the term also applies to cutting free lumber. Crosscut saws may be smaller traditional carpentry saws and larger saws used for forestry and logging work.

   The teeth of crosscut saws are designed to cut wood at a right angle to the direction of the wood grain. The cutting edge of each tooth is angled back and has a beveled edge. This design allows each tooth to act like a knife edge and slice through the wood. This is in contrast to a rip saw that tears along the grain.
- **Cross-sectional properties:** Dimensions and other geometric properties of structural components.

Crown: Most lumber is not chalkline straight but will bow slightly along its length. The upward bow is called the Crown of the board. See **camber**.

   Also, the live branches and foliage of a tree and the upper part of a tree.

Crown molding: A molding used on cornice or wherever an interior angle is to be covered, especially at the roof and the wall corner.

Crowned stud: When a wall is assembled on the floor, the framer places the crown of the stud upward. When the wall is stood in place, the convex or crown side of the stud is the "front," and the convex side of the stud is the "back."

Crowning: Arranging all framing members so that all crowns are in the same direction. See crowned stud.

Crows nest: See **Cupola**.

Crushed stone: Aggregate of stone that has irregular sides usually of one dimension or sieve size.

Crust: The lithosphere, the outer 80 km of the earth's surface made up of crustal rocks, sediment, and basalt. The general composition is silicon-aluminum-iron.

Crutch-pattern screwdriver: This screwdriver has a large, flattened-oval handle enabling the user to deliver tremendous torque to stubborn screws, especially in tight corners.

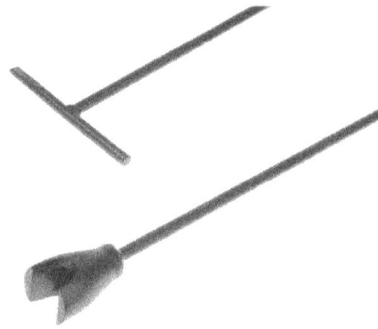

Crutch-pattern screwdriver handle.

Cryptoflorescence: Hidden salt crystallization that occurs within the pores below the masonry surface.

   The fine pores cannot accommodate the increasing accumulations of salts and are eventually broken apart by the expansive forces of the crystal growth, thereby causing the surface to decay.

Cubic foot (CF): A three-dimensional volume measurement equal to the amount contained by a cube that is one foot wide, one foot long, and one foot high.

Cubic foot of gas: This is the amount of gas that will occupy one cubic foot at a temperature of 60° F, and under a pressure equivalent to that of 30 inches of mercury.

Cubic yard (CY): A three-dimensional volume measurement equal to the amount contained by a cube that is one yard wide, one yard long, and one yard high. Each cubic yard contains 27 cubic feet.

Cul de sac: A street or alley that is closed at one end.

Culinary water: Water that is fit for human consumption. Also called potable water.

Cultural landscape: A geographic area, including both cultural and natural resources and the wildlife or domestic animals therein, associated with a historic event, activity, or person or exhibiting other cultural or aesthetic values. There are four general kinds of cultural landscape that are not mutually exclusive: historic sites, historic designed landscapes, historic vernacular landscapes, and ethnographic landscapes.

Cultural resource: An aspect of a cultural system that is valued by or significantly representative of a culture or that contains significant information about a culture. A cultural resource may be a tangible entity or a cultural practice.

Culture: A system of behaviors (including economic, religious, and social), beliefs (values, ideologies), and social arrangements.

Cultured countertop: A type of solid plastic material usually mixed with a pattern that imitates a type of stone. Culture marble countertops are probably the most common type but cultured granite is also common. Cultured materials are also used to make tub and shower surrounds. See **Plastic laminate countertop**, **Solid surface countertop**, **Solid plastic countertop**, **Wood block countertop**, **Cultured marble countertop**, **Stone countertop**, and **Tile countertop**.

Cultured marble countertop: The most common type of cultured countertop contains swirls and color variations that imitate the look of a marble countertop. See **Plastic laminate countertop**, **Solid surface countertop**, **Solid plastic countertop**, **Wood block countertop**, **Cultured marble countertop**, **Stone countertop**, and **Tile countertop**.

Cultured stone: Masonry units made from man-made materials such as plaster or plastic. Cultured stone is shaped and colored to resemble natural stone but is much lighter weight.

Culvert: A usually round, corrugated drain pipe.

Cunit: A measurement equal to 100 cubic feet of solid wood.

Cup: A deviation in the face of a piece of lumberfrom a straight line drawn from edge to edge of a piece of lumber (and also across the width of a door).Warping that causes boards to curl up at their edges is called **cupping**.

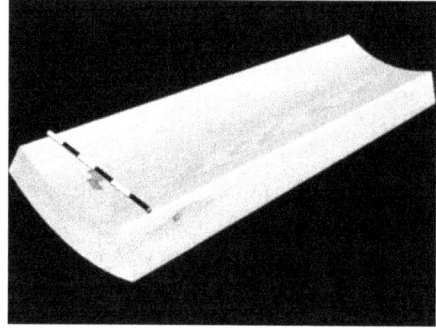

Cup.

Cupola: A vent positioned at the ridgeline. Cupolas are often in the shape of a small house or dome, topped with a weathervane. Some cupolas are installed for decoration only and are nonfunctional.

Also, a relatively small roofed structure set on the ridge of a main roof area. Also referred to as a **crow's nest**.

Curable defect: A deficiency in a property that is easy or inexpensive to fix, such as chipping paint.

Curb: The short elevation of an exterior wall above the deck of a roof.

Also, a raised member used to support skylights, HVAC equipment, hatches and other pieces of mechanical equipment above the level of the roof surface. Generally a curb should be a minimum of eight inches (8") in height.

Curb appeal: The first impression of a house as seen from the street.

Curb cock shutoff: A valve normally used with water meters set between the meter and the building.

Curb stop: Normally a cast iron pipe with a lid that is placed vertically into the ground, situated near the water tap in the yard, and where a water cut-off valve to the home is located (underground). A long pole with a special end is inserted into the curb stop to turn off/on the water.

Cure: To change the properties of a material (rubber, plastic, concrete, etc.) by a chemical reaction such as condensation, polymerization, or vulcanization. Curing usually means hardening and includes the action of heat or a catalyst with or without pressure.

Curing: The hardening of concrete, paint or other wet materials through evaporation, hydration, or chemical reaction.

A **curing agent** is the material additive that alters chemical activity between the components resulting in a change in the rate of cure.

Curing compound is material that is sprayed over newly placed concrete forming a film that retards the dehydration of the concrete, thereby allowing it to cure over a longer period of time.

Current: The flow of electrons through a conductor, measured in amperes (amps). If the current flows back and forth through a conductor, it is called alternating current (AC). If the current flows in one direction only, as in a car battery, it is called direct current (DC). AC is most widely used because it is possible to increase ("step up") or decrease ("step down") the current through a transformer. For example, when current from an overhead power line is run through a pole-mounted transformer, it can be stepped down to normal household current. Also, alternating current can travel enormous distances with little loss of voltage, or power.

Curtain drain: A ditch sometimes filled with gravel or drainage tile which diverts storm water away from a structure.

Curtain wall: A non-load bearing exterior closure wall assembly that is attached to the structural frame of a building.

Curtilage: An area of land attached to a house and forming one enclosure with it.

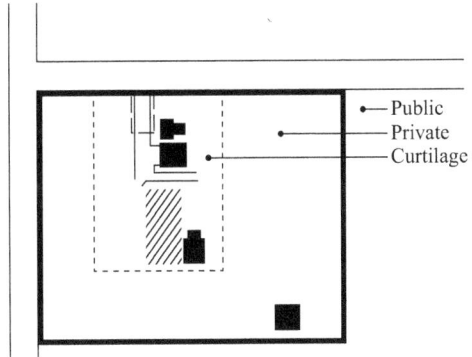

Each enclosed area represents a curtilage.

Curvature: The rotation per unit length of a member due to bending forces.

Curve sawing: To saw on the contour or sweep of the log. Also referred to as **sweep sawing** or **shape sawing.**

Curved-tooth file: This tool is widely used on aluminum and sheet metal. It features deeply cut, curved teeth for fast cutting and reduced clogging of soft material.

Curved-tooth file.

Custom builder: A builder who constructs a home or building based on plans created by the owner.

Custom cabinets: Milled cabinets that are built for a specific kitchen to match specified dimensions and design. See **Built-in cabinets, Milled cabinets,** and **Mass produced cabinets.**

Custom home: A structure designed by an architect hired by the owner.

Custom keying: An option for how door locks are keyed, e.g., deadbolt, entry set, etc. Custom keying indicates that the owner will supply either a copy of a key or a key-code for the new lock.

Cut: Terms that begin with cut:

- **Cut and cover:** A method of tunnel construction that involves digging a trench, building a tunnel, and then covering it with fill.

- **Cut pile:** Carpet pile in which the ends are looped, both ends are attached to the carpet backing, then the centers of the loops are cut. See **Pile** and **Loop pile**.
- **Cut rate:** The amount of material removal by the abrasive per unit of time.
- **Cutback:** Bitumen thinned by solvents that is used in cold-process roofing adhesives, roof cements, and roof coatings.
- **Cutoff:** Refers to the smooth cutting of wood, plywood, chipboard, paneling, pressboard, etc. Below is also **cut-off** that is completely different.
- **Cutout:** The open area between shingle tabs. Also referred to as a throat.

Three-tab shingles. Cutouts are between the tabs.

- **Cutout box:** An enclosure designed for surface mounting that has swinging doors or covers secured directly to and telescoping with the walls of the equipment.
- **Cut-in brace:** Nominal 2-inch-thick members, usually 2 by 4's, cut in between each stud diagonally.
- **Cut-list:** A list of components with dimensions used for fabrication and accounting purposes. See **Bill of materials**, **Bill of lading**, and **Shipping list**.
- **Cut-off:** A detail designed to seal and prevent lateral water movement in an insulation system.
  A cut-off is also used to separate different sections of a roofing system.
- **Cutting circle:** The circle described by the outer rim or extremity of the teeth of a circular saw.

Cycle: When alternating current flows back and forth through a conductor, it is said to cycle. In each cycle, the electrons flow first in one direction, then the other. In the United States, the normal rate for power transmission is 60 cycles per second, or 60 Hertz (Hz).

Cyclical maintenance: Maintenance that is performed less frequently than annually. Cyclical maintenance may involve replacement or at least mending of material. See **Routine maintenance**.

Cylinder: With regard to a locking mechanism, the cylinder shaped mechanism that contains the actual keying slot and locking code for locking a door.

Cylinder.

With regard to hydraulic equipment, a device that converts fluid power or air into linear mechanical force and motion. It usually consists of a movable element such as a piston and piston rod, plunger rod, plunger, or ram, operating within a cylindrical bore.

Also, the outermost lining of a hydraulic jack.

D-cracks: A series of cracks in concrete near and more-or-less parallel to joints, edges, and structural cracks. D-cracks are usually associated with distress from freezing and thawing of saturated aggregate particles in concrete pavements.

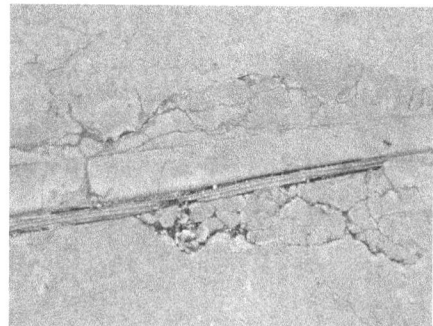

D-cracking concrete.

Dado: A groove cut into a board or panel intended to receive the edge of a connecting board or panel. A dado is three-sided and cut into a board, usually across the grain, as opposed to a rabbet, which has two sides and is at the edge of the board.

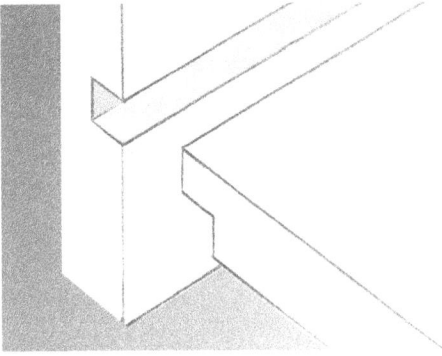

Rabbet and dado joint.

Daily construction report: A written document and record that furnishes information to concerning important details of events as they occur daily and hourly, and provides historical documentation that might later have a legal bearing in cases of disputes. Daily reports should: be numbered to correspond with the working days established on the progress schedule; include a description of the weather; record the total number of employees; list subcontractors by name; identify work started and completed; list the major equipment on the job site; list names and titles of visitors; note all accidents and/or safety meetings, and; in a factual and business-like manner provide comments on the progress of work and other job-related information.

The report should be made even on no-work days, e.g., when work is halted due to rain, strike, etc. The report includes a description of the weather, a record of the total number of employees, subcontractors by name, work started, and completed today, equipment on the job site, job progress today, names and titles of visitors, accidents and/or safety meetings, and a remarks column for other job-related information.

Dalle glass: A decorative composite glazing material made of individual pieces of glass that are embedded in concrete or epoxy.

Damp proofing: A black, tar like water-proofing material applied to the exterior of a foundation wall to inhibit the passage of moisture. Also spelled **dampproofing**.

Also, the application of an impermeable material (usually lead-based, bitumen-based, or plastic-based, although new materials are constantly coming to market) built into a wall near the ground to prevent rising dampness.

Also, heavy duty polythene, incorporated within floors built on the ground to prevent rising dampness entering through the slab.

Damper: A metal "door" placed within the fireplace chimney. Normally closed when the fireplace is not in use.

Also, a mechanical shutter or flap that controls the flow of gasses through a circuit.

Damping: The rate at which natural vibration decays as a result of the absorption of energy. In buildings it is an inherent nature to resonate inefficiently to vibration depending on structural connections, kinds of materials, and nonstructural elements used. "Damping" design measures can reduce the magnitude of seismic forces.

Critical damping is the minimum damping that will allow a displaced system to return to its initial position without oscillation.

Date of agreement: The date of the agreement can usually be found on the front page of the agreement or by the signatures. Sometimes the date of agreement is considered to be the day it was recorded or actually awarded to the contractor.

Date of commencement of work: The date established in a written notice to proceed from the owner to the contractor.

Date of substantial completion: The date certified by the owner, or more commonly the architect, when the work or a designated portion of it is substantially complete.

Daylight: The end of a pipe (the terminal end) that is not attached to anything.

Daylight factor: The ratio of exterior illumination to interior illumination, expressed as a percentage.

Daylight glazing: A vertical window area located 7'6" above the floor. Glazing at this height is the most effective at distributing daylight deep into the interior space.

Daylight opening (DLO): The width and the height of the visible glass.

Daylighting: The controlled admission of natural light into a space through glazing to reduce or eliminate electric lighting.

Dead-blow hammer: A tool that enables a user to strike blows without damaging the work's surface; in auto repair they are commonly used for chassis work, dislodging stuck parts and sometimes for hubcap installation and removal.

Dead-blow hammers are usually constructed of solid orange or black polyurethane. Composite heads and fiberglass handle models are also available. Options include shock absorbent rubber grips, head cavities filled with shot and replaceable faces with metal screws for quick replacement.

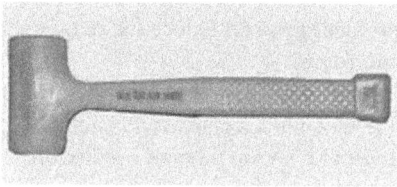

Dead-blow hammer.

Dead bolt: An exterior security lock installed on exterior entry doors that can be activated only with a key or thumb-turn. Unlike a latch, which has a beveled tongue, dead bolts have square ends. The use of dead bolts is regulated by most fire codes.

Dead flat: No gloss or sheen.

Dead front: As regards electrical equipment, equipment without live parts exposed to a person on the operating side of equipment.

Dead knots: Openings where a portion of the wood knot has dropped out or where cross checks have occurred to present an opening. Also referred to as **open knots**.

Dead level: Refers to a roof with no slope or pitch.

Dead-level asphalt: A roofing asphalt conforming to the requirements of ASTM Specification D 312, Type I. This asphalt is used on roofs that do not exceed a 1/4 in 12 slope (2%).

Dead light: The fixed, non-operable window section of a window unit.

Dead load: A permanent load consisting of all building parts and built-in fixtures that will be supported by a structural part.

Debark: To remove bark from trees or logs. The machine used to debark is, naturally, called a debarker.

Debris flow: A moving mass of rock fragments, soil, and mud, more than half of the particles being larger than sand size.

Decay: The decomposition of wood substance caused by the action of bacteria and wood-destroying fungi, resulting in softening, loss of strength, weight, and often in change of texture and color.

Buildings and cities also decay, i.e., they fall into disrepair and deteriorate.

Deceleration device: Any mechanism, such as a rope grab, rip-stitch lanyard, specially-woven lanyard, tearing or deforming lanyard, automatic self-retracting lifeline/lanyard, etc., which serves to dissipate a substantial amount of energy during a fall arrest, or otherwise limit the energy imposed on an employee during fall arrest.

Deceleration distance: The additional vertical distance a falling employee travels, excluding lifeline elongation and free fall distance, before stopping, from the point at which the deceleration device begins to operate. It is measured as the distance between the location of an employee's body belt or body harness attachment point at the moment of activation (at the onset of fall arrest forces) of the deceleration device during a fall, and the location of that attachment point after the employee comes to a full stop.

Deciduous: A type of tree with broad leaves that usually are shed annually, such as maple. Commonly referred to as Hardwood.

Decimalized feet: An expression of distance in feet and decimal portions of feet rather than feet and inches. For example, 6.25 feet is the decimalized expression of 6 feet 3 inches.

Deck: A structural component of the roof of a building. The deck must be capable of safely supporting the design dead and live loads, including the weight of the roof systems, and the additional live loads required by the governing building codes. Decks are either non-combustible (e.g., corrugated metal, concrete, or gypsum) or combustible (e.g., wood plank or plywood), and provide the substrate to which the roofing or waterproofing system is applied.

Also, a deck may be a floor or roof covering made out of gage metal attached by welding or mechanical means to joists, beams, purlins, or other structural members. Such decking can be galvanized, painted, or unpainted and may be specified, e.g., Type "B" Wide Rib, Type "F" Intermediate, Type "N" Deep Rib, Type "A" Narrow Rib, Composite, Cellular, etc.

Also, lumber that forms a floor's surface. Decking fastens directly to the floor joists.

Also, to install the decking material on floor joists, rafters, and trusses.

Deck mounted faucet: A faucet that is mounted on the deck of the bathtub enclosure, rather than on the rim of the bathtub or on the wall. Also referred to as a **Roman spoutfaucet** and a **sunken tub**. The deck-mount tub is usually mounted on a platform and has no apron or decorative side.

Deck mounted faucet with gooseneck spout.

Declarant: A LEED project team member who is technically qualified to verify the content of a LEED credit submittal template, and is authorized by the project administrator to sign the template and upload it to LEED Online. The Declarant must have had a significant degree of responsibility for the credit, such as participation in or oversight of the implementation or verification. The declarant for credits may be restricted or nonrestricted. For example, for Sustainable Sites Credit 4, only the property manager or facility manager may submit verification; for others, any team member, including contractors or consultants, can prepare the submittal documentation.

Decompression: The release, usually at a slow, controlled rate, of confined fluid to reduce pressure on the fluid.

Dedicated circuit: A circuit that consists of a home run that connects to a single device. An electrical circuit that serves only one appliance (i.e., dishwasher) or a series of electric heaters or smoke detectors.

Deep well: Usually a well that is more than 25 ft. deep.

Default: Breach of a mortgage contract (not making the required payments).

Defect: A shortcoming, imperfection, or lack. A characteristic that makes a product or thing either less desirable or completely unsuitable for the intended purpose.

Checks, splits, open joints, knotholes, cracks, loose knots, wormholes, gaps, voids, or other opening interrupting the smooth continuity of the wood surface are called open defects.

Deflection: The degree to which a structural element is displaced under a load. It may refer to an angle or a distance. Measured deflection usually takes into account both elastic and creep deformation.

Building codes usually specify the maximum deflection under various conditions, usually as a fraction of the span e.g., 1/400 or 1/600. Either the allowable stress or the serviceability limit state including deflection considerations among others may govern the minimum dimensions of the designed member.

Deformation: A change in the original shape of a material.

Deformations: Deformations of members in a metal structure that may be caused by fabrication or erection out-of-plumbness, lack of fit or slip at connections, settlement or failure of supports, overstressing, inadequate bracing, changes in temperature, removed members or missing connectors, and torsional effects unaccounted in design.

Deformability: Deformation and strength properties of masonry are measured using thin, bladder-like flatjack devices installed in cut mortar joints in the masonry, to measure load-deformation (stress-strain) properties. Boundary effects of the collar joint behind the wythe tested and adjacent masonry are neglected. In the case of multi-wythe masonry, deformability is estimated only in the wythe in which the flatjack is inserted. Deformability of other wythes may be different.

Deformed reinforcing bar: A steel reinforcing bar that is manufactured with surface deformations to provide a locking anchorage with surrounding concrete bars. The required spacing, height, and other physical features of the deformations on steel bars that are used for concrete reinforcement are referred to as **deformation requirements**. These deformations affect bond strength.

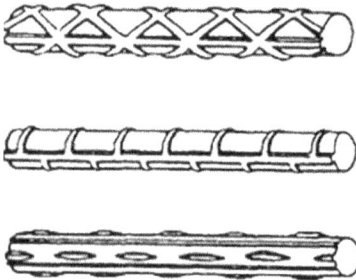

Examples of deformed bars.

Degradation: A deleterious change in the chemical structure, physical properties, or appearance of a material due to natural or artificial exposure (e.g., exposure to radiation, moisture, heat, freezing, wind, ozone, oxygen, etc.).

Degree of freedom: A displacement quantity which defines the shape and location of an object. In the two dimensional plane, a rigid object has three degrees of freedom: two translations and one rotation. In three-dimensional space, a rigid object has six degrees of freedom: three translations and three rotations.

Delamination: A horizontal splitting, cracking, or separation of a concrete member in a plane roughly parallel to and generally near the surface. Delamination of concrete is frequently caused by corrosing of reinforcing steel. Bridge decks and parking deck slabs are susceptible to delamination.

Delamination can also be found in the separation of the other materials, e.g., a panel due to failure of the adhesive. This is usually caused by excessive moisture.

Delta (Δ): The amount of change in a number, size, or position. In other words, a mathematical variation of a variable or function.

Also, a triangular tract of sediment deposited at the mouth of a river, typically where it diverges into several outlets, e.g., the Mississippi River Delta.

Demand factor: The ratio of the maximum demand of a system, or part of a system, to the total connected load of a system or the part of the system under consideration.

Demand flow technology (DFT): A strategy to define and deploy business processes in a flow, driven in response to customer demand. DFT is based on a set of applied mathematical tools that are used to connect processes in a flow and link it to daily changes in demand. DFT represents a scientific approach to flow manufacturing for discrete production. In the early years, DFT was regarded as a method for **just-in-time (JIT)** that advocated manufacturing processes driven to actual customer demand.

Demising wall: The boundaries that separate the space assigned to the contractor from neighboring space.

Also, demise is a term used to describe a conveyance of an estate in real property. It is most commonly used as a synonym for **let** in a lease.

Densely occupied space: An area that has a design occupant density of 25 people or more per 1,000 square feet (40 square feet or less per person). Most codes specify ventilation requirements for these spaces.

Densifier: A substance that is applied to a surface, e.g., a concrete floor, to fill pores and increase surface density.

Design development phase: The second phase of the architect's basic services in which the architect prepares drawings and other presentation documents to fix and describe the size and character of the entire project. It includes architectural, structural, mechanical and electrical systems, materials, and other essentials as may be appropriate. A statement of probable construction cost is also developed.

On architectural projects, the design development phase serves as the transitional phase from the schematic phase to the contract document phase during design.

Density: The weight (mass) per unit of volume; density measures the degree of compactness of a substance, e.g., soil. It is commonly expressed as grams per cubic centimeter (g/cm^3), kilograms per cubic meter (kg/m^3), or pounds per cubic foot (lb/ft^3), although kilograms per liter (kg/l) and pounds per gallon ($lb/gallon$) are also sometimes used.

The density of some common materials is as follows:
- Air (at sea level), 0.0013 g/cm^3
- Gasoline, 0.8 g/cm^3
- Gold, 19 g/cm^3
- Ice, 0.92 g/cm^3
- Steel, 7.9 g/cm^3
- Water, 1 g/cm^3

Average density (D) is computed by the formula $D = m/V$ where "m" is the mass of a substance and "V" is its volume. Relative density (also called specific gravity) is the ratio of the density of a substance at 20°C or 68°F to the density of water.

In computing, density is a measure of the amount of information on a storage medium, e.g., a disc. In urban planning density is the number of people in a given area or space. And, in photography, the opacity of a photographic image is referred to as its density.

Density factor (kd): A coefficient used in calculating the Landscape Coefficient; it modifies the Evapotranspiration Rate to reflect the water use of a particular plant or group of plants, particularly with reference to the density of the plant material.

Dentils: A small square block used in series. Often found in cornices.

Dentils.

Depth of focus: The depth of the focus or hypocenter beneath the earth's surface commonly classifies earthquakes as: shallow (0–70 km); intermediate (70–300 km); and, deep (300–700 km).

Depth of the jamb: The point where the exterior casing ends to the point where the interior casing begins. On clad units, the point from the backside of the nailing fin to the interior of the frame.

Depth of joist: The out-to-out distance from the top of the top chord to the bottom of the bottom chord taken at some reference location, usually at the midspan of the joist or joist girder.

Derrick: A kind of fixed crane with a movable pivoted arm for moving or lifting heavy weights, especially., on a ship.

Also, the framework over an oil well or similar boring that holds the drilling machinery.

As a bit of trivia, derrick also the gallows and was the surname of a hangman in London, England.

Desanco fitting: A type of compression adapter that connects tubular brass fittings to the PVC pipe.

Desanco fitting.

Descriptive specification: Specifications that describe, often in intricate detail, the materials, workmanship, manufacture methods, and installation of the obligatory goods.

Desiccant: A material used to absorb moisture. Often used in shipping containers and other sealed airspaces.

Design: The combination of elements that create the form, plan, space, structure, and style of a human endeavor.

Other terms that begin with design:
*   **Design acceleration:** The anticipated ground acceleration at a site used for earthquake-resistance design of a structure.
*   **Design check:** The evaluation of a design to determine whether it conforms with the design brief and can be expected to provide a safe engineered solution. Design checks are often conducted by objective third parties. Also referred to as peer review.
*   **Design development phase:** Usually there are five distinct stages of the design process: (1) research, strategy or feasibility; (2) design concepts; (3) detailed design development; (4) implementation; and (5) supervision of production.
      During the design development phase the selected concept is worked up with

all details implemented. This stage may seem confused as many things go on simultaneously. A detailed specification of the design for production planning and final costing is usually created during this stage.
*   **Design documents:** The plans, details, sections, specifications, etc., prepared by the designer.
*   **Design earthquake:** Generally defined as 2/3 of the maximum considered earthquake.
*   **Design forces:** The loads that act on the structural system. For example, dead load, live load, wind load, snow load, seismic load, and other dynamic loads are all design forces.
*   **Design heat loss:** The calculated values, expressed in units of Btu per hour (abbreviated Btuh), for the heat transmitted from a warm interior to a cold outdoor condition, under some prescribed extreme weather conditions. The values are useful for selecting heating equipment and for estimating seasonal energy requirements. Infiltration heat loss is a part of the design heat loss.
*   **Design intent:** The creative objectives of a designer, architect, landscape architect, engineer, or artist that were applied to the development of a project.
*   **Design length:** With regard to joists the span of a joist or joist girder in feet minus 0.3333 feet.
*   **Design light output:** The light output of lamps at 40% of their useful life.
*   **Design loads:** Loads that are specified in building codes or standards published by federal, state, county, or city agencies, or in owners' specifications to be used in the design of a structure.
*   **Design pressure (DP):** A measurement of the structural performance of a window or door. Usually specified as one-and-a-half times greater than necessary based on expected building, wind and weather conditions.
*   **Design strength:** The resistance provided by a structure, member, or connection to the forces imposed on it.

- **Design-build contract:** A building contract in which the builder is responsible for all or some of the design. Government agencies are tending more and more to this form of contract.
- **Desigh-XCM:** A variation of the extended services form of CM, where the A/E also provides the CM function.

Designated person: See **Authorized person**.

Designer: A person who recommends aesthetically pleasing combinations of shapes and shades for the interior or the exterior of a building.

Desktop geographic information system (GIS): GIS software programs that support a wide variety of functions, queries, and mapping capabilities for personal computer-based applications. It is intended especially for visual presentation and descriptive analyses of geo-coded data.

Detail: One of the five basic views found on a plan. A detail is a close-up (i.e., large scale) view of some part of a section view, used to show exactly how parts connect together.

Detail drawing: A shop drawing that defines the exact shape, dimensions, bolt hole patterns, etc., of a piece[s] of steel. The piece[s] of steel may stand alone in the structure or may be one of many pieces in an assembly or shipping piece.

Detail piece: A single piece of steel that may stand alone in the structure or that is one of many pieces in an assembly or shipping piece.

Detailer: A person who is charged with the production of the advanced bill of materials, final bill of materials, and the production of all shop drawings needed to purchase, fabricate, and erect structural steel.

The production of different types of shop drawings needed to fabricate and erect structural steel is called **detailing**.

Detailing file: This tool is basically a half-round combination file. It is handy for working small flats, rounds and hollows. Its ends are tapered for work in confined spaces.

Detailing file.

Deterioration: The impairment of usefulness in materials.

Development footprint: The area affected by project activity. Hardscape, access roads, parking lots, non-building facilities, and the building itself are all included in the development footprint.

Dew point temperature: The temperature of the air at which the water vapor in the air starts to condense in the form of liquid or as frost.

De-energize: To free from any electric connection and/or electric charge.

Degree-day: A measure of heating demand, based on the difference between the mean daily outdoor temperature and 65°F. Cumulative totals for the month or heating season are used by engineers for estimating heating energy requirements.

De-humidistat: A control mechanism used to operate a mechanical ventilation system based upon the relative humidity in the home.

De-superheater: A device for removing the excess heat in gas as its pressure is reduced. Its most common application is the reduction of temperature in a steam line through the direct contact and evaporation of water.

Diagonal bracing: A reinforcing member that is attached at an angle to provide lateral strength to another member such as a rafter.

Diagonal bridging: Two angles or other structural shapes connected from the top chord of one joist to the bottom chord of the next joist to form an "X" shape whose l/r ratio cannot exceed 200. The bridging members are almost always connected at their point of intersection.

Diagonal wood floor installation: An installation where wood strips are installed in a pattern that runs diagonally to at least one of the walls. See **Straight wood floor installation** and **Herringbone wood floor installation**.

Dial plate: As regards faucets, a trim piece found behind a single-control wall-mount faucet handle. Also known as a **face plate**.

Dial plate.

Also, any number of plates that are mounted on equipment, e.g., the letters and numbers on a telephone.

Diagramming: The process of drawing a floor plan and sometimes elevations that include dimensions and other important information such as the location of doors, windows, outlets, switches, etc.

Diameter: A straight line segment that passes through the center of a circle and terminates at the outer edges of the circle. It is the longest line segment that will fit inside a circle. The diameter is equal to twice the length of the radius.

Diamond polishing: A process that uses diamond tooling to polish a surface, e.g., a concrete floor, to a high reflectivity.

Diaphragm: A type of structural roof deck capable of resisting shear that is produced by lateral forces such as wind or seismic loads.

Diaphragm action: The resistance to a racking affect or in-plane shear forces offered by roof decking, panels, or other structural members when properly attached to a structural frame.

Diaphragm valve: A valve with a flexible membrane that deflects down onto a rigid area of the valve body to regulate water flow from the supply lines. This eliminates the possibility of debris build-up within the valve. Diaphragm valves are used on shut-off and throttling service for liquids, slurries, and vacuum/gas.

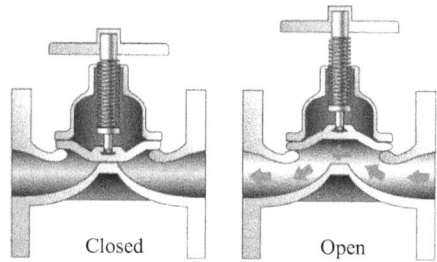

Diaphragm valve.

Diatomite: A hydrous form of silica. Diatomite deposits are made up of shells of diatoms (algae). It is used for filtration, in a brick or powder form for the insulation of furnaces and refrigerators, for sound-proofing and as an abrasive in metal and concrete polishes. Also referred to as diatomaceous earth and kieselguhr.

Dichroic glass: A thin metallic coating on any type of glass. This coating is applied at a high temperature in a vacuum chamber.

Dichroic glass on the exterior of a building.

Die: A device that cuts external threads on a rod or pipe. Special designs are used for cleaning up existing threads of rust or rolled over threads, called rethreading dies.

Die stock: A two-handle, adjustable tool that holds and turns dies.

Die stock.

Dielectric fitting: In a home or commercial building's water supply system, a special type of adapter (such as a union) used to connect a pipe containing copper with a pipe containing iron. Dielectric fittings should always be used between dissimilar metals to prevent galvanic action from causing corrosion failure.

Differential settlement: Relative movement of different parts of a building or home caused by uneven sinking of the structure.

Diffuser: A mechanism, usually louvered, for the distribution of air at the terminus of a HVAC circuit.

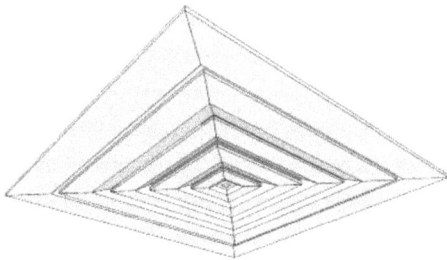

HVAC diffuser.

Also, a device for reducing the velocity and increasing the static pressure of a fluid passing through a system.

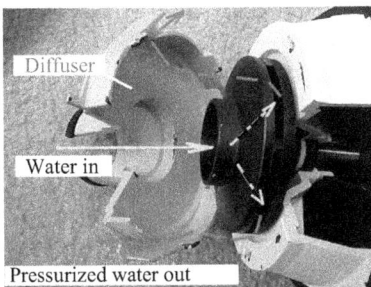

Diffuser pump.

Diffusion: The movement of moisture from areas of high to low concentration or temperature. The spreading of light from a light source usually to reduce glare and harsh shadows. The intermingling of substances by the natural movement of their particles.

Digital imaging: The creation, manipulation, and production of images using computer technology.

Digital protractor: Essentially four tools in one. A digital protractor features an angle finder, compound cut calculator, protractor, and level.

Dilatancy: As pertains to rock, the increase in volume of rock due to a change in strain.

Diluent: A liquid used in coatings to reduce the consistency and make a coating flow more easily. The water in latex coatings is a diluent. A diluent may also be called a reducer, thinner, reducing agent, or reducing solvent.

Dimension(s): Dimensions are the measurements of a room or object. From plans, dimensions can be determined using the measurement on the drawing and the scale of the drawing or by reading the dimensions printed on the drawing.

Some materials are characterized by standard measurements, for example:

- **Dimensional beam:** A beam that is made from dimensional lumber and is usually available in rough-cut or surfaced finishes.
- **Dimensional lumber:** The wood that is commonly used in construction. Dimensional lumber is manufactured in a large variety of sizes for interior and exterior applications. Dimensional lumber is sold in a nominal dimension, not its finished dimension. For example, a 2 × 4 is lumber with an actual finished size of 1.5" thick by 3.5" wide.
- **Dimensional shingles:** See **Architectural shingles**.
- **Dimensional stability:** The ability of a material to retain its current properties and to resist a change in size resulting from exposure to temperature changes and moisture.

Dip: The angle of a fault or other planar geologic feature relative to horizontal.

Dip-slip fault: See **Fault, dip-slip**.

Direct chemical attack: Deterioration of materials by attack of chemical solutions with either low or high acidity (pH).

Direct costs: When applied to construction costs that can be specifically identified with a project or with a unit of production within a project; direct labor would be a direct cost.

Direct glaze: Refers to a window with no sash. The glass is glazed directly into the frame and is stationary.

Direct glaze window.

Directional control valves: Valves that allow fluid to flow into different paths from one or more sources.

Directivity: Directivity is an effect of a fault rupturing whereby earthquake ground motion in the direction of rupture propagation is more severe, than that in other directions from the earthquake source.

Discharge head: The difference in elevation between the liquid level of the discharge tank and the centerline of the pump. Also includes any additional pressure head that may be present at the discharge tank fluid surface.

Discoloration: A departure of color from the normal. Discoloration is usually the result of stains or chemical changes.

Disconnect: A large (generally 20 Amp) electrical on-off switch.

Disconnecting means: A device, or group of devices, or other means by which the conductors of a circuit can be disconnected from their source of supply.

Discount rate: A mortgage interest rate that is lower than the current rate for a certain period of time, e.g., 2.00% below variable rate for 2 years.

Also, The percentage rate at which money or cash flows are discounted. The discount rate reflects both the market risk-free rate of interest and a risk premium. See **Opportunity cost**.

Discretionary change order: Owner directed change order for work that is not required for completion of the project, for the projects' intended use, or in accordance with the construction contract.

Discrimination: With regard to a business, bias, or prejudice that results in the denial of opportunity, or unfair treatment regarding a selection, promotion, or transfer. Discrimination is practiced commonly on the grounds of age, disability, ethnicity, religion, sex, etc., that are irrelevant to competence or suitability. Also, unequal treatment provided to one or more parties on the basis of a mutual accord or some other logical or illogical reason. Ability is nothing without opportunity.

Also, differences in two rates that are not explainable or justifiable by economic considerations such as cost.

Disintegration: Deterioration into small fragments or particles due to weathering, chemical attack, erosion, and other factors.

Dismantle: To take apart or remove any components, device, or piece of equipment that would not be taken apart or removed by a homeowner in the course of normal home maintenance.

Displacement: Lateral movement of the structure caused by lateral force.

Also, in geology, it is the permanent offset of a reference point across a fault.

Distortion: Warping and deforming due to overloading, poor design, ground movement, expansion, poor manufacturing quality, and other factors.

Specifications often state tolerable limits of distortion of materials.

Distributed load: An external force that acts over a region of length, surface, or area: essentially any external force that is not a concentrated force.

Diverter: A diverter is used on a shower/tub combination, multiple shower sprays, or similar to divert water from one source to another. A diverter is often attached to wall tub filler (a knob on top that can be raised or lowered), or a handle on a shower wall that is turned to switch the water flow from one devise to another.

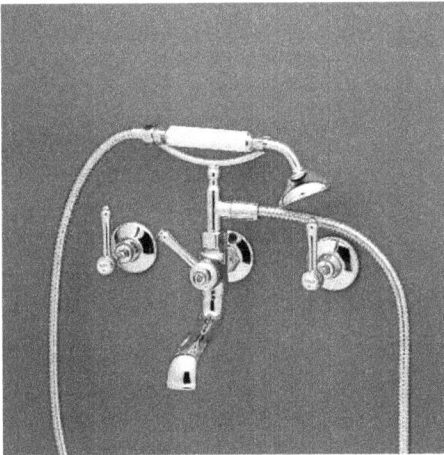

Faucet with a diverter.

Diversion channel: A bypass created to divert water around a dam so that construction can take place. See **Cofferdam**.

Divided lites: Separately framed pieces or panes of glass. A double-hung window, for instance, often has several lites divided by muntins in each sash. These designs are often referred to as six-over-six, eight-over-one, etc., to indicate the number of lites in each sash. Designs simulating the appearance of separately framed panes of glass are often referred to as SDLs or simulated divided lites. Designs using actual separate pieces of glass are sometimes referred to as TDLs or true divided lites.

Dog-leg chisel: This chisel's skewed blades are ideal for trimming joint work in furniture making and for undercutting.

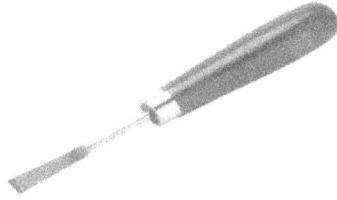

Dog-leg chisel.

Dome: A roof with a partial-spherical shape; imagine a three-dimensional arch.

Domestic hot water: Water heated for residential washing, bathing, etc.

Doming head: Equipment used to form a dome (chamfer) on the end of roundwood posts. A doming head consists of a rotating cutterhead.

Door: A hinged, sliding, or revolving barrier at the entrance to a building, room, or vehicle, or in the framework of a cupboard.

Other terms that begin with door:
- **Door, bi-fold:** A door that is hinged so as to fold against the door jamb. Bifold doors are normally classified as either two- or four-leaf units.
- **Door buck:** The frame of a door consisting of the head and jamb sections connected together as a single assembly (usually a hollow metal section).
- **Door, combination:** A door assembly of stiles and rails that will include multiple door types within a single door. These door types would typically include combinations of flat or raised panels, lites and/or louvers.
- **Door frame:** A group of components (wood, aluminum, or steel) that are assembled to form an enclosure and support for a door. Also referred to as door jambs.
- **Door, French:** A door assembly of stiles and rails (and possibly muntins and bars) surrounding a single or multiple glazed opening.

Interior French door.

- **Door gasket:** Insulating material placed around the door jamb where the slab meets the jamb or door stop when closed. The door gasket is used on exterior doors.
- **Door gibs:** Devices at the bottom of horizontal sliding door panels that stick into sill grooves and eliminate door panels swinging in or out. Some building codes require that exterior elevator doors have gibs so that the doors cannot be pushed in.
- **Door handle plate:** A large decorative plate that covers any bored hole in a door with an attached knob or lever.

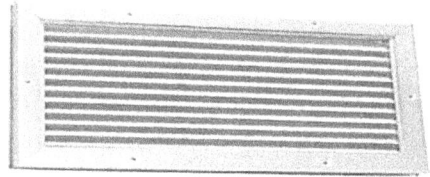

Door handle plate

Door handle plate.

- **Door hanger:** A rolling assembly fastened to the top of a door panel which supports and allows horizontal sliding movement of the door panel. The door track on which the hanger rolls is part of the door hanger assembly.
- **Door hardware:** The latch, door knob, and striker plate.
- **Door jamb:** The surrounding case into which and out of which a door closes and opens. It consists of two upright pieces, called side jambs, and a horizontal head jamb. These three jambs have the **door stop** installed on them. Also called a **doorpost**. Also spelled **doorjamb**.
- **Door louver:** A set of angled slats or flat strips fixed or hung at regular intervals in a door, shutter, or screen to allow air or light to pass through. There is also a **louvered door**.

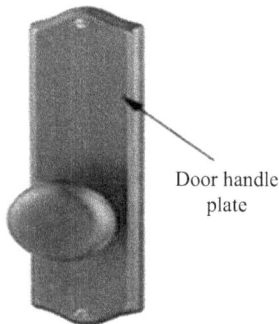

Door louver.

Louvered door.

- **Door operator:** An automatic garage door opener.
- **Door, panel:** A door assembly of stiles, rails, and one or more panels. Intermediate rails or mullions are used to separate panels. Panels can be raised or flat.

Panel door.

- **Door protective device:** Any type of device used with automatic power operated doors that detects obstructions to the normal closing of the elevator doors and either causes the doors to reopen or go into some other mode of operation, such as nudging. A safe edge, a safety astragal, a photoelectric device (safe ray), and electrostatic field device are examples of door protective devices.
- **Door sill:** The threshold of a door opening with grooves to guide the bottom of the car door.
- **Door slab:** A rectangular door without hinges or frame.
- **Door stop:** The wooden style that the door slab will rest upon when it's in a closed position.

- **Door swing:** The direction the door opens while standing on the outside.

    If you stand on the outside of the door and you push it away from yourself to open, it is in-swing.

    If you stand on the outside of the door and you pull it toward you to open, it is out-swing.
- **Door thickness:** The actual measurement of the physical thickness of your door.

Dope: A pasty lubricant used on seal pipe threads prior to making a threaded pipe connection.

Doric: A Doric column features a fluted, tapered shaft, and a square abacus capital, the Doric column is the simplest of the architectural orders.

Doric   Ionic   Corinthian

A Doric column is the simplest of the architectural orders.

Dormer: An opening in a sloping roof, the framing of which projects out to form a vertical wall suitable for windows or other openings.

Double acting cylinder: A cylinder in which fluid force is applied to the movable element in both directions. In order to connect the piston in a double-acting cylinder to a crank shaft there must be a hole in one end of the cylinder for the piston rod. This hole is fitted with a gland or "stuffing box" to prevent the working fluid from escaping. Double-acting cylinders are common in steam engines but unusual in other engine types.

Double bull nose: A specialty tile trim piece with rounded corners on both sides. **See Cove molding/piece** and **Tile base**.

Double coverage: Roofing that is installed so that there is a double layer of roofing.

Double curvature: When end moments on a structural member produce a bending effect that causes the member to form an S shape or has a reversal in curvature.

Double cylinder door locks: Door locks that require a key to lock/unlock the door from both the outside and inside. The double cylinder function is an option for door locks like deadbolts and entry sets. Double cylinder locks are usually not recommended because they make it more difficult to exit in an emergency.

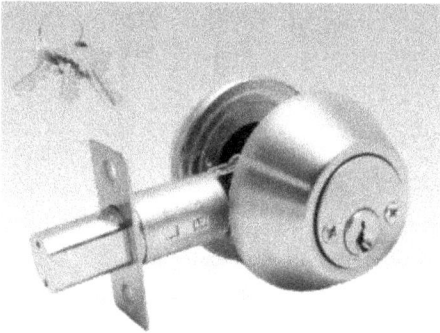

Double glazed window.

Double hung window: A window with two vertically sliding sashes, both of which can move up and down.

Double pull: A method for using the ladder to climb a lower roof section, then pulling the ladder onto the roof for use when climbing a second higher roof section.

Double square: This unique tool is designed for checking squareness of a board's edge after planing and jointing. It fits easily in a work apron pocket.

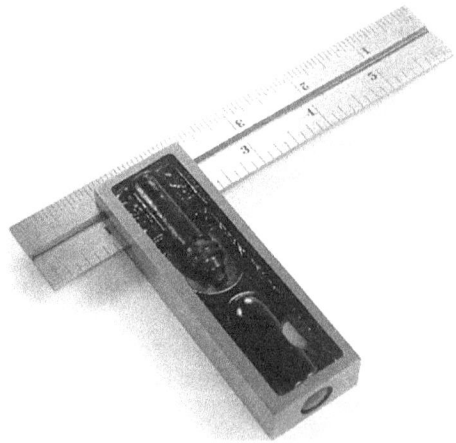

Double cylinder deadbolt.

Double glass: A window or door in which two panes of glass are used with a sealed air space between. The use of two panes of glass in a window increases energy efficiency and provides other performance benefits, although, if not properly sealed the unit may not provide much insulation. Also known as **insulating glass, energy panel (EP)**, and **double glazing**.

Double graveling: Installing one layer of gravel in a flood coat of hot bitumen, removing the excess gravel and then installing a second layer of gravel in another flood coat of hot bitumen.

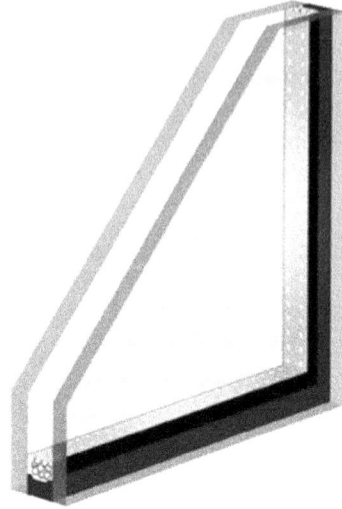

Double square.

Double-strength glass: Glass that is between 0.115 and 0.133 inch thick.

Double threshold shower: A shower that is set in a corner; one threshold is used as an entrance and the other has a glass wall.

Double threshold shower.

Double window: Two windows separated by a mullion, forming a unit. Also called a **coupled window**.

Antique double window.

Also, a regular window plus a storm sash. The exterior storm sash is also known as **a storm window**.

Double windows.

Also, a window with insulating air space between panes may be referred to as a **double window**. Also known to as **insulating glass** and **double glazing**. See **Double glass** and **Energy panel**.

Dovetail joint: A joint where a mortise and tenon combine to form a solid structure. See **Mortise and tenon**.

Dovetail joint.

Dovetail saw: Many woodworkers consider this tool the most important of the back saws (and one of the smallest). Dovetail saws are dedicated almost entirely to cutting dovetails. Dovetail saws have three handle designs: closed, pistol-grip and a turned spindle-like handle.

Dovetail Saw

Dovetail square: This tool is used to lay out and marks dovetail joints with consistent accuracy.

Dowel: A stick of wood, plastic, or metal that fits into corresponding holes to attach two pieces of material together.

In terms of concrete work, a short piece of reinforcing rod used to join two separate parts of a concrete assembly together as a whole.

Doweled construction: A method of construction of stile and rail wood doors where holes are machined into, but not through, the stiles and where matching holes are machined into the ends of the rails. Glue and dowels are inserted into these holes to attach the rail to the stile.

Down payment: The difference between the sales price and the mortgage amount. A down payment is usually paid at closing.

Downstanding leg: The leg of a structural angle that is projecting down from you when viewing it.

Down-flow furnace: A furnace which forces air down and out the bottom of it.

Downspout: A pipe, usually of metal, for carrying rainwater down from the roof's horizontal gutters.

Downspout adapters: There are no existing specifications or standards on dimensions for downspouts; one manufacturer's 3 × 4 downspout may not be the same size as another's 3 × 4 downspout, and thus adapters may be required.

Downstream face: The side of the dam that is not against the water.

Doze: A form of incipient decay characterized by dull and lifeless appearance of the wood, accompanied by a lack of strength and softening of the wood.

Drain tile: A perforated, corrugated plastic pipe laid at the bottom of the foundation wall and used to drain excess water away from the foundation. It prevents ground water from seeping through the foundation wall. Sometimes called perimeter drain.

Draft hood: A device placed in and made part of the vent connector, chimney connector, or smokepipe, from an appliance, or in the appliance itself. It is designed to ensure the ready escape of the products of combustion.

Drafter: A person who translates the ideas provided to them by an architect or designer into accurate plans that may be used for construction. In this age of computers fine drafting has become a lost art.

Dragline: This excavating machine is generally used for surface mining for coal, sand, and gravel. The dragline is attached to cables and pulls the bucket back to the machine.

Drain: An outlet or other device used to collect and direct the flow of runoff water from an area, e.g., a roof drain.

Drain field: A series of pipes through which waste is released into the soil after it has been treated in the septic tank. The size of the drain field depends on the number of people being serviced by the system and the ability of the soil to absorb liquid.

Drainage system: Generally, all the piping within a public or private premises that conveys sewage, rain water, or other liquid wastes to a legal point of disposal. The definition does not include the mains of a public sewer system or private or public sewage-treatment or disposal plant.

Draw: The amount of progress billings on a contract that is currently available to a contractor under a contract with a fixed payment schedule.

Dressed lumber: Lumber that has been trimmed and planed at the sawmill. The **dressed size** of lumber after being surfaced with a planing machine differs from the nominal size of the lumber. For example, a nominal 2- by 4-inch stud actually measures a dressed size of about 1-1/2 by 3-1/2 inches.

Driers: Various compounds added to coatings to speed the drying.

Drift: The horizontal displacement of basic building elements due to lateral earthquake forces.

Also, the lateral movement or deflection of a structure.

Drift index: The ratio of the lateral deflection to the height of the building.

Drift pin: A tapered steel rod used by ironworkers to align bolt holes in structural steel for eventual fastening.

Drill press: This versatile tool is used for hole boring a variety of holes accurately and easily. It can also be used to sand, rout, polish, saw, shape, grind, sharpen, and mortise. Drill presses are used for both metalworking and woodworking.

Drill press.

Drip: A member of a cornice or other horizontal exterior finish course that has a projection beyond the other parts for throwing off water.

Also, a groove in the underside of a sill or drip cap to cause water to drop off on the outer edge instead of drawing back and running down the face of the building.

A molding or metal flashing placed on the exterior topside of a door or window frame to cause water to drip beyond the outside of the frame is referred to as a **drip cap** and **drip edge**.

A steel flashing bent at a 90° angle that is placed along the outer perimeter of steep sloped buildings is also a **drip edge**. Drip edges help direct runoff water away from the building. A drip edge resembles **nosing** except that it has an outwardly-angled bottom edge (preferably hemmed).

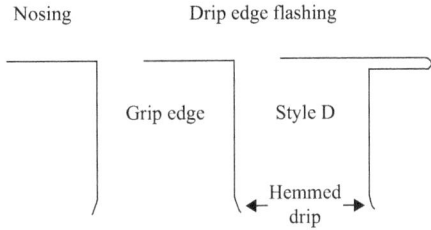

Showing the difference between drip edges and nosing.

Drip irrigation: A high-efficiency method in which water is distributed at low pressure through buried mains and submains. From the submains, water is distributed to the soil from a network of perforated tubes or emitters. Drip irrigation is a type of **microirrigation**.

Drip leg: A stub end pipe placed at a low point in the gas piping to collect condensate and permit its removal.

Drip leg.

Drip loop: A loop that is made in the drop wires just in front of the weather head. The drip loop prevents water from dripping down the wires and into the weather head.

Drip loop.

Drip pan: A fabricated sheetmetal pan used to collect leakage and condensate from a piece of mechanical equipment and drain it away from other assemblies, equipment, or electrical circuits.

Drop-in sink: A washbasin fitted into a worktop surface from above.

Drop plug: See **button plug**.

Drop siding: Shiplap siding with special shaping on its face; for example, it may have a rounded face to look like logs.

Drop wire: A wire that connects the transformer to the structure. The drop wire for a residential structure usually contains two current carrying leads and a neutral.

Dry bulb: An ordinary exposed thermometer bulb, especially as used in conjunction with a wet bulb that is a drybulb with a wet sock pulled over it to measure the cooling effect of moisture evaporation from its surface.

The difference in wetbulb and drybulb temperatures is called a wetbulb depression.

Dry colors: Powder-type colors to be mixed with water, alcohol, or mineral spirits and resin to form a paint or stain.

Dry film thickness: The thickness in mils (thousandths of an inch), of a dried coating or mastic.

Dry glazing: A form of glazing in which the glass is secured in the frame with a dry gasket, wood stops, or metal stops, instead of by a glazing compound. See **Reglet**.

Dry in: To install the black roofing felt (tar paper) on the roof.

Also, making a low-slope roof watertight.

Dry looper: A machine used in the manufacture of composition shingles that is designed to allow air to circulate around roofing material to dry the material.

Dry lumber: The moisture content of wood is calculated by the formula:

$$\text{Moisture content} = M_g - M_{od}/\,M_{od}$$

Here, M_g is the green mass of the wood and M_{od} is its oven-dry mass. The equation can also be expressed as a fraction of the mass of the water and the mass of the oven-dry wood rather than a percentage.

Softwoods are generally considered as "dry" or "kiln dried" at 19% or lower moisture content, as specified by the American Softwood Lumber Standards. Hardwoods are generally considered dry when at 10% or lower moisture content, although there is no definitive standard as with softwood species.

Dry pipe valve: A valve that automatically controls the water supply to a sprinkler system so that the system beyond the valve is normally maintained dry.

Dry rot: A term loosely applied to any dry, crumbly rot in wood but especially to rot that is in an advanced stage and permits the wood to be crushed easily to dry powder.

Drywall: A term used to describe gypsum board wall construction.

Drywall hammer: This hammer is designed for installing drywall. It has a milled striking face opposite a hatchet blade. The hatchet blade is used to score the drywall to permit snapping it along the scored line of cut.

Drywall hammer.

Drywall (or Gypsum Wallboard (GWB), Sheet rock or Plasterboard: Wall board or gypsum. A manufactured panel made out of gypsum plaster and encased in a thin cardboard. Usually 1/2" thick and 4' × 8' or 4' × 12' in size. The panels are nailed or screwed onto the framing and the joints are taped and covered with a "joint compound." "Green board" type drywall has a greater resistance to moisture than regular (white) plasterboard and is used in bathrooms and other wet areas.

Dry weight of wood: The oven-dry weight, or simply dry weight of wood after drying to a constant weight at a temperature slightly above the boiling point of water (215–220°F).

Drying: A process used to remove moisture from lumber. The moisture content in the lumber is reduced to the average amount it will maintain when used in building construction. After drying, many lumber mills seal the lumber with stain or wax to prevent them from absorbing moisture again. Surface drying and kiln drying are two methods used for drying lumber.

Drying oil: An oil that when exposed to air will dry to a solid through chemical reaction with air, e.g., **linseed oil, tung oil, perilla, fish oil,** and **soybean oil**.

Drying shrinkage: The change in linear dimension of a test specimen due to drying from a saturated condition to an equilibrium mass and length under specified accelerated drying conditions. For example, concrete masonry units are routinely subjected to such a test.

Dual level drain: An drain that will permit drainage at two different levels.

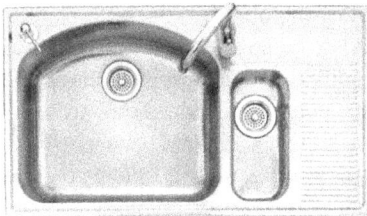

Dual level drain (perforated side walls allows ingress at lower level).

Duckbill snips: Snips used to cut curves in either direction. Can be used for straight cutting but will require slightly more effort to cut with than straight pattern snips.

Duckbill snips.

Duct: A fabricated hollow conduit for the containment of a HVAC or electrical circuit.

Earlywood: The portion of the annual growth ring that is formed during the early part of the growing season. Earlywood is usually less dense and weaker mechanically than **latewood**.

Earnest money: A sum paid to the seller to show that a potential purchaser is serious about buying.

Earth pigments: Pigments that are obtained from the earth, including barytes, ocher, chalk, and graphite.

Earthquake fault terms
- **Fault:** A fracture plane in the earth's crust across which relative displacement has occurred. (Location of slippage between the earth's plates).
- **Normal fault:** A fault under tension where the overlying block moves down the dip or slope of the fault plane.
- **Strike-slip fault (or lateral slip):** A fault whose relative displacement is purely horizontal.
- **Thrust (reverse) fault:** A fault under compression where the overlying block moves up the dip or slope of the fault plane.
- **Oblique-slip Fault:** A combination of normal and slip or thrust and slip faults whose movement is diagonal along the dip of the fault plane.

Earthquake: The shaking or vibrating of the ground caused by the sudden release of energy stored in rock beneath the earth's surface.

Other terms that begin with earthquake:
- **Earthquake hazard:** Any physical phenomenon associated with an earthquake that may produce adverse effects on human activities.
- **Earthquake hazard reduction:** A broad term used for the process of lessening the impact of earthquakes on society.
- **Earthquake loss:** The actual or anticipated damage to life and property caused by earthquakes.
- **Earthquake risk:** The social and economic consequences of anticipated earthquakes

expressed in economic loss or casualties. Risk may be expressed as the probability that these will equal or exceed specified values in an area during a specified interval of time.

- **Earthquake source:** The origination point of earthquake energy release.
- **Earthquake strap:** A metal strap used to secure gas hot water heaters to the framing or foundation of a house. Intended to reduce the chances of having the water heater fall over in an earthquake and causing a gas leak.

Easemen: A formal contract which allows a party to use another party's property for a specific purpose. For example, a sewer easement might allow one party to run a sewer line through a neighbors property.

Eave: The horizontal exterior roof overhang; the line along the sidewall of a building formed by the intersection of the plane of the roof and the plane of the wall.

The vertical distance from finished floor to the eave is the **eave height**.

A structural member located at the eave of a building that supports a roof and/or wall panels is an **eave strut**.

An additional layer of roofing material applied at the eaves to help prevent damage from water back-up is called **eaves flashing**.

A channel (usually sheet metal) installed along the downslope perimeter of a roof to transport runoff water from the roof to the drain leaders or downspouts is an **eaves-trough** that is also referred to as a **gutter**.

Eccentric: The condition that exists when a load is applied on a line of action that does not pass through the centroid of the body to which it is applied.

Eccentricity: The distance between a line of action of force and the centroid of the member to which it is applied.

Eccentric braced frame: A steel frame that has diagonal bracing arranged eccentricly to column/beam joints.

Ecological restoration: The process of assisting in the recovery and management of

ecological integrity, including biodiversity, ecological processes and structures, regional and historical content, and sustainable cultural practices.

Ecologically appropriate site features: Natural site elements that maintain or restore the ecological integrity of the site. Examples include native or adapted vegetation, waterbodies, exposed rock, unvegetated ground, or other features that are part of the historic natural landscape within the region and provide habitat value.

Ecological restoration: The process of assisting in the recovery and management of ecological integrity, including biodiversity, ecological processes and structures, regional and historical context, and sustainable cultural practices.

Edge: The edge of a roof, property, waterway, etc., border, boundary, extremity, fringe, margin, side; lip, rim, brim, brink, verge; perimeter, circumference, periphery, limits, bounds.

Other terms that begin with edge:
- **Edge angle:** A structural angle that is connected around the edge of a joist extension or other member.

    Also, an angle used around the sides of a floor to contain the concrete when it is being poured. Also referred to as a pour stop.
- **Edge band:** A strip along the outside edges of the two sides and/or top and bottom of the door.
- **Edge bore:** The hole that is drilled into the edge of the door, connecting from the edge of the door to the larger main bore.

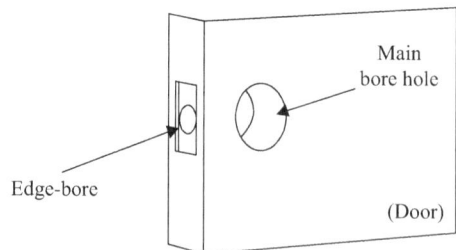

Edge bore.

- **Edge distance:** The distance from the center of a hole to the edge of a connected part.
- **Edge effect:** Heat transfer at the edge of an insulating glass unit due to the thermal properties of spacers and sealants.
- **Edge grain shingles:** Shingles that are made from lumber whose annual rings form at least a 45° angle at the face.
- **Edge strip:** The width or region around the edges of a building where uplift values are higher than in the interior of the roof.
- **Edge stripping:** Membrane flashing strips cut to specific widths used to seal/ flash perimeter edge metal and the roof membrane.
- **Edge venting:** The installation of vent material along a roof edge (e.g., **starter vent**) as part of a ventilation system. Edge vent material should be used in conjunction with other venting material (e.g., a **ridge vent**) as it not intended for use by itself.

An edge vent provides intake ventilation for homes with little or no overhang. It's a roof-top installed, shingle-over intake vent.

Edger: A machine used to square-edge wany lumber (the absence of square wood on the edge of a board) and also to rip lumber. It consists of a frame supporting an arbor on which is mounted one to several saws and transmission gear.

Edging: An edge molding.

Also, a plain or molded strip of metal, wood, or other material used to protect edges of a panel or hide the laminations as in plywood or roof sheathing, an edging strip.

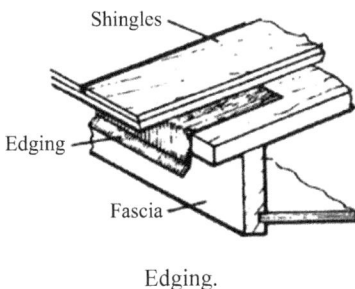

Edging.

Also, in concrete finishing, the process of rounding the exposed edges of slabs to reduce the possibility of chipping or spalling.

Also, boards that are nailed along the eaves and rakes after cutting back existing wood shingles to provide secure edges for reroofing with asphalt shingles.

Effective depth: As applied to a concrete beam or slab, it is the distance of the centroid of the reinforcement from the compression face of the concrete.

Effective length: The equivalent length, KL, used in compression formulas. This method estimates the interaction effects of the total frame on a compression member by using K factors to equate the strength of a framed compression member of length L to an equivalent pin-ended member of length KL subject to axial load only. See **Effective length factor (K)**.

Buckled shape of column shown by dashed line						
Theoretical K value	0.5	0.7	1.0	1.0	2.0	2.0
Recommended design value K	0.65	0.80	1.2	1.0	2.10	2.0
End condition key		Rotation fixed and translation fixed				
		Rotation free and translation fixed				
		Rotation fixed and translation free				
		Rotation free and translation free				

Examples of effective length.

Effective length factor (K): The ratio between the effective length and the unbraced length of a member measured between center of gravities of the bracing members. K values are given for several idealized conditions in which joint rotation and translation are realized. See **Effective length**.

Effective moment of inertia: The moment of inertia of the cross section of a member that remains elastic when partial plastification takes place. See **Moment of inertia**.

Effective peak acceleration: A coefficient shown on National Earthquakes Hazards Reduction Program (NEHRP) maps used to determine seismic forces.

Effective width: The transverse distance indicating the amount of slab that acts in conjuction with the supporting member.

Efficiency: The ratio of the useful work performed by a machine or in a process to the total energy expended or heat taken in; the ratio of output to input. Efficiency is usually expressed as a percent.

When thinking about organizational performance it is worthwhile asking: Are we doing this as best possible? That is, are we as efficient as possible? This as opposed to **effectiveness** that answers the question: Is what we are doing necessary?

Efflorescence: Deposits of white salts, usually calcium carbonate, on the surface. Efflorescense is found on many surface types, including brick, block, tile, grout, slate, stone, concrete work, pavers, limestone, marble, granite, etc. Two conditions must be present to create efflorescence: a source of water soluble salts, and water moving through the material to carry the salts to the surface. The water evaporates and leaves the white powder behind.

Egg and dart: A decorative pattern consisting of alternating shapes of a void and arrow, usually used for molding. See **Moldings**.

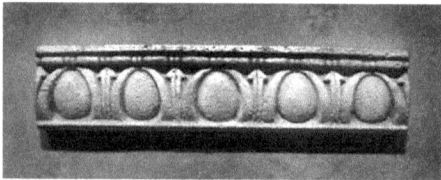

Egg and dart molding.

Eggshell: With regard to paint, gloss that lies between semigloss and flat.

Egress window: A means of exiting the home. Most codes require an egress window in every bedroom and basement. Normally a 4'× 4' window is the minimum size required. Also referred to as **emergency exit window**.

Elastic: When we think of something that is elastic we usually think of something that is: stretchy, elasticized, stretchable, springy, flexible, pliant, pliable, supple, yielding, plastic, resilient. See **Elasticity**.

Other terms that begin with elastic:
- **Elastic analysis:** The analysis of a member that assumes that material deformation disappears on removal of the force that produced it and the material returns to its original state.
- **Elastic design:** See **Allowable stress design** and **Working stress design**.
- **Elastic deformation:** A nonpermanent deformation in which a solid returns to its original size and shape after an external deforming force is removed.
- **Elastic energy:** The energy stored in deformed elastic material (e.g., a watch spring). Elastic energy equals $(k\Delta^2)/2$ where k is the stiffness, and Δ is the associated deflection. Elastic energy is sometimes called **elastic potential energy** because it can be recovered when the object returns to its original shape; see **Potential energy**.
- **Elastic limit:** The maximum stress or force per unit area within a solid material just prior to the onset of permanent deformation. When stresses up to the elastic limit are removed, the material resumes its original size and shape. Stresses beyond the elastic limit cause a material to yield or flow. For such materials the elastic limit marks the end of elastic behavior and the beginning of plastic behavior.

  For most brittle materials, stresses beyond the elastic limit result in fracture with almost no plastic deformation.

  The elastic limit is in principle different from the proportional limit that marks the end of the kind of elastic behavior that can be described by Hooke's law, in which the stress is proportional to the strain (relative deformation) or, said differently, that in which the load is proportional to the displacement. The elastic limit nearly coincides with the proportional limit for some elastic materials, so that at times the two are not distinguished; whereas for other materials a region of nonproportional elasticity exists between the two. The proportional limit is the end point of what is called linearly elastic behavior.

- **Elastic rebound:** A theory of earthquakes that envisages gradual deformation of the fault zone without fault slippage until friction is overcome at which time the fault suddenly slips to produce the earthquake; the release of strain energy by the abrupt movement of a fault with a resultant earthquake.

Elasticity: The property of a body or substance that enables it to resume its original shape or size when a distorting force is removed.

Elastomer: Natural or synthetic material that at room temperature can be stretched under low stress and, upon immediate release of the stress or force, will return quickly to its approximate original dimensions.

Elastomeric: A rubber like material that has a predictable size and shape within its stress level.

Elastomeric coating: A coating system that when fully cured is capable of being stretched at least twice its original length (100% elongation) and recovering to its original dimensions.

Elastomeric roof system: Stretchable single membrane roof system made from either plastic Polyvinyl chloride (PVC) or rubber ethylene propylene diene monomer (EPDM).

Elastoplastic: The total range of stress (deformation), including expansion beyond elastic limit into the plastic range. In the plastic range deformation is permanent.

Elbow (ell): A plumbing or electrical fitting that lets you change directions in runs of pipe or conduit. See **Ell.**

Electric/Electrical: Something that is operated by electricity: electric-powered, electrically operated, battery-operated.

Other terms that begin with electric and electrical:
- **Electric lateral:** The trench or area in the yard where the electric service line (from a transformer or pedestal) is located, or the work of installing the electric service to a home.
- **Electric operator:** An electrically operated device that will open and close windows that otherwise use a crank handle. This is used in lieu of a roto gear crank or pole crank.
- **Electric resistance coils:** Metal wires that heat up when electric current passes through them and are used in baseboard heaters and electric water heaters.
- **Electrical circuit:** A group of outlets or other electrical devices that connect to a fuse box or breaker panel through a common lead.
- **Electrical device:** A unit of an electrical system that is intended to carry but not utilize electric energy.
- **Electrical engineer:** An engineer who is concerned with electrical devices and systems and with the use of electrical energy.
- **Electrical entrance package:** The entry point of the electrical power including: (1) the **strike** or location where the overhead or underground electrical lines connect to the house; (2) the **meter** that measures how much power is used; and (3) the **panel** or **circuit breaker box** (or **fuse box**) where the power can be shut off and where overload devices such as fuses or circuit breakers and located.
- **Electrical rough:** Work performed by the electrical contractor in conjunction with or after the plumber and heating contractor are completed with their work. Normally all electrical wires, and outlet, switch, and fixture boxes are installed before insulation is put in.
- **Electrical trim:** Work performed by the electrical contractor when the house is nearing completion. The electrician installs

all plugs, switches, light fixtures, smoke detectors, appliance pig tails, bath ventilation fans, wires the furnace, and "makes up" the electric house panel. The electrician does all work necessary to get the home ready for and to pass the municipal electrical final inspection.

- **Electrically supervised:** As applied to a control circuit, in the event of an interruption of the current supply or in the event of a break in the circuit, a specific signal will be given.

Electricity: The flow of an atom's electrons through a conductor.

Electrode: A conductor through which an electric current enters or leaves an electrolyte, an electric arc, or an electronic valve or tube. It is the device through which current is conducted to the arc or base metal during the process of welding.

Electro-hydraulic servo valve: Valves that are electro-hydraulicly operated by transforming a changing analogue or digital input signal into a smooth set of movements in a hydraulic cylinder.

Servo valves are directional type valves that receive a variable or controlled electrical signal and that control or meter hydraulic flow. Servo valves can provide precise control of position, velocity, pressure and force with good post movement damping characteristics.

Electrolytic or electrochemical corrosion: Oxidation of materials due to chemical reaction between (usually) a metal and oxygen in the environment. This is the most common cause of deterioration of unprotected iron and steel (rust).

Electronic level: This tool makes use of blinking lights and a beeping signal to indicate level, plumb, pre-selected and unknown angles.

Electrostatic filter: A filter that magnetically attracts dust to its surface.

Elemental mercury: Pure mercury (rather than a mercury-containing compound), the vapor of which is commonly used in fluorescent and other lamp types.

Elevation: One of the five basic views found on a plan. Elevation is an eye-level view of a surface on the building.

Elevation sheet: The page on the blue prints that depicts the house or room as if a vertical plane were passed through the structure.

Elevator: A hoisting and lowering mechanism equipped with a car or platform that moves in guides in a substantially vertical direction, and that serves two or more floors of a building.

Elevatoring: Applying available elevator technology to meet the traffic demands in a building. Elevatoring involving calculations and simulations of the elevator system's performance on the basis of estimations of population and patterns.

Ell: An extension of a building at a right angle to its length. See **Elbow (ell)**.

Elongated toilets: Toilets that have longer bowls from front to back. To some people they are sleeker than their round front counterparts.

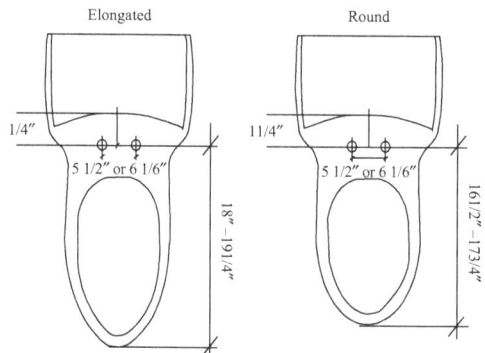

Elongated and round toilets.

Elongation: The increase in length of a specimen measured after fracture as a percentage of the original length. Elongation is a measure of ductility. See also **Reduction of area**.

Embankment dam: A dam composed of a mound of earth and rock.

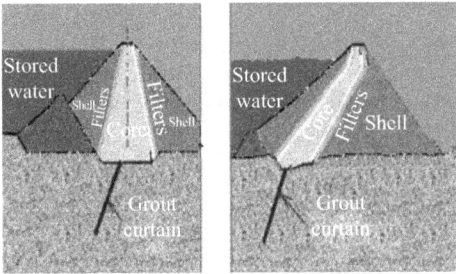

Very simplified versions of embankment dams.

Embedment: A steel member such as a plate, bolt, stud, or bar cast into a concrete structure that is used to transmit applied loads to the concrete.

With regard to roofing, the process of installing or pressing-in a reinforcement felt, fabric, mat, or panel uniformly into bitumen or adhesive; the process of pressing granules into coating during the manufacture of factory-prepared roofing; and, the process whereby ply sheet, aggregate, or other roofing components settle into hot- or cold-applied bitumen via the force of gravity.

Emergency board-up: The process of securing a structure from weather and unwanted entry. Damaged doors and windows or other easily accessible openings are typically covered with plywood and opening in the roof are typically covered with plastic. This work is usually done under contract with a city or county agency.

Emergency generator: An engine using internal combustion to drive a generator that supplies (usually in limited amounts) electrical power to a building during period of outage.

Emergency interlock release switch: With regard to elevators, a device to make inoperable, in the case of emergency, door or gate electric contacts or door interlocks.

Emergency power: Dedicated electrical circuits within a building that are connected to the main electrical supply as well as the emergency generator.

Emergency switch: A switch that is separate from any piece of mechanical or electrical equipment and that is used to break an electrical circuit to that piece of equipment. The switch must be located in plain sight at or near the piece of equipment.

Emissivity: The ratio of the radiation emitted by a surface to the radiation emitted by a black body at the same temperature.

Employee, construction, per OSHA: Every laborer or mechanic, regardless of the contractual relationship which may be alleged to exist between the laborer and mechanic and the contractor or subcontractor who engaged him. "Laborer" generally means one who performs manual labor or who labors at an occupation requiring physical strength; "mechanic" generally means a worker skilled with tools.

Emulsion: A mixture of solids suspended in a liquid. As examples: **Latex paint**s an emulsion that has resins suspended in water. The resins are brought together with the aid of an emulsifier. Asphalt emulsion is a combination of three basic ingredients, asphalt, water, and a small amount of an emulsifying agent. The emulsifier keeps the asphalt droplets in a stable suspension and controls the breaking time. The result is a liquid product that can be used in cold processes for road construction and maintenance.

Enamel: A broad classification of paints that dry to a hard, usually glossy finish. Most equipment-coating enamels require baking. Enamels for walls do not.

Encapsulant: A substance that forms a barrier between lead-based paint and the environment using a liquid-applied coating or an adhesively bonded covering material.

Encapsulation is the application of an encapsulant.

Enclosed: As pertains to electrical equipment, a case, housing, fence or wall(s) that prevents people from accidentally contacting energized parts.

End: Extremity, furthermost part, limit; margin, edge, border, boundary, periphery; point, tip, tail end, tag end, terminus.

Other terms that begin with end:
- **End bay:** The bay that is located from the end of a building to the first interior main frame.
- **End body:** A faucet valve body configuration in which both the inlet and outlet are at the end of the valve body.

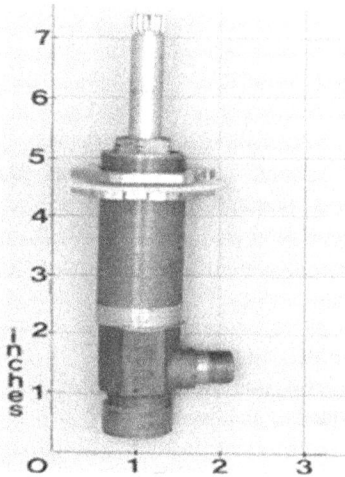

End body assembly.

- **End diagonal or web:** The first web member on either end of a joist or joist girder which begins at the top chord at the seat and ends at the first bottom chord panel point.
- **End distance:** The horizontal distance from the first top chord panel point at the end of a joist to the first bottom chord panel point.
- **End drain:** The side of a tub where the drain is found. When facing a tub, if the drain is on the right, this is a Right-Hand tub (RH). If on the left, it is a Left Hand tub (LH). On whirlpools, the pump is located opposite the drain.
- **End lap:** The lap at the end of a sheet of deck that bears over the primary support (joist or beam).

  With regard to roofing, the distance of overlap where one ply, panel, or piece extends beyond the end of the immediately adjacent underlying ply, panel, or piece.
- **End moment:** A moment that is generated at one end or both ends of a joist, joist girder, or beam due to continuous frame action. An end moment can be caused wind, live load, or dead load moment.
- **End panel:** The distance from the panel point at the joist seat to the first top chord panel point toward the interior.
- **End wall:** An exterior wall that is perpendicular to the ridge of the building.
- **End-matched board:** End-matched boards have tongues and grooves along their sides and the ends.

Energize: To direct electric current through a conductor. Power lines and wires can be intentionally energized (or de-energized) to carry current to an electrical device. But conductive surfaces which are unintentionally energized, like the metal case of a tool, the metal housing of a circuit box, or a metal object such as an aluminum ladder, present a danger of electrocution.

Something that is electrically connected to a source of voltage is said to be **energized**.

Energy: A property of a body related to its ability to move a force through a distance opposite the force's direction. It is the product of the magnitude of the force times the distance. Energy may take several forms: see **Kinetic energy, Potential energy**, and **Elastic energy**.

Other terms that begin with energy:
- **Energy audit:** An audit that identifies how much energy is used in a building, for what purposes, and identifies opportunities for improving efficiency and reducing costs.
- **Energy dissipation:** The reduction in intensity of earthquake shock waves with time and distance, or by transmission through discontinuous materials with different absorption capabilities.
- **Energy panel (EP):** Formerly called a removable double glazing (RDG), an EP is a piece of glass annealed or tempered, and finished on the edges by a surround. EPs are applied to windows or doors and rest on the glazing stop. EPs offer the homeowner added energy efficiency. Think of it as an interior storm window.

• **ENERGY STAR®️ rating:** A measure of a building's energy performance compared with buildings with similar characteristics, as determined by use of the ENERGY STAR®️ Portfolio Manager. A score of 50 represents average building performance.

Engaged column: A half column that is set against a wall or into a wall.

Engaged columns.

Engineer: A person licensed to practice the profession of engineering under state law.

Engineer of Record: A sole practioner or member of a firm who is in "responsible charge" of a design and who ultimately seals the drawings and specifications with his or her professional seal.

The term **responsible charge** is used by many states to establish who may seal or stamp engineering documents. The definitions offered for the term are usually open to a degree of interpretation since it is difficult to write a rule or regulation for each conceivable situation.

Engineered wood products: A composite wood product using glued fiber, lumber, and/or veneer to meet specific design criteria. Such products include **laminated veneer lumber (LVL), parallel strand lumber (PSL),** and **structural I-beams**. Products

under development include various molded, extruded, and other structural and nonstructural composites. Also referred to as **engineered materials**.

Large-sized engineered wood members can be made from small diameter trees; this is advantageous. But engineered wood products are generally more costly and require more energy to produce than lumber.

Engineered veneers: These are first peeled, normally from Obeche (the wood of an African obeche tree that is used especially for veneering) or Poplar logs. The peeled veneer leaves are dyed to a specified color, and then glued together in a mold to produce a large laminated block. The shape of the mold determines the final grain configuration. The block is then sliced into leaves of veneer with a designed appearance that is highly repeatable.

Engineering brick: A type of brick that is manufactured at extremely high temperatures, thereby forming a dense and strong brick. Engineering bricks offer excellent load bearing capacity, damp-proofing characteristics and chemical resisting properties. The traditional product was blue in color, but other colors are now available.

Engineering bricks.

English bond: Alternate rows of bricks that consist of all headers and all stretchers. Traditionally considered to be the strongest bond, it is often found in engineering works like bridges and retaining-walls.

English bond.

English bow saw: This tool can accurately saw curves and straight. The blade can be rotated 360° to a convenient working position. Historically it has been used to cut chair legs, arm, and arches. Nowadays, the jigsaw or the bandsaw substitute for the bowsaw.

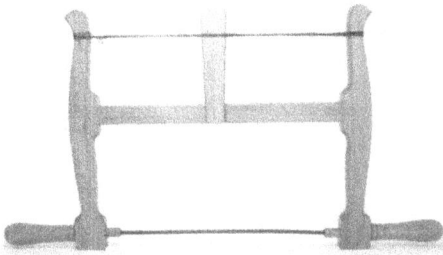

Traditional English bow saw.

English garden wall bond: Most brickwork bonds are constructed so that one side of the wall is "fair-faced" (suitable for viewing as finished work) while the other side, the hidden side, is left rough. Garden walls, however, are seen from both sides, so Garden Wall bond is constructed so that both sides can be built fair-faced.

English garden wall bond.

Entablature: A horizontal, continuous lintel on a classical building supported by columns or a wall, comprising the architrave, frieze, and cornice. See **Cornice**.

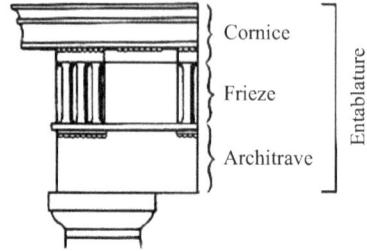

Entablature (Same picture is at cornice).

Entablement: A platform supporting a statue, above the dado and base.

Entablement.

Envelope: A continuous seal for preventing bitumen from leaking down into or off a building. The envelope is constructed by extending the base sheet or other non-porous ply of felt beyond the edge of the field plies. It is then turned back onto the top of the system and adhered. See **Bitumen-stop**.

Also, the physical separator between the interior and the exterior environments of a building. The building's envelope serves as the outer shell to help maintain the indoor environment and facilitate its climate control.

Environment: The surroundings or conditions in which a person, animal, or plant lives or operates. The aggregate of physical, chemical, and biological factors that act upon an organism or an ecological community and ultimately determine its form and survival.

Also, the natural world as a whole or in a particular geographical area. Frequently considered in the ways that human beings affect the world or area.

Environmental: Of, relating to, or associated with the environment. Someone who is concerned with the ecological impact of altering the environment is said to be an environmentalist. Other terms beginning with environmental:

- **Environmental attributes of green power:** The environmental attributes of green power include the emissions reduction benefits that result from the substitution of renewable energy sources for conventional power sources.
- **Environmental conditions:** Features or state of the physical environment and the surroundings, factors, or forces that influence or modify that environment.
- **Environmental hazards:** Any physical or natural condition or event that possesses a risk to humans.
- **Environmental impacts:** The repercussions of an activity or specific land use on the physical/social environment as a consequence of emissions, waste disposal, water, and power usage, etc. (Shouldn't really be **environmental effects**?)
- **Environmental tobacco smoke (ETS), or secondhand smoke:** Airborne particles that are emitted both directly from cigarettes, pipes, and cigars and indirectly, as exhaled by smokers.

Environmentally assisted cracking (EAC): A process that sometimes occurs with the use of high strength steel fasteners. Cracks occur in the fastener at a comparatively low stress level due to interactions with the environment. Hydrogen is suspected of causing EAC in high strength steel fasteners, the hydrogen being produced as a result of chemical reactions (galvanic corrosion in a moist environment) or being present from a plating process that may have been applied to the fastener.

Epicenter: The point of the earth's surface directly above the focus or hypocenter of an earthquake.

Epoxy: A two-part resin that, when mixed, forms a tight cross-linked polymer. Epoxy forms a hard, tough surface that is highly resistant to corrosion.

A concrete finish made by spreading an epoxy adhesive over cured concrete, then placing aggregate over the epoxy which glues the aggregate to the underlying concrete is an **epoxy finish**.

The process of injecting epoxy resin into a crack is referred to as an **epoxy injection technique**. A good epoxy joint is usually stronger than the material it replaces.

Equilibrium: An object is in equilibrium if the resultant of the system of forces acting on it has zero magnitude. See Static equilibrium and Dynamic equilibrium.

Equations of equilibrium: Equations of equilibrium are the equations relating a state of static equilibrium of a member or structure when the resultant of all forces and moments are equal to zero. Three equations must be fulfilled simultaneously: sum of the forces in the X-direction must equal zero, sum of the forces in the Y-direction must equal zero, and the sum of the moments about any point must equal zero for a two dimensional structure.

Equilibrium moisture content (EMC): The moisture content at which wood neither gains nor loses moisture when surrounded by air at a given relative humidity and temperature. This, however, is a dynamic equilibrium and changes with relative humidity and temperature.

Equipment screen: A nonstructural wall or screen constructed around rooftop equipment such as HVAC units, curbs, etc., to hide the look of the equipment and make the structure more aesthetically pleasing.

Equity: With regard to real property, the valuation that one has invested, i.e., the property value less the mortgage loan outstanding.

An **equity lease**s a type of joint venture arrangement in which an owner enters into a contract with a user who agrees to occupy a space and pay rent as a tenant, but at the same time, receives a share of the ownership benefits such as periodic cash flows, interest and cost recovery deductions, and perhaps a share of the sales proceeds.

The return on the portion of an investment financed by equity capital is the **equity yield rate**.

Equivalent, per OSHA: Alternative designs, materials, or methods to protect against a hazard which the employer can demonstrate will provide an equal or greater degree of safety for employees than the methods, materials, or designs specified in the standard.

Equivalent lateral force (ELF): The representation of earthquake forces on a building by a single static force applied at the base of a building; also referred as base shear (V).

Equivalent uniform load: A uniform load (in plf) derived from the maximum reaction (in lbs) or the maximum moment (in inch-lbs) of a member carrying various loads.

Equiviscous temperature (EVT): A measure of viscosity used in the tar industry, equal to the temperature in degrees Celsius at which the viscosity of tar is 50 s as measured in a standard tar efflux viscometer.

EVT is the temperature at which a bitumen attains the proper viscosity for use in built-up roofing. There is usually a 25°F variance permitted above and below the recommended EVT. The EVT is measured in application equipment just prior to application using a standard thermometer or it can be measured just after application using a laser thermometer.

Here are some sample EVT temperatures:

Asphalt – ASTM D 312

Asphalt Type	Mop Application	Mechanical Spreader
Type I Dead Level	350	375
Type II Flat	400	425
Type III Steep	425	450
Type IV	450	475

Temperatures in °F
Temperatures may vary ± 25°F

Erection: The process of installing structural members, e.g., joists, joist girders, beams, bridging, deck, etc., to construct a structure.

A primary shop drawing that illustrates to the field crew how to assemble the shipping pieces is an **erection drawing**. For example, ironworkers match piece marks on the actual shipping pieces to the piece marks noted on the erection drawings.

Floor or roof plans that identify individual marks, components, and accessories furnished by the structural members provider[s] in a detailed manner to permit proper construction of the structure are **erection plans**. See **Framing plan** and **Placing plan**.

Erector: The person or company that actually erects the structure.

Ergonomic design: Design intended to provide optimum comfort and to avoid stress or injury. Generally if a product is physically comfortable, it is well-designed ergonomically.

Erosion: The process by which the materials of Earth's surface are loosened, dissolved, or worn away, and transported by natural agents.

As regards concrete, erosion is the progressive disintegration of concrete by the abrasive or cavitation action of gases, fluids, or solids in motion.

Escalators: A power driven inclined continuous stairway used for raising or lowering passengers.

Escrow: The handling of funds or documents by a third party on behalf of the buyer and/or seller.

Escutcheon: An ornamental plate that fits around a pipe, keyhole, door handle, or light switch to hide the cut out hole.

Also, a shield or emblem bearing a coat of arms.

Espagnolette: Tilt-Turn hardware that houses the gear mechanism for the Tilt-Turn, inswinging casement and hopper handles.

Espagnolette.

Estimate: The amount of labor, materials, and other costs that a contractor anticipates for a project as summarized in the contractor's bid proposal for the project. Estimates of cost-to-complete and final cost are frequently prepared during construction.

A calculation of costs is referred to as an **estimate of construction cost**. Estimates of construction cost vary in their degree of detail, ranging from very detailed analyses of materials, labor and schedule for all items of work down to "ball-park" estimates base on area, volume or similar unit and the estimator's experience.

An estimate of the cost yet to be expended on a work-scope in order to complete it is an **estimated cost-to-complete**. The estimated cost-to-complete is the difference between the cost-to-date and the estimated final cost.

An estimate of the final cost of a work item based on its cost-to-date and the estimated cost to complete it is an **estimated final cost**; it is the sum of the **cost-to-date** and the **estimated cost-to-complete**.

Estimating: The process of calculating the cost of a project. This can be a formal and exact process or a quick and imprecise process.

Etched glass: Hardened glass that has been sandblasted or otherwise engraved to form a pattern or design in the glass.

Ethnographic landscape: Areas that contain a variety of natural and cultural resources that associated people define as heritage resources, including plant and animal communities, geographic features, and structures, each with their own special local names.

Ethylene Propylene Diene Monomer (EPDM): A synthetic rubber material used to make elastomeric roof membranes.

European style cabinet door hinge: A cabinet hinge that is mortised into the back of the door on one side and attached to the cabinet box on the other side. Once installed, it can be adjusted in a variety of directions. The European style cabinet door hinge was designed for use on the frameless style cabinets that were developed in Europe after World War II, but is also found on some higher quality framed style cabinet. Also called a **six way adjustable hinge** and a **recessed hinge**. See **Standard cabinet door hinge**.

European style cabinet door hinge.

European wallpaper roll: A roll of wallpaper that contains about 28 square feet of material.

Eutrophication: Artificially high levels of nitrogen and phosphorus in the water promote excessive growth of microscopic or macroscopic plants, in a process called eutrophication. When these plants accumulate, die, and decay, they cause low oxygen content in the water, thereby effecting both flora and fauna in the local ecosystem.

Evaluation of historic sites: The process by which the significance of a property is judged and eligibility for National Register of Historic Places (or other designation) is determined.

Evaporate: To turn from liquid into vapor.

Also, to lose or cause to lose moisture or solvent as vapor.

Evaporative cooler: System which cools by drawing air through moist filters that transfer moisture into the air. Evaporative coolers are effective only in regions with relatively low humidity.

Evaporator coil: The part of a cooling system that absorbs heat from air in your home. See **condensing unit**.

Evapotranspiration: Water lost through transpiration through plants plus water evaporated from the soil.

The amount of water lost from a vegetated surface in units of water depth, expressed in millimeters per unit of time is the **evapotranspiration rate (ET)**.

Excavation: Any man-made cut, cavity, trench, or depression in an earth surface formed by earth removal.

Excavator: A contractor or piece of equipment that moves the soil out of the area where footings, foundations, or utility lines will be placed and backfills the soil around these parts once they are in place.

Excavators are heavy construction equipment that consist of a boom, stick, bucket, and cab on a rotating platform (known as the house). The house sits atop an undercarriage with tracks or wheels. A cable-operated excavator uses winches and steel ropes to accomplish the movements.

Exceedance probability: The probability that a specified level of ground motion or social or economic impact in an area will be exceeded in a specified time.

Excess air: The quantity of air supplied that exceeds the minimum necessary to support the combustion chemistry. To ensure complete combustion of the fuel used, combustion chambers are usually supplied with excess air to increase the amount of oxygen and thus the probability of combustion of all fuel.

Combustion efficiency increases with increased excess air until the heat loss in the excess air is larger than the heat provided by more efficient combustion. When fuel and oxygen in the air are in perfectly balance the combustion is said to be stoichiometric (stoikē-ō-metrik).

Typical excess air to achieve highest efficiency for different fuels are:
- 5–10% for natural gas
- 5–20% for fuel oil
- 15–60% for coal

Exhaust air: The air removed from a space and discharged outside the building by means of mechanical or natural ventilation systems.

Exhaust vent: A device used to vent air from the roof cavity with vents that are installed on or near the higher portions of the roof such as the ridge.

Exhaust vent.

Existing building commissioning or **retro-commissioning:** Commissioning involves developing a building operating plan that identifies current building operating requirements and needs, conducting tests to determine whether the building and fundamental systems are performing optimally in accordance with the plan, and making any necessary repairs or changes.

Exit: A means of egress from the interior of a building to an open exterior space. Exits may include, either singly or in combination, vertical exits, exit passageways, horizontal exits, interior stairs, and fire towers or fire escapes. If you are a designer be sure to learn local exiting requirements.

Expanded foam: A foam that contains small beads with air voids around the beads. When used as insulation, expanded foam should only be installed above grade.

Expanded metal: An intermediate raw material that allows light and air to flow freely through it and that has more rigidity that the sheet from which it was produced.

Expanded metal is formed in an expanded metal press and produced from any malleable metal sheet product. The plate, sheet, or coil is simultaneously slit and stretched into diamond-shaped openings.

Expansion: Enlargement, increase in size, swelling, dilation; lengthening, elongation, stretching, thickening. As a contractor you should always be aware of expansion and contraction.

Other terms that begin with expansion:
- **Expansion cleat:** A cleat designed to handle thermal movement of the metal roof panels.
- **Expansion factor:** With regard to chips, bark, sawdust, and shavings, the ratio of volume occupied in one of these forms to the volume of solid wood before conversion. With respect to soils and rock, one cubic of soil or rock in-situ expands and does not translate into one cubic of fill in the bed of a truck or placed and compacted on the site; this is extremely important when estimating earthwork.

Thermal expansion is the tendency of matter to change in volume in response to a change in temperature; all materials have this tendency. Materials that contract with increasing temperature are rare; this effect is limited in size, and only occurs within limited temperature ranges. The degree of expansion divided by the change in temperature is called the material's **coefficient of thermal expansion** and generally varies with temperature.

There are several types of coefficients: volumetric, area, and linear.

For example, when measuring or laying out distances using a surveyors' tape, there is always a change in temperature, especially when the taping operation requires. Nowadays lasers are mostly used, but if measurement is with a tape, to avoid an introduced error due to temperature, tapes should be standardized and a standard temperature for the tape determined.

The correction of the tape length due to change in temperature is given by

$$C_f = C \times L(T - T_s)$$

C_f is the correction to be applied to the tape due to temperature; T is the observed temperature or average observed temperature at the time of measurement; T_s is the standard temperature, the temperature at which the tape was standardized; C is the coefficient of thermal expansion of the tape; L is the length of the tape or length of the line measured.

The correction C_f is added to L to obtain the corrected distance:

$$D = L + C_f.$$

Usually, for common tape measurements, the tape used is a steel tape with coefficient of thermal expansion C equal to 0.0000116 units per unit length per degree Celsius change. This means that the tape changes length by 1.16 mm per 10 m tape per 10°C change from the standard temperature of the tape. This may seem minimal, but over the course of a long distance it adds up.

• **Expansion joint:** A constructed joint in an assembly allowing for movement of that assembly in a controlled manner. See **Control joint**.

Also, material that is installed to permit movement between structural members.

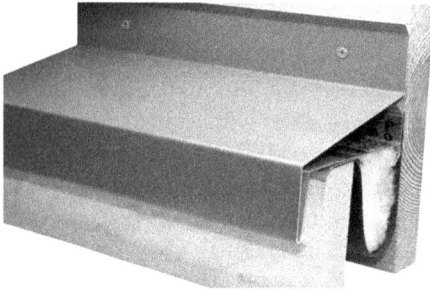

Roof to wall expansion joint

• **Expansion slots:** Laser cut slots on a saw blade. The slots are designed to prevent distortion of the rim of larger diameter blades. They provide relief areas for expansion of materials from heat and prevent blade distortion.

Expansive bit: Taking the place of many larger bits, expansive bits are adjusted by moving the cutting blade in or out by a geared dial or by a lockscrew to vary the size of the hole.

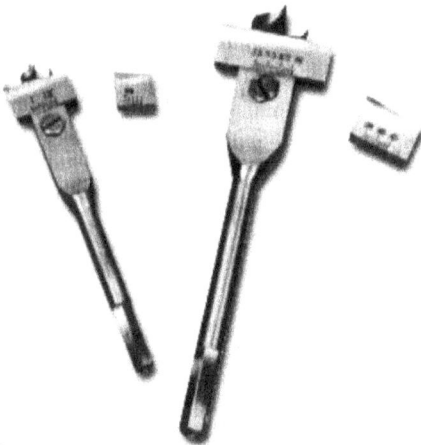

Expansive bits.

Expansive soils: Earth that swells and contracts depending on the amount of water that is present. ("Betonite" is an expansive soil).

Exposed/exposure: Live electrical parts that are capable of being inadvertently touched or approached nearer than a safe distance by a person are considered to be exposed. It is applied to parts that are not suitably guarded, isolated, or insulated.

Wiring that is on or attached to the surface or behind panels designed to allow access are also considered to be exposed.

Concrete that had its top layer of cement/sand washed off, thereby exposing, usually, gravel is said to have an **exposed aggregate finish**.

The **exposed-nail method** is a method of asphalt roll roofing application in which all nails are driven into the adhered, overlapping course of roofing. Nails are exposed to the weather.

The condition where concrete reinforcement has been visibly exposed by the loss of its protective cover is called **exposure**.

The potential casualties or economic loss to all or to specific subsets of a population or structures from one or more earthquakes in an area is also called as **exposure**.

The traverse dimension of a roofing element or component not overlapped by an adjacent element or component in a roof covering is also **exposure**.

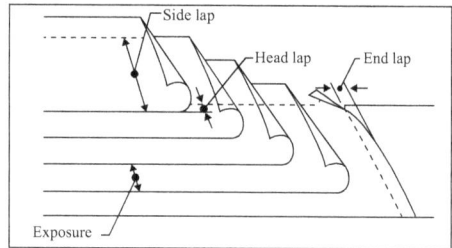

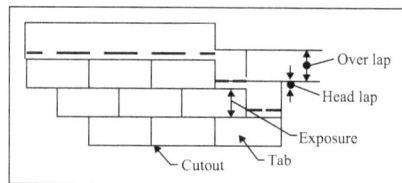

The personal air monitoring of an employee's breathing zone to determine the amount of contaminant, e.g., lead, to which he/she is exposed, is **exposure monitoring**.

And, a type of plywood that has been approved by the American Plywood Association for exterior use is **exposure 1 grade plywood**.

Extended end: The extended part of a joist top chord with also the seat angles extended from the end of the jost extension back into the joist maintaining the standard 2 1/2 inch end bearing depth over the entire length of the extension.

Extended services contract – CM: A form of construction management contract that adds other services such as design, construction, and contracting to the CM firm's contract.

Extender: Ingredients added to paint to increase coverage, reduce cost, achieve durability, alter appearance, control rheology and influence other desirable properties. Less expensive than prime hiding pigments such as titanium dioxide. Examples are: barium sulfate, calcium carbonate, clay, gypsum, silica, talc. An extender may improve coating performance.

Exterior insulation and finishing system (EIFS): A type of building exterior wall cladding system that provides exterior walls with an insulated finished surface and waterproofing in an integrated composite material system.

Extractive: Substances in wood that are not an integral part of the cellular structure and that can be removed by solution in hot or cold water, ether, benzene, or other solvents that do not react chemically with wood components.

Extension: An attachment to a drill that extends the overall length of the bit to facilitate long reaches.

Extension jamb: A board or trim component that extends from the interior of the window frame to the interior wall. It is used to increase the depth of the jambs of a window to fit a wall of any given thickness.

External blower: A blower unit that is mounted outside of the home on the wall or roof.

External force: A surface force or body force acting on an object. External forces are sometimes called applied forces.

Extrados: The upper or outer curve of an arch. Often contrasted with intrados.

Extras: With regard to contracts, additional work requested of a contractor, not included in the original plan, that will be billed separately.

Extruded foam: Foam that is smooth, with no beads or voids. Because extruded foam will not absorb water, it can be installed as insulation above or below grade.

Extrusion: The process, in which a heated material is forced through a die, used to produce aluminum, vinyl (PVC) and other profiles or components used in the production of windows and doors. This term also is used to refer to the profiles or lineals manufactured by this process and used to make window and door components.

Extrusive rock: The cooling of magma on the earth's surface creates an igneous rock.

Eye: The hole in the center of a circular saw blade so it can be fitted on the arbor.

Eyebrow: A dormer, usually of small size, whose roof line over the upright face is typically an arched curve, turning into a reverse curve to meet the horizontal at either end. Also, a small shed roof projecting from the gable end of the larger, main roof area.

Eyebrow dormer.

F rating: A rating of effectiveness of firestop material that measures its resistance to flames and water.

Fabricated frame scaffold: A scaffold consisting of platforms supported on fabricated end frames with integral posts, horizontal bearers, and intermediate members.

Fabricated frame scaffold.

Fabrication: The act of changing steel from the mill or warehouse into the exact configuration needed for assembly into a shipping piece or directly into a structural frame. Fabrication includes material handling, template making, cutting, bending, punching, welding, and grinding.

Façade: The principal front portion of a building; the façade looks onto a street or open space.

Face-nailing: Driving a nail perpendicularly through the width side board. See **Toe nailing**.

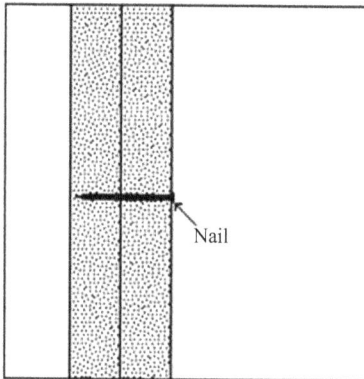

Face-nail.

Face width: With regard to doors, the total width of the stile, rail, or panel minus the width of the molding patterns. The most common way of showing dimensions on a stile and rail door elevation.

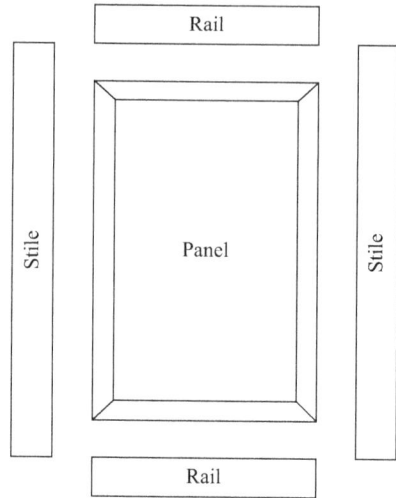

Stile and rail door.

Faceplate: The small, rectangular-shaped piece of metal that surrounds the door latch on the edge of the door.

Faced concrete: To finish the front and all vertical sides of a concrete porch, step(s), or patio. Normally the "face" is broom finished.

Faces: The vertical or inclined earth surfaces formed as a result of excavation work.

Facility alterations and addition: Building work that is done on an existing building. Facility alterations refer to changes made to a building that do not interfere with the original design character of the building, such as replacement flooring, painting, etc. Facility additions are structures added to the original building structure that are subordinate in scale to the main building and typically located to the side or rear of the original structure. An example of an addition would be an extra floor added to the building.

Facility manager: Belongs to "a profession that encompasses multiple disciplines to ensure functionality of the built environment by integrating people, place, process, and technology" according to the International Facility Management Association (IFMA).

Facing brick: The brick that is used and exposed on the outside of a wall.

Factory built fireplace: A pre-built firebox assembly that includes a heat exchanger, air movement equipment, and the flue assembly. Also called a **zero clearance fireplace**.

Factor of safety: The ratio of the ultimate load for a member divided by the allowable load for a member and must always be greater than unity.

Factored load: The product of the nominal load and a load factor.

Factory lumber: Lumber intended to be re-manufactured after it leaves the sawmill, e.g., for making sashes and door.

Factory seam: A splice/seam made in the roofing material by the manufacturer.

Fading: Losing brightness or brilliance.

Failure: Load refusal, breakage, or separation of component parts. Load refusal is the point where the ultimate strength is exceeded.

The manner in which a structure fails (column buckling, overturning of structure, etc.) is referred to as **failure mode**.

Fairtrade: A product certification system overseen by FLO-CERT International. See **FLO-CERT GmbH.**

Fallback: A reduction in the softening point temperature of asphalt that occurs when asphalt is overheated for pro-longed periods of time. See **Softening point drift.**

False tread and riser: A non-structural, decorative tread and riser assembly that is often placed over the structural tread and riser when carpet runs down the center of a stair. False tread and risers are typically stain-grade wood that gives the appearance of higher grade wood treads and risers at a lower cost. See **Balustrade.**

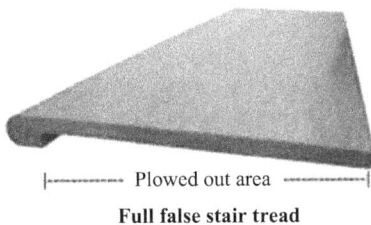

|-------------- Plowed out area --------------|
Full false stair tread

Falsework: A temporary structure used to support a permanent structure while it is not self-supporting.

Fan light: A semicircular window with ribbed bars, normally found over a door or another window. Also spelled **fanlight.**

A fan light.

Also, a ceiling fan with a light.

A ceiling fan light.

Fan room: Any room or space that is dedicated to air handling equipment that contains blowers and or fans.

Fascia: Horizontal boards attached to rafter/truss ends at the eaves and along gables. Roof drain gutters are attached to the fascia.

Fast saw: A circular saw that wobbles, weaves, or snakes because the rim is too long for the speed at which it runs.

Fast track construction: A method of construction management that involves a continuous design-construction operation. This usually involves a prime or main contractor starting the construction work before the plans and specifications are complete. Fast track construction is fraught with danger, as it is not uncommon for work to have to be undone.

Fasteners: Any of a wide variety of mechanical securement devices and assemblies, including nails, screws, cleats, clips, and bolts, which may be used to secure various components of a roof assembly.

Fatigue: The continuous stressing of a material usually in bending leading to the eventual failure of that material.

The tendency of a material to fracture due to progressive brittle cracking under repeated alternating of cyclic stresses of an intensity below the normal strength is called **fatigue failure** and **fatigue cracking**. Initially these cracks usually propagate slowly depending on the intensity and frequency of the stress cycles; the number of cycles needed to cause fatigue failure decreases as the stress is increased. If detected in time, remedial action can be taken to prevent failure. If the cracks are allowed to propagate unrestricted they frequently lead to brittle fracture.

Mechanical properties such as cyclic counting, endurance limit, strain-frequency curve, hardness, and stress-frequency curve that characterizes resistance of the metals to fatigue crack initiation and propagation under low amplitude high frequency cyclic stresses or strains are referred to as **fatigue properties**.

Faucet: A faucet is an entire fixture that is composed of a spout and handle(s).

Fault: A fracture in the earth along which the two sides have been displaced relative to each other.

Other terms that are associated with earthquake faults:
• **Fault, active:** A fault along which displacement has occurred in recent geologic time or along which earthquake foci are

located. Active faults are assumed to be capable of producing earthquakes.
• **Fault, capable:** An active fault or fault zone that has the potential to cause surface displacement during the lifetime of a project under consideration.
• **Fault, dip-slip:** A fault with the major component of relative displacement along the direction of dip of the fault.
• **Fault, left-lateral:** A strike-slip fault on which displacement of the block opposite the observer is to the left.
• **Fault, normal:** A sloping faulting on which the block above the fault has moved downward relative to the block below.
• **Fault, oblique-slip:** A fault with both strike-slip and dip-slip movement.
• **Fault plane:** A plane that approximates the rupture surface of a fault.
• **Fault-plane solution:** A way of showing the fault and the direction of slip on it from an earthquake, using circles with two intersecting curves that look like beach balls. Also called a **focal-mechanism solution**.
• **Fault, right-lateral:** A strike-slip fault on which displacement of the block opposite the observer is to the right.
• **Fault trace:** The intersection of a fault with the land surface. Commonly plotted on geologic maps to represent the location of the fault.
• **Fault scarp:** Steplike linear landform caused by the geologically recent offset of the land surface by a fault.
• **Fault segment:** A discrete section of a fault, separated by recognizable boundaries that tends to rupture independently.
• **Fault, stick-slip:** The fast movement that occurs between two sides of a fault when the two sides become unstuck. The rock becomes distorted (bent), but holds its position until the rock snaps back (elastic rebound) into an unstrained position. Stick-slip displacement on a fault radiates energy in the form of seismic waves that create an earthquake.

- **(Fault) strain release:** Movement along an earthquake fault plane; can be gradual or abrupt.
- **(Fault) stress drop:** The difference between the stress across a fault before and after an earthquake.
- **Fault, strike-slip:** A fault on which the movement is parallel to the fault's strike.
- **Fault, thrust or overthrust:** A dip-slip fault in which the upper block moves over the lower block. The dip of some thrust faults is low and the displacement may be tens of miles.
- **Fault, transform:** The plane along which the break or shear of a fault occurs.
- **Fault zones:** The area surrounding a major fault, consisting of numerous interlacing small faults.

Also, a fault is an insulation failure that exposes electrified conductors, thereby causing current to leak and possibly resulting in electric shock.

Faulting: The movement that produces relative displacement of adjacent rock masses along a fracture. Displacement of the ground or sea floor surface by a fault movement is called **surface faulting**. See **Fault**.

Feasibility analysis: The process of evaluating a proposed project to determine if that project will satisfy the objectives set forth by the agents involved (including owners, investors, developers, and lessees). Also referred to as a **feasibility study**.

Feasibility phase: The conceptual phase of a project preceding the design phase. The feasibility phase is used to determine whether a project should be constructed or not.

Feather-edge board: A board that is thicker on one side than the other. Feather-edge boards are used for fencing, where they are fixed vertically and overlapping. Feather-edge boards are sometimes found in tiled roofs, fixed horizontally, with the thicker edge at the top to provide a hanging point for tiles.

Feathering strips: Tapered wood filler strips placed along the butt ends of old wood shingles to create a relatively smooth surface when reroofing over existing wood shingle roofs. Referred to in some regions of the country as "horse feathers" or leveling strips.

Feed rate: The speed at which a drilling tool or saw is fed or pushed into a work piece.

Feeder: Connects the meter base to the breaker panel(s) or fuse box(es.) Usually contains four cables that are twisted together.

Feedwater: Any water that is fed into the water tanks for use in the boiler. **Feedwater control valves** are actuated valves that throttleflow into the tanks; **feedwater control valves** may be operated manually or electronically.

Fahrenheit (°F): The temperature scale in which water freezes at 32°F and boils at 212°F under normal atmospheric conditions.

$$°F = (°C \times 1.8) + 32$$

Feathering strips: Tapered wood filler strips placed along the butts of old wood shingles to create a level surface when reroofing over existing wood shingle roofs. Also referred to as **horsefeathers**.

Feature: With regard to historic property, a prominent or distinctive aspect, quality, or characteristic of the property.

Also, simply a historic property.

A property's expression of the aesthetic or historic sense of a particular period of time is referred to as its **historic feeling**.

Felt: A flexible sheet manufactured by the interlocking of fibers through a combination of mechanical work, moisture, and heat. Roofing felts may be manufactured principally from wood pulp and vegetable fibers (organic felts), asbestos fibers (asbestos felts), glass fibers (fiberglass felts or ply sheet), or polyester fibers.

Felt machine/Felt layer: A mechanical device that is used for applying bitumen and roofing felt or ply sheet simultaneously.

Female threads: Faucet threads that are in place on the inside of a fitting.

Fenestratio: Originally, an architectural term for the arrangement of windows, doors, and other glazed areas in a wall. This has evolved to become a standard industry term for windows, doors, skylights, and other glazed building openings. From the Latin word, "fenestra," meaning window.

Ferrule: Metal tubes used to keep roof gutters open. Long nails (ferrule spikes) are driven through these tubes and hold the gutters in place along the fascia of the home.

Ferrules.

Also, a short ring for reinforcing or decreasing the interior diameter of the end of a tube.

A ferrule.

Also, a ring or cap, usually of metal, put around the end of a post, cane, or the like, to prevent splitting.

Also, a short metal sleeve for strengthening a tool handle at the end holding the tool.

Tool handle with ferrule.

Also, a bushing or adapter holding the end of a tube and inserted into a hole in a plate in order to make a tight fit, used in boilers, condensers, etc.

Ferrule.

Also, a short plumbing fitting, covered at its outer end and caulked or otherwise fixed to a branch from a pipe so that it can be removed to give access to the interior of the pipe.

Ferrule fitting.

Also, cylindrically shaped copper tubes crimped to the ends of stranded wire to create a secure, reliable connection. Similar ferrules are used with fiber optic fibers.

Ferrules.

Festoon lighting: A string of outdoor lights that is suspended between two points.

Fiber and cement shingle: A shingle made from a combination of wood fiber and Portland cement. Fiber and cement shingles can be made to resemble slate, tile, or wood.

Fiber glass mat: An asphalt roofing base material manufactured from glass fibers.

Fiber saturation point (fsp): A term used in wood mechanics and especially wood drying, to denote the moisture content at which moisture is saturated within the cell walls of wood and the cell cavities are free of water. Further drying of the wood results in strengthening of the wood fibers, and is usually accompanied by shrinkage. Wood is normally dried to a point where it is in equilibrium with the atmospheric moisture content or relative humidity. Because this varies so does the equilibrium moisture content. This averages around 20% moisture content. Below FSP water is held in wood as bound water within the cell cavities or lumen. (The central cavity of a tubular or other hollow structure in an organism or cell.)

Fiberboard: A generic term applied to sheet materials of widely varying densities. Fiberboard is made of manufactured or refined or partly refined wood or other vegetable fibers. Bonding agents and other materials may be added to increase strength, resistance to moisture, fire, and decay, and to improve some other property.

Fiberglass: Glass filaments that are formed by pulling or spinning molten glass into random lengths. As an example of how fiberglass is used, fiberglass fibers are condensed into strong, resilient mats for use in roofing materials.

Fibercsope: An inspection device with an eye piece at one end of fiber optics enabling viewing of otherwise non-visible features. Normally equipped with a lamp.

Fibrous admixture: Short lengths of fiber added to concrete to increase it's tensile strength.

Fick's Law: In chemistry and physics, an observed law stating that the rate at which one substance diffuses through another is directly proportional to the concentration gradient of the diffusing substance. Fick's law is a function of time, the chloride concentration at the surface, and a diffusion constant.

Fiddleback: A unique figure on the face of a wood, giving it a washboard effect.

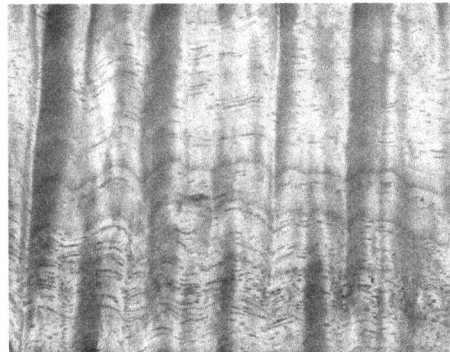

Fiddleback on wood.

Field: Other terms that begin with field:
• **Field measure:** To take measurements (cabinets, countertops, stairs, shower,

doors, etc.) on-site instead of using the blueprints.

- **Field of the roof:** The central or main portion of a roof, excluding the perimeter and flashing.
- **Field order (FO):** A written order issued to a contractor by the owner, or owner's representative effecting a minor change or clarification with instructions to perform work not included in the contract for construction. The work will eventually become a change order. A field order is an expedient process used in an emergency or need situation that in many cases does not involve an adjustment to the contract sum or an extension of the contract time. See also **Construction change directive**.
- **Field photography:** Photography intended for producing documentation.
- **Field records:** Notes of measurements taken, field photographs, and other recorded information intended for producing documentation.
- **Field schedule:** A graphic, tabular or narrative representation or depiction of the construction portion of the project. The schedule shows field activities and durations in sequential order. Field personnel develop their relatively short interval field schedule on month-to-month, week-to-week, or day-to-day basis from the project's milestone schedule.
- **Field seam:** A splice or seam made in the field (not factory) where overlapping sheets are joined together using an adhesive, splicing tape, or heat- or solvent-welding.
- **Field total:** The amount of material, including waste that will be required in the field in order to complete the job.
- **Field weld:** The term used for the welding of structural members on the jobsite and not in a fabricator's shop.
- **Field work order (FO):** A written request/directive to a subcontractor or vendor, usually from the general or main contractor, for services or materials.

Fill: A repair to an open defect usually made with fast drying plastic putty; often referred to as **putty repairs**. The repairs should be made with non-shrinking putty of a color matching the surrounding area of the wood, to be flat and level with the face and panel, and to be sanded after application and drying.

Also, to build up the level of low-lying land or to replace excavated earth with material such as earth or gravel. Also the material that is used for this.

Filler: A relatively inert ingredient added to modify physical characteristics.

Also, a rod, plate, or angle welded between a two angle web member or between a top and bottom chord panel to tie them together. Fillers are usually located at the middle of the member. See **Tie** or **Plug**.

Filler joist floor: An obsolete but commonly found form of floor comprising a concrete slab reinforced with steel I-beams known as rolled steel joists.

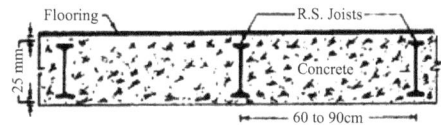

Filler joist floor.

Fillet: A sealant material installed at horizontal and vertical planes to remove 90° angles; a concave junction formed where two surfaces meet.

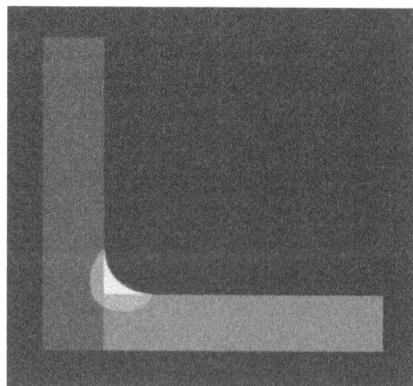

Fillet.

Also, a small piece of decorative wood that fills the space between balusters in a bottom rail. See **Balustrade**.

Film build: The amount of thickness produced in an application. Millimeters (mils) of dry film per mils of applied wet film.

Film thickness: The depth or thickness of the dry coating in millimeters.

Filter fabric: A woven synthetic fabric that is used as a screen to prevent the passage of sand and silt, while allowing for the passage of water.

Filtration rate: The rate at which the water is traveling through the filter, measured as GPM (gallons per minute) per sq. ft.

Fin: A sharp protrusion in a roof deck that can damage roof components.

Final acceptance: The last stage in the commissioning of a building for its intended use. Final acceptance is given by the owner of the project to the general contractor.

Final inspection: A review by the building inspector after the interior and exterior construction is complete, to check for any problems that may endanger the health or safety of the building occupants. In most areas a certificate indicating that the final inspector has been successfully completed is required before the project can be considered completed.

Financial leverage: The use of borrowed funds to acquire an investment.

Financial and management control system: A manual or computerized management control system that is used by the project team to guide the course of a project and record its status and progress.

Financial risk: The possible change in an investment's ability to return principal and income.

Fine aggregate: Aggregates whose size is less than 4.75 mm, e.g., sand that is used in the preparation of concrete and cement mortar.

Fine mineral-surfacing: Water-insoluble, inorganic material, more than 50% of which passes through a No. 35 sieve. Used on the surface of various roofing materials and membranes to prevent sticking.

Finger joint: A manufacturing process of interlocking two shorter pieces of wood end to end to create a longer piece of dimensional lumber or molding. Often used in jambs and casings and are normally painted (instead of stained).

Finish: May refer to the plumbing, electrical, carpentry, or HVAC work that is visible when construction is complete.

Also, the visible coating on walls, trim, deck sheet, etc. Painted, unpainted and galvanized are examples of finishes.

Other terms that begin with finish:

• **Finish coat:** Final coat of any material on a surface.

   With regard to stucco, the third coat of common stucco is the **finish coat**. It contains the texture and may contain the pigment. If the finish coat does not contain pigment, the surface of the stucco must be painted when dry. See **Finish**. For stucco, see **Scratch coat** and **Brown coat**.

• **Finish electrical:** Any electrical part that will be installed after the walls and ceiling are finished.

• **Finish nail:** A small headed nail that can be recessed below the surface of the wood, leaving only a small hole that can be puttied easily.

• **Finish plumbing:** As with finish electrical, finish plumbing is the installation of the plumbing fixtures, faucets, and other visible parts of the plumbing system.

• **Finish strip:** A roof deck accessory made out of gage metal for finishing out runs of deck for small areas of coverage where full sheet coverage is impractical.

• **Finish water:** Water that has completed a purification or treatment process.

Finpipe/Fin tube: Pipe with added materials, e.g., circular steel coils, that increase the surface area of the pipe to aid heat transfer.

Fire: The most common secondary hazard resulting from an earthquake.

Other terms beginning with fire:

- **Fire alarm:** A system, either automatic or manual, that is arrange to give a signal that indicates a fire emergency.
- **Fire area:** A floor area that is enclosed by fire divisions. See **Fire divisions**.
- **Fire box:** The interior of a fireplace system built of heat-resistant materials which contains the fire and radiates heat in the room. Can be made from a variety of materials including special fire brick, prefabricated masonry panels or metal. Also spelled **firebox.**
- **Fire block:** Short horizontal members sometimes nailed between studs, usually about halfway up a wall. See **Fire Stop**.
- **Fire canopy:** A solid horizontal projection that extends beyond the exterior face of a building wall and located over a wall opening so as to retard the spread of fire through openings from one story to another.
- **Fire damper:** A damper inserted in ductwork to prevent the passage of smoke and fire through a fire rated assembly.
- **Fire division:** Any construction that has the required fire-resistance rating and structural stability under fire conditions to provide a fire barrier between adjoining buildings or between adjoining or superimposed fire areas or building sections within the same building.
- **Fire door:** A door that is made of non-combustible material, the purpose of which is to prevent a fire from spreading within a building
  Also, a door that leads to the outside of a building and that can be easily opened from inside; an emergency exit. See Fire rated doorb and Fire resistance.
- **Fire hose closet:** Usually a compartment built into an interior wall and covered by a glass door. It holds a coiled fire hose that is connected to a charged water line.
- **Fire proofing/Fireproofing:** The process of coating (usually) a structural steel member with a fire retardant material to make the member resistant to fire.

Also, the noncombustible material that is applied directly or indirectly around a structural steel element to inhibit the passage of excessive heat caused by a fire.

- **Fire protection plan:** A report that contains a narrative description of the life and fire safety systems and evacuation system for a structure.
- **Fire rated door:** A door that has been constructed in such a manner that when installed in an assembly will pass a fire test under neutral (UL 10B) or positive (UL 10C) pressure criteria and can be rated as resisting fire for 20 min (1/3 hour), 45 min (3/4 hour), 1 hour, or 1-1/2 h. The door must be tested and carry an identifying label from a qualified inspection agency.
- **Fire resistance:** With regard to paint, the ability of a coating to withstand fire or to protect the substrate to which it is applied from fire damage.

  With regard to doors, doors whose core typically incorporate minerals rather than wood fiber as the primary component. They are designed to improve fire resistance and thermal transmission.

  With regard to structural members, the ability to resist fire because of the type of protection they have, e.g., membrane protection and spray on protection. There are hundreds of floor-ceiling and roof-ceiling assemblies with their fire-resistance ratings given in the **Underwriters Laboratory Fire Directory**.

  A **fire-protection rating** is the time in hours or fractions of an hour that an **opening protective** and its assembly will withstand fire exposure as determined by a fire test that has been made in conformity with specified standards of the local responsible agency. (An **opening protective** is an assembly of materials and accessories, including frames and hardware that are installed in an opening in a wall, partition, floor, ceiling or roof to prevent, resist, or retard the passage of flame, smoke, and hot gasses.)

A **fire-resistance rating** is the time in hours or fractions of an hour that materials or their assemblies will withstand fire exposure as determined by a fire test that is in conformity with a specified standard of the local responsible agency.

- **Fire retardant:** A coating that will reduce flame spread, resist ignition when exposed to high temperature and insulate the substrate and delay damage to the substrate.
- **Fire retardant chemical:** A chemical or preparation of chemicals used to reduce the flammability of a material or to retard the spread of flame.
- **Fire retardant treated wood:** The wood that has been pressure impregnated with fire retardant chemicals so as to reduce its combustibility.
- **Fire safety plan:** A description of the fire drill and evacuation procedures for a structure that is required to be submitted to the local fire department in accordance with local rules.
- **Fire section:** A sprinklered area within a building that is separated from other areas by noncombustible construction having at least a 2-h fire resistance rating.
- **Fire separation:** Any construction that has the fire-resistance rating to provide a fire barrier between adjoining rooms or spaces within a building, building section, or fire area.
- **Fire stand pipe:** A main water line that feeds the water sprinkling system or fire hose connections. This is separate from the lines feeding restrooms or other fixtures.
- **Fire stop:** A material used to block the passage of smoke and fire in and around a penetration through a fire rated assembly.

  Also, a solid, tight closure of a concealed space, placed to prevent the spread of fire and smoke through such a space. In a frame wall, this will usually consist of 2 by 4 cross blocking between studs. Work performed to slow the spread of fire and smoke in the walls and ceiling (behind the drywall) includes stuffing wire holes in the top and bottom plates with insulation, and installing blocks of wood between the wall studs at the drop soffit line. This is integral to passing a rough frame inspection. See **Fire block**.

- **Fire suppression piping system:** Any system, including any and all related equipment and material, with the purpose of controlling, containing, suppressing, or extinguishing fire.
- **Fire surround:** A shelf and side elements framing a fireplace. Also referred to as chimneypiece or mantelpiece. Also spelled **firesurround**.

Fire surround.

- **Fire tape:** The process in which drywall is finished to provide fire protection only and not to provide a smooth finish wall. Fire taped drywall has tape embedded along all joints which are then covered with one additional layer of mud. Fasteners are also covered with mud.
- **Fire wall:** A wall which has been designed to resist the spread of fire. Fire walls in homes are typically required between the garage and living space. Fire walls are usually rated by the hours they are designed to resist the spread of fire.
- **Fire-resistive/Fire rated:** Materials that are not combustible in the temperatures of ordinary fires and will withstand such fires for a rated period. Building codes establish required fire ratings.

- **Fireclay:** Fireclay predates the widespread use of cast iron. It has a lustrous, glossy finish. Fireclay is less susceptible to warping and, therefore, often used for large sinks.

A large, handpainted fireclay sink.

- **Fired brick:** A masonry unit made from clay that is formed and then baked at a high enough temperature to cause a partial melting or glazing on the surface. This glaze provides a seal that protects the brick from moisture. Used in a fireplace and boiler.
- **Fireman telephone jack:** A special phone jack that can be built into an elevator car operating panel or hall station that enables firefighter communication.
- **Fireplace chase flashing pan:** A large sheet of metal that is installed around and perpendicular to the fireplace flue pipe. Its purpose is to confine and limit the spread of fire and smoke to a small area.
- **Firewood processor:** High production equipment used to produce firewood as a business.

Firmwood: Solid wood free of decay and voids. Firmwood is a term used in log scaling that by its very nature is subjective and requires training.

Firring: A piece of timber cut as a wedge and fixed to the top of a joist. Firring is used to give flat roofs a fall for drainage, or to level up uneven floors, etc.

First draw: Water that has been sitting in pipes or plumbing fixtures overnight and is first drawn when taps are opened in the morning. Where lead is present in plumbing, this water would have the highest levels of lead contamination. Many buildings, especially schools, require that drinking fountains be flushed daily prior to the start of the work day.

First fix: The fixing of the wires (**electrical first fix**) and pipes (**plumbing first fix**) in a structure, before closing the walls. A carpentry first fix is the provision of joists, studs and rafters.

First motion: The direction of ground motion as the P-wave arrives at a seismometer. Used to compute fault-plane solutions.

Fish tape: A long strip of spring steel used for fishing cables and for pulling wires through conduit.

Fishmouth: With regard to roofing, a half-cylindrical or half-conical shaped opening or void in a lapped edge or seam, usually caused by wrinkling or shifting of ply sheets during installation; in shingles, a half-conical opening formed at a cut edge. Also referred to as an **edge wrinkle**.

Shingle fishmouth curling.

Fishplate: A wood, plywood or metal piece that is used to fasten the ends of two members together at a butt joint with nails or bolts. Sometimes used at the junction of opposite rafters near the ridgeline. Also called a **gusset plate** and **gang nail plate**.

Fissuring: A method of imparting a set of ragged depressions into the face of acoustical tile or panels during manufacture for appearance and acoustical performance.

Five-quarter: A term used for exterior decking material. Refers to the one and one-quarter or **five-quarter** thickness of the material.

Fixed: As used here, fastened, attached, affixed, secured; joined, connected, coupled, linked; installed, implanted, embeded; stuck, glued, pinned, nailed, screwed, bolted, clamped, clipped.

Other terms beginning with fixed:
- **Fixed collar:** A collar that is firmly attached to the saw arbor. A fixed collar is distinguished from a loose collar that is held to the arbor with a nut.
- **Fixed connection:** In two dimensions, a fixed connection between two members restrains all three degrees of freedom of the connected member with respect to one another. A fixed connection is sometimes referred to as a rigid connection or moment-resisting connection.
- **Fixed fee:** A set contract amount for all labor, materials, equipment, and services. The fixed fee includes the contractor's overhead and profit for all work being performed for a specific scope of work.
- **Fixed limit of construction costs:** A construction cost ceiling agreed to between the owner and architect or engineer for designing a specific project.
- **Fixed lite:** A non-venting or non-operable window.
- **Fixed pane:** A non-operable door usually combined with operable door unit.
- **Fixed price contract:** A contract with a set price for the work. See **Time and materials contract**.
- **Fixed rate:** A loan where the initial payments are based on a certain interest rate for a stated period. The rate payable will not change during this period regardless of

changes in the lender's standard variable rate.
- **Fixed rate mortgage:** A mortgage with an interest rate that remains the same over the years.
- **Fixed support:** In two dimensions, a fixed support restrains three degrees of freedom: two translations and one rotation.
- **Fixed window:** A window unit that does not open. Also called a **picture window**.
- **Fixed-end support:** A condition where no rotation or horizontal or vertical movement can occur at that end. This type of support has no degrees of freedom. Three reactive forces exist at the rigidly fixed end. See **Rigid connection**.

Fixture: The visible parts of the plumbing and electrical systems. Fixtures are attached to floors, ceilings, and walls; toilets, sinks, wall sconces, and ceiling lights are fixtures.

Flagging tape: A non-adhesive, weatherproof tape made of polyvinyl chloride that is used in construction, landscaping, excavating, identification of trees or hazards, and in locating underground cables or pipe. Flagging tape comes in a variety of bright colors for outdoor use.

Flagman/Flagperson: A person who warns of construction operations, danger, caution, and general hazardous conditions by the use of flags or signs.

Flagstone: Flat stones (1 to 4 inches thick) used for walks, steps, floors, and vertical veneer (in lieu of brick). Also called **flagging** or **flags**.

Flakeboard: A manufactured wood panel made out of 1–2" wood chips and glue. Often used as a substitute for plywood in the exterior wall and roof sheathing. Also called OSB or wafer board.

Flaking: The detachment of a uniform layer of a coating or surface material, usually related to internal movement, lack of adhesion, or passage of moisture.

Flame retardant: A substance that is added to a polymer formulation to reduce or retard its tendency to burn.

Flame retention burner: An oil burner, designed to hold the flame near the nozzle surface.

Flame spread classes: The Unified Building Code (UBC) and Building Officials and Administrators International, Inc.'s (BOCA) codes use I-II-III designation and the Standard code uses A-B-C. The flame-spread categories are per ASTM E-84/UL 723:

- **Class A or I:** flame spread 25 or less.
- **Class B or II:** flame spread 26 to 75.
- **Class C or III:** flame spread 76 to 200.

Fire retardant treated wood (FRTW) must have a flame spread of 25 or less in the 10-min ASTM E-84/UL 723 test, and then the test is continued for 20-min more during which there must be no evidence of significant progressive combustion and the flame front may not progress more than 10.5 feet from the burner. This is considerably more severe than the 10-min test used for fire retardant surface coatings and other building materials.

Flame spread rating: A measure of the time it takes for fire to spread across the surface of a combustible material based on ASTM E84.

Flammability: The characteristics of a material to burn or support combustion.

Flange: The projecting edge of a rigid or semi-rigid component, such as a metal edge flashing flange, skylight flange, flashing boot, structural member, etc.

Flange.

Flanges: Parallel edges that are perpendicular to the center web of a steel beam (girder).

I-BEAM

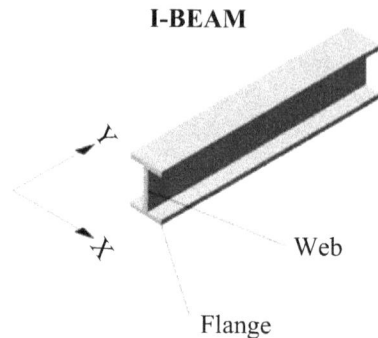

Web

Flange

Flanges on a steel I-beam.

Flanker: A former term used to describe a side or lateral part. Also previously used to describe a 3-wide picture unit or bay.

Flare fitting: A fitting used to connect flex copper pipe, usually gas pipe.

Flaring tool: A tool that is used to create a cone-shaped enlargement at the end of a piece of tubing to accept a flare fitting.

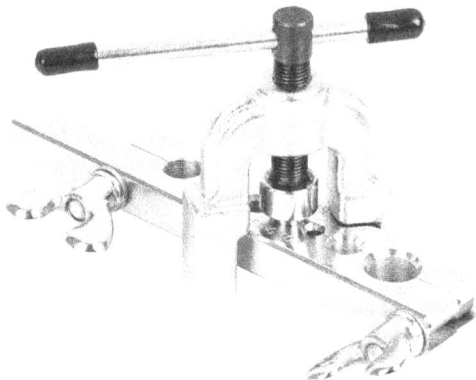

Flaring tool.

Flash point: The lowest temperature of a liquid at which it gives off vapors sufficient to form an ignitable mixture with air near its surface.

Flashing: Any piece of material installed to prevent water from penetrating into the structure around doors, windows, chimneys, and roof edges. Flashing may be made of metal, plastic, rubber, or impregnated paper.

The purpose of a flashing is usually to divert water to the outside. Therefore, the material that is selected for the flashing must be water tight and long lasting. It should also be workable and easy to install or else there is a high chance of improper installation.

An important aspect of longevity for flashing materials are their reaction with other building materials. Corrosion may be caused by galvanic action between materials or from corrosive elements in one or both of the materials; see the table below.

Metal corroding \ Contact metal	Magnesium & alloys	Zinc & alloys	Aluminium & alloys	Cadmium	Steel-carbon	Cast iron	Stainless steels	Load, tin and alloys	Nickel	Brasses, nickel silvers	Copper	Bronzes, cupro-nickels	Nickel copper alloys	Nickel-Chrome-Mo alloys Titanium, silver, graphite Graphite, gold, platinum
Magnesium & alloys		X	X	X	X	X	X	X	X	X	X	X	X	X
Zinc & alloys			X	X	X	X	X	X	X	X	X	X	X	X
Aluminium & alloys				X	X	X	X	X	X	X	X	X	X	X
Cadmium					X	X	X	X	X	X	X	X	X	X
Steel-carbon						X	X	X	X	X	X	X	X	X
Cast iron							X	X	X	X	X	X	X	X
Stainless steels					X	X		X	X	X	X	X	X	X
Load, tin and alloys							X		X	X	X	X	X	X
Nickel										X	X	X	X	X
Brasses, nickel silvers						X	X	X	X		X	X	X	X
Copper								X	X	X		X	X	X
Bronzes, cupro-nickels													X	X
Nickel copper alloys														X
Nickel-Chrome-Mo alloys Titanium, silver, graphite Graphite, gold, platinum														

X = Galvanic corrosion risk

Also, a flaw in drywall that occurs when reflected light reveals a difference between areas where drywall mud was applied over tape or fasteners and areas where mud was not applied.

Flashing cement: With regard to roofing, an ASTM D 2822 Type II roof cement that is a trowelable mixture of solvent-based bitumen and mineral stabilizers that may include asbestos or other inorganic or organic fibers. Generally, flashing cement is characterized as vertical-grade, which indicates it is intended for use on vertical surfaces. See **Asphalt Roof Cement** and **Plastic Cement**.

Flashing collar: An accessory flashing used to cover and/or seal soil pipe vents and other penetrations through the roof. Also referred to as a **roof jack** or **flashing boot**.

Flat: A surface that scatters or absorbs the light falling on it so as to be substantially free from gloss or sheen (0–15 gloss on a 60° gloss meter).

Also, a surface that is horizontal, level and smooth.

Other terms beginning with flat:
- **Flat bit:** See **Spade bit**.
- **Flat casing:** Flat, surfaced on four sides, pieces of pine of various widths and thicknesses for trimming door and window openings. The casing serves as the boundary molding for siding material and also helps to form a rabbet for screens and/or storm sash or combination doors.
- **Flat grain shingles:** Shingles made from lumber whose annual rings form less than a 45° angle at the face.
- **Flat lock:** A type of interlocking two separate metal panels by folding one panel over on top itself and the folding the other down under itself and then hooking the panels together.
- **Flat mold:** Thin wood strips installed over the butt seam of cabinet skins.
- **Flat paint:** An interior paint that contains a high proportion of pigment and dries to a flat or lusterless finish.
- **Flat panel:** A flat, thin plywood panel used in a frame and panel cabinet door. See frame and panel cabinet panel, slab cabinet door and raised panel.
- **Flat roof:** A roof style that appears flat but actually has a slight slope to allow drainage of precipitation. As a minimum, roofs should be sloped 1/4 inch per foot. Do not forget to consider deflection due to loading when designing a roof.
- **Flat tiles:** A tile shingle with a flat surface. The surfaces of flat tiles often have a grain simulation and the sides are usually rabbeted and grooved.

- **Flat top grind:** A term used to describe a saw tooth filed square on top. The teeth may also be ground with various combinations of beveled tops. For example, alternate top bevels are often used for crosscutting saws. Tops may also be pointed or chamfered and alternated with flat tops on very hard, dense woods to break up the chip to reduce tooth load.
- **Flat truss:** A roof or floor truss with horizontal top and bottom chords reinforced with diagonal members between them.
- **Flat-laid countertop:** Type of plastic laminate countertop with no integral backsplash, a flat smooth surface, and usually a square front edge. See plastic laminate countertop and post-formed countertop.
- **Flatwork:** A common word for concrete floors, driveways, basements, and sidewalks.

Fleck, ray: A portion of a ray as it appears on the quartered or rift cut surface. Fleck can be dominant appearance in oak and is sometimes referred to as flake.

Quarter sawn surface, showing ray fleck.

(Quarter-sawing is so named because the log is first quartered lengthwise. This results in wedges with a right angle ending at approximately the center of the original log. Each quarter is then cut separately by tipping it up on its point and sawing boards successively along the axis, resulting in boards with the annual rings mostly perpendicular to the faces and, especially with oaks, a distinctive ray and fleck figure.

Quartersawn boards can also be produced by cutting a board from one flat face of the quarter, flipping the wedge onto the other flat face to cut the next board, and so on.)

Fleece: With regard to roofing, mats or felts composed of fibers (usually non-woven polyester fibers), often used as a membrane backer.

Flemish bond: Probably the most common bond in brickwork, it consists of alternating headers and stretchers, with each header being in the middle of the stretchers above and below.

Flemish bond.

Fletton brick: A common type of machine-made yellow/orange frogged brick used in the south-east of England and London. Named after Fletton, near Peterborough.

Fletton brick.

Flexible conduit: A conduit, either electrical or mechanical, that allows for differential movement between two different pieces of equipment or circuits while maintaining the integrity of that circuit.

Flexible connector: A braided hose that connects a faucet or toilet to the water supply stop valve. It serves as a riser but is much more flexible and easier to install. Flexible connectors are usually smde of stainless steel or PVC/Polyester reinforced hose.

Flexible system: A structural system that will sustain relatively large displacements without failure.

Flexibility: Flexibility is the inverse of stiffness. When a force is applied to a structure, there is a displacement in the direction of the force; flexibility is the ratio of the displacement divided by the force. High flexibility means that a small load produces a large displacement.

Flexure: Bending deformation, i.e., deformation by increasing curvature.

Flint: A hard, gray colored stone that is sometimes used as a building stone.

Flint-lime brick: A kind of calcium silicate brick.

Flitch: A large piece of lumber cut out of a log that is then sawn into boards or veneer strips. A flitch is usually cut from the outside of a tree trunk and very likely has wane (the absence of square wood on the edge of a board from any source) on one or both edges.

Also, a complete bundle of veneer sheets laid together in sequence as they are cut from a given log or section of log.

Also, a timber beam strengthened with one or more steel plates bolted or screwed to it, often sandwiched between timbers. Also referred to as a flitch beam.

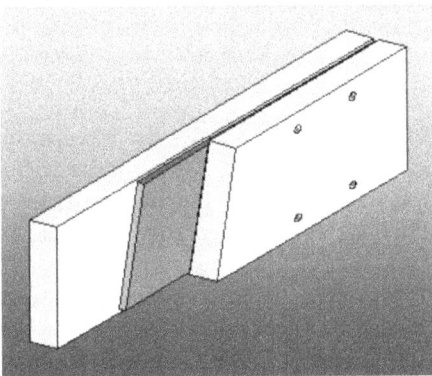

Flitch beam.

Float: A hand tool used to provide an even texture to concrete or plaster surfaces before they set.

Also, a scheduling term indicating that an activity or a sequence of activities does not necessarily have to start or end on the scheduled date to maintain the schedule on the critical path. The difference between the early start and late finish of an activity, minus the activities duration. See **Critical Path Method**.

Float glass: Glass that is produced by a process in which the ribbon is floated across a bath of molten tin. The vast majority of flat glass is now produced using this method. The terms **plate glass** and **sheet glass** refer to older manufacturing methods still in limited use.

Floating: The next-to-last stage in concrete work, when you smooth off the job and bring water to the surface by using a hand float or bull float.

Other terms beginning with floating:
- **Floating earthquake:** Earthquakes in regions of moderate to high seismicity that cannot be correlated with known geologic structures and thus are assumed to be likely to occur at any location in the region.
- **Floating panels:** Floating panels are utilized in stile and rail wood door construction where the panel is not glued to the frame; it is left to "float" to accommodate and compensate for natural wood thermal expansion and contraction due to atmospheric conditions. This allows for the seasonal movement of the wood comprising the panel, without cracking the panel or distorting the frame.
- **Floating type flange joint:** A conventional flanged joint in which a gasket is compressed by bolts; the gasket is not rigidly located.

Floating flange rubber expansion joint.

• **Floating wall:** A non-bearing wall built on a concrete floor. It is constructed so that the bottom two horizontal plates can compress or pull apart if the concrete floor moves up or down. Normally built on basements and garage slabs.

Flood coat: Heavy, smooth asphalt coating mopped over the cap sheet of a multiple ply membrane roof to provide a smooth surface. The flood coat must be protected from sun damage by painting it with a UV coat or by covering it with aggregate.

A flood coat is generally thicker and heavier than a glaze coat, and is applied at ~45–60 pounds per square (2–3 kilograms per meter). Also referred to as pour coat.

Flood plane structure: A structure that sits atop columns that raise the main floor above the flood plane.

Flood test: The procedure where a controlled amount of water is temporarily retained over a horizontal surface to determine the effectiveness of the waterproofing.

Floor: A story or level of a building. A **floor area** is the projected horizontal area inside walls, partitions, or other enclosing construction. And when used to determine the occupant load of a space the **net floor area** is the horizontal occupiable area within the space, excluding the thickness of walls, and partitions, columns, furred-in spaces, fixed cabinets, equipment, and accessory spaces such as closets, machine and equipment rooms, toilets, stairs, halls, corridors, elevators, and similar occupied spaces. Establishing net floor area may become important when you are seeking a building permit for a building that is just squeeking into a zoning regulation.

Other terms beginning with floor:
• **Floor joists:** Subfloor framing members that support the floor span. Joist are usually made of 2×8 (or larger) lumber.
• **Floor plan:** One of the five basic views found on a plan. A floor plan is a view as

though you are looking directly down on a building with the top removed so you can see the layout of the floor including walls and fixtures. Floor plans may include fixtures and other key elements of the floor it is depicting.

Also referred to as a plan view.
• **Floor plate:** The entire floor area and or assembly of any one story of a building.
• **Floor sink:** A receptacle that is usually made of enameled cast iron and is located at floor level. It is connected to a trap to receive the discharge from indirect waste and floor drainage. Cleaner and more sanitary than regular floor drains. Floor sinks are usually used in hospitals and restaurants.

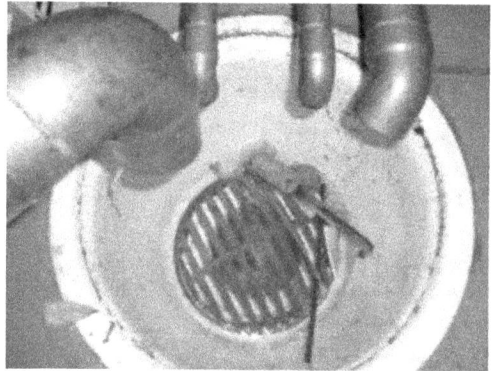

Floor sink.

• **Floor system:** Includes the framing support members such as floor joists or floor trusses and the sheathing that provides the floor system surface. This is a structural part.
• **Floor truss:** A floor truss may be used instead of regular joists or I-joists. They are generally placed on wider centers, and are deeper and more expensive than other joists. Floor trusses are designed to allow plumbing, electrical, and heating runs to be placed inside of them instead of below them like is often required in other joists.

Floor trusses.

Foot valve: A special type of check valve located at the bottom end of the suction pipe on a pump. It opens when the pump operates to allow water to enter the suction pipe but closes when the pump shuts off to prevent water from flowing out.

Flow control valve: A valve which controls the rate of liquid flow.

Flow rate: The volume, mass, or weight of a fluid passing through any conductor per unit of time.

Flow restrictor: A device that restricts flow in showerheads. By law all showerheads in the United States have flow restrictors that limit water consumption to 2.5 gallons per minute. There are also air flow restrictors.

Flue: An enclosed passageway in a chimney to carry products of combustion to the outer air.

Other terms with the word flue:
• **Flue baffle:** A device to deflect, check, or regulate flow through a pipe.
• **Flue cap:** A cap placed on the top opening of the flue in such a way as to permit proper ventilation of the inner chambers of the flue pipe and at the same time prevent moisture or small animals from entering the flue. Also known as a **flue collar**.
• **Flue damper:** An automatic door located in the flue that closes it off when the burner turns off; purpose is to reduce heat loss up the flue from the still-warm furnace or boiler.
• **Flue exhauster:** A fan that is installed in the vent pipe to provide a positive induced draft. Also referred to as a draft inducer.

• **Flue lining:** Fire clay or terra-cotta pipe (round or square) and usually made in all ordinary flue sizes. Used for the inner lining of chimneys with the brick or masonry work done around the outside. Flue linings in chimneys usually run from one foot below the flue connection to the top of the chimney.

Fluid: A liquid or gas with no fixed shape. Fluids yield to external pressure.

Fluid mechanics: The science of the properties and motion of liquids and gases.

Fluid-applied elastomer: A liquid elastomeric material that cures after application to form a continuous waterproofing membrane.

Fluorescent lighting: A gas-filled glass tube with a phosphor coating on the inside. The gas inside the tube is ionized by electricity that causes the phosphor coating to glow. Lamps normally have two pins that extending from each end.

Fluorescent tube: A glass tube that radiates light when phosphor on its inside surface is made to fluoresce by ultraviolet radiation from mercury vapor. Also referred to as a fluorescent light and fluorescent bulb.

Fluoropolymer binders (PVDF): Fluoropolymers have non-stick properties, chemical resistance, and resistance to UV damage. Polyvinylidene Fluoride (PVDF) is a premier resin in this category. PVDF components are used extensively in the high purity semiconductor market, pulp and paper industry (chemically resistant to halogens and acids), nuclear waste processing (radiation and hot acid applications), and the general chemical processing industry (chemical and temperature applications). Fluoropolymers have also met specifications for the food and pharmaceutical processing industries.

Currently, PVDF is at least 70% of the binder that is used in superior performance coatings. Acrylic is usually the other 30%.

Fluoro-carbon thread coating: A low friction coating applied to bolt threads. Fluoro-carbon thread coating is frequently used to prevent thread fouling especially when an assembly containing threaded fasteners is painted. Unless masked before painting, electro deposited primers can cover the threads resulting in difficulties, unless the expensive chore of cleaning the threads is completed. A fluoro-carbon thread coating will not adhere to the coating. This type of coating also prevents problems caused by weld splatter obstructing the threads of weld nuts during their placement. Fluoro-carbon thread coating also has the property of reducing the torque-tension scatter during tightening.

Flush door slab: Door with a flat, smooth face with no panels or decoration. Flush doors may have a solid or hollow core.

Flush pointing: Pointing that is flush with the surface of the bricks. See Bucket handle pointing.

Flush tile: Flat ceiling tiles with neither embossed designs nor recessed edges.

Flushometer: A toilet valve that automatically shuts off after it meters a certain amount of water flow.

Flute: (Not the musical instrument) Wave shapes pressed into a corrugated medium. They form the columns and arches that give corrugated sheets their strength. Flutes are categorized by the size of their wave.

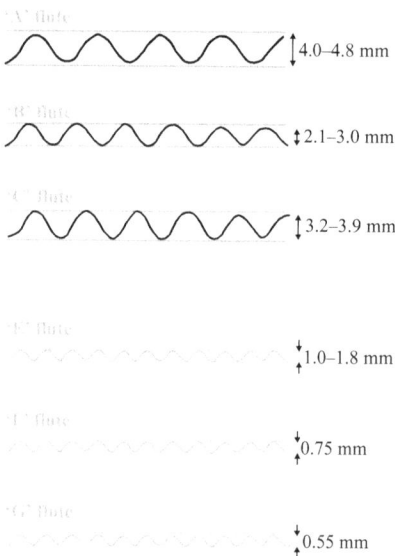

Flutes.

Also, the fold or bend in a sheet of deck that forms a groove or furrow.

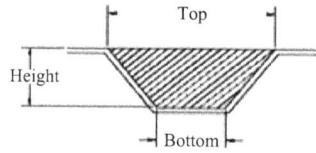

Typical flute section.

Also, on a drill bit, the exit path for chips.

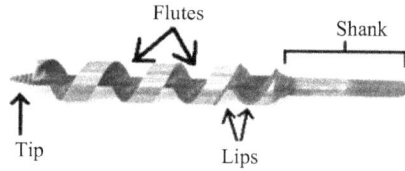

Flutes on a drill bit.

Fluted casing: A casing that contains a series of round reliefs called flutes along its length. Fluted casings are designed to look like fluted columns.

Fly rafters: A gable rafter that is located under the overhang part of the roof sheathing on the gable end. They are not directly supported by the exterior wall. Also sometimes referred to as the barge rafter or barge board.

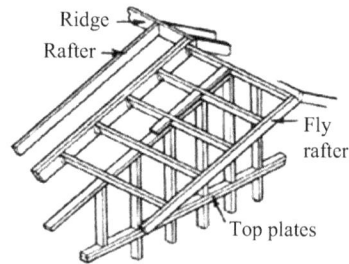

Fly rafters.

Foam foundation sill plate sealer: Thin foam strip placed on top of the foundation and under the foundation sill plate. This strip is actually pushed down over the bolts and fills in any gaps between the foundation and the foundation sill plate.

Foam stop: The edge metal used to terminate sprayed polyurethane foam.

Foaming: The formation of a foam layer on the surface of the boiler water. It naturally occurs in a body of impure water subject to rapid circulation currents. The greater the amount of impurities dissolved in the water the greater the propensity there is for foaming.

Focal depth: The depth of an earthquake (or hypocenter) below the ground surface.

Focus or Hypocenter of an earthquake: The point at which the rupture occurs (it marks the origin of the kinetic waves of an earthquake).

Fogging: A deposit or film left on an interior surface of a sealed insulating glass unit due to extreme conditions or failed seals.

Folding partition: A moveable wall on a track suspended from a joist or beam that usually folds like an accordion. Depending on their size folding walls may be stored against the end wall, in a closet or pocket in a wall. Large folding partitions are generally motorized. Also referred to as folding wall.

Foot: Unit of length in the Imperial equal to 12 inches or a third of a yard (0.3048 m).

Footbolt: A locking rod device installed vertically in the stile or astragal of a door or screen that when activated secures the panel or screen in a stationary position.

Footer/Footing: The base upon which the structure will stand, it rests on the soil. A footing ultimately supports all of the weight of the structure, it is a structural part.

Footing drain: A drain, usually a porous or perforated pipe that extends around a building's footings for the purpose of draining ground water away from the substructure.

Force: In physics, an influence tending to change the motion of a body or produce motion or stress in a stationary body. It is measured with the SI unit of newtons and represented by the symbol F. A, force can cause an object with mass to change its velocity (including from a state of rest), or can cause a flexible object to deform. Force can also be described by intuitive concepts, e.g., any push or pull measured in units of weight.

In hydraulics, total force is expressed by the product P (force per unit area) and the area of the surface on which the pressure acts, i.e., $F = P \times A$.

In structures, force is a directed interaction between two objects that tends to change the momentum of both. Since a force has both direction and magnitude, it can be expressed as a vector.

Force system: See **System of forces**.

Forced air system: A common form of heating with natural gas, propane, oil, or electricity as a fuel. Air is heated in the furnace and distributed through a set of metal ducts to various areas of the house.

Forced convection: A heat transfer process, aided by mechanical circulation of a liquid (such as water) or a gas (such as air). This applies to natural wind flow over a window.

Forced dry: Baking the paint between room temperature and ~150° F to speed the drying process.

Forearc: The forearc is the region between the subduction zone and the volcanic chain (volcanic arc).

Foreshocks: Smaller earthquakes preceding the main earthquake in a series.

Forged: A blacksmithing technique that shapes metal by hammering, usually while red or white hot.

Form: A temporary structure erected to contain concrete during placing and initial hardening. A form may be constructed of wood, plastic, steel, or soil.

Form ties: Steel rods that hold two separate forms apart to facilitate the placement of concrete. Form ties usually remain in place through the curing process or until the forms are stripped.

Formal region: A region identified by political jurisdiction or on the basis of the presence or absence of one or more distinguishing features or characteristics.

Formaldehyde: A component of resin used to manufacture plywood and panel products, and a naturally occurring component of wood. It is a colorless pungent gas in solution made by oxidizing methanol. The molecular formula for formaldehyde is CH_2O.

Formaldehyde is commonly used as a preservative in medical laboratories and mortuaries. Formaldehyde is also found in many products such as chemicals, particle board, household products, glues, permanent press fabrics, paper product coatings, fiberboard, and plywood. It is also widely used as an industrial fungicide, germicide, and disinfectant.

Formaldehyde is a sensitizing agent that can cause an immune system response upon initial exposure. It is also a cancer hazard.

Formica: A brand name of a common type of plastic laminate material. The term Formica is often used in the industry when referring to plastic laminate. See **Plastic laminate countertop**.

Formwork: Either temporary or permanent molds into which concrete or similar materials are poured.

Fossiloferous: A rock containing fossils.

Fouling: Anything that adheres to boiler surfaces causing localized overheating and disruption of water circulation. In bad water areas fouling also occurs in water tanks, pipework, valves, injectors, clacks, etc.

Foundation: The supporting portion of a structure below the first floor construction, or below grade, including the footings.

Other terms beginning with foundation:
• **Foundation pier:** A foundation element that consists of a column embedded into the soil below the lowest floor to the top of a footing or pile cap. Where a pier bears directly on the soil without intermediate footings or pile caps, the entire length of the column below the lowest floor is usually considered as a foundation pier. Foundation piers should be able to be inspected, otherwise they are deemed to be piles.
• **Foundation sill plate:** A piece of lumber (some codes require treated lumber) that is used between the foundation and the framing. It is attached to the foundation with anchor bolts.
• **Foundation ties:** Metal wires that hold the foundation wall panels and rebar in place during the concrete pour.
• **Foundation walls:** The vertical walls of a substructure that transfer their loads to footings. Foundation walls extend below grade.
• **Foundation waterproofing:** A high-quality below-grade moisture protection that is used for below-grade exterior concrete and masonry wall damp-proofing to seal out moisture and prevent corrosion.

Four way inspection: An inspection of the rough-in of four trades including framing, plumbing, HVAC, and electric. This inspection must be completed before the walls or ceilings are closed-in.

Fracture properties: Mechanical properties such as the ductile-brittle transition temperature (DBTT), nil ductility temperature (NDT), or nil-ductility transition temperature that characterize resistance of the metals to brittle fracture, especially at low temperatures.

The ductile-brittle transition temperature (DBTT), nil ductility temperature (NDT), or nil ductility transition temperature of a metal represents the point at which the fracture energy passes below a pre-determined point for a standard **Charpy impact test**. DBTT is important since, once a material is cooled below the DBTT, it has a much greater tendency to shatter on impact instead of bending or deforming.

Fractured composition shingles: A composition shingle that has been torn by the impact from a hailstone. The fractures often radiate out from the center of the hailstone impact in a spider web pattern. Also see **Bruised composition shingles** and **Granular loss**.

Fracture cracking: Brittle cracks that occur with little or no preceding plastic deformation. Low temperature, stress or strain concentrations, and metallurgical composition are important factors influencing fracture cracking. Fracture cracks are oftened triggered by impact or a sudden increase in load.

Frame: A system of assembled structural elements.

With regard to windows, the stationary portion of a window that encloses either the glass (direct glaze) or the sash (operating or stationary) and consists of the head jamb (top), sill (bottom), sub-sill, side jambs, jamb extension, brick mold or flat casing, and blindstop.

Frame and panel cabinet door: A cabinet door that consists of a frame that surrounds a panel. The panel may be glass, a veneered plywood flat panel, or a solid wood raised panel. See **Slab cabinet door**.

Frame clamp or cramp: A metal component screwed to the window or door frame and built into the masonry wall.

Frame clamp/cramp.

Frame expander: A flat aluminum extrusion used in conjunction with the 90° frame expander to provide a flat casing appearance for clad units.

Frame inspection: The act of inspecting the home's structural integrity and it's compliance with local municipal codes.

Framed opening: A structurally-framed opening in a roof of a building for use in installing large items such as HVAC units, skylights, or ventilators.

Framed style cabinets: A cabinet style in which a face frame is attached to the cabinet box. See **Frameless style cabinets**.

Framed to square: A term to denote that the building framing has been completed to the point that it is ready for the roof frame to be built.

Frameless style cabinet: A cabinet style that has no face frame attached to the cabinet box. Often called a **European style cabinet** because it was developed in Europe during the reconstruction following World War II as an alternative to the more labor-intensive framed style cabinet. See **Framed style cabinets**.

Framer: The carpenter contractor that installs the lumber and erects the frame, flooring system, interior walls, backing, trusses, rafters, decking, installs all beams, stairs, soffits, and all work related to the wood structure of the home. The framer builds the home according to the blueprints and must comply with local building codes and regulations.

Framing: Lumber used for the structural members of a building, such as studs, joists, and rafters.

Floor or roof plans that identify individual marks, components, and accessories furnished by the structural materials manufacture[s] is called a **framing plan**. The framing plan is detailed to permit proper erection of the building's frame. See **Erection plan** and **Placing plan**.

Members that connect the bottoms of opposing rafters together to prevent them from moving outward are **framing tie**s. Ceiling joists are commonly used as framing ties.

Free along side (FAS): A commercial delivery term that signifies that the seller must at the seller's own risk and expense deliver goods to the side of the transporting medium in the usual manner and obtain and tender a receipt for the goods in exchange for which the carrier must issue a bill of lading.

Free and clear: Unincumbered.

Free fall: The act of falling before a personal fall arrest system begins to apply force to arrest the fall.

The vertical displacement between onset of the fall and just before the fall arrest system begins to apply force to arrest the fall is termed the **free fall distance**. This distance excludes deceleration distance, and lifeline/lanyard elongation, but includes any deceleration device slide distance, or self-retracting lifeline/lanyard extension before they operate and fall arrest forces occur.

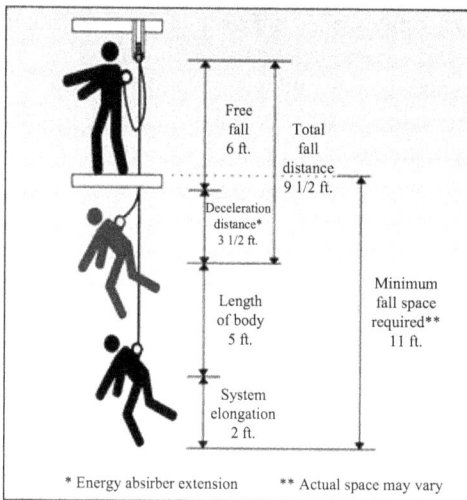

Fall distances.

Free field: Ground motion measurements that are not influenced by constructed structures.

Free-body diagram: A diagram on which all of the external forces acting on a body are shown at their respective points of application.

Free-tab shingles: Shingles that do not contain factory-applied strips or spots of self-sealing adhesive.

Freeze thaw damage: Surface damage to concrete resulting from the cycle actions of freeze thaw.

Deterioration of concrete from freeze thaw actions may occur when the concrete is critically saturated. This occurs when ~91% of its pores are filled with water. When water freezes to ice it expands to 9% more volume than that of liquid water. Because there is no space for this expansion in concrete, the freezing causes distress that continues with successive freeze thaw cycles resulting in repeated loss of concrete surface. To protect concrete from freeze/thaw damage, it should be air-entrained by adding a surface active agent to the concrete mixture. This creates a large number of closely spaced, small air bubbles in the hardened concrete that serve to relieve the pressure build-up caused by ice formation by acting as expansion chambers. About 4% air by volume is needed and the air-bubbles should be well distributed and have a distance between each other of less than 0.25 mm in the cement paste.

Concrete with high water content and a high water to cement ratio is less frost resistant than concrete with lower water content.

Freezeless hose bibb: A faucet designed to supply water to the outside of the structure without danger of freezing in cold temperatures. The faucet is located on the outside of the structure but the valve portion is located inside the heated structure.

Freezing and thawing: The loss in mass as a percentage of the original mass of a dried specimen when subjected to standard cyclic testing of alternations of freezing and thawing.

French door: Generally refers to a pair of hinged doors that open from the middle. French doors incorporate wider stile-and-rail components around the glass than typical glazed doors.

French window: Two casement sash hinged on the sides to open in the middle; the sash extends to the floor and serves as a door to a porch or terrace.

Frenchman: A tool for forming the shape of pointing. Also referred to as **tucking irons**, they are made of hard steel and come in many sizes and widths to match previous work.

Tucking irons (Frenchmen).

Frequency: The number of times an action occurs in a unit of time or in a given sample; the inverse of period.

Also, the ratio of the number of actual to possible occurrences of an event.

The rate at which a vibration occurs that constitutes a wave, either in a material (as in sound waves), or in an electromagnetic field (as in radio waves and light), usually measured per second. (Symbol: f or $\mathbf{v}$).

A pump or motor's basic frequency is equal to its speed in revolutions per second multiplied by the number of pumping chambers.

The number of cycles per second of alternating current (example: 60 cycles per second or 60 hertz per second).

Friction loss: In pipes, it is the loss of energy or head that occurs in pipe flow due to viscous effects generated by the surface of the pipe. The shear stress of a flow is dependent on whether the flow is turbulent or laminar. For turbulent flow, the pressure drop is dependent on the roughness of the surface, while in laminar flow, the roughness effects of the wall are negligible.

Friction-weld: A process that uses high-speed vibrations to join materials together.

Frieze: The central horizontal band of an entablature, found below the cornice and above the architrave. See **Entablature**.

Frieze board: In house construction a horizontal member connecting the top of the siding with the soffit of the cornice.

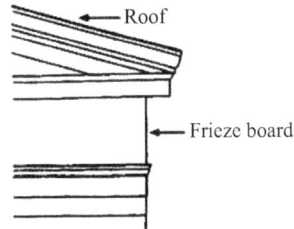

Frieze board.

Frog: A hollow on one or both of the larger faces of a brick or block. Frogs reduce the weight of the brick or block. They may be filled with mortar. Also referred to as panels.

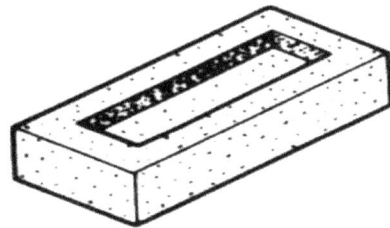

Brick with a frog.

Also, a device that permits train wheels on one rail of a track to cross the rail of an intersecting track.

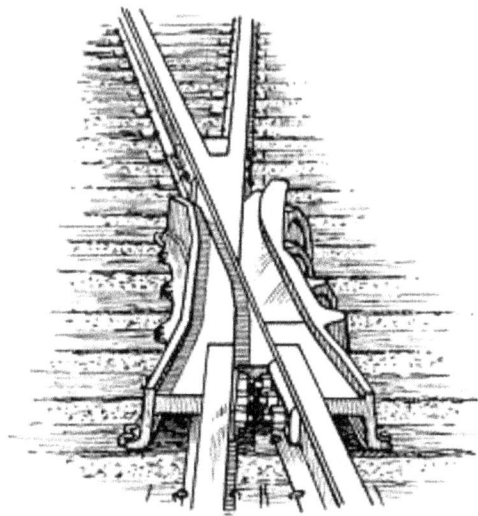

Railroad frog.

Front end loader: A mobile machine mounted on a wheeled or tracked chassis, equipped with a grapple, tuck, bucket, or fork-lift device, and employed in the loading, unloading, stacking, and sorting of materials.

Frost lid: A round metal lid that is installed on a water meter pit.

Frost line: The depth of frost penetration in soil and/or the depth at which the earth will freeze and swell. This depth varies in different parts of the country.

FROTH-PAK™: A foam sealant that is a two-component, quick-cure polyurethane foam that fills cavities, penetrations, cracks, and expansion joints.

Fuel-efficient vehicles: Vehicles that have achieved a minimum green score of 40 on the American Council for an Energy Efficient Economy annual vehicle-rating guide.

Fuhgedd about it: What a contractor says when an owner wants to add work to a contract without wanting to pay for it.

Full: Crowded, packed, crammed, congested, wall-to-wall, stuffed, etc., that lead to numerous construction terms:

- **Full backsplash:** A backsplash that runs from the countertop to the bottom of the upper cabinet. See **Backsplash** and **Block backsplash**.
- **Full basement structure:** A structure that has a basement level, the floor usually positioned below ground level beneath the main level.
- **Full cutoff:** A luminaire that has a light distribution in which the candela per 1000 lamp lumens does not numerically exceed 25 (2.5%) at or above an angle of 90° above nadir, and 100 (10%) at or above a vertical angle of 80° above nadir. Light pollution is become a major issue in urban areas and one that you may wish to familiarize yourself with.
- **Full disclosure:** For products that are not formulated with listed suspect carcinogens full disclosure has two components:

(1) disclosure of all ingredients (both hazardous and non-hazardous) that make up 1% or more of the undiluted product and (2) use of concentration ranges for each of the disclosed ingredients. Full disclosure for products that are formulated with listed suspect carcinogens has three components: (1) disclosure of listed suspect carcinogens that make up 0.1% or more of the undiluted product, (2) disclosure of all remaining ingredients (both hazardous and non-hazardous) that make up 1% or more of the undiluted product, and (3) use of concentration ranges for each of the disclosed ingredients.

Previously fully disclosure was not a concern to contractors, but nowadays both clients and workers are requiring that this information be available on all job sites.

- **Full extension glide:** Hardware attached between a cabinet drawer and the cabinet box that allows the drawer to be pulled completely out of the cabinet box. See **Glide**.
- **Full flow:** A condition where all the fluid must pass through the filter element or medium. For example, a full-flow oil filter is a device that filters the oil passing through the engine before it reaches the bearings.

In England or Australia people might say that an activity is in full flow, if it is happening fast and with energy.

- **Full height cabinet:** Any cabinet that runs the full height from the floor to the level of the upper unit. See **Lower unit, Vanity cabinet**, and **Upper unit**.
- **Full sawn:** Lumber cut, in the rough, to its full nominal size.
- **Full-time equivalent (FTE):** Generally, a 40 h per week job. Part-time and overtime positions have FTE values based on their hours per week divided by 40. Multiple shifts are included or excluded depending on the intent and requirements of the credit. Estimators frequently develop labor costs based on FTEs, but beware as fringe

benefits are often different for part-time and overtime hours.

- **Fully shielded:** An exterior light fixture that is shielded or constructed so that its light rays project below the horizontal plane passing through the lowest point on the fixture from which light is emitted.

Fume hoods: A ventilated enclosure in a laboratory that evacuates harmful volatile fumes.

There are two basic types of laboratory fumehoods: constant air volume and variable air volume. The constant air volume (CAV) fumehood exhausts the same amount of air all the time, regardless of sash position. As the sash is lowered and raised, the velocity at the face of the hood changes. This change in face velocity can result in less than optimal hood performance.

Variable air volume (VAV) fumehoods are fitted with a face velocity control that varies the amount of air exhausted from the fumehood to respond to changing the sash opening area. This allows for optimal hood performance regardless of the sash position. VAV hoods also provide energy savings by reducing the flow rate from the hood when the sash is closed but have higher initial cost.

A VAV system should tie the fumehood exhaust to the room's air supply and exhaust. Then the VAV system can ensure that the required room air changes per hour are being met, that sufficient make-up air is provided to balance the air that's being exhausted, and that appropriate room temperatures are maintained.

When designing a fume hood, consideration should be given to maintaining it. Laboratory fume hoods may exhaust hazardous, caustic and even explosive materials. Eventually the hood and its entire exhaust system will have to be maintained and cleaned. While laboratory personel are trained in the hoods proper use, maintenance workers may not be, with serious consequences.

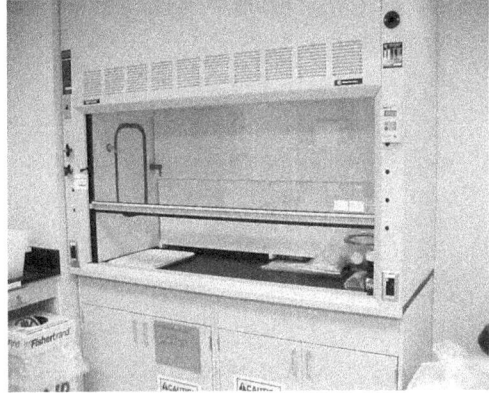

A laboratory fume hood.

Function: The normal and characteristic purpose or action of a system, component, or device.

Functional drainage: The ability to empty a plumbing fixture in a reasonable time.

Functional flow: The flow of the water supply at the highest and farthest fixture from the building supply shutoff valve when another fixture is used simultaneously.

Fundamental/Natural period: The elapsed time, in seconds, of a single cycle of oscillation. The inverse of Frequency.

Fungal decay: Wood decay fungus is a variety of fungus that digests moist wood, causing it to rot. Some wood-decay fungi attack dead wood and some are parasitic and colonize living trees. Fungi that not only grow on wood but actually cause it to decay, are called lignicolous fungi. They do not necessarily need to decay lignin in the wood to be termed lignicolous.

Wood-decay fungi can be classified according to the type of decay that they cause:

- **Brown rot:** Brown-rot fungi break down hemicellulose and cellulose. As a result of this type of decay, the wood shrinks, shows a brown discoloration, and cracks into roughly cubical pieces; hence the name brown rot or cubical brown rot.
- **Soft rot:** Soft-rot fungi secrete an enzyme that breaks down cellulose in the wood. This leads to the formation of microscopic

cavities inside the wood, and sometimes to a discoloration and cracking pattern similar to brown rot.

• **White rot:** White-rot fungi break down the lignin in wood, leaving the lighter-colored cellulose behind. Some white-rot fungi break down both lignin and cellulose. White-rot fungi are able to produce enzymes, such as laccase, needed to break down lignin and other complex organic molecules and are therefore being studied for other uses.

Funicular: A funicular shape is one similar to that taken by a suspended chain or string subjected to a particular loading.

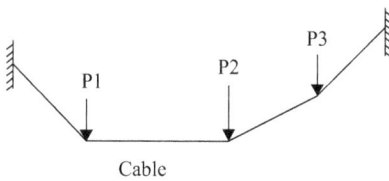

A structure in pure tension having the funicular shape is termed a cable.

Furlong: A unit of length in the Imperial system; ten chains, or 660 feet, one-eight of a mile, equal to 201.168 m.

Furnace: A mechanical piece of equipment used to produce hot air through combustion of a fuel, heat exchanger, and or radiant electric coils.

Furniture, fixtures, and equipment (FFE): Includes all items that are not base building elements, such as lamps, computers and electronics, desks chairs, and tables. If your contract requires that you, the contractor, must provide FFE, make sure there is full understanding with the owner as to exactly what this means. Also, make sure all purchase orders make clear work items such as delivery, e.g., sidewalk or installed, and time of delivery, e.g., normal work hours or at night so as not to interfere with construction.

Furring strips: Strips of wood, often 1 × 2 and used to shim out and provide a level-fastening surface for a wall or ceiling.

Fuse: A device often found in older homes designed to prevent overloads in electrical lines.

A fuse allows a piece of metal to become part of a circuit. The metal melts under heat created by excessive current, thereby interrupting the circuit and preventing the flow of electricity from exceeding the circuit's current-carrying capacity. See Circuit breakers.

Fused glass: Glass that has been heated in a kiln to the point where two separate pieces are permanently joined as one without losing their individual color.

Fusible link: A metal tie in a fire warning, prevention, or suppression system that will melt at a predetermined temperature initiating that systems operation.

Fusion-weld: A term for a type of corner construction, used with vinyl and other types of windows and doors, in which a small amount of material on the ends of two pieces are melted or softened, then pushed together to form a single piece. This also is referred to simply as a welded corner.

Future value (FV): The amount to which money grows over a designated period of time at a specified rate of interest.

For example, the FV of $1,000 invested at 7% for one year with compounding monthly is: $1,072.29. See **Annual percentage rate (APR), Annual percentage yield (APY),** and **Compound interest**.

Gable: The end, upper, triangular area of a home, beneath the roof. Other terms associated with the word gable:

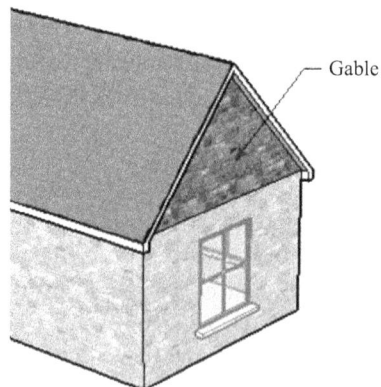

Gable.

- **Gable end truss:** A truss used at the ends of a gable roof. It has vertical members that are spaced to allow convenient attachment of the exterior wall sheathing.

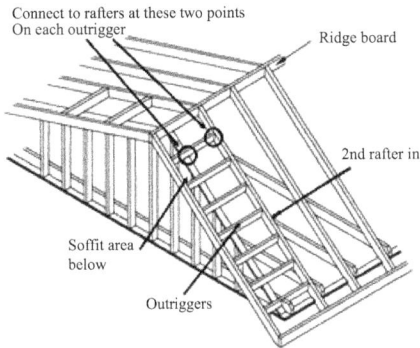

Gable end truss.

- **Gable-on-hip roof:** A roof configuration with hips coming up from the eave corners that terminate into a gable roof.

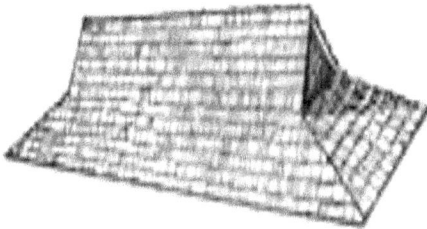

Gable-on-hip roof

- **Gable joist:** A non-standard type of joist where the top chord is double pitched at an extreme pitch (say 3/12) and the bottom chord is straight or level.
- **Gable roof:** A roof style consisting of two sides that slope in opposite directions down from the peak or ridge. The roof end forms an inverted V and is filled in with triangular shaped gable end wall. Also referred to as gable-shaped roof.
- **Gable vent:** Vents placed in the gable ends of the roof. Gable vents facilitate the flow of air in the attic while protecting it from insects and the weather.

Galling: A severe form of adhesive wear that occurs during sliding contact of one surface relative to another. Clumps of one part stick to the mating part and break away from the surface.

Gallon: A unit of volume. One US gallon is equal to 4 quarts or 231 cubic inches (~3.79 liters). One British imperial gallon is equal to four quarts or 4.55 liters.

Galvalume®: A trade name for a coating, used over metal, that is composed of aluminum zinc for corrosion protection.

Galvanic action: An electroylic reaction between dissimilar metals in the presence of an electrolyte.

Galvanized: A generic term used to describe a steel coated with zinc applied in an electro-galvanizing or dipping process. The galvanic series provides a list of metals with those on the top of the list being attacked by those lower down in the list. The father apart on the list, the faster the attack.

Steel coated with zinc for corrosion resistance is called **galvanized steel.**

Gambrel roof: A roof style consisting of two sides that meet at the ridge and slope in opposite directions. Each side has two sections, the lower section having a steeper slope than the upper section. The Gambrel roof is often used on barns.

Gambrel roof.

Gambrel truss: A truss used to make a Gambrel roof. It functions in the same way as a gable truss.

Gandy: A tool that uses a screen to press the course aggregate downward while leaving the fine aggregate at the surface. A flat gandy is dropped lightly over the entire surface. A rolling gandy uses a screen shaped like a barrel and is rolled across the entire surface. Widely used in residential construction, its use is discouraged by many structural engineers because it can severely damage the concrete unless used skillfully.

Gang: More than one switch installed in a single electrical box.

Gang nail plate: A steel plate attached to both sides at each joint of a truss. Sometimes called a fishplate or gussett.

Ganger: The leader (foreman) of a gang of laborers.

Gaps: Open slits in the inner ply or plies or improperly joined veneer when joined veneers are used for inner plies.

Garret: A top-floor or attic room; usually pictured as a cramped, dismal room with sloping ceilings.

Terms beginning with gas:
- **Gas cock:** A plug valve installed the main gas line and an appliance.
- **Gas distribution piping:** All piping from the house side of the gas meter piping that distributes gas supplied by a public utility to all fixtures and apparatus used for illumination or fuel in a building.
- **Gas lateral:** The trench or area in the yard where the gas line service is located, or the work of installing the gas service to a home.
- **Gas meter piping:** The piping from the gas service line valve to the outlet of the meter regulator set or the meter it no regulator is required.
- **Gas piping systems:** The gas service piping, meter piping, and distribution piping.
- **Gas service line valve:** The valve that is usually located at or below grade on the supply side of the meter or service regulator, if the service regulator is required. If a plug type valve is used it should be constructed so as to prevent the core from being blown out by the pressure of the gas. In addition, it should be of a type that is capable of being locked in the off position by the local gas utility.

- **Gas service piping:** The supply piping from the street main up to and including the gas service line valve.

Gasket: Soft pliable material used to prevent joint leakage.

Gate valve: A valve that lets you completely stop, but not modulate the flow within a pipe.

Gauge: The thickness of metal or wire. A heavier gauge means that the metal is thicker, but is denoted by a smaller number. For example:

GAGE NO.	STEEL	ALUM.
7	0.179	-
8	0.164	-
9	0.150	-
10	0.135	-
11	0.120	-
12	0.105	-
13	0.090	0.093
14	0.075	0.079
15	0.067	0.071
16	0.060	0.064
17	0.054	0.058
18	0.048	0.052
19	0.042	0.046
20	0.036	0.040
21	0.033	0.037
22	0.030	0.034
23	0.027	0.031
24	0.024	0.028
25	0.021	0.025
26	0.018	0.022
27	0.016	0.019
28	0.015	-
29	0.014	-
30	0.012	-
31	-	-

Sheet metal gauge chart.

Also, the thickness of a saw blade. Saw blade thickness is usually expressed in decimals of an inch or millimeters.

Also, the height of brickwork, specified as the number of courses per foot or per 300 mm.

Also, the distance from centerline hole to centerline hole across a set of holes; the distance between the two rails of a railroad; the distance between two wheels on an axle.

Also, an instrument for measuring or testing. When attached to a boiler, a **gauge glass or liquid level** allows the water level to be seen or determined.

Boiler gauge glass. The glass is attached to a gauge frame.

Gauge pressure: A pressure scale that ignores atmospheric pressure. Its zero point is 14.7 psi absolute.

General conditions: A written portion of the contract documents in which the owner stipulates the contractor's minimum acceptable performance requirements including the rights, responsibilities, and relationships of the parties involved in the performance of the contract. General conditions are usually included in the book of specifications but are sometimes found in the architectural drawings.

General condition items: Usually include purchases, services, and materials required to facilitate construction at the site. The construction budget will usually show that these are financial obligations of the owner and the logistic responsibility of the CM; a GC may be fully responsible.

The general conditions of the contract for construction also prescribes the rights, responsibilities, and relationships of the parties signing the agreement and outlines the administration of the contract for construction. AIA's document A201 provides an example of the general conditions of the contract for construction.

General contractor (GC): A contractor who enters into a contract with the owner of a project for the construction of the project and who takes full responsibility for its completion. The contractor may enter into subcontracts with others for the performance of specific parts or phases of the project.

A GC firm usually performs work with its own employees who are supplemented by specialty subcontractors. Or a GC may perform the project work as an independent contractor, providing services to owners through the use of subcontractors while under a contract normally use by the general contracting system. In the latter case, the GC is referred to as a **Paper Contractor**; owners would be wise to avoid such a situation.

Geodesic dome: A geodesic dome uses a pattern of self-bracing triangles in a pattern that gives maximum structural advantage, thus theoretically using the least material possible. (A "geodesic" line on a sphere is the shortest distance between any two points.)

Geodesic dome.

Geodesy: The study of the shape and size of the earth.

Geodimeter: An instrument used to measure distance between points on the surface of the earth.

Geographic Information System(s) (GIS): System(s) (usually computer-based) that are used for capturing, handling, storing, retrieving, managing, manipulating, and displaying geographic information or geo-coded data.

Geologic hazard: A geologic feature or process that has the potential to have an adverse effect on people or structures.

Geology: Geology is the study of the planet earth: the materials it is made of, the processes that act on those materials, the products formed, and the history of the planet and its life forms since its origin.

Geometry of a structure: The configuration of all the structural elements noted on the engineering and shop drawings that depict the relationship of one structural element to the next. The geometry is controlled by the drawings produced by the engineer of record. The detailer uses working points and dimensions on these drawings to produce erection, detail, and other shop drawings. Should the geometry not close, i.e., not come together exactly, then one or more dimensions are wrong.

The actual or nominal dimensions of structural components are referred to as the **geometry of structural components**, and the actual or nominal dimensions that define the geometry of a structure are referred to as the **geometry of structure**.

Geomorphology: The study of the origin and character of landforms.

Geophysics: Geophysics is the branch of earth science that employs physical measurements and mathematical models to explore and analyze the structure and dynamics of the solid Earth and similar bodies and their fluid envelopes.

Geotechnical: Related to the use of scientific methods and engineering principles to acquire, interpret and apply knowledge of earth materials to solving engineering problems.

Geotechnical engineer: See **Soils engineer**.

Geothermal energy: Electricity generated by converting hot water or steam from within the Earth.

A system that uses pipes to transfer heat from underground for heating, cooling, and hot water is a **geothermal heating system**; the system retrieves heat from the Earth during cool months and returns heat in summer months.

Gingerbread: Ornate scroll-sawn wood applied to gothic-revival homes.

Gingerbread on a house.

Ginny wheel: Pulley used for hoisting things up a scaffold.

Ginny wheel.

Girder: A large or principal beam of wood or steel used to support loads along its length.

Girt: A horizontal beam that is placed between support columns that is used for attaching wall cladding.

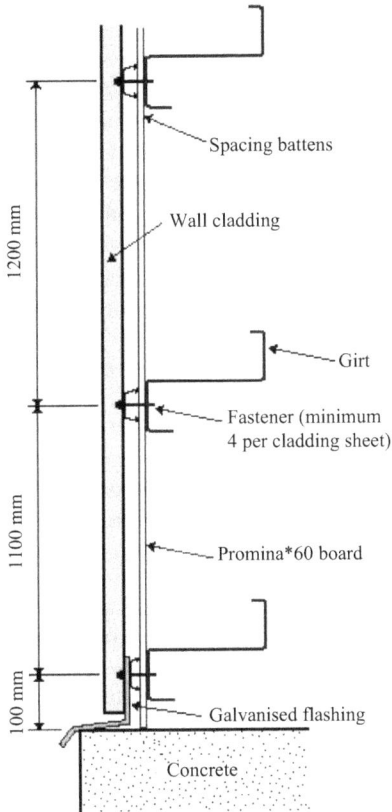

Girts.

Also, a heavy horizontal beam located above the posts in timber framed homes. These beams often support the floor joists. Nowadays, this form of construction is rare.

Girts.

Gland: In a pump, the part that holds one half of the mechanical seal and attaches to the stuffing box.

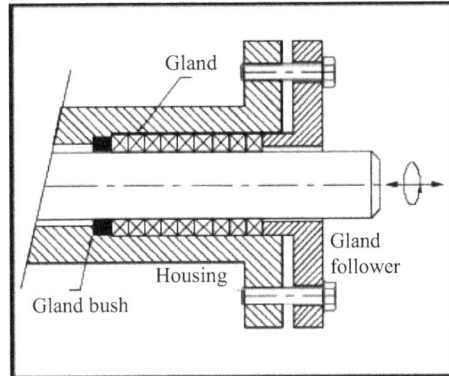

Stuffing box packing.

Glare: Any excessively bright source of light within the visual field that creates discomfort or loss in visibility.

Glass felt: A sheet composed of bonded glass fibers, suitable for impregnation and coating in the manufacture of bituminous roofing and waterproofing materials, and shingles.

Glass stop/Glass bead: Strips of profiled wood or vinyl used to hold the glass in position in the sash. Wood glazing bead is attached to the rails and stiles of the sash using staples, small nails, or vinyl barbs. A vinyl bead is held in place by extruded barbs positioned in the kerf. Aluminum caps may be used over the vinyl bead in some cases. Also referred to as **glazingstop/bead**.

Glaze coat: The top layer of asphalt on a smooth-surfaced built-up roof membrane.

Also, a thin protective coating of bitumen applied to the lower plies or top ply of a built-up roof membrane when application of additional felts or the flood coat and aggregate surfacing are delayed. See **Flood coat**.

Glazed tile: A tile shingle with a color glaze compound put on its surface that produces a smooth and shiny face. **Glazed clay tiles** are baked. **Glazed concrete tiles** dry chemically. Glaze usually adds significantly to the cost of clay tiles, but adds only moderately or not at all to the cost of concrete tiles. **Glazed wall tiles** come in a variety of sizes but are usually about 4"× 4" and typically have a high gloss or matte glaze applied to the finish surface. See **Ceramic mosaic tile** and **Quarry tile**.

Glazing: The process of installing glass that commonly is secured with glazier's points and glazing compound.

Also, glass (and other materials) in a window or door.

Glazing factor: The ratio of interior illuminance at a given point on a given plane (usually the work plane) to the exterior illuminance under known overcast sky conditions. LEED uses a simplified approach for its credit compliance calculations. The variables used to determine the daylight factor include the floor area, window area, window geometry, visible transmittance (Tvis) and window height.

Glazing tape: A two-sided adhesive tape placed between the glass rabbet and the glass and/or the glazing bead and glass of some unit types.

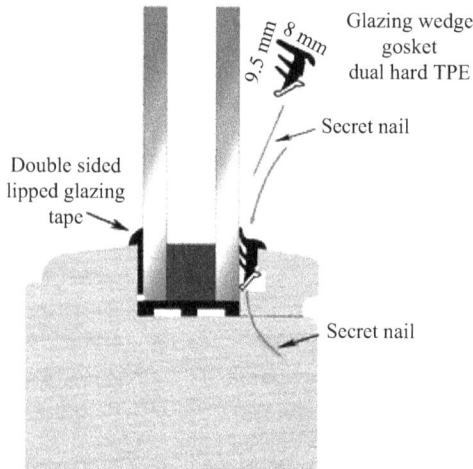

One type of glazing tape.

Glide: Hardware attached between a cabinet drawer and the cabinet box that holds the drawer in a level position as the drawer is pulled out of and pushed into the cabinet box. See **Full extension glide**.

Glider: A window with a movable sash that slides horizontally. Also referred to as a **Horizontal sliding window**.

Globe valve: A valve that lets you adjust the flow of water to any rate between fully on and fully off. See **Gate valve**.

Gloss: The luster or shininess of paints and coatings. Different types of gloss are frequently arbitrarily differentiated, such as sheen, distinctness-of-image gloss, etc. Trade practice recognizes the following gloss levels, in increasing order of gloss: flat (or matte), practically free from sheen, even when viewed from oblique angles (usually less than 15 on 60° meter); eggshell, usually 20–35 on 60° meter; semi-gloss, usually 35–70 on 60° meter; full-gloss, smooth and almost mirror-like surface when viewed from all angles, usually above 70 on 60° meter.

Gloss enamel: A finishing paint material. Forms a hard coating with maximum smoothness of surface and dries to a sheen or luster (gloss).

Gloss meter: A device for measuring the light reflectance of coatings. Different brands with the same description (such as semi-gloss or flat) may have quite different ratings on the gloss meter.

Glue: An adhesive substance used for sticking objects or materials together.

Glue laminating (Glulam): Production of structural or non-structural wood members by bonding two or more layers of wood together with adhesive. A **glued laminated beam** is a structural beam that is composed of wood laminations or lams. The lams are pressure bonded with adhesives to attain a typical thickness of 1 ½".

Glueline: The layer of adhesive that attaches two adherends. Also called a **Bondline**.

Glue-up tile: A roof tile that is glued into place. Glue-up tile is usually 12" by 12".

Goose neck: A long, curving handrail piece that is used to step down and make a long vertical transition between handrail parts on a stair balustrade. See **Top rail, volute, One-quarter turn**, and **Balustrade**.

A goose neck handrail piece.

Also, a faucet that has a high, arched spout. Cooks who must fill large pots with water find goose neck faucets especially handy to use.

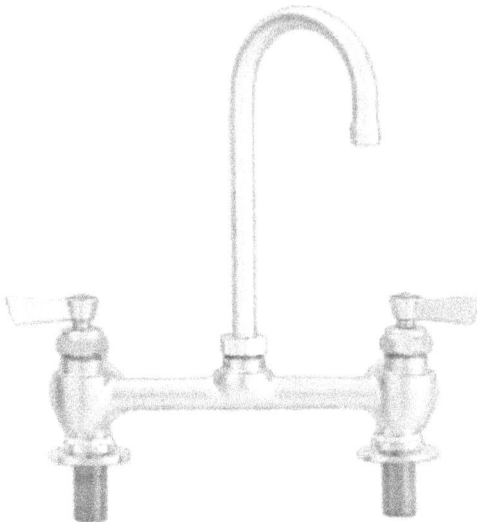

Goose neck faucet.

Gouge: Rock crushed in a fault zone.

Government incentives: Concession given or measures taken by local or regional government to attract firms or investment dollars to a given locality for the purposes of promoting economic growth and encouraging development.

Governor: Besides the guy who works in your state's capital, a mechanical speed control mechanism. For an elevator, it is a wire rope-driven centrifugal device used to stop and hold the movement of its driving rope. This initiates the activation of the car safety device. It opens a switch, which cuts off power to the drive motor and brake if the car travels at a preset overspeed in the down direction. Some types of governors will also open the governor switch and cut off power to the drive motor and brake if the car overspeeds in the up direction.

Also, a device automatically regulating the supply of fuel, steam, or water to a machine, ensuring uniform motion, or limiting speed.

Graben: Long, narrow trough bounded by one or more parallel normal faults. These down-dropped fault blocks are caused by tensional crustal forces. Also referred to as rift valley.

Grade: The elevation at ground level or at any given point.

Also, the work of leveling dirt, as in to grade.

Also, the designated quality of a manufactured piece of wood, steel, or other material.

Grade beam: A part of a building's foundation.It is a reinforced concrete beam that transmits the load from a bearing wall into spaced foundations such as pile caps or caissons.

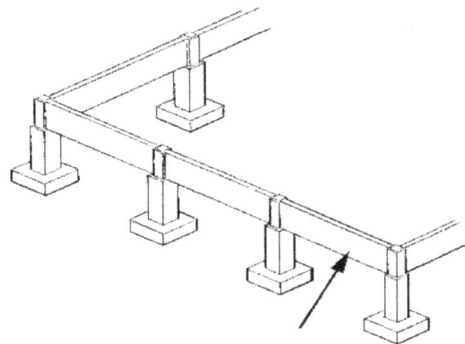

Grade beams.

A grade beam differs from a wall footing: while a grade beam is designed for bending and typically spans between pile caps or caissons, a wall footing bears directly on soil to transmit the weight of the wall into the ground.

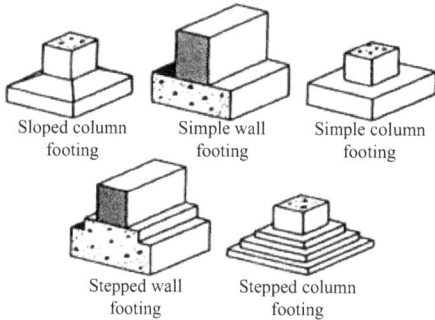

Sloped column footing   Simple wall footing   Simple column footing

Stepped wall footing   Stepped column footing

Wall footings.

A grade beam also differs from a strap beam: while a grade beam is reinforced to distribute the weight of a wall to separate foundations, a strap beam is designed to redistribute the weight of a column between footings.

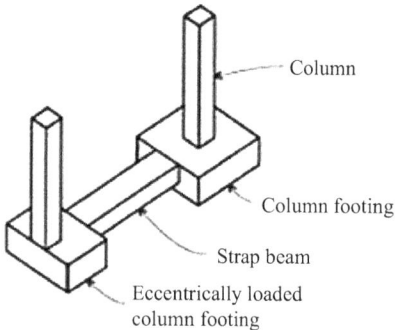

Column

Column footing

Strap beam

Eccentrically loaded column footing

Strap beam.

Grade stake: A stake that is placed in the ground and marked at the point where the grade should be found once the building part is in place. Grade stakes are often placed and marked to indicate where the top of the concrete will be located once the pour is complete. Grade stakes used as a guide for establishing the final level of the concrete are usually pulled out and their holes filled with wet concrete once the wet concrete has been leveled at the proper grade.

Grade stamp: A **grade stamp** is printed on lumber that is produced in North America. As a natural product most lumber has some defects, e.g., large knots or splits, that may reduce its strength. Because of these and other less obvious defects, lumber needs to be appraised by trained inspectors and assigned a grading stamp when it leaves the mill.

A grade stamp is put on lumber to show its important characteristics and to provide mill information. A grademark is a stamp or symbol applied to a piece of lumber, by the grader at a planermill, to designate grade. Building inspectors may ask to see grademarks to ensure lumber quality.

Grader: A construction machine with a long blade that is used to create a flat surface. Typical models have three axles, with the engine and cab situated above the rear axles at one end of the vehicle and a third axle at the front end of the vehicle, with the blade in between.

Graders are used to bring the surface to **finish grade**; the **rough grading** having been performed by heavy equipment or engineering vehicles such as scrapers and bulldozers.

In areas that receive little snow, graders are sometimes used to remove the occasional snow accumulations.

Gradient: Inclination of a road, piping, or the ground, expressed in percent.

Graduated payment mortgage (GPM): A fixed-rate, fixed-schedule loan. It starts with lower payments than a level payment loan; payments rise annually, with the entire increase being used to reduce the outstanding balance. The increase in payments may enable the borrower to pay off a 30-year loan in 15–20 years, or less.

Grain: With regard to wood, the direction, size, arrangement, appearance, or quality of the fibers in wood. To have a specific meaning the term must be qualified.

• **Close:** Narrow, inconspicuous annual rings. The term is sometimes used to designate wood having small and closely spaced pores, but in this sense the term "fine-textured" is preferred.

• **Coarse:** Wide, conspicuous annual rings in which there is considerable difference between springwood and summer-wood.

The term is sometimes used to designate wood with large pores, such as oak, ash, chestnut, and walnut, but in this sense the term "coarse-textured" is preferred.

- **Cross:** Fibers that deviate from a line parallel to the sides of the piece. Cross grain may be either diagonal or spiral grain or a combination of the two.
- **Curly**: Fibers that are distorted so that they have a curled appearance, as in "birdseye" wood. The areas showing curly grain may vary up to several inches in diameter.
- **Diagonal:** Has annual rings that are at an angle with the axis of a piece as a result of sawing at an angle with the bark of the tree or log. Diagonal grain is a form of cross grain.
- **Edge:** Sawed so that the wide surfaces extend approximately at right angles to the annual growth rings. Lumber is considered edge-grained when the rings form an angle of 45–90° with the wide surface of the piece.
- **End:** The grain as seen on a cut made at a right angle to the direction of the fibers, e.g., on a cross section of a tree.
- **Flat:** Sawed parallel to the pith and approximately tangentially to the growth rings. Lumber is considered flat-grained when the annual growth rings make an angle of less than 45° with the surface of the piece.
- **Interlocked:** Fibers that for several years slope in a right-handed direction, and then for a number of years slope to a left-handed direction, and so on. Interlocked wood is very difficult to split radially, though tangentially it may split fairly easily.
- **Open:** The common classification for woods with large pores, such as oak, ash, chestnut, and walnut. Also known as "coarse-textured."
- **Spiral:** Fibers that take a spiral course around the trunk of a tree instead of the normal vertical course. The spiral may extend in a right-handed or left-handed direction around the tree trunk. Causes slope of grain in lumber. Spiral grain is a form of cross grain.

- **Straight:** Fibers that run: parallel to the axis of a piece.
- **Vertical:** Another term for edge-grained lumber.
- **Wavy:** Fibers that collectively take the form of waves or undulations.

Also, the individual mineral pieces or crystals that make up a rock.

Also, a unit of measure for the mass of moisture: a unit of weight equal to 0.002285 ounces or 0.036 dram.

Grain slope: Regarding doors, expression of the angle of the grain to the long edges of the veneer component.

Grain sweep: Regarding doors, expression of the angle of the grain to the long edges of the veneer component over a 12 inch (300 mm) length from each end of the door.

Gram (g): A unit of mass in the SI system of weights.

Granite: A very hard, granular, crystalline, igneous rock consisting mainly of quartz, mica, and feldspar and often used as a building stone. It is common practice in the industry to classify architectural granite as either of two types:

- **Building granite:** Granite used either structurally or as a veneer for exterior or interior wall facings, steps, paving, copings, or other building features.
- **Masonry granite:** Granite used in larger blocks for retaining walls, bridge piers, abutments, arch stones, and similar purposes.

Gram: A metric unit of weight equal to one thousandth of a kilogram. One ounce is ~28 g.

Granular loss: Granular loss occurs when mineral granules embedded in a composition shingle are loosened by the impact from a hailstone when the hailstone does not bruise or fracture the shingle. See **Fractured composition shingles** and **Bruised composition shingles**.

Granule: Opaque, natural, or synthetically colored aggregate commonly used to surface cap sheets, shingles, and other granule-surfaced roof coverings. Also referred to as **mineral** or **ceramic granule**.

Grass cloth wallpaper: Wall covering made of loosely woven vegetable fibers.

Grating: A framework of parallel or crossed bars, typically preventing access through an opening while permitting communication or ventilation.

Also, protection over a channel, trench, etc.

Gravel: Aggregate resulting from the natural erosion of rock.

Gravel stop: A low profile upward-projecting metal edge flashing with a flange along the roof side, usually formed from sheet or extruded metal. Installed along the perimeter of a roof to provide a continuous finished edge for roofing material. A gravel stop acts as a bitumen-stop during mop application of hot bitumen along a perimeter edge.

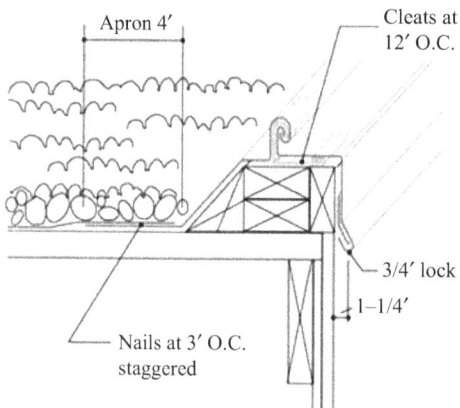

Gravel stop.

Gravity: An attractive force between two objects; each object accelerates at a rate equal to the attractive force divided by the object's mass. Objects near the surface of the earth tend to accelerate toward the earth's center at a rate of 32.2 ft/s^2 (9.81 m/ s^2); this value is often called the gravitational constant and denoted as g. See **G** or **g**.

Gravity dam: A dam constructed so that its weight resists the force of water pressure. Gravity holds the dam in place against the

push from the water. The water presses laterally (downstream) on the dam, tending to overturn it by rotating about its toe (a point at the bottom downstream side of the dam). The dam's weight counteracts that force.

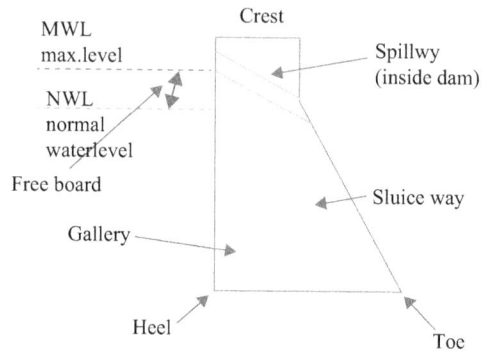

A gravity dam.

Gravity-fed flush: A common gravity-fed flush toilet that quietly removes waste by dropping the water from the tank to the bowl thereby creating a siphonic action that pushes waste out of the bowl.

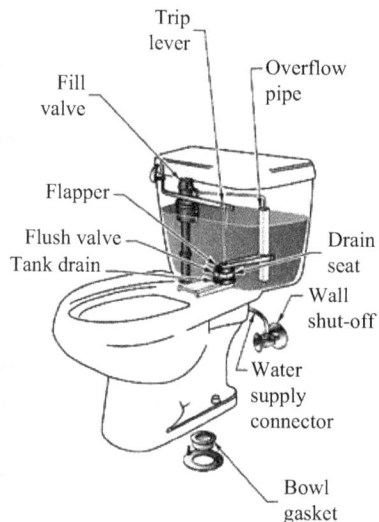

A gravity-fed water closet.

Graywater or gray water: Defined by the Uniform Plumbing Code (UPC) in its Appendix G, titled "Gray Water Systems for Single-Family Dwellings," as "untreated household waste water which has not come into contact with toilet waste. Gray water includes used water from bathtubs, showers, bathroom wash basins, and water from clothes-washer and laundry tubs. It shall not include waste water from kitchen sinks or dishwashers." The International Plumbing Code (IPC) defines graywater in its Appendix C, titled "Gray Water Recycling Systems," as "waste water discharged from lavatories, bathtubs, showers, clothes washers, and laundry sinks." Some state and local authorities allow kitchen sink wastewater to be included in graywater. State and local codes may differ from the UPC and IPC definitions in other respects as well; designers must comply with the definitions established by the authority having jurisdiction in their areas.

Great circle: The shortest path between two points on the surface of a sphere lies along a great circle.

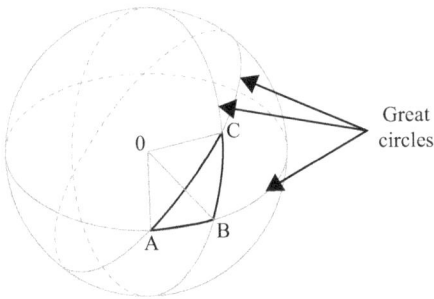

Great circles.

Green: As used here, favoring or supporting environmentalism or something or someone not mature or ripe. For example:

- **Green building:** A movement in architectural and building circles aimed at creating structures that are occupant and environmentally friendly. Criteria such as sustainability, energy efficiency, and healthfulness are considered.
- **Green cleaning:** The use of cleaning products and practices that have less adverse environmental affects than conventional products and practices. Green cleaning may add cost to your project.
- **(The) Green Globes™ system:** In 2004, the Green Building Initiative finalized an agreement to bring the Green Globes™ environmental assessment and rating tool into the US market. The Green Globes system is a green management tool that includes an assessment protocol, rating system, and guide for integrating environmentally friendly design into commercial buildings. It also facilitates recognition of the project through third-party verification.
- **Green lumber:** Freshly sawed or undried wood.
- **Green roof:** A flat roof that is deliberately covered with growing material.

A green roof.

- **Green Seal:** Green Seal is a not-for-profit organization that uses science-based programs to empower consumers, purchasers, and companies to create a more sustainable world. Green Seal is used by purchasers as a way to determine the environmental preferabilty of a product or service. With organizations and governments of all levels requiring the purchase of environmentally preferable products (EPPs), the **Green Seal certification** ensures that the product meets sustainability requirements.
- **Green strength:** The immediate holding power achieved by a sealant or adhesive.
- **Green target sizes:** The size to which green lumber must be cut to allow for

sawing deviation, shrinkage in drying and allowance for fiber removed in the finishing process.

Greenfield conduit: A type of flexible metal conduit.

Greenhouse gasses: Gases that provide an insulating effect in the earth's atmosphere. Some of these gasses are accumulating and some are being destroyed because of human activity, thereby leading to global climate change. These gases include carbon dioxide, methane, nitrous oxide, ozone, and water vapor.

Grid: The completed assembly of main and cross tees in a suspended ceiling system before the ceiling panels are installed. Also the **decorative slats (muntin)** installed between glass panels.

Grillage beam: A short beam used like a bearing plate to distribute large reactive loads to a wall such as the load from a joist girder.

Grille: A term referring to window pane dividers or muntins. It may be a type of assembly fitted to the interior of the window or door unit that can be detached for cleaning. Also a grille can be fitted inside the sealed insulating glass unit, when it also is referred to as a **grid**.

Grilles between glass (GBG's): Aluminum flat or contour bars divide the glass visually. Because the grille is between the glass, grilles do not become misplaced or damaged. Cleaning is easy without small panes of individual lites.

Grilles between glass.

Grip length: The total distance between the underside of the nut to the bearing face of the bolt head; includes washer, gasket thickness, etc.

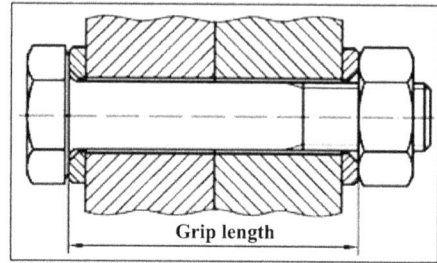

Grip length.

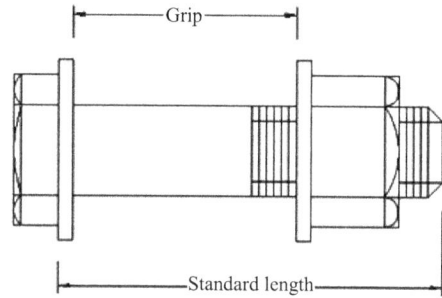

As opposed to standard length.

Grommet: A circular eyelet which reinforces a hole that has been punched into a piece of material.

Gross: As used here, consisting of an overall total exclusive of deductions:

• **Gross area:** The entire floor area of a building or the total square footage of a floor. See **Habitable space**.

• **Gross capacity:** The maximum weight that a machine is designed to lift.

• **Gross leasable area (GLA):** The total floor area designed for tenant occupancy and exclusive use, including basements, mezzanines, and upper floors, and it is measured from the center line of joint partitions and from outside wall faces. GLA is that area on which tenants pay rent; it is the area that produces income.

• **Gross lease:** A lease in which all expenses associated with owning and operating the property are paid by the landlord. See **Net lease**.

- **Gross operating income:** The total income generated by the operations of a property before payment of operating expenses. It is calculated from potential rental income, plus other income affected by vacancy, less vacancy and credit losses, plus other income not affected by vacancy. The Annual Property Operating Data form or the Cash Flow Analysis Worksheet can be used to calculate a property's gross operating income.

Ground (terms that begin with ground and relate to earthquakes):

- **Ground acceleration:** Acceleration of the ground due to earthquake forces.
- **Ground displacement:** The distance that ground moves from its original position during an earthquake.
- **Ground motion:** A general term that refers to the quantitative or qualitative motion of the earth's surface produced by earthquakes or explosions. Also referred to as **ground response**.
- **Ground movement:** A general term that includes all aspects of motion: acceleration, particle velocity, and displacement. (The plates of the earth's crust move slowly relative to one-another thereby accumulating pressure or strain resulting in slippage and complex vibration inducing forces in a building.)
- **Ground velocity:** The velocity of the ground during an earthquake.

Ground (terms that begin with ground and relate to electricity): Ground refers to electricity's habit of seeking the shortest route to earth. Neutral wires carry it there in all circuits. An additional grounding wire or the sheathing of the metal-clad cable or conduit protects against shock if the neutral leg is interrupted.

- **Ground wireman** electrical conductor that leads to an electric connection at the earth.
- **Ground-fault:** A fault, or insulation failure, in the wire used to create a path to ground.
- **Grounding:** To prevent the buildup of hazardous voltages in a circuit by creating a low-resistance path to earth or some other ground plane.

- **Grounding rods:** Metal rods that are driven into the soil to a predetermined depth for the purpose of grounding the building against lightning strike.

Ground (other terms that begin with the word ground):

- **Ground failure:** A situation in which the ground does not hold together such as landslides, mud flows, and liquefaction.
- **Ground iron:** The plumbing drain and waste lines that are installed beneath the basement floor. Cast iron was once used, but black plastic pipe (ABS) is now widely used.
- **Ground joint:** A joint that consists of pressing two mating surfaces together without a gasket or supplemental sealant.

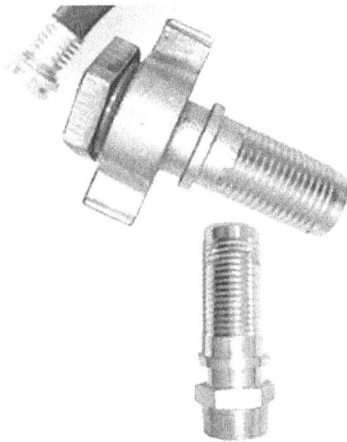

Ground joint.

- **Ground lease:** A lease of the land only. Usually the land is leased for a relatively long time to a tenant who constructs a building on the property. A land lease separates ownership of the land from ownership of buildings and improvements constructed on the land.
- **Ground thread:** A class of fit or how tightly a fastener will fit into a threaded hole. A ground thread is the ultimate in tap accuracy. Class of fit is specified according to H limits. One H limit equals .0005" over the basic pitch diameter.
- **Groundmass:** The main part of an igneous rock made up of finer grains in which the larger crystals are set.

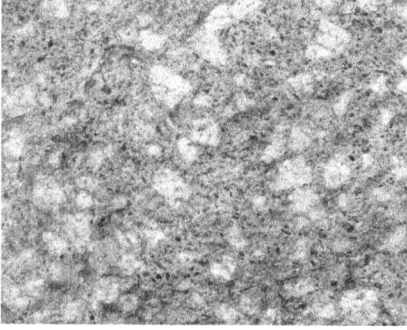

The finer grains make up the groundmass.

- **Groundskeeper:** A qualified professional with relevant and sufficient expertise who oversees and is responsible for the establishment and maintenance of landscaping, vegetation, and pest control on the project building's grounds.
- **Groundwater:** Water from an aquifer or subsurface water source. The level of this water in the soil is called the **water table**.
- **Groundwork:** Foundations, drainage, leveling, and other building operations involving digging.
- **Ground-fault circuit interrupter (GFCI):** A fast-acting circuit breaker designed to shut off electric power in the event of a ground-fault within as little as 1/40 of a second. It works by comparing the amount of current going to and returning from equipment along the circuit conductors. When the amount going differs from the amount returning by ~5 milliamperes, the GFCI interrupts the current.

  The GFCI is rated to trip quickly enough to prevent an electrical incident. If it is properly installed and maintained, this will happen as soon as the faulty tool is plugged in. If the grounding conductor is not intact or of low-impedance, the GFCI may not trip until a person provides a path. In this case, the person will receive a shock, but the GFCI should trip so quickly that the shock will not be harmful.

  The GFCI will not protect a person who is holding two hot wires, a hot and a neutral wire in each hand, or contacting an overhead power line. However, it protects against the most common form of

electrical shock hazard, the ground-fault. It also protects against fires, overheating, and destruction of wire insulation. Also known as **residual-current device (RCD)** and **residual-current circuit breaker (RCCB)**.

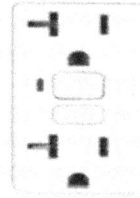

Ground-fault circuit interrupter (GFCI).

Group multi-occupant spaces: Includes conference rooms, classrooms, and other indoor spaces used as places of congregation for presentations, training sessions, etc., where workers engage in a common task and share the lighting and temperature controls. Group multi-occupant spaces do not include open office plans that contain individual workstations.

Group velocity: The velocity with which most of the energy in a wave train travels.

Grousers: A protrusion on the surface of a wheel or continuous track segment, Grousers increase traction in soil, snow, or other loose material. They are commonly used on bulldozers, loaders, and excavators, agricultural vehicles, and snowmobiles.

Grousers on a bulldozer track.

Grout: A wet mixture of cement, sand, and water that flows into masonry or ceramic crevices to seal the cracks between the different pieces. Also, mortar made of such consistency (by adding water) that it will flow into the joints and cavities of the masonry work and fill them solid.

Non-shrink grout is a cementitious material that is used to partially fill penetration pockets (pitch pans). A pourable sealer is used afterward.

Grooved fitting: A gasketed mechanical pipe fitting made of ductile iron that is clamped onto a groove in a pipe rather than screwed onto threads. It is mostly used in commercial applications on pipes larger than 1 1/2" in diameter.

Grooved fitting.

Growth ring: One year's growth increment of a tree composed of one band of springwood (earlywood) and one band of summerwood (latewood). Also called an Annual ring.

Guarded: As relates to electrical, something that is covered, shielded, fenced, enclosed, or otherwise protected by means of suitable covers, casings, barriers, rails, screens, mats, or platforms to remove the likelihood of approach or contact by persons or objects to a point of danger. See **Guarding**.

Guarding: Protection against people or things falling off the edge of stairs, landings, or balconies.

Also, placement of live parts of electrical equipment where they cannot accidentally be contacted, such as in a vault, behind a shield, or on a raised platform, to which only qualified persons have access.

Guardrail system: A barrier erected to prevent employees from falling to lower levels. See **Guarding**.

Guaranteed maximum price-construction management (GMP-CM): A form of CM contract whereby the construction manager guarantees, in addition to providing additional construction management services, a ceiling price to the owner for the cost of construction.

Gudgeon: A socket for the pintle of a hinge.

Gudgeon hinge pack.

Guide arm: A valve that controls water flow into the toilet tank.

Guide blocks: The arms of the saw-guide mechanism which hold the guide pins on a bandsaw.

Guide rails: A device or mechanism to direct products, vehicles or other objects through a channel, conveyor, roadway, or rail system. There are numerous types of guide rails, among them the following:

- **Factory or assembly line rails.**
- **Accessories to power tools:** such as a straight, swivel, or angle jig for a circular saw.
- **Elevator or lift shafts:** SteelT-, round, or formed sections with guiding surfaces installed vertically in a hoistway to guide and direct the course of travel of an elevator car and elevator counterweights.

- **Roadway and bridge guide rails:** Sometimes roadway and bridge **guide rails** are referred to as **guardrails**. But there is a legal distinction between a guide rail and a guard rail. While guide rail systems are intended to guide vehicles back onto the road guardrails are intended to guard them from going off the road and into potential danger.
- **Central rail:** A rail that guides the rubber-tired train of a rubber tired metro.

Guide shoes: Devices that are used mainly to guide an elevator car and counterweight along the path of the guide rails. They also assure that the lateral motion of the car and counterweight is kept at a minimum as they travel along the guide rails.

Also, guiding projections mounted on the bottom edge of horizontally sliding elevator doors or gates, or on the sides of vertically sliding doors or gatesto guide them.

Gullet: The area of the saw tooth in which the sawdust is carried. Formuli are available to determine feed speed. Feed speed should not allow the gullet fill to exceed 100% (woodproductsonlineexpo.com.)

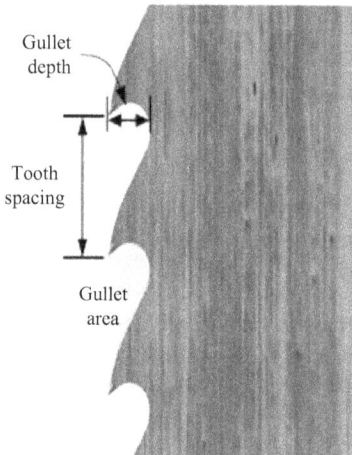

Gullet area (from wood products online expo, bandmill saws – gullet fill percentage).

Gully: A water-worn ravine.

Also, a deep artificial channel serving as a gutter or drain.

Gum pockets: Well-defined opening between rings of annual growth containing gum or evidence of prior gum accumulations. Found mainly in cherry wood.

Gum spots: Accumulation of resinous material often found on panel surfaces. Usually removed by sanding.

Gun nails: Most modern framers use nail guns and rarely use a hammer. Gun nails come in strips or coils so they can easily be loaded in the nail gun. See **Common nails, Box nails**, and **Sinkers**.

Gusset: A flat wood, plywood, or similar type member used to provide a connection at the intersection of wood members. Most commonly used at joints of wood trusses. They are fastened by nails, screws, bolts, or adhesives.

Also, a steel plate used to connect structural steel members or to reinforce members. It is usually inserted between the top or bottom chord of a joist or joist girder.

Gutter: A shallow channel or conduit of metal or wood set below and along the (fascia) eaves of a house to catch and carry off rainwater from the roof.

Also, an open channel for receiving and carrying away rain water.

Gutter brace: A place where the gutter is attached to the structure's fascia.

Gymnosperm: A plant that has seeds unprotected by an ovary or fruit. Gymnosperms include the conifers, cycads, and ginkgo. Compare with angiosperm.

Gyp board, drywall, wall board, and gypsum board: A panel (normally 4' × 8', 10', 12', or 16') made with a core of Gypsum (chalk-like) rock that covers interior walls and ceilings.

Gypsum-based underlayment: An underlayment made from fiber reinforced gypsum which is easy to cut and install and is highly resistant to indentation. See **Underlayment, particleboard underlayment, Plywood underlayment, Lauan plywood underlayment, Cement board underlayment**, and **Untempered hardboard underlayment**.

Gypsum board: A construction board of gypsum (hydrous calcium sulfate) sandwiched between two layers of craft paper or other fibrous material used as interior or exterior surfacing in a wall or ceiling assembly. Also called **drywall** and/or **plasterboard**.

Gypsum plaster: Gypsum formulated to be used with the addition of sand and water for base-coat plaster.

H Clip: Small metal clips formed like an "H" that fits at the joints of two plywood (or wafer board) sheets to stiffen the joint. Normally used on the roof sheeting.

H Clip

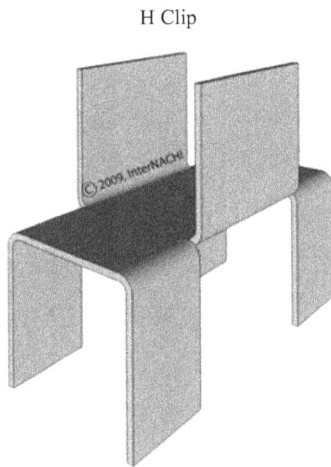

H-section: A structural member shaped in cross-section like an H.

H section.

H-series joist: A series of joist adopted in 1961 so proportioned that the allowable tension or bending stress does not exceed 22,000 psi or 30,000 psi depending on whether 36 ksi or 50 ksi yield steel is used.

Habitable space: A space in a building for living, sleeping, eating, and cooking. Most codes do not consider bathrooms, toilet rooms, closets, halls, storage, utility spaces, etc., as habitable spaces.

Hairline: A thin, perceptible line showing at the joint of two pieces of wood.

Half bond: A course of brick in which the vertical joint between bricks is half-way across the length of the brick in the course below it. Also referred to as **running bond** and **stretcher bond**.

Half gable roof: See **Shed roof**.

Half laced valley: A pattern formed in the valley of a roof by overlapping the valley with shingles from one side of the valley and cutting shingles from the other side so they end at the center of the valley.

　　Also, a molding whose profile is half a circle. May be used as a screen molding or bead, shelf edge, or panel mold.

Half timbered: A descriptive term for a traditional timber-framed house. It is a method of construction that uses timber frames (post and beam) for internal and external walls. Brick and plaster are normally used to fill the gaps between timbers.

Half timbered.

Half-round: A method of veneer cutting similar to rotary cutting, except that the piece being cut is secured to a "stay log" a device that permits the cutting of the log on a wider sweep than when mounted with its center secured in the lathe to produce rotary sliced veneer. A type of half-round cutting is used to achieve plain-sliced or flat-cut veneer.

Halogen burner: An electrical burner which instantly becomes hot when the burner is turned on. See **Ceran top**.

Halons: Substances used in fire-suppression systems and fire extinguishers. These substances deplete the stratospheric ozone layer and are outlawed in some areas.

Hammerbeam roof: A form of historical roof truss, usually comprising a central truss section spanning between two cantilevers.

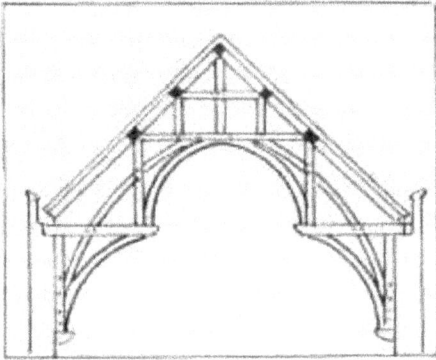

Hammerbeam roof.

Hand as a preface:

- **Hand brace installer bit:** An installer bit with a taper square shank. The taper square fits into and is held by a hand brace. See **Brace**.

Hand brace installer bit.

- **Hand hole:** An oval opening (approximately hand sized) in the boiler plate allowing access for inspection and washing out.

   A **hand hole door** is a solid oval piece of metal with a threaded shaft extending from the middle. The door fits through a

hand hole (approximately hand sized) and butts against the inside of the outer boiler plate. A clamp fits over the shaft that is bolted down sealing the door against the boiler plate.

Hand hole and hand hole door.

- **Hand rail:** Used as a hand support in a stairwell.
- **Hand shower:** A shower-head attached to a flexible shower hose. Also referred to as a hand held.
- **Hand texture:** Any texture that is applied to drywall by hand, without the use of a machine. This include brush textures and hock textures.
- **Hand-feet rule:** A rule for safe use of a ladder. The Hand/Feet Rule states that when climbing a ladder you should always have either one foot and two hands or two feet and one hand on the ladder at all times.
- **Hand-planned finish:** A distressing treatment by which a new floor or board is scraped with blades by hand to give an undulating and worn effect.
- **Hand-sealing:** The method to assure and enhance sealing of shingles with roofing cement (ASTM D-4586) on very steep slopes, in high wind areas, and when installing in cold weather.

   To hand-seal shingles the roofer uses a tube of roof cement in a caulking gun, placing a quarter-sized dollop of roof cement under the front corners of each shingle or shingle tab and then presses the down; the cement shouldn't be visible below the shingle.

- **Hand-split and resawn shake:** Wood shake with a rough, split face, and a sawn back.
- **Hand-tabbing:** Applying spots of adhesive to shingle tabs.
- **Handing:** A term used to describe the right or left hand operation of a window or door.

  To determine handing:

  1. Stand on the outside of the door. For example, if you are getting the handing for the front entry door of a house, stand outside of the house; if you are getting the handing for a bathroom door stand outside of the bathroom, etc.
  2. Facing the door, see on what side the hinges are located. If the hinges are on the left side of the door, it is a left hand door. If the hinges are on the right side of the door, it is a right hand door.

- **Handles:** With regard to faucets, handles typically come in three styles: knob, lever, and cross. Some manufacturers allow a customer to mix and match components to create a custom handle consisting of handle bases and handle inserts.
- **Handrail:** A length of timber or metal at hand height at the side of a staircase or landing.

Hang a door: The procedure for attaching a door to its frame and adjusting it to open, shut, latch, and lock properly.

Hangers: Any device that is used to hang a mechanical piece of equipment, pipe, duct, and ceiling, off the underside of a structural frame or floor assembly above.

Hanger wires: Wires that are used to suspend the acoustical ceiling from the existing structure. The standard material is 12 gauge galvanized, soft annealed steel wire, but heavier gauge wire is available for higher load carrying installations, or situations where hanger wire spacing exceeds 4 feet on center.

Stainless steel wire and nickel-copper alloy wire are often used in severe environment designs. Seismic designs or exterior installations subject to wind uplift may require supplemental bracing or substantial hanger devices such as metal straps, rods, or structural angles.

Hard as a preface:
- **Hard costs:** All items of expense directly incurred by or attributable to a specific project, assignment, or task. Also referred to as **direct costs** and **construction costs**.

  On the other hand, **soft costs** or **indirect costs** refer to costs incurred in addition to the direct construction cost such as marketing, taxes, finance charges, insurance, interest payments, and general administration costs.
- **Hard hat:** A hat constructed of a rigid material and designed to take a predetermined impact load for the protection of the head. It must be ASTM and OSHA approved.
- **Hard joint:** A joint in which the plates and material between the nut and bolt bearing surfaces have a high stiffness when subjected to compression by the bolt load. A joint is usually defined as hard if the bolt is tightened to its full torque and it rotates through an angle of 30° or less after it has been tightened to its snug condition.
- **Hard water:** Natural water containing impurities in various proportions. Traditional hardness is a measure of calcium or dissolved solids in a solution, measured in parts per million. Hard water generally ranges from 100 to 250 ppm.
- **Hardened washers:** The force under the head of a bolt or nut can exceed, at high preloads, the compressive yield strength of the clamped material. If this occurs excessive embedding and deformation can result in bolt preload loss. To overcome this hardened washers under the bolt head can be used to distribute the force over a wider area into the clamped material. A more modern alternative is to use a flange headed nuts and bolts.

Hex flange head bolt.

- **Hard-coat glass:** A glass product that is coated during the manufacturing process at the molten glass stage. Also known as a pyrolytic coating, this type of coating offers a surface that is generally as durable as an ordinary glass surface, and therefore requires no special handling and does not need to be used in an insulating glass unit. The other type of glass coating is a sputter-coat, which is applied in a secondary process. Sometimes referred to as a soft-coat, these types of coatings generally require some additional care in handling and fabrication and must be used within an insulating glass unit.

- **Hardboard:** Generally this term refers to one of the botanical groups of trees that have broad leaves in contrast to the conifers or softwoods. The term has no reference to the actual hardness of wood. Also called **Angiosperms** or **Deciduous**.

- **Hardness:** Hardness is usually defined as the resistance of metal to plastic deformation, usually by indentation. The term may also refer to stiffness or temper, or to resistance to scratching, abrasion, or cutting. It is the property of a metal that gives it the ability to resist being: permanently, deformed (bent, broken, or have its shape changed), when a load is applied. The greater the hardness of the metal, the greater resistance it has to deformation. For example, for rivets and threaded fasteners, sufficient hardness ensures that the surface of the fasteners have the capacity to transmit load by bearing. Hardness is

usually a specified requirement for rivets, nuts, and washers.

In metallurgy hardness is defined as the ability of a material to resist plastic deformation.

In mineralogy hardness is the property of matter commonly described as the resistance of a substance to being scratched by another substance; hardness is based on the Mohs scale of mineral hardness. This scale is based on the ability of one natural sample of matter to scratch another mineral. The samples of matter used by Mohs scale are as follows:

Mineral	Mohs Hardness	Absolute Hardness
Talc	1	1
Gypsum	2	3
Calcite	3	9
Fluorite	4	21
Apatite	5	48
Orthoclase Feldspar	6	72
Quartz	7	100
Topaz	8	200
Corundum	9	400
Diamond	10	1600

- **Hardscape:** Consists of the inanimate elements of the building landscaping, including pavement, roadways, stone walls, concrete paths, and sidewalks, and concrete, brick, or tile patios.

- **Hardware:** All of the "metal" fittings that go into the home when it is near completion. For example, door knobs, towel bars, handrail brackets, closet rods, house numbers, door closers, etc., usually the interior trim carpenter installs the "hardware."

- **Hardwood:** A general term used to designate lumber or veneer produced from temperate zone deciduous or tropical broad-leaved trees in contrast to softwood that is produced from trees that are usually needle bearing or coniferous. The term does not infer hardness in its physical sense, for example, balsa wood is a hardwood although very weak and soft. (Efforts should be made to ensure that the timber is from renewable sources.)

Harmonic tremor: Describes continuous rhythmic earthquakes that can be detected by seismographs. Harmonic tremors often precede or accompany volcanic eruptions.

Hasp: A simple bar and ring piece of hardware that allows for the locking of a door or gate by inserting a padlock through the hardware.

Hatch: A unit used to provide access to a roof from the interior of a building.

Haunch: An extension, knee like protrusion of the foundation wall that a concrete porch or patio will rest upon for support.

Hazard insurance: Protection against damage caused by fire, windstorms, or other common hazards. Many lenders require borrowers to carry it in an amount at least equal to the mortgage.

Hazardous atmosphere: An atmosphere that by reason of being explosive, flammable, poisonous, corrosive, oxidizing, irritating, oxygen deficient, toxic, or otherwise harmful, may cause death, illness, or injury.

Hazardous substance: A substance that by reason of being explosive, flammable, poisonous, corrosive, oxidizing, irritating, or otherwise harmful, is likely to cause death or injury.

Hazardous waste: Any waste as defined in 40 CFR 261.3 (RCRA). RCRA's definition means a solid waste, or combination of solid wastes, that, because of its quantity, concentration, or physical, chemical, or infectious characteristics may pose a substantial present or potential hazard to human health or the environment. See **RCRA**.

Head: Main horizontal frame member at the top of a window or door. Also referred to as **head jamb**.

Also, an often used preface:

- **Head joint:** The vertical joint of mortar in all masonry bonding.
- **Headbolt:** A locking rod device installed vertically in the stile or astragal of a door or screen which when activated secures the door in a stationary position.

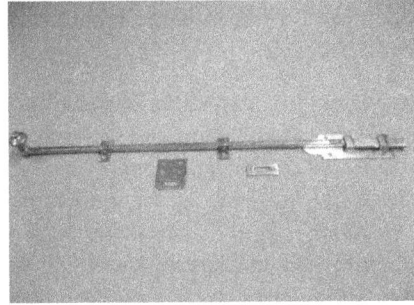

Brass head bolt.

- **Headlap:** The distance of overlap measured from the uppermost ply or course to the point that it laps over the undermost ply or course.
- **Header:** A beam placed perpendicular to joists and to which joists are nailed in framing for a chimney, stairway, or other opening.

    A header is placed between two joists or between a joist and a wall that carries another joist or joists. A header is usually made up of an angle, channel, or beam with saddle angle connections on each end for bearing.

    Also, the horizontal structural member over an opening (for example over a door or window).

    Also, A brick whose "head" or short end is visible on the surface of the wall. See **Stretcher**.

- **Header board:** Terminology used in deck construction. When a deck that juts out from a house is tied into the house structurally a header board or "ledger" is bolted to the house. The header board supports the house-side of the deck. Usually this board will be the same width as the deck joists that will be "hung" from the header board using joist hangers.

The joists meet the header board at a right angle, and the tops of the joists are at the same level as the top of the ledger.

• **Header jamb:** Specific name for the jamb found on the top of the inside of window and door openings.

Hearth: The fireproof area directly in front of a fireplace. A hearth is the inner or outer floor of a fireplace.

Heartwood: The non-active center of a tree generally distinguishable from the outer portion (sapwood) by its darker color.

Heartwood is found at the center of the tree. Generally it is of higher quality than sapwood, with less and tighter knots and more resistant to decay.

Heat: Energy that creates warmth and increases the temperature of a substance. Any energy that is wasted or used to overcome friction is converted to heat. Heat is measured in calories or British thermal units (Btus). One Btu is the amount of heat required to raise the temperature of one pound of water one degree Fahrenheit.

Many other terms include the word heat:

• **Heat blister:** A bubble that forms in a shingle when the asphaltic coating does not properly bond to the mat.

• **Heat exchanger:** A device that transfers heat from a source, such as a flame, to a conductor, such as air or water.

• **Heat gain:** The transfer of heat from outside to inside by means of conduction, convection, and radiation through all surfaces of a house.

• **Heat gun:** A hand-held device used to emit a stream of hot air. Temperatures usually vary between 200°F and 1000°F, with some hotter models running as high as 1400°F (for removing lead paint temperatures below 1100°F) are used to minimize vaporization.

Typically they are rated between 600 and 1750 watts. Heat guns usually have the form of a handgun, hence the name.

While heat guns are usually electrically heated some use a gas flame. They have a mechanism to move the hot air, such as an electric fan; a nozzle directs the air. Heat guns are commonly used to thaw pipes, remove paint, and for soldering, bending PVC pipe and heat shrinking tubing. Although not called such, a hair dryer is a form of low-temperature heat gun.

• **Heat island effect:** Refers to the absorption of heat by hardscapes, such as dark, nonreflective pavement, and buildings, and its radiation to surrounding areas. Particularly in urban areas, other sources may include vehicle exhaust, air-conditioners, and street equipment; reduced airflow from tall buildings and narrow streets exacerbates the effect.

• **Heat line:** A distinct line left on walls by superheated smoke that was stopped at the ceiling. The bottom edge of this superheated smoke often leaves a line on the walls.

• **Heat loss:** The transfer of heat from inside to outside by means of conduction, convection, and radiation through all surfaces of a house.

• **Heat meter:** An electrical municipal inspection of the electric meter breaker panel box.

• **Heat pump system:** A mechanical system that uses compression and decompression of gas to either take heat out of the structure or bring heat into the structure.

• **Heat rough:** Work performed by the heating contractor after the stairs and interior walls are built. This includes installing all ductwork and flue pipes. Sometimes, the furnace and fireplaces are installed at this stage of construction.

• **Heat tightening:** Heat tightening takes advantage of the thermal expansion characteristics of a bolt. A bolt is heated and expands and then allowed to cool. As the bolt attempts to contract it is constrained longitudinally by the clamped material and a preload results. Methods of heating include direct flame, sheathed heating coil, and carbon resistance elements. The process is slow, especially if the strain in the bolt is to be measured, since the system must return to ambient temperature for each measurement. This is not a widely used method and is generally used only on very large bolts.

- **Heat trim:** Work done by the heating contractor to get the home ready for the municipal Final Heat Inspection. This includes venting the hot water heater, installing all vent grills, registers, air conditioning services, turning on the furnace, installing thermostats, venting ranges and hoods, and all other heat-related work.
- **Heat vent:** Holes on a circular saw blade that prevent blade distortion and provide relief areas for expansion of materials as a result of friction-generated heat. See **Expansion slot**.
- **Heat welding:** A method of melting and fusing together the overlapping edges of separate sheets or sections of polymer modified bitumen, thermoplastics or some uncured thermoset roofing membranes by the application of heat that is usually in the form of hot air or an open flame and pressure. Also referred to as **heat seaming**.
- **Heating degree-days:** The sum, on an annual basis, of the difference between 65°F and the mean temperature for each day as determined from "NOAA (National Oceanic and Atmospheric Administration) Annual Degree Days to Selected Bases Derived from the 1960–1990 Normals" or other weather data sources acceptable to code officials.
- **Heating load:** The amount of heating required to keep a building at a specified temperature during the winter, usually 65° F, regardless of outside temperature.
- **Heating zone:** Room or group of rooms that is heated or cooled as a unit, usually controlled through a single thermostat.

Heavy construction equipment: (This list is from the Alberta Labour Relations Board. Roadbuilding and Heavy Construction, 1 December 2003).

Roadbuilding equipment:

- **Asphalt planers:** A machine used to strip or plane asphalt from existing road during road repairs.

- **Asphalt or Concrete Paver:** A machine that is used to lay down concrete or asphalt. Also known as a **lay-down machine**.
- **Backfillers:** Equipment such as dozers, hoes, and cranes that have special attachments for backfilling trenches and excavations.
- **Base spreader/Jersey spreader:** A machine that is used to spread soil cement or lime on a road surface. It is also used in the operation of hardening the base portion of a road or a parking lot.
- **Boring machine (horizontal):** An auger-type machine that is used to bore under roads and other surface obstructions.
- **Boring machine (vertical):** An auger-type machine that is used for drilling small and large diameter holes for piling for building foundations and bridges, etc. It stabilizes ground.
- **Cat D-2 to D-11:** Commonly known as a bulldozer, its prime function is to move materials such as dirt, sand, gravel, etc.
- **Clam/Dragline:** A conventional crane with a clam bucket that is used to move material from hard to reach locations.
- **Compressor:** A pump or other machine that increases the pressure of a gas. Compressors come in a wide range types and capacities and have many uses.
- **Concrete pump:** A truck-mounted or mobile pump that is used to pump and place concrete.
- **Crane:** Under this category are all types and sizes of rubber-tired and tracked conventional and hydraulic cranes that are used to lift, move, and lower equipment and materials.
- **Directional drilling machine:** Drills that can drill on an angle, used to drill under rivers, etc.
- **Ditch-Witch:** A small trencher that is used to dig trenchs for smaller cables.
- **Double-drum hoist:** A hoist with double drums that are used for hoisting and lowering materials. Both single- and double-drum hoists are usually stationary and air- or motor-driven.
- **Excavators:** All sizes of rubber tire and track shovels (old cable) and backhoes

(new, hydraulic) that are used to excavate and move various materials (sand, dirt, etc.).

- **Farm tractors:** Two- or four-wheel rubber-tired farm tractors of various types and sizes. They come with attachments such as buckets, forks, ploughs, used for towing any equipment such as rone plough, trailers, discs, and compaction equipment.
- **Feller buncher:** A machine that is used to cut, delimb, and pile timber. Also known as a **Tree farmer**.
- **Forklift:** A machine with forks used to raise, lower, and transport materials. Used primarily to load and unload trucks.
- **Front ender:** These machines come in many sizes and configurations, i.e., they may be either track (crawler) or rubber-tired. They are used primarily to load trucks, transport materials, and backfill.
- **Gradall:** A rubber-tired or tracked machine that is used to dig, grade, and slope trenches, etc.
- **Gravel and chip spreader:** A machine that is used to spread gravel chips on road surface.
- **Hydro axes:** A machine that is used to cut bush and shrub along right of ways.
- **Motor scraper:** A motorized earth mover that is used to move dirt. Picks up and releases dirt though opening in the bottom of body. Also called a **buggy**.
- **Motor grader:** A machine that is used to cut, slope, level roads, ditches, and building sites. Sometimes pressed into service for snow removal.
- **Off highway vehicles:** All types and sizes of off road vehicles, e.g., snow mobiles, and all 4 × 4, 6 × 6, and 8 × 8 trucks and vehicles.
- **Piledriver:** A machine that is used to drive steel or wooden piles into the ground to stabilize soil.
- **Power mounter drill:** A drilling unit where the prime mover is an integral part of the drill.
- **Power dozer:** A dozer that has a unique mechanized blade; picture a conveyor belt that travels horizontally around the blade. The blade's conveyor-action permits

simultaneous loading and unloading of material at a rapid rate. Efficiently casts fill into trenches.

- **Push cat:** A dozer that is used to push other equipment, usually a scraper.
- **Ripper cat:** A dozer with a ripper that is used to rip hardpan or frozen soil.
- **Rock drills:** Air track drills that run off compressors. They are used to drill holes for blasting.
- **Rubber-tired rollers:** A macine that packs gravel, sand, asphalt, or soil for hardening of road, etc.
- **Scraper cat:** A dozer with an attached scrapper. It is used to remove soil or material from areas on which a motor scraper cannot travel.
- **Screed (asphalt or concrete):** An attachment or part of a paver that controls the width and depth of material being placed.
- **Self-propelled compaction equipment:** Mobile rubber-tired or steel rollers with or without attachments. This equipment is used to compact materials on road, etc.
- **Side loader:** The same as front-ender loader, but loads from side.
- **Sideboom:** A machine that is used for hoisting and lowering pipe and materials into ditch.
- **Single-drum hoist:** A hoist with a single drum for hoisting and lowering of materials.
- **Skid-steer loader:** As that is mall loader on rubber tires. It may have several attachments. Commonly known as a **bobcat**.
- **Skidder:** A machine that may have a blade and/or grapple. It is used to move (skid) logs.
- **Slip form paver:** A machine that lays down the proper width and depth of asphalt or soil cement on roadways and parking lots. Also known as an **extruder**.
- **Soil cement and lime travel mixer:** A machine that is used to mix and spread soil cement and lime to harden surfaces.
- **Steel roller:** The roller follows the paving machine to pack down the newly placed asphalt. Also known as a **breakdown**.

- **Steel roller (base):** A steel compaction roller used to pack base material.
- **Steel roller (finish):** This roller packs down and smoothes asphalt behind a paving machine.
- **Sweeper:** Self-propelled sweepers come in many types and sizes. They are used to clean surface of foreign objects.
- **Tow cat:** A dozer with a winch that is used to tow other equipment and vehicles.
- **Towed-compaction equipment:** Compactors that are not self-propelled but must be towed by farm tractors or other equipment.
- **Tower and hammerhead cranes:** Stationary cranes that are used to hoist, move, and lower material.
- **Trencher:** A ditching machine that is used to dig trenches for pipe and cable.
- **Zoom boom:** Similar to forklift, but has a telescopic boom to reach distances. Also known as a **material handler**.

Trucks that are used in roadbuilding and heavy construction:

- **"A" Frame:** A truck that is used to hoist and transport equipment or material. (Being replaced by picker truck). Also known as a **winch or bed truck**.
- **Distributor:** A truck with a tank, pump, and spray bar for application of bitumen or other fluids to road surfaces. Also referred to as **black top, maintenance**, and **asphalt distributors**.
- **Dump:** A truck that is used to transport materials such as sand, gravel, and asphalt. Also known as a **Belly dump**.
- **Flat deck:** A truck that has no without lifting device. It is used to transport small equipment and material.
- **Fuel:** A truck delivers gas or diesel fuel toon-site equipment.
- **Hi-Boy:** A truck that is used to move parts, equipment, and materials.
- **Hy-Ab:** A flat-deck truck with a knuckle-type lifting device that is used to transport small equipment (pumps) and materials.
- **Hydro-vac:** A truck that is used to locate buried cables and oil, gas, and water lines.

- **Low boy:** A truck that is used to move equipment.
- **Picker truck:** A truck that has a crane mounted on it and is used to hoist and transport materials and equipment. Also known as **boom, stinger**, and **pitman** (A pitman is also a worker who is employed inside a pit such as a coal mine.)
- **Pressure:** A truck that is used to steam or wash equipment. Also known as a **steam and wash truck**.
- **Pump truck:** A truck that is used to pump concrete.
- **Service:** A truck that is operated by service technicians who maintain the on-site equipment.
- **Trans-mix:** A truck that is used to transport concrete or other material.

Hectare (ha): A metric unit of area, 100 meters by 100 meters (10,000 square meters). A Hectare is equivalent to 2.471 acres.

Heel cut: The vertical cut portion of a rafter that sets on a cap plate. A heel cut and seat cut combined notch the end of a rafter to permit it to fit flat on a wall and on the top, doubled, exterior wall plate.

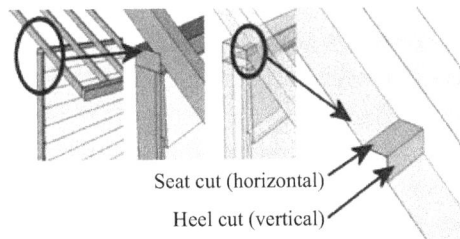

Seat cut (horizontal)
Heel cut (vertical)

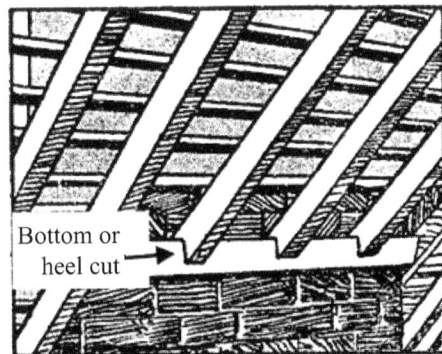

Bottom or heel cut

Heel cut.

Heel height: The vertical measurement from the outside top of a plate or beam to the top of the rafter.

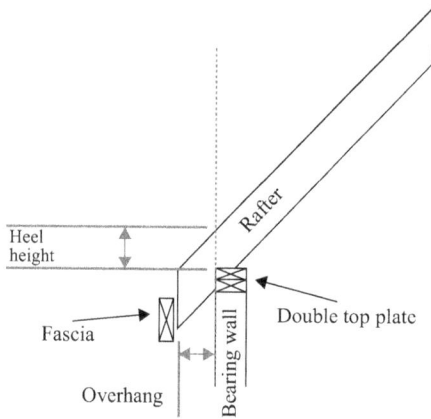

Helical: Spiral-shaped: in the shape of a helix or spiral.

Hem: The edge created by folding metal back on itself.

Herringbone wood floor installation: An installation where wood strips are installed in a zigzag pattern. See **Straight wood floor installation** and **Diaginal wood floor installation**.

Hertz (Hz): A unit of frequency; equal to one cycle per second (cps).

Heterogeneous: Stone formed from several types of material. See **Homogeneous**.

Hex key adjusting screw: Used to adjust the pressure applied at jaws of a tool. The screw can be adjusted by hand or with appropriate size Hex.

Hex key adjusting screws.

HEX shingles: Shingles that have the appearance of a hexagon after installation.

Hex shingles.

High alumina cement (HAC): Concrete made with this type of cement cures faster than concrete made with Portland cement. Unfortunately, it has the disadvantage that it tends to become weaker over time especially in a moist atmosphere and is not used for structural use.

High-density fiberboard (HDF): No longer defined under ANSI A208.2; HDF is a marketing term to define MDF grades above Grade 150. Medium-density fiberboard is denser than plywood. It is made up of separated fibers. It is stronger and much denser than normal particle board and can be used as a building material similar in application to plywood.

High-density urethane foam pad: A urethane foam pad that looks much like a thin wrestling pad. Unlike most other types of pad, water can be extracted from high-density urethane foam pads. See **Synthetic felt pad, Waffle type sponge rubber**, and **Rebound pad**.

High efficiency toilets (HETs): Toilets that use 1.28 gallons per flush, ~20% less water than regular toilets.

High-efficiency particulate air (HEPA) filters: Filters that remove virtually all (99.97%) 0.3-micron particles.

High gloss paint: As the name suggests high gloss paint has a high gloss. High-gloss paint can be water-based (also called latex or acrylic) or alkyd-based, commonly referred to as oil-based paint. High-gloss alkyd paints are very hard, scrub able, and resistant to abrasions.

High pressure decorative laminate: A high impact resistant surface material consisting of decorative surface paper impregnated with melamine resins pressed over multiple craft paper layers saturated with phenolic resins, thermoset at high pressure and temperature.

High nailing: When shingles are nailed or fastened above the manufacturer's specified nail location; it is an improper placement of nails.

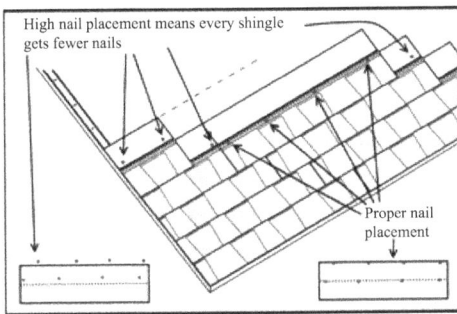

High nailing.

High speed steel (HSS): A formula of metals that has sufficient alloys to withstand frictional heat up to 1000° F without softening.

High-strength bolt: A structural steel bolt having a tensile strength greater than 100,000 pounds per square inch, usually A325 or A490; a steel bolt used to connect steel structural members together through friction.

Until relatively recently, structural steel connections were either welded or riveted. Now, high-strength bolts have completely replaced structural steel rivets. Indeed, the latest steel construction specifications published by AISC (the 13th Edition) no longer covers their, i.e., rivets, installation.

High strength friction grip bolts (HSFG): Bolts that are of high tensile strength. HSFG are used in conjunction with high strength nuts and hardened steel washers in structural steelwork. The bolts are tightened to a specified minimum shank tension so that transverse loads are transferred across the joint by friction between the plates rather than by shear across the bolt shank.

High strength friction grip bolt, nut, and washers.

High tensile steel: A grade of steel that is stronger than mild steel. High tensile steel is used in structural steelwork and concrete reinforcement.

Highest and best use: With regard to land use, the reasonably probable and legal use of vacant land or an improved property that is physically and financially feasible, and that results in the highest value. All possible use scenarios, including renovation, rehabilitation, demolition, and replacement should be analyzed.

Highlights: A light spot, area, or streak on a painted surface.

Hinge: A jointed device or flexible piece that a door, gate, shutter, lid, or other attached part turns, swings, or moves. A hinge has one degree of freedom. It can freely rotate about its axis but it cannot displace in any direction. Two mutually perpendicular reactive forces exist at the hinge and their lines of action pass through the center of the hinge. See **Pin connection and Support**.

Hip: A roof with four sloping sides. The line where two adjacent sloping sides intersect is called the hip. Also called a **hip roof**.

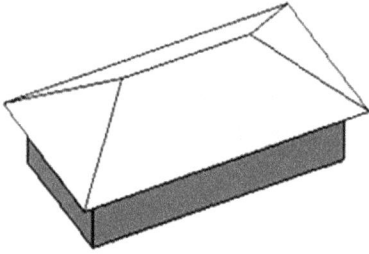

Rectangular hip roof.

A roof with hips and valleys.

Also, the external angle formed by the meeting of two sloping sides of a roof.

A **hip jack rafter** runs from the hip rafter to the wall top plate. The down-slope ridges on hip roofs are called **hip legs**. And, a rafter that forms the hip line of the roof from the ridge to the outside corner of the exterior walls is the **hip rafter.**

Hip rafter and hip jack rafter.

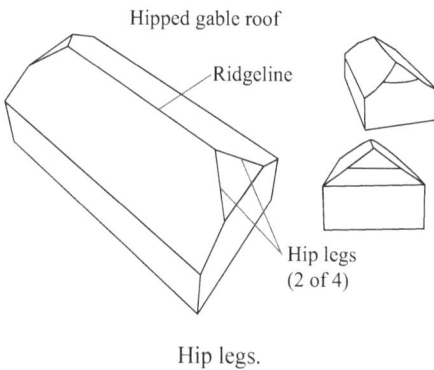

Hip legs.

Hip and valley: A system of roof framing where support members form valleys and ridges.

Hip shingles: Shingles that are used to cover the inclined external angle formed by the intersection of two sloping roof planes.

Historic American Buildings Survey (HABS)/Historic American Engineering Record (HAER): Architectural and engineering documentation programs that produce a thorough archival record of buildings, engineering structures, and cultural landscapes.

Terms that begin with historic and historical:

* **Historic character:** The sum of all visual aspects, features, materials, and spaces associated with a property's history.
* **Historic designed landscape:** A landscape significant as a design or work of art that meets the following criteria was consciously designed and laid out by a professional or amateur according to a recognized style or tradition and has a historical association with a significant person, trend or movement in landscape gardening or architecture, or a significant relationship to the theory or practice of landscape architecture.
* **Historic district:** A local or national geographically definable area, urban, or rural, possessing a significant concentration, linkage, or continuity of sites, landscapes, structures, or objects, united by past events or aesthetically by plan or physical developments. A district may also be composed of individual elements separated geographically but linked by association or history.
* **Historic document:** Any recorded information in any medium that has a direct, physical association with a past human

event, activity, observation, experience, or idea.

- **Historic landscape:** A cultural landscape associated with events, persons, design styles, or ways of life that are significant in American history, landscape architecture, archeology, engineering, and culture; a landscape listed in or eligible for the National Register of Historic Places.

- **Historic preservation:** Historic preservation is defined by the Secretary of the Interior's Standards for the Treatment of Historic Properties, 1995, as the act or process of applying measures necessary to sustain the existing form, integrity, and materials of an historic property. Work, including preliminary measures to protect and stabilize the property, generally focuses upon the ongoing maintenance and repair of historic materials and features rather than extensive replacement and new construction. New exterior additions are not within the scope of this treatment; however, the limited and sensitive upgrading of mechanical, electrical, and plumbing systems and other code-required work to make properties functional is appropriate within a preservation project.

- **Historic property:** A district, site, structure, or landscape significant in American history, architecture, engineering, archeology, or culture; an umbrella term for all entries in the National Register of Historic Places.

- **Historic significance:** The meaning or value ascribed to a structure, landscape, object, or site based on the National Register criteria for evaluation. It normally stems from a combination of association and integrity.

- **Historic site:** The site of a significant event, prehistoric, or historic occupation or activity, or structure or landscape whether extant or vanished, where the site itself possesses historical, cultural, or archeological value apart from the value of any existing structure or landscape. See **Cultural landscape**.

- **Historic vernacular landscape:** A landscape whose use, construction, or physical layout reflects common traditions, customs, beliefs, or values, which over time is manifested in physical features and materials and their interrelationships, and which reflect the customs and everyday lives of people.

- **Historical architect:** A specialist in the science and art of architecture with specialized advanced training in the principles, theories, concepts, methods, and techniques of preserving prehistoric and historic structures.

- **Historical context:** An organizing structure created for planning purposes that groups' information about historic properties based on common themes, time periods, and geographical areas.

- **Historical landscape architect:** A specialist in the science and art of landscape architecture with advanced training in the principles, theories, concepts, methods, and techniques of preserving cultural landscapes.

Hob: A projection or shelf at the back or side of a fireplace, used for keeping food warm.

Hod: A three-sided container mounted on a pole that is used to carry bricks or mortar up a ladder.

A Hod.

Hod carrier: A bricklayer's laborer.

Hoe: Simplistically, a long-handled gardening tool with a thin metal blade, used mainly for weeding and breaking up soil.

Hoes are among mans' oldest agricultural tools and they can be found with varing shapes around the world. They are used to move small amounts of soil, control weeds, pile soil around the base of plants create furrows and shallow trenches, etc. In construction hoes are often used to mix small batches of mortar and concrete.

Hog fuel: Wood chips or shavings, residue from sawmills, etc., used for fuel, landfill, animal feed, and surfacing paths and running tracks.

Hoist: A mechanical lifting device. Usually a chain or electric lifting device usually attached to a trolly that travels along a monorail or bridge crane.

Hoistway: Any shaftway, hatchway, well hole, or other vertical opening or space in which an elevator, dumbwaiter, or material lift is designed to operate. The space is enclosed by fireproof walls and elevator doors. The hoistway includes the pit and terminates at the underside of the overhead machinery space floor or grating, or at the underside of the roof where the hoistway does not penetrate the roof. (Hoistway is sometimes called "hatchway" or "hatch.")

Hold-down: Used to connect the outside of the framing to the foundation. The hold-downs are placed in the foundation while the concrete is still wet.

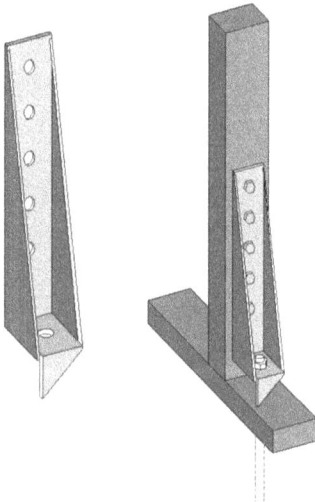

Framing hold-down.

Also, a type of pipe anchor consisting of a U-strap bolted at either end onto a horizontal plate.

Hold-down pipe clamp.

Holding capacity, kitchen range hoods: The interior space of a canopy. Generally speaking, the more powerful the cooking equipment beneath it, the larger the holding capacity that is required.

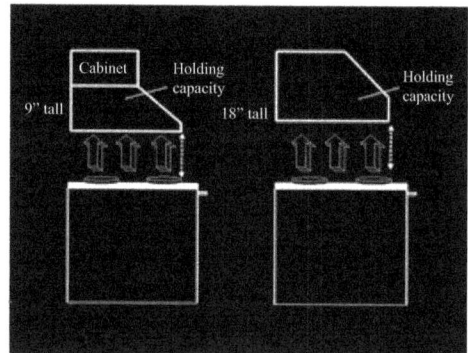

Holding capacity.

Hole saw: A small, cylindrical attachment for a power drill consisting of a circular saw blade for cutting holes.

Hole, per OSHA: A gap or void 2 inches (5.1 cm) or more in its least dimension, in a floor, roof, or other walking/working surface.

Holes, worm: Holes resulting from infestation by worms. By definition, worm holes are greater than 1/16 inch (1.6 mm) in diameter and do not exceed 5/8 inch (16 mm) in length.

Holidays: Bubbles that result from the separation of plies in multiple ply membrane roofs, usually due to improper installation; an area where a liquid-applied material is missing or absent.

Hollo-bolt: A proprietary expanding bolt that can be used in making bolted connections to hollow sections, and other situations where lack of access prevents a nut being used.

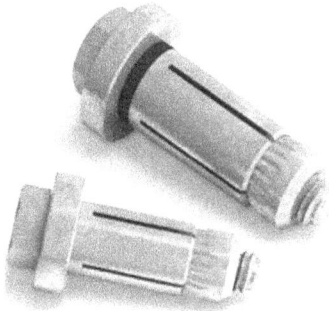

Type HB hollo-bolt.

Hollow metal: The term used to describe the cold formed sheet steel used in the construction of doors, door frames, borrow light frames, and some window frames.

Hollow section: A tubular structural steel member. Available are circular, rectangular, square, and elliptical hollow sections.

Hollow-core door: A flush door constructed with two skins or door faces separated by stiles and rails at the perimeter. Generally, a honeycomb-type support is used inside the door between the two faces.

Home run: The electrical cable that carries power from the main circuit breaker panel to the first electrical box, plug, or switch in the circuit.

Homogeneous material: A material having the same engineering design properties throughout, for example, stone formed from just one material. See **Heterogeneous**.

Holocene: The past 10,000 years of geologic time.

Honeycombs: The appearance concrete makes when rocks in the concrete are visible and where there are void areas in the foundation wall, especially around concrete foundation windows. Honeycombs are usually due to

lack of adequate vibration, inappropriate low slump, or congestion of the reinforcing steel.

Excessive areas of voids and or air pockets present in concrete placement that will diminish the strength of that placement area is referred to as **honeycombing**.

Honeycombs are also a cellular separation in the interior of a wood piece, usually along the wood grain. Wood honeycombing is a result of internal stress. It normally occurs when too much heat is applied too rapidly during kiln drying. White and Red Oak are particularly susceptible to honeycombing.

Brickwork built with gaps between the bricks is called **honeycomb brickwork**.

Honeycomb brickwork.

Also spelled **Honey comb**.

Hood: With regard to a fireplace, the canopy overhanging a fireplace to increase the draft.

Fireplace hood.

Hook: A semicircular bend in a reinforcing bar that enables it to resist a pulling loads anchoring it securely in concrete. Used to tie two separate concrete placements together.

Hook angle: The angle of the teeth on a saw blade that determines how fast or aggressive the saw will cut through materials.

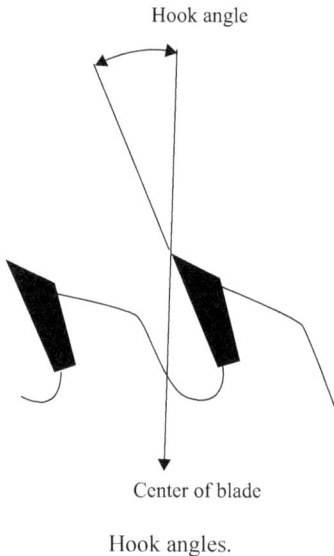

Hook angle

Center of blade

Hook angles.

Hooke's law: Discovered by the English scientist Robert Hooke in 1660, the law states that, for relatively small deformations of an object, the displacement or size of the deformation is directly proportional to the deforming force or load. Under these conditions the object returns to its original shape and size upon removal of the load.

Hopper: A bin or container that holds a reserve of dry material such as premixed mortar for controlled release as required.

Hopper window: A window unit that opens by moving the top of the window sash inward. The bottom of the window sash is attached with hinges.

Horizontal bridging: A continuous angle or other structural shape connected to the top and bottom chord of a joist horizontally, whose l/r ratio cannot exceed 300.

Horizontal shear stress: Horizontal shear stress is zero at the outer fibers of a section and is maximum at the neutral axis. It tends to cause one part of the section to slide past the other.

Beam shear is defined as the internal shear stress of a beam caused by the shear force applied to the beam:

$$\tau = VQ/It,$$

where

V = total shear force at the location along the beam where we wish to find the horizontal shear stress.

Q = statical moment about the neutral axis of the entire section of that portion of the cross-section lying outside of the cutting plane.

Q = the cross-sectional area, from the point where we wish to find the shear stress at, to an outer edge of the beam cross section (either top or bottom).

t = the width of the beam at the point that we wish to determine the shear stress.

I = Moment of Inertia of the entire cross sectional area.

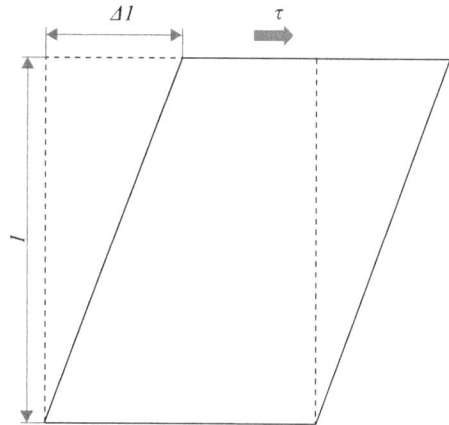

Since the area at the outer edge of the beam is zero horizontal shear stress is also zero there. The horizontal shear stress is (normally) a maximum at the neutral axis of the beam. This is the opposite of the behavior of the bending stress that is maximum at the outer edge of the beam, and zero at the neutral axis.

Horizontal slider: A window with a movable sash that slides horizontally. Also referred to as a **gliding window**.

Horizontal truss reinforcing: A wire truss that is placed in a bed masonry joint of mortar for the purpose of reinforcing the masonry assembly at a predetermined regular interval. Usually every six courses of common brick and or every two courses of block receive horizontal truss reinforcing.

Horsefeathers: See **Feathering strips**.

Horsepower (hp): The power required to lift 550 pounds one foot in 1 s or 33,000 pounds one foot in 1 min. A horsepower is equal to 746 watts or to 42.2 British thermal units per minute.

Horst: An elongated block of the earth's crust uplifted relative to surrounding rocks along bounding faults.

Hose bibb: An exterior water faucet. Sometimes spelled **hose bib**. Also referred to as a **sill cock**.

Hot or Hot stuff: The roofer's term for hot bitumen.

Hot mop installation method: An installation method where bonding materials are heated and mopped onto roofing materials to form a bond between layers, overlapping seams, or flashing. On modified bitumen roof systems, a method whereby SBS-type modified bitumen roofing is adhered to the base sheet.

Hot-rolled shapes: Structural steel sections that are formed by rolling mills from molten steel. Hot-rolled shapes include W-, M-, S- and HP-shapes, channels (C- and MC-shapes), angles (L), and structural tees (WT-, MT- and ST-shapes).

A hot rolled shape with symbol HP is used for bearing piles that have essentially parallel flanges and equal web and flange thickness.

Hot spot: A volcanic center, 100 to 200 km across and persistent for at least a few tens of millions of years, that is thought to be the surface expression of a rising plume of hot mantle material.

Hot-water heater: A heater and storage tank to supply heated water. Easy enough, except there are many types of hot-water heaters. While not a complete list, here are the more common types of water-heaters:

- **Traditional tank water heaters:** With these heaters all incoming water must be heated to the desired temperature and then continuously maintained at this temperature; the water is first heated to the set point and then allowed to cool to the point at which the thermostat calls for another heating cycle.

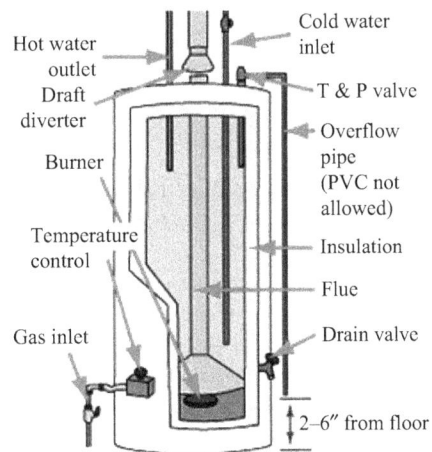

Traditional gas hot water-heater.

- **Condensing gas storage water-heaters:** A draft-inducing fan pushes air and fuel into a sealed combustion chamber that is inside the storage tank. As the fuel burns, combustion gas is exhausted through a secondary heat exchanger of coiled stainless steel tubing that is submerged within the tank's water supply. Both the combustion chamber and heat exchanger have large surfaces that help to maximize the heat transfer to the water.

These heaters are very efficient, so efficient in fact that the combustion gases actually cool to the point where the water vapor in the exhaust condenses, releasing its latent heat that is transferred to the stored water.

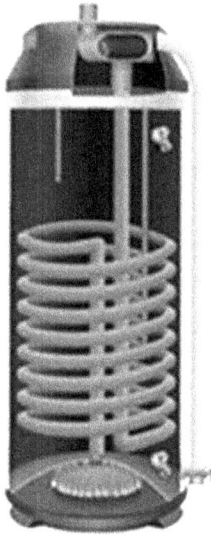

Gas tankless hot water-heater.

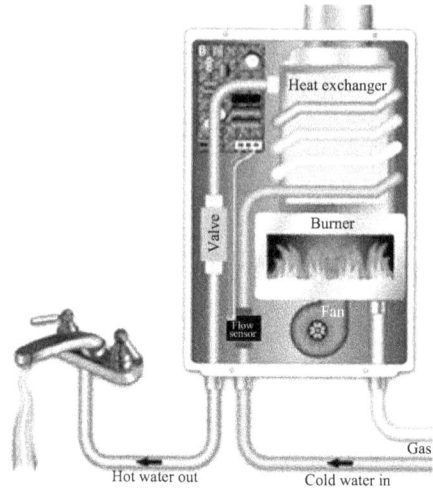

Condensing gas storage water-heater.

- **Gas tankless water-heaters:** The intent of these water-heaters is to eliminate the losses that are due to keeping tanks constantly hot. They are marketed in large measure because they do this.

In reality the efficiency of these heaters depends very much on the consumption patterns of the user. In most cases higher efficiencies are gained at steady-state high flow rates. Smaller intermittent flows reduce efficiency because the unit has to continuously go through both the shutdown and startup for each event. Each startup requires the heat exchanger that was cooled by fans at shutdown, to be reignited and reheated. Unfortunately, most residential hot water events are of relatively low flow and short duration.

- **Electric tankless water-heaters:** These heaters can generally be broken down into one of four categories: whole house, point of use, booster and extender. They work well with preheated water, help to reduce scald potential, and can support other water heating technology such as solar, heat pump, and geothermal heat recovery systems.

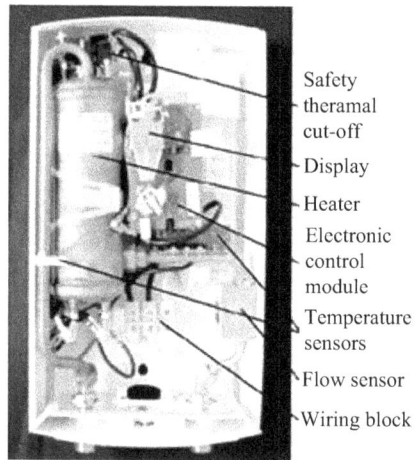

Electric tankless hot water-heater.

- **Heat pump water-heaters:** These heaters work in the same manner as an air conditioner, except in reverse. That is, an air conditioner takes heat from the compressor and expels it from the cooled space, a heat pump puts the heat generated from a compressor into water that is in a tank.

Heat pump water-heaters have the added advantage for some situations of cooling air that can be used to condition nearby space.

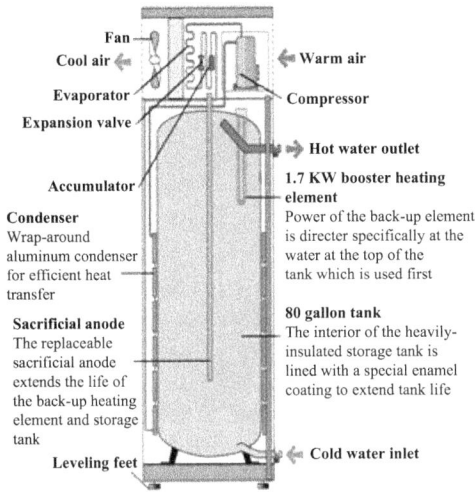

Fan
Cool air ←
Warm air ⇐
Evaporator
Compressor
Expansion valve

Hot water outlet ⇒
1.7 KW booster heating element
Power of the back-up element is directer specifically at the water at the top of the tank which is used first

Accumulator

Condenser
Wrap-around aluminum condenser for efficient heat transfer

80 gallon tank
The interior of the heavily-insulated storage tank is lined with a special enamel coating to extend tank life

Sacrificial anode
The replaceable sacrificial anode extends the life of the back-up heating element and storage tank

Leveling feet
Cold water inlet ←

Heat pump water-heater.

- **Solar water-heaters:** In general function in one of two ways: direct and indirect. In a direct system water is pumped directly through the solar collector and into a storage tank. In an indirect system, a freeze resistant heat transfer fluid is pumped through the solar collector and passed through the heat exchanger where the heat is transferred to the potable water. Indirect systems are generally used in colder climates where the water in the solar collectors may freeze.

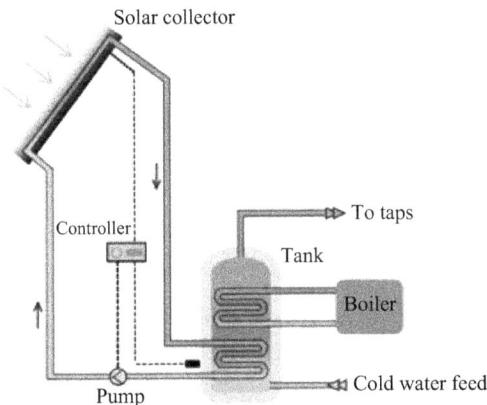

Solar collector

Controller

To taps ⇒

Tank

Boiler

Pump

Cold water feed ←

Solar hot-water system.

Hot wire: The wire that carries electrical energy to a receptacle or other device—in contrast to a neutral, which carries electricity away again. Normally it is the black wire. See **Ground**.

Hub: Regarding piping, hub typically refers to a solvent weld joint that will fit over a pipe or the spigot of another fitting.

Humidifier: An appliance normally attached to the furnace, or portable unit device designed to increase the humidity within a room or a house by means of the discharge of water vapor.

Humidistat: An instrument used for measuring and controlling moisture in the air.

Humidity: The amount of moisture contained in the atmosphere. Generally expressed as percent relative humidity (the ratio of the amount of moisture [water vapor] actually present in the air, compared to the maximum amount that the air could contain at the same temperature.)

Hundredweight: In the British imperial units system, a weight of 112 pounds. A hundredweight is also equivalent to eight stone, or one twentieth of a ton.

Hung out: A home is said to be "hung out" after the drywall installation has been completed. See **Scrap out**.

Hurricane clip: Metal straps that are nailed to secure the roof rafters and trusses to the top horizontal wall plate. Sometimes called a **Teco clip and hurricane tie**.

Hurricane clip.

Husk: A term used for the parts of the sawing system that support the arbor, saw, saw guide, and splitter, usually on a circular saw headrig.

Hydration: The chemical process that occurs when water and cement combine to form the adhesive paste that holds the aggregate together and makes concrete harden. The correct mixture of water, cement, and temperature is needed for proper hydration to occur.

Hydraulic: To move or convey by fluid.

Hydraulic balance: A condition of equally opposed hydraulic forces acting on a part in a hydraulic component.

Hydraulic cement: Cement that sets under water, e.g., Portland cement.

Hydraulic control: A control that is actuated by hydraulically induced forces.

Hydraulic head: A measurement of liquid pressure above a geodetic datum. It is usually measured as a liquid surface elevation, expressed in units of length, at the entrance (or bottom) of a piezometer (an instrument for measuring the pressure of a liquid or gas, or something related to pressure (such as the compressibility of liquid). Piezometers are often placed in boreholes to monitor the pressure or depth of groundwater). Hydraulic head can also be measured in a column of water using a standpipe piezometer by measuring the height of the water surface in the tube relative to a common datum. The hydraulic head can be used to determine a hydraulic gradient between two or more points.

In fluid dynamics, head is a concept that relates the energy in an incompressible fluid to the height of an equivalent static column of that fluid. The total energy at a given point in a fluid is the energy associated with the movement of the fluid, plus energy from pressure in the fluid, plus energy from the height of the fluid relative to an arbitrary datum. Head is expressed in units of height such as meters or feet. (For more on this you should learn about Bernoulli's Principle.)

The static head of a pump is the maximum height (pressure) it can deliver. The capability of the pump can be read from its Q-H curve (flow vs. height). Head is equal to the fluid's energy per unit weight. Head is useful in specifying centrifugal pumps because their pumping characteristics tend to be independent of the fluid's density. There are four types of head used to calculate the total head in and out of a pump:

- **Velocity head:** Due to the bulk motion of a fluid (kinetic energy).
- **Elevation head:** Due to the fluid's weight, the gravitational force acting on a column of fluid.
- **Pressure head:** Due to the static pressure, the internal molecular motion of a fluid that exerts a force on its container.
- **Resistance head**: Due to the frictional forces acting against a fluid's motion by the container. Also referred to friction head and head loss.

Hydraulic shock: The instantaneous pressure caused when a closed plumbing valve stops flowing water.

Hydraulics: The branch of science and technology concerned with the conveyance of liquids through pipes and channels.

Also, hydraulic systems, mechanisms, or forces that use pressurized oil to move its components.

Hydro jetting: The process of clearing blocked pipes using high pressure water as a boring and flushing mechanism.

Hydrochlorofluorocarbons (HCFCs): Refrigerants that are used in building equipment and cause depletion of the stratospheric ozone layer but are less damaging than CFCs.

Hydrodynamics: The branch of science pertaining to liquid flow and pressure.

Hydro energy: Electricity produced from the downhill flow of water from rivers or lakes.

Hydrofluorocarbons (HFCs): Refrigerants that do not deplete the stratospheric ozone layer but may have high global warming potential and thus are not environmentally benign.

Hydrology: The branch of geology that studies water on the earth and in the atmosphere: its distribution and uses and conservation.

Hydrostatic pressure: A uniform pressure exerted on a surface by ground or standing water.

Hydrostatics: The branch of science pertaining to the energy of liquids at rest.

Hygrometer: An instrument for measuring the humidity of the air or a gas.

Hygroscopic: Tending to absorb moisture from the air to be in equilibrium with the atmosphere. Also, the ability of a material to absorb moisture.

Hypalon™: A registered trademark of E.I. duPont de Nemours, Inc., for "chlorosulfonated polyethylene." Hypalon is a single-ply roofing material. See **Chlorosulfonated polyethylene** (CSPE).

Hypocenter: The point below the epicenter at which an earthquake actually begins. Also referred to as the **focus**.

Hypotenuse: The long side on a right triangle found opposite the 90° angle.

Hysteresis: A term that describes the behavior of a structural member subjected to reversed, repeated load into the inelastic range whose plot of load verses displacement is characterized by loops. The amount of energy dissipated during inelastic loading is indicated by the enclosed area within these loops.

Elastic hysteresis of an idealized rubber band. The area in the center of the hysteresis loop is the energy dissipated due to material plasticity.

I-beam: A steel beam with a cross section resembling the letter I. It is used for long spans as basement beams or over wide wall openings, such as a double garage door, when wall and roof loads bear down on the opening. Also referred to as an **I-section**.

I-joist: A manufactured structural building component that resembles the letter "I." Used as floor joists and rafters. I-joists include two key parts: flanges and webs. The flange of the I joist may be made of laminated veneer lumber or dimensional lumber, usually formed into a 1 1/2" width. The web or center of the I-joist is commonly made of plywood or oriented strand board (OSB). Large holes can be cut in the web to accommodate duct work and plumbing waste lines. I-joists are available in lengths up to 60 feet long.

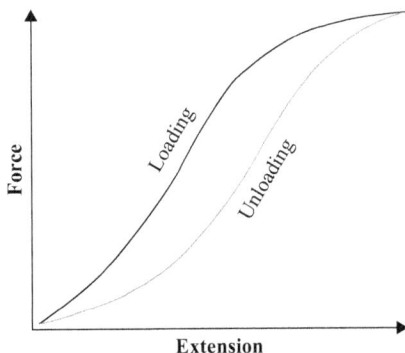

I-joist.

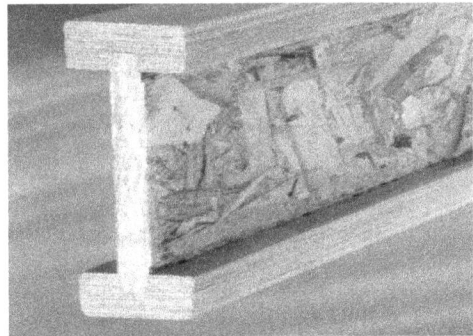

Ice and water shield: A self-adhering membrane that is specifically designed to be used in heavy rain and snow areas where leaks can be a problem. In most cases an ice and water shield would be installed on the first three feet of a roof in addition to underlayment. Building codes will require this in certain areas.

Ice-damming: A condition that occurs when snow melts on the heated portion of an improperly ventilated roof. The water drips down to the unheated portion where it freezes into ice. Eventually the build up of ice will cause the roof to leak.

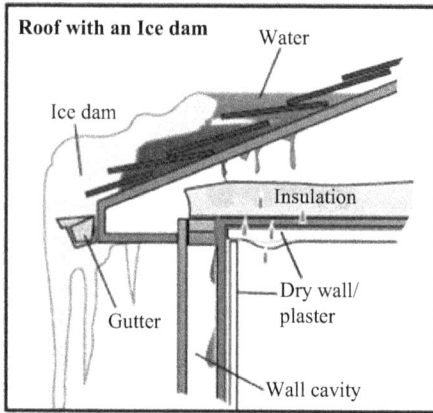

Ice-damming.

Igneous rock: One of the three main rock types (the others being **sedimentary and metamorphic rock**). Igneous rock is formed through the cooling and solidification of magma or lava. It may form with or without crystallization, either below the surface as intrusive (plutonic) rocks or on the surface as extrusive (volcanic) rocks.

Ignition temperature: The minimum temperature at which a material will combust.

Illuminance: The density of luminous flux incident of a surface in lumens per unit area.

Terms beginning with the word impact:

• **Impact factor:** The factor by which the static weight is increased by dynamic application.
• **Impact load:** A weight that is dropped or a dynamic load generated by movement of a live load such as vehicles, craneways, etc.
• **Impact noise:** Impact noise occurs when an object collides with another object, impact noise is usually heard from the floor above. It could be footsteps, a chair sliding across a wood or tile floor, or an object falling on the floor. Impact noise travels freely through a structure and through air pockets.

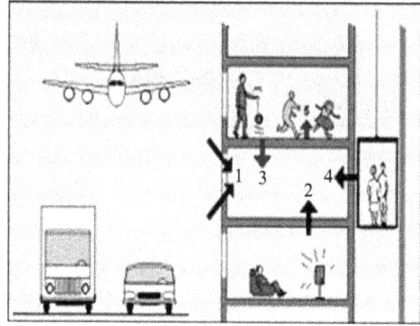

Impact noises.

• **Impact strength:** The ability of a material to absorb the energy of a load delivered rapidly to a member.
• **Impact wrench:** A pneumatic tool that is used to tighten nuts on bolts.

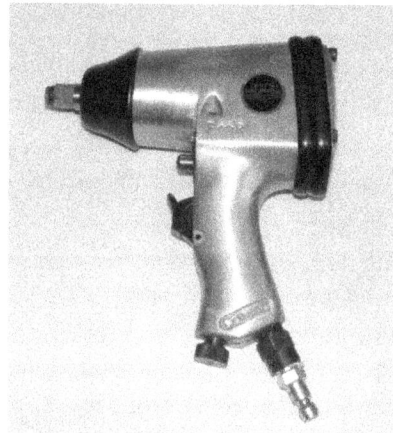

Air powered impact wrench.

• **Impact wrench auger bit:** An auger bit with a 7/16" or 5/8" quick-change shank for use in impact wrench drills.

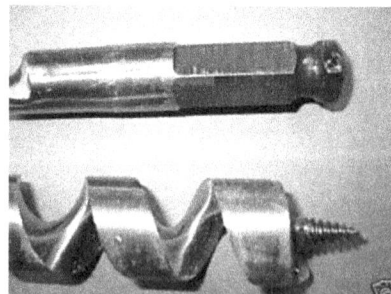

Impact wrench auger bit.

- **Impact-resistant:** Term used to describe window and door products that have passed established tests for resistance to windborne debris. Such products are typically used in coastal areas that are prone to hurricanes.

  Also, the ability of a roofing material to resist damage (e.g., puncturing) from falling objects, application equipment, foot traffic, etc. The impact resistance of the roofing assembly is a function of all of its components, not just the membrane itself.

Impedance: Opposition to the flow of alternating (AC) electric current. See **Resistance**.

Impeller: The rotating wheel with vanes that is found inside a centrifugal pump. As it spins at high speed it draws fluids in and thrusts them under pressure to the discharge outlet.

Imperial system: The traditional system of weights and measures used in many English-speaking countries until superseded by SI units metrication. The principal Imperial elements are yards (with their subdivisions of feet and inches) and pounds (divided into ounces and multiplied into hundredweights and tons).

Imperviousness: Resistance to tpenetration by a liquid and is calculated as the percentage of an area covered by, say, a paving system that does not allow moisture to soak into the ground.

  In the case of ceramic glazed masonry, imperviousness measures the resistance to staining when subjected to a standard test.

Impervious surfaces: Impervious surfaces promote runoff of precipitation instead of infiltration into the subsurface. The imperviousness or degree of runoff potential can be estimated for different surface materials.

Implode: Collapse or cause to collapse violently inward.

Impossibility: A legal defense to non-performance of a contract. It arises when performance is impossible due to the destruction of the subject matter of the contract (as, for example, by fire) or the death of a person necessary to perform it; performance is then excused and the contract is terminated. In civil law, impossibility is an excuse for non-performance of a contract where the promised performance has become illegal.

  Generally contractors are not excused for failure to perform because of unforeseen difficulties. To be excused the work must truly be impracticable due to some extreme, or unreasonable difficulty, expense, injury or loss; however, to get a ruling of impossibility a contractor may not have proven scientific or actually impossibility.

Impregnate: To saturate. in roofing, asphalt impregnated fiber glass roofing felts are fiber glass mats that have been completely permeated with asphalt bitumen.

In situ: Something being in its original position; not having been moved.

In situ stresses: The sum of stresses that are in a member resulting from manufacturing, fabrication, and erection residual stresses and the stresses resulting from in situ loads and deformations.

Inactive door, in a double door: The door in a double-door set that does not contain a latch-set, but instead is bolted at the top and bottom to hold it stationary when shut. This door receives the latch or bolt of the active door.

Incandescent lamp: A lamp employing an electrically charged metal filament that glows at white heat. A typical light bulb.

Inch: A unit of length equal to one-twelfth of a foot (2.54 cm).

Incinerator: A furnace or container for burning waste materials.

Incline: The slope of a roof expressed either in percent or in the number of vertical units of rise per horizontal units of run. See **Slope**.

Inclusions: Nonmetallic material that is entrapped in sound metal.

Income statement: In the time intervals between the issuance of balance sheets (that present snap shots at an instant of time), two major types of transactions of interest to stockholders and creditors occur. Many involve the acquisition and financing of productive resources; we can call these financing-investing or capital activities. The other activities of major concern relate to the production and sale of goods and services; we call these operating activities. The income statement summarizes the major changes in the resources of the business as a result of the operating activities. It attempts to report on the performance of the business during a particular period.

See also **Balance sheet** and statement of **Changes in financial position**.

Inconspicuous: With regard to construction imperfections, generally defined as barely detectable with the naked eye at a distance of 6 ft. to 8 ft. (1.8 m to 2.4 m).

Indentations: With regard to doors, areas in the face that have been compressed as the result of residue on the platens during pressing or handling damages through the factory.

Independent contractor: Someone who makes an agreement with another to do work while retaining in him/herself control of the means, methods, and manner of producing the end result. Neither party has the right to terminate the contract at will.

Index: The interest rate or adjustment standard that determines the changes in monthly payments for an adjustable rate loan.

Index lease: A lease in which the rental amount adjusts accordingly to changes and/or movements in a price index, commonly the consumer price index.

Indirectcost: Costs for items and activities other than those directly incorporated into the project but considered necessary to complete the project. A contractor's or consultant's overhead expense; expenses indirectly incurred and not chargeable to a specific project or task. Also referred to as **Soft costs**. See **Hard costs**.

Indirect wastes pipe: A waste pipe that is used to convey gray water by discharging it into a plumbing fixture such as a floor drain.

Individual occupant workspaces: Areas where workers use standard workstations to conduct individual tasks. Examples are private offices and open office areas with multiple workers.

Indoor air quality (IAQ): The nature of air that affects the health and well-being of building occupants.

Indoor-outdoor carpet: A type of carpet that may be used in interior or exterior applications. Originally, indoor-outdoor carpet was made to imitate grass, but today it imitates many types of traditional interior carpet. See **Wool carpet, nylon carpet, Berber carpet, sculptured carpet**, and **shag carpet**.

Induction motor: The most common type of electric motor used in industry. It has a slippage of 2–5% compared to synchronous motors.

Industrial property: Commercial properties that are used for the purposes of production, manufacturing, or distribution.

Industrial lift: A hoisting and lowering mechanism of a nonportable power-operated type for raising or lowering material vertically, and that operates entirely within one story of a building.

Industrial waste: Liquid, gaseous or solid substances, or a combination of these that are the result of any process of industry, manufacturing, trade or business, or from the development or recovery of any natural resource.

Inelastic: The behavior of an element beyond its elastic limit, having permanent deformation. See **Yield point**.

Inert: A material that will not react chemically with other ingredients.

Inertia: The tendency of an object at rest to remain at rest, and of an object in motion to remain in motion.

Inertial force: A fictitious force used for convenience in visualizing the effects of forces on bodies in motion. For an accelerating body, the inertial force is considered as a body force whose resultant acts at the object's center of gravity in a direction opposite the acceleration. The magnitude of the force is the mass of the object times the magnitude of the acceleration.

Inertial forces: With regard to earthquakes, earthquake generated vibration of the building's mass causing internally generated inertial forces and building damage. Inertial forces are the product of mass times acceleration ($F = m \times a$).

Infeasible, per OSHA: Impossible to perform the construction work using a conventional fall protection system (i.e., guardrail system, safety net system, or personal fall arrest system) or technologically impossible to use any one of these systems to provide fall protection.

Infeed: The direction a workpiece is fed into a saw blade or cutter. See **Outfeed**.

Infiltration: The passage of air from indoors to outdoors and vice versa; term is usually associated with drafts from cracks, seams, or holes in buildings.

Inflection point: A point of zero moment in a structural member.

Influence line: A curve whose ordinates give the values of some particular function (shear, moment, reaction, etc.) in an element due to a unit load acting at the point corresponding to the particular ordinate being considered. Influence lines for statically determinate structures are straight lines and for statically indeterminate structures the lines are curved and their construction involves considerable analysis.

Infrared emittance: Infrared emittance indicates the ability of a material to shed infrared radiation, whose wavelength is roughly 5–40 μm. Infrared emittance is measured on a scale of 0–1.

Infrared thermography: The use of an infrared camera to detect moisture in roof insulation.

Inglenook: A recess for a bench seat or two next to a fireplace. Popular in Shingle style and Craftsman homes.

An inglenook.

Initial rate of absorption (IRA): The mass of water that is absorbed when a brick or structural clay unit is partially immersed in water for 1 min, expressed in $kg/m^2/min$. Also referred to as suction.

Injected insulation: Insulation that is injected into place. There are two types of injected insulation: (1) foam that is injected into holes and cracks through tubes and (2) insulation that is injected through mesh into a framing cavity.

Inlay: Embedding pieces of a different material into another. An inlay is usually made flush with the surface into which it is inlaid.

A floor inlay.

Inorganic: Any chemical or compound that is derived from minerals, does not contain carbon, and is not classified as organic; being or composed of materials other than hydrocarbons and their derivatives; not of plant or animal origin.

Input motion: A term representing seismic forces applied to a structure.

Insect screen: A screen which will prevent the passage of insects into a building assembly.

Insert fitting: Fittings that are equipped with external, annular rings for gripping. They are inserted into flexible rubber or plastic tubing as a connection.

Insert fitting.

Inserts: Faucet handles.

Inside casing: The inside visible molding surrounding the interior of the window frame. Also referred to as **interior casing, interior finish**, and **interior trim**.

Inside corner: The point at which two walls form an internal angle, as in the corner of a room.

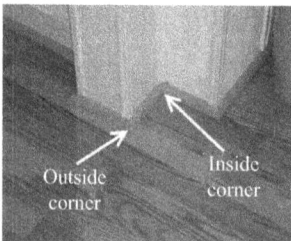

Inside corner vs. outside corner.

Instability: An important branch of structural mechanics that examines alternate equilibrium states associated with large deformation. Briefly, instability is reached when a structure or structural member is loaded so that continued deformation results in a decrease in its load-resisting capacity.

The importance of the subject is recorded in the history of structural collapses caused by neglect or misunderstanding of the stability aspects of design; perhaps the most famous among these is the collapse of the Tacoma Narrows Bridge in 1940, due to aerodynamic instability.

Instantaneous water heater: A type of water heater that heats water as it flows through a heat exchanger coil.

Insulated flue: A flue consisting of an inner pipe and an outer pipe with the space between the two filled with heat-resistant insulation.

Insulating glass (IG): Two or more lites of glass with a hermetically sealed airspace between the lites. The sealed space may contain air or be filled with an inert gas such as argon. Also known as double glass.

Insulation: With regard to the transfer of heat, any material high in resistance to heat transmission that, when placed in the walls, ceiling, or floors of a structure, and will reduce the rate of heat flow.

Also, non-conductive materials used to cover or surround a conductor, permitting it to be handled without danger of electric shock.

An example of insulation is **rigid insulation board**, that is a structural building board made of coarse wood or cane fiber in 1/2- and 25/32-inch thicknesses. It can be obtained in various size sheets and densities. Insulation board is characterized by an integral bond produced by interfelting of the fibers, to which other materials may be added during manufacture to improve certain properties, but which is not consolidated under heat and pressure. Generally insulation board has a density of more than 10 pounds per cubic foot (pcf) (specific gravity 0.16) but less than 31 pounds per cubic foot (specific gravity 0.50).

With regard to the flow of electrical current, any material, such as glass or rubber, that prevents the flow of electric current.

Insulation stop: Material placed in the roof system to prevent insulation from falling through the space between the top of the exterior wall and the bottom of the roof sheathing.

Insurances: There are several types of insurance that may be a contractor's responsibility: **workman's compensation; public liability and property damage; contingent liability;** and, **hold harmless.** Workman's compensation, and public liability and property damage rates vary according to the trades involved and the state that work is being performed. States can vary by as much as 100%. Rates also vary based on the contractor's safety record, which is one reason it is important to have a good safety record. Standard rates are usually applied to contingent liability and hold harmless insurances.

Inswing French door: A French door with panels that swing to the inside. One, two, three, and four panel units available as stationary or operating.

Intake ventilation: The fresh air that is drawn into a passive ventilation system through vents typically installed in the soffit or eave of a roof.

Integral apron: One-piece construction that streamlines the look of built-in whirlpools and baths. The exposed front of the tub is incorporated into the construction.

An integral apron.

Integral drawer: A type of drawer in which the face serves as the front piece of the drawer. The drawer sides attach directly to the drawer face. See **Attached drawer**.

Integral stops: Stops that allow the owner to shut off the water supply at the faucet for maintenance without shutting off the water supply to the entire house.

Integral vacuum breaker: A device typically used in a sink or shower sprayer to keep water from backflowing into the fresh water supply.

Integrated pest management (IPM): The coordinated use of knowledge about pests, the environment, and pest prevention and control methods to prevent unacceptable levels of pest infestation and damage by the most economical means while minimizing hazards to people, property, and the environment.

Integrated sash: A sash unit in which the insulating glass spacer profiles are integrated into the sash profiles. Separate IG construction is eliminated as the two lites of glass are applied and sealed directly to the sash, creating one assembly.

Integrity: With regard to historic sites, authenticity of a property's historic identity, evidenced by the survival of physical characteristics that existed during its historic or prehistoric period, the extent to which a property retains its historic appearance.

Intensity: A subjective measure of the force of an earthquake at a particular place as determined by its effects on persons, structures, and earth materials. Intensity is a measure of energy. The principal scale used in the United States today is the **Modified Mercalli, 1956 version**. MM (or Modified Mercalli) scale is based on observation of the effects of the earthquake MM-I thru MM-XII (MM-I = not felt, MM-XII = damage nearly total).

Intensive survey, historic sites: A systematic, detailed examination of an area that is designed to gather information about historic properties. An intensive survey should be sufficient to evaluate the properties against predetermined criteria of significance within specific historic contexts.

Intercepter: A device for separating grease and oil from the drainage system. See **Grease trap**.

Grease interceptor
(drawing not scale)

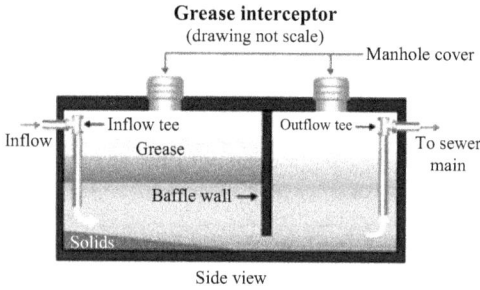

Side view

Interconnect device: A device that can be assembled into an entry set that allows a person to disengage both the deadbolt and the lower latch at the same time just by operating the handle. The door will open with this single action, as opposed to the usual two actions of first unlocking the top deadbolt before operating the handle.

Interconnect device.

Interest: The cost paid to a lender for borrowed money.

Interest-only loan: A method of loan amortization in which interest is paid periodically over the term of the loan and the entire original loan amount is paid at maturity.

Interlayment: A felt, metal, or membrane sheet material used between courses of steep-slope roofing to improve the weather- and water-shedding characteristics of the primary roof covering during times of wind-driven rain and snow. Typically used with wood See **Neutral pressure shakes**.

Interlocking shingles: Shingles with interlocking edges; designed so that the wind cannot lift them. The most common type of interlocking shingle is the **T-lock**.

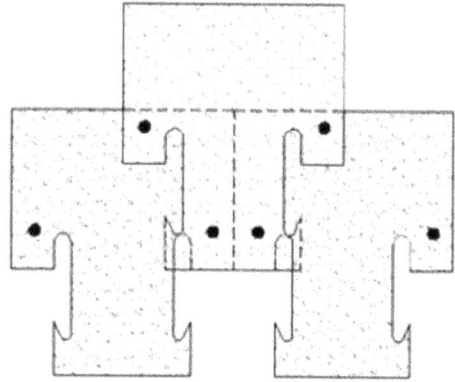

Three t-lock shingles

Interlocking siding: Siding made from metal or vinyl with edges that interlock as the pieces are installed forming a weather tight seam.

Internal hinge: See **Pin connection**.

Internal pressure: Pressure inside a building that is a function of ventilating equipment, wind velocity, and the number and location of openings and air leaks.

Terms beginning with the word interior:

- **Interior bearing:** Bearing supports that are interior to two exterior supports.

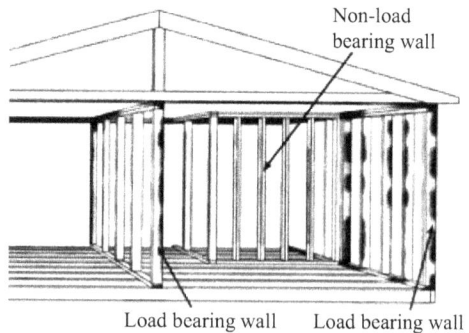

The center wall is an interior (load) bearing wall.

- **Interior casing:** The casing trim used on the interior perimeter of the window or door. Generally supplied by others except in the case of a round top casing that is usually factory supplied.

- **Interior finish:** Material used to cover the interior framed areas of walls and ceilings.
- **Interior glazes:** Glazing installed from inside of building.
- **Interior mullion casin:** The inside trim between adjacent windows.
- **Interior plaster:** Plaster that is used as a wall or ceiling finish inside the structure.

Intermittent weld: A weld that is not continuous. It is broken by recurring unwelded spaces.

Internal pressure: The pressure inside a building that is a function of the wind velocity and the number and locations of openings.

Internal rate of return (IRR): The discount rate often used in capital budgeting that makes the net present value of all cash flows from a particular project equal to zero; determining this would be an iterative process. Generally, the higher a project's internal rate of return, the more desirable it is. Therefore, IRR can be used to rank several prospective projects that are under consideration. Assuming all other factors are equal among the various projects, the project with the highest IRR would probably be considered the best and undertaken first. This comparative method is sometimes referred to as the internal rate of return method.

IRRs can also be compared against prevailing rates of return in the securities market. If a firm can't find any projects with IRRs greater than the returns that can be generated in the financial markets, it may simply choose to invest its retained earnings into the market until better opportunities arise.

IRR is sometimes referred to as the **economic rate of return (ERR)**.

Internally cured concrete: A fairly new method of designing concrete specifically to provide internal curing (actually, the concept of using lightweight aggregate to improve the hydration of the cement paste was observed back in the 1950s). The American Concrete Institute (ACI) has defined internal curing as "supplying water throughout a freshly placed cementitious mixture using reservoirs, via pre-wetted lightweight aggregates, that readily release water as needed for hydration or to replace moisture lost through evaporation or self-desiccation."

While lightweight aggregate is the most common method used as a water reservoir, researchers throughout the world are also investigating the use of superabsorbent polymers and natural fibers. While external curing water is applied at the surface and its depth of penetration is influenced by the quality of the concrete, internal curing enables the water to be distributed more equally throughout the cross section. You can expect to hear a great deal more on this subject.

International 1/4-inch Log Rule: Developed in 1906 this rule is based on a reasonably accurate mathematical formula. It allows for a 1/4-inch saw kerf (the width of the cut made by a saw blade) and a fixed taper (the difference in diameter between the top end and the butt end of a log) allowance of 1/2 inch per 4 feet of log length. Deductions are also allowed for shrinkage of boards and a slab thickness that varies with the log diameter. The International 1/4-Inch Log Rule is often used as a basis of comparison for log rules.

International Building Code (IBC): The IBC is one of the family of codes and related publications published by the International Conference of Building Officials (ICBO) and other organizations, such as the International Association of Plumbing and Mechanical Officials (IAPMO) and the National Fire Protection Association (NFPA). The IBC is designed to be compatible with these other codes, as together they often make up the enforcement tools of a jurisdiction. More and more governing authorities are accepting the IBC (with edits) as their standard.

International Fire Code (IFC): A model code that regulates minimum fire safety requirements for new and existing buildings, facilities, storage, and processes. The IFC addresses fire prevention, fire protection, life safety, and safe storage and use of hazardous materials in new and existing buildings, facilities, storage, and processes. The IFC provides a total approach of controlling hazards in all buildings and sites, regardless of the hazard being indoors or outdoors.

Interplate coupling: Interplate coupling is the ability of a fault between two plates to lock and accumulate stress. Strong interplate coupling means that the fault is locked and capable of accumulating stress, whereas weak coupling means that the fault is unlocked or only capable of accumulating low stress.

Interplate earthquake: An earthquake that occurs on the boundary between two tectonic plates.

Interrupting rating: The highest current at rated voltage that a device is intended to interrupt under standard test conditions.

Intrados: The underside of an arch. See **Extrados**.

Intraplate earthquake: An earthquake that occurs within the interior a tectonic plate.

Intrusive: Igneous rock formed by the cooling of magma inside the earth's crust.

Intumescence: A mechanism whereby fire-retardant paints protect the substrates to which they are applied. An intumescent paint puffs up when exposed to high temperatures, forming an insulating, protective layer over the substrate.

A material that expands when exposed to extreme heat or fire to fill any gap, e.g., between a door and frame and pipe penetrations is said to be **intumescent**.

Invasive plants: Both indigenous and exotic species that are characteristically adaptable and aggressive, have a high reproductive capacity, and tend to overrun an area. Collectively, they are one of the great threats to biodiversity and ecosystem stability.

Inventory: A complete list of items such as property, goods in stock, or the contents of a building.

Also, a quantity of goods held in stock.

Also, in accounting, the entire stock of a business, including materials, components, work in progress, and finished products.

Also, with regard to cultural resources, a list of cultural resources, usually of a given type and/or in a given area.

Invert: The elevation of the inside diameter of a pipe.

Inverted Roof Membrane Assembly (IRMA®): A variation of the "Protected Membrane Roof Assembly" in which Styrofoam® brand insulation is used. IRMA® and Styrofoam® are registered trademarks of the Dow Chemical Company.

In-place management: A series of steps used as an alternative to lead-based paint removal. In-place management improves the condition of intact lead-based paint to reduce and/or eliminate hazards without total removal.

Environmental Protection Agency (EPA) studies, Department of Housing and Urban Development (HUD) studies, and independent studies have shown that in-place management of lead-based paint is the most cost-effective way to significantly reduce the hazard of lead-contaminated dust. In-place management makes a house lead-safe, rather than lead-free. Costs to perform in-place management activities is pennies per square foot, rather than several dollars per square foot to perform the other methods. These activities involve the repairing of areas where paint is chipping, cracking, chalking or peeling, and performing specialized cleaning on a regular basis with a lead-specific detergent.

In-kind: In the same manner or with something equal in substance having a similar or identical effect. Beware, many owners will not accept what you believe is an in-kind substitute.

Ionic order: One of the ancient Greek orders of architecture, characterized by a fluted column and a capital consisting of four volute scrolls. Named after Ionia in Greece, where it was first used.

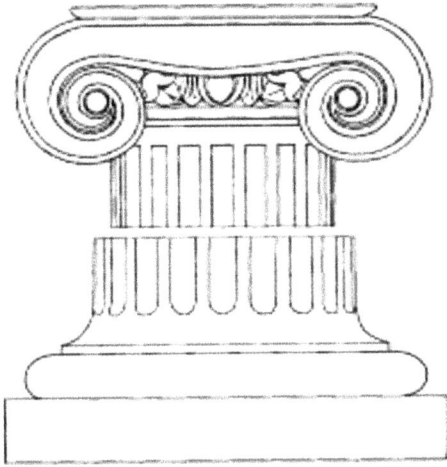

Ionic order.

Iron (Fe): The second most common metal after aluminum. The most important iron ores can be divided into four groups: Magnetite, Hematite, Limonite, and Siderite.

Ironworker: A skilled mechanic who erects and installs structural steel building frames.

Irrigated land: An area that has water delivered through artificial methods, i.e., other than rain. Conventional methods utilize pressure to deliver and distribute water through sprinkler heads. Other more efficient methods include drip irrigation, a highly efficient type of microirrigation that delivers water at a low pressure through buried mains and sub-mains of perforated tubes or emitters.

Irrigation: A lawn sprinkler or other similar system.

Isocyanate: A highly reactive organic chemical containing one or more Isocyanate groups. A basic component in **sprayed polyurethane foam (SPF)** systems and some **polyurethane coating** systems.

Isolated: As applied to location, not readily accessible to persons unless special means for access are used.

Isolation membrane: A protective layer of material that is installed between dissimilar materials. In a tile floor, an isolation membrane protects the tile from movement as the underlying system absorbs water that

penetrates the tile surface. See **Thinset tile, Mortar bed**, and **Cement board**.

Isoseismals: Map contours drawn to define limits of estimated intensity of shaking for a given earthquake.

Isostasy: A condition of hydrostatic equilibrium with the rigid part of the earth's crust floating on a denser and more mobile sublayer. Thus extensive areas of high topography are underlain by "roots" of relatively low density and the adding or removal of material at the surface causes vertical adjustments to maintain equilibrium.

Isotropic: Exhibiting the same properties in all directions. For example, a solid is said to be isotropic if it expands equally in all directions when thermal energy is provided to it.

J bend: The trap section, with a 180° bend, or a multi-piece P-trap. Also called a **Return bend**.

J channel: Metal edging used on drywall to give the edge a finished appearance when a wall is not "wrapped." Generally, basement stairway walls have drywall only on the stair side. J Channel is used on the vertical edge of the last drywall sheet.

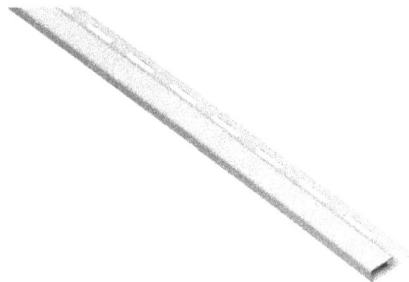

White vinyl siding J Channel
shown is 5/8-in × 12-ft 6-in

J hook: A pipe hanger in the shape of a J.

J-series joist: A series of joists adopted in 1961. J-series joists are made of A36 steel and are proportioned so that the allowable tension or bending stress does not exceed 22,000.

Jack hammer: A pneumatic tool that combines a hammer directly with a chisel. They are usually used to break up rock, pavement, and concrete. Jack hammers may be hand held or rig mounted. Hand held jack hammers are typically powered by compressed air, but some use electric motors. The larger rig mounted jack hammers are usually hydraulically powered.

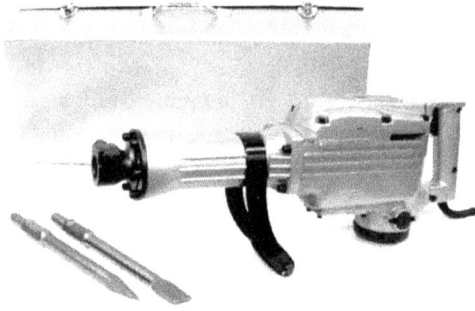

Electric demolition jack hammer.

Jack post: A type of structural support made of metal, which can be raised or lowered through a series of pins and a screw to meet the height required. Basically used as a replacement for an old supporting member in a building. See **Monopost**.

Jack rafter: See **Rafter, jack**.

Jack truss: Any of a number of trapezoidal trusses that are used to support those areas of a hiproof that are not beneath the peak or ridge, parallel to the truss or trusses that meet at the peak or ridge.

Also, a joist girder that is supporting another joist girder.

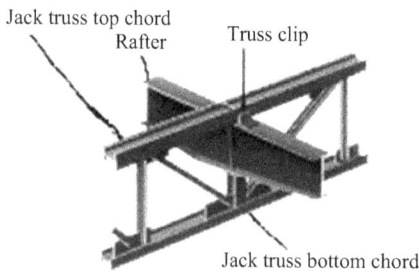

Jack truss.

Jacket: A metal covering around a pipe or piece of mechanical equipment.

Jacketing: The encasement of existing columns with steel or Kevlar.

Jacking operation: The task of lifting a slab (or group of slabs) vertically from one location to another (e.g., from the casting location to a temporary (parked) location, or to its final location in the structure), during the construction of a building/structure where the lift-slab process is being used.

Lift slab construction.

Jackson turbitity unit (JTU): A quantitative unit of turbidity originally based on the comparison of a liquid (such as water) with a suspension of a specify type of silica, using the turbidity measure in a **Jackson candle turbidimeter**.

Jal awning: A window with several outswinging, awning-type windows that pivot near the top of the glass and operate in unison. Also referred to as awning window.

Jalousie: A shutter or blind with fixed or adjustable slats that exclude rain and provide ventilation, shade, and visual privacy.

A **jalousie window** is a window that has parallel glass, acrylic, or wooden louvers that are set in a frame. The louvers are locked together onto a track, so that they may be tilted open and shut in unison, to control airflow through the window; they are usually controlled by a crank mechanism.

Jalousie window.

Jamb: The side and head lining of a doorway, window, or other opening. It includes studs as well as the frame and trim.

Jamb depth: Width of a window or door from the interior to the exterior of the frame.

Jamb extension: A jamb-like member, usually surfaced on four sides, which increases or extends the depth of the exterior or interior window or door frame; jamb extensions imply a larger depth than "wood jamb liners."

Jambliner: The track installed inside the jambs of a double-hung window, on which the window sash slide.

Also, thin strips of wood attached to the head jamb, side jambs, and sill to accommodate various wall thicknesses. Common jamb depths are: 4 9/16", 4 13/16", 5 1/16" and 5 3/16".

Jet: An orifice or other feature of a toilet that is designed to direct water into the trapway quickly to start the siphon action.

Jetty: In traditional timber-framed buildings, the projection of an upper story over the story below. The reason for this form of construction seems originally to have been simply to increase the floor area of the upper stories.

A double jettied building.

Also, a landing stage or small pier at which boats can dock or be moored.

A jetty.

Also, a **breakwater** constructed to protect or defend a harbor, stretch of coast, or riverbank.

A jetty (breakwater).

Jib: The arm on a crane. It adds length to the boom.

A jib crane is a cantilevered boom or beam with a hoist and trolly used to pick up loads in all or part of a circle around which it is attached.

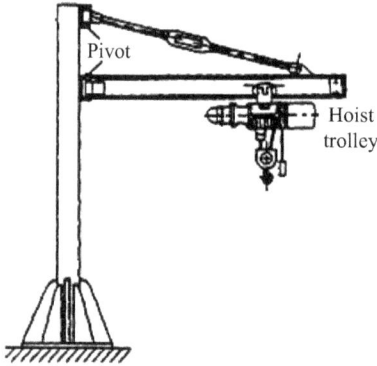

Jib crane.

Jig: A device that holds work or pieces of material in a certain position until rigidly fastened or welded during the fabrication process.

Also, while this has nothing to do with construction, in sailing a jib is a triangular staysail set forward of the forwardmost mast.

Jobsite: The specific location where a construction project is occurring.

Johnny bolts: Closet bolts that are used to mount toilet bowls to the closet flange. Also referred to as toilet bowl bolts.

Joinery: Finished timber fixtures of buildings such as doors, windows, paneling, cupboards, etc.

Joint: Besides the place you go to after work, a joint is the location between the touching surfaces of two members or components joined and held together by nails, glue, cement, mortar, or other means.

Also, one length of pipe.

Some traditional woodworking joints include the following:

- **Butt joint:** The ends of two pieces of wood are butted against one another. This is the simplest and weakest joint.
- **Miter joint:** Similar to a butt joint, except that both pieces are at a 45° angle.

- **Lap joint:** One piece of wood overlaps another.
- **Box joint:** Used for the corners of boxes. It involves several lap joints at the ends of two boards. Also called a finger joint.
- **Dovetail joint:** A form of box joint where the fingers are locked together by diagonal cuts.
- **Dado joint:** A slot is cut across the grain in one piece of wood for another piece to set into.
- **Groove joint:** The slot is cut with the grain.
- **Tongue and groove:** Each piece has a groove cut all along one edge, and a thin, deep ridge (the tongue) on the opposite edge. If the tongue is unattached, it is considered a spline joint.
- **Mortise and tenon:** A stub (the tenon) will fit tightly into a hole cut for it (the mortise). This is a traditional method of jointing frame and panel members in doors, windows, and cabinets. Mission Style furniture traditionally use mortise and tenon joints.
- **Birdsmouth joint:** Used in roof construction. A V-shaped cut in the rafter connects the rafter to the wall-plate.

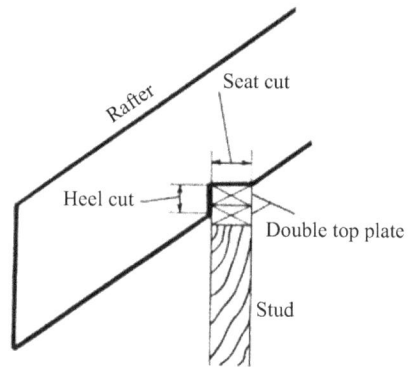

Birdsmouth joint

- **Comb Joint:** A joint used as a way of conserving timber, as a means of joining random lengths of timber to be machined to a finished piece.

Other terms that begin with joint:

- **Joint cement/Joint compound:** A powder that is usually mixed with water and used for joint treatment in gypsum-wallboard finish. Often called **spackle** or **drywall mud**.

- **Joint, edge:** A joint running parallel to the grain of the wood.
- **Joint, open:** A joint in which two adjacent pieces of veneer do not fit tightly together.
- **Joint penetration:** The minimum depth the weld metal extends from its face into a joint.
- **Joint runner:** Collar like device that keeps molten lead in place while sealing a joint in cast iron pipe.

Planer head.

Joint runner.

- **Joint sealer:** Caulking compound.
- **Joint tape:** Tape used to seal joints between insulation boards.
- **Joint tenancy:** A form of ownership in which the tenants own a property equally. If one dies, the other automatically inherits the entire property.
- **Joint trench:** When the electric company and telephone company dig one trench and "drop" both of their service lines in.
- **Joint venture:** A business undertaking by two or more parties in which profits, losses, and control are shared. A joint venture is usually more limited in scope and duration than a partnership, although the terms are often considered synonymous and both indicate similar types of joint liability for debts and torts.
- **Jointing:** The act of reducing the points of all knives on a planer head to coincide with the circumference of a circle when the head is rotated.

Joist: A structural load-carrying member with an open web system that supports floors and roofs utilizing hot-rolled or cold-formed steel and is designed as a simple span member.

Looking at the definition of a "joist" and "joist girder," they appear similar. The difference is that a joist girder is the main horizontal beam. The joists, which are usually smaller, are connected to the girder and are supported by the girder. See **Joist girder**.

Also, wooden 2 × 8's, 10's, or 12's that run parallel to one another and support a floor or ceiling, and supported in turn by larger beams, girders, or bearing walls.

Other terms that begin with joist:

- **Joist designation:** A standard way of communicating the joist safe uniformly distributed load-carrying capacities for a given span such as 16K5 or 24K10 where the first number is the nominal joist depth at midspan in inches and the last number is the chord size. See **Longspan designation** and **Joist girder designation**.
- **Joist girder:** A primary structural load-carrying member with an open web system designed as a simple span supporting equally spaced concentrated loads of a floor or roof system acting at the panel points of the joist girder and utilizing hot-rolled or cold-formed steel.
- **Joist girder designation:** A standard way of communicating the girder design loads, e.g., 48G6N10.5K where the first number is the nominal girder depth at midspan in

inches, 6N is the number of joist spaces on the span of girder, and 10.5K is the kip load on each panel point of the girder. The approximate dead load weight of the member is included in the kip load. See **Joist designation** and **Longspan designation**.

- **Joist hanger:** A metal "U" shaped item used to support the end of a floor joist and attached with hardened nails to another bearing joist or beam.
- **Joist manufacturer:** The producer of joists or joist girders who is Steel Joist Institute (SJI) approved.
- **Joist spacing:** The distance from one joist to another.
- **Joist substitute:** A structural member that is intended for use at very short spans (10 feet or less) where open web steel joists are impractical. They are usually used for short spans in skewed bays, over corridors, or for outriggers. It can be made up of two or four angles to form channel sections or box sections. See **Angle unit**.
- **Joist, trimmer:** A crosspiece fixed between full-length joists (and often across the end of truncated joists) to form part of the frame of an opening in a floor or roof.

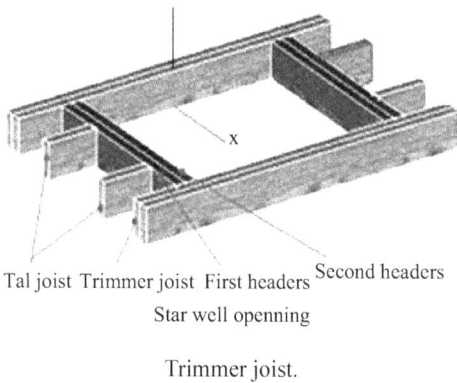

Tal joist Trimmer joist First headers Second headers
Star well openning

Trimmer joist.

Jost effect: The reduction in the frictional resistance that occurs in a direction different to that in which slip is occurring. This effect is used in many applications including the removal of corks from bottles. If the cork is first rotated the force needed to pull the cork from the bottle is significantly reduced. It is also the fundamental reason why threaded fasteners experience self-loosening. Frictional resistance

is first overcome in the transverse direction by slip occurring on the joint resulting in the frictional resistance in the circumferential direction reducing to a small value. The torque acting on the fastener in the loosening direction (as a result of its preload) coupled with the Jost Effect results in self-loosening occurring.

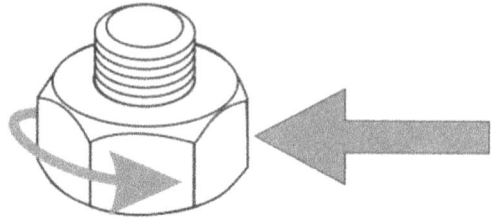

Jost Effect: the nut slides and rotates.

The term is named after the Institute that completed research into this effect, the Jost Institute of Tribotechnology at the University of Central Lancashire in the UK.

Jumpers: Water pipe that is installed in a water meter pit (before the water meter is installed), or electric wire that is installed in the electric house panel meter socket before the meter is installed. This is sometimes illegal.

Junction box: A box that protects splices in electrical wires and provides access. Switches, outlets, and boxes for light fixtures are junction boxes.

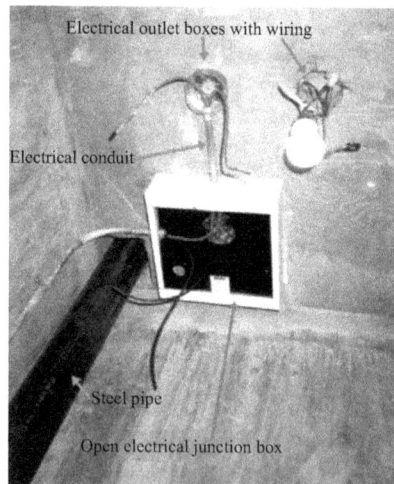

Electrical junction box.

Juvenile wood: The innermost rings of wood adjacent to the pith (The small cylinder of primary tissue of a tree stem around which the annual rings form), formed during the initial development of that part of the stem by the live crown. As the crown moves higher with growth, "mature" wood rings are formed. Juvenile wood differs from mature wood in cell structure and size.

K-series joist: A series of joist adopted in 1986 based on a load/span type of determination.

Kafer fitting: Cast iron drainage pipe fitting with a threaded-on hub that is used to attach to existing cast iron lines.

Kafer insertable fitting.

Kaizen: A Japanese management philosophy typically translated as "continuous improvement." Using this concept, employees are given the authority and resources to solve problems to make improvements. In kaizen events, teams of managers, employees, and others are brought together to improve an entire process. These process-improvement methods are often introduced into a company in conjunction with a lean manufacturing philosophy.

Keenes cement: A cement that is composed of finely ground, anhydrous, calcined gypsum, the set of which is accelerated by the addition of other materials. Keenes cement is used in areas subjected to moisture. It is a hard, strong finishing plaster that is made from gypsum and maintains a high polish because of its density. It excels for use in bathrooms and kitchens and is also widely used for the finish coat in auditoriums, public buildings, and other places where walls may be subjected to unusually hard wear or abuse. Also called **Anhydrous Calcined Gypsum**.

Keeper: The metal latch plate in a door frame into which a doorknob plunger latches.

Kelly ball: A test for the consistency of concrete using the penetration of a half sphere; a 1-inch (2.5-cm) penetration by the Kelly ball corresponds to about 2 inches (5 cm) of slump.

Kerf: The width of the cut made by a saw blade. Also, the width of the saw tooth at its outermost widest point.

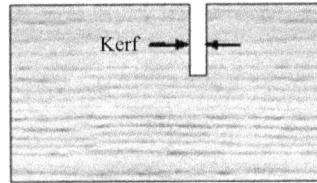

Kerf.

Also, a slot cut into the edge of a piece, e.g., granite, with a saw blade for insertion of an anchor.

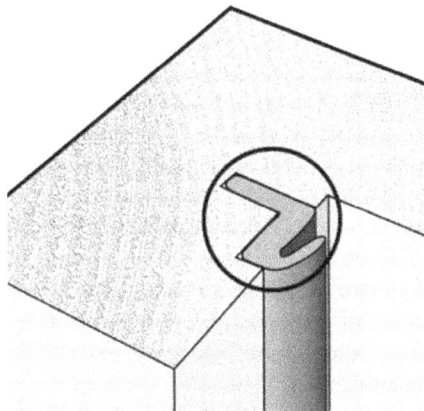

Kerf. (The kerf shown is called a rebated kerf. It has an additional cut that countersinks the kerf from the back edge of the kerf.)

Key: A small piece of shaped metal with incisions cut to fit the wards of a particular lock, and that is inserted into a lock and turned to open or close it.

Also, a small, parallel-sided piece, flat or tapered on top, for securing pulleys and other parts to shafts. The key is placed in a keyway that is a groove or channel for a key, as in a shaft or the hub of a pulley; a **Keyseat**.

Also, a formed slot depression in concrete interlocking the joining of two separate concrete placements.

Other terms that begin with key, keyed, etc.:

Key federal laws: With respect to

- the handling of hazardous materials, they are important laws or statutes enacted to enforce the responsible handling of materials to minimize the danger to human beings and/or the environment.
- **Key plan:** A small reference plan or outline of the whole building on each plan sheet divided into smaller areas for which each sheet is drawn. It can also show different sequences, phases, sheet number that area is drawn on, etc.
- **Keyed alike, keyed differently:** Keyed alike is a keying option in which all the cylinders in an order operate by the same key, whereas with keyed differently the cylinder for one particular door lock operates by its own unique key.
- **Keyed cylinder lock:** A lock providing an exterior entry and locking convenience.
- **Keying:** An option for door locks such as deadbolts or entry door handle sets. Keying is used to specify how the cylinder in the lock should be keyed. See **keyed alike, keyed differently keyed differently**.
- **Keyless:** A plastic or porcelain light fixture that operates by a pull string. Generally found in the basement, crawl space, and attic areas.
- **Keystone:** A central stone at the summit of an arch, locking the whole together. (A voussoir is also a wedge-shaped or tapered

stone used to construct an arch, but smaller than a keystone.)

Keystone and voussoirs.

- **Keyway:** A slot formed and poured on a footer or in a foundation wall when another wall will be installed at the slot location. This gives additional strength to the joint/meeting point.

    Also, a slot cut in a part of a machine or an electrical connector to ensure correct orientation with another part that is fitted with a key.

Sprocket with an internal parallel keyway.

- **Key-in knobs or levers:** Knobs or levers that contain a cylinder on the exterior side that can be locked/unlocked by a key. The interior side has a thumbturn to operate the locking function. Key-in door knobs and levers are often combined with a deadbolt installed above it for best security.

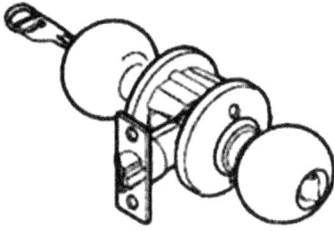

Single key-in function.

Kick back: A work piece that is thrown back by a cutter. Kick back is prevented by using anti-kick back devices on power tools such as table saws. Also spelled **kickback**.

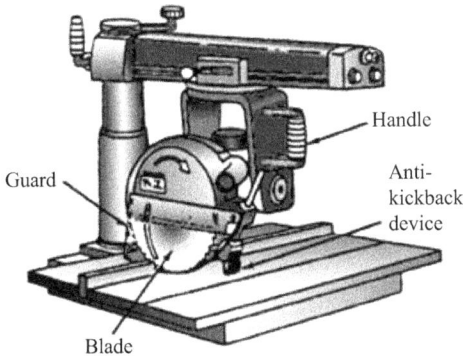

An anti-kick back device.

Kick board: A temporary board used at the perimeter of a building or opening for protection from falling objects which might be kicked off a slab or floor plate onto workers below.

Kicker: A structural member used to brace a concrete form, column, etc.

Adjustable kickers.

Also, in concrete construction, a low plinth at the base of a column to help locate the formwork for a wall or column.

Kicker.

Kilowatt (kw): One thousand watts. A kilowatt-hour is the base unit used in measuring electrical consumption. See **Watt**.

Kiln: A chamber having controlled airflow, temperature, and relative humidity for drying lumber, veneer, and other wood products.

Lumber that is dried in a closed chamber in which the removal of moisture is controlled by artificial heat and usually by controlled relative humidity is referred to as being **kiln-dried**. Kiln drying is about 10 times faster than surface drying.

Kiln-drying schedule: A specified set of dry- and wet-bulb temperatures and air velocities employed in drying a kiln charge of lumber or other wood products.

Kilo: SI prefix for 10^3 or 1000.

Kilogram (kg): The basic unit of mass in the SI system, equal to 1,000 g ($\sim$ 2.2 lbs).

Kilometer (km): A measure of length equal to 1,000 m or 0.62 miles.

Kilonewton (Kn): One thousand newtons, a unit of force in the SI system. Newtons are very small, and the kiloNewton is the practical unit most often used by engineers. In imperial terms it is approximately equivalent to the weight of two hundredweights.

Kinematic: The general movement patterns and directions of the earth's rocks that produce rock deformation.

Kinetic energy: The work, i.e., the amount of energy transferred by a force acting through a distance, needed to accelerate a body of a given mass from rest to its stated velocity. Having gained this energy during its acceleration, the body maintains this kinetic energy unless its speed changes. The same amount of work is done by the body in decelerating from its current speed to a state of rest.

Kingpost/King post truss: A roof truss with a central vertical member. See **Queen truss**.

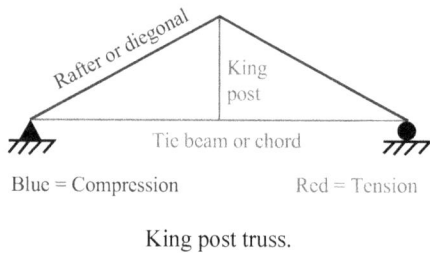

Blue = Compression   Red = Tension

King post truss.

King stud: The vertical 2 x's frame lumber (left and right) of a window or door opening, and that runs continuously from the bottom sole plate to the top plate.

Kip: A unit of weight equal to 1000 pounds.

Knee brace: A structural brace positioned diagonally between a beam or column and a joist panel point.

Knee cap: Sheet metal trim that fits over a panel rib after it has been cut and bent.

Knee wall: Any short wall assembly.

Knife cuts per inch (KCPI): A measure of the smoothness of machined lumber. It can be determined by holding the surfaced board at an angle to a strong light source and counting the visible ridges per inch, usually perpendicular to the profile. The surface is smoother with more knife marks per inch. KCPI is generally used when describing the result of molded profiles or S4S materials.

Knife marks: Very fine lines that appear across the panel veneer or wood solids that can look as though they are raised resulting from some defect in the lathe knife that cannot be removed with sanding.

Knife plate seat: A vertical plate used as a joist seat whose width is small for bearing purposes. It is used for hip and valley bearing conditions, canted seat conditions, and extreme skewed conditions.

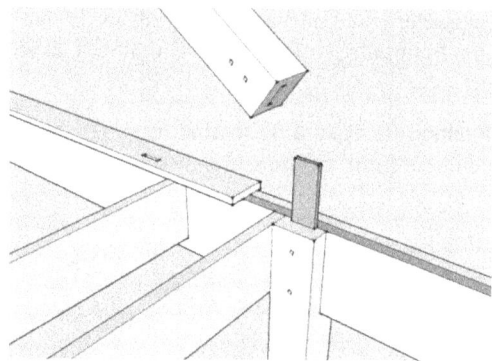

Knife plate.

Knob and tube wiring: The wiring system used before 1945. Two strands of copper wire are run along framing by connecting them to porcelain knobs and through framing inside porcelain tubes.

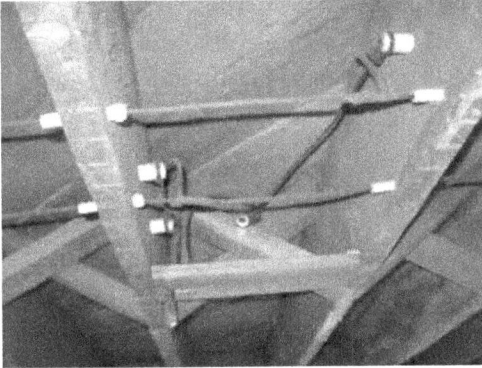

Floor joists with knob and tube wiring.

Knot holes in old wooden fence.

Knock down texture: Any type of drywall texture that is flattened or smoothed. Texture may be knocked down when it is semi-dry with a drywall knife or it can be sanded after it is dry.

Knockout: A key used to remove a drill from a collet (a holding device that exerts a clamping force on the object when it is tightened).

Knockout plug: APVC test plug.

Knot: In lumber, the portion of a branch or limb of a tree that appears on the edge or face of the piece.

A knot is an imperfection in a piece of wood that affects the technical properties of the wood, usually for the worse, but may be exploited for visual effect. In a longitudinally sawn plank, a knot will appear as a usually dark roughly circular node around which the grain of the rest of the wood parts and rejoins. Within a knot, the direction of the wood (grain direction) is up to 90° different from the grain direction of the regular wood.

In grading lumber and structural timber, knots are classified according to their form, size, soundness, and the firmness with which they are held in place. This firmness is affected by, among other factors, the length of time for which the branch was dead while the attaching stem continued to grow.

Other terms that begin with the word knots:

- **Knot holes:** Voids that are produced when knots drop from the wood in which they were originally embedded.

- **Knots, blending pin:** Sound knots 1/4 inch (6.4 mm) or less that generally do not contain dark centers. Blending pin knots are barely detectable at a distance of 6 ft. to 8 ft. (1.8 m to 2.4 m), do not detract from the overall appearance of the panel, and are not prohibited from appearing in all grades.

- **Knots, conspicuous pin:** Sound knots 1/4 inch (6.4 mm) or less in diameter containing dark centers.

- **Knots, loose:** A knot in timber that is not sound and may end up becoming dislodged over time.

- **Knots, open (knot holes):** Openings where a portion of the wood substance of the knot was dropped out, or where cross checks have occurred to produce an opening.

- **Knots, sound tight:** Knots that are solid across their face and fixed by growth to retain their place.

Kollar Kaps: Styrofoam forms used to protect floor drains while concrete is being poured around them.

Kraft paper: A heavy brown paper made of a sulfate pulp that is often used to face batt insulation. Kraft paper is not resistant to fire, so Kraft-faced batt insulation must be covered with a fire-resistant material.

Krypton gas: An inert gas known for its ability to provide insulating properties in a small air space.

L flashing: L-shaped flashing that is made from one continuous piece of metal.

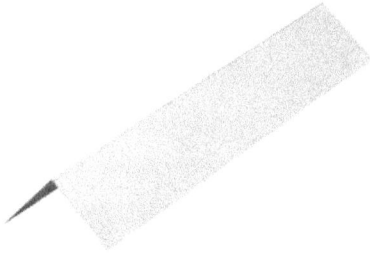

L flashing.

L tubing: An industry standard for copper tubing that is defined by the tube wall thickness and identified by a "blue" strip. Type "L" copper tube wall is ~ 50% greater thickness than Type "M."

Labeled: Equipment or materials to which has been attached a label, symbol, or other identifying mark of an organization that is acceptable to the authority having jurisdiction and concerned with product evaluation, that maintains periodic inspection of production of labeled equipment or materials, and by whose labeling the manufacturer indicates compliance with appropriate standards or performance in a specified manner.

Labor and material payment bond: Security from a surety (bonding) company to the owner, on behalf of the prime contractor or subcontractor. A labor and material payment bond guarantees payment to the owner should the contractor fail to pay for all labor, materials, equipment, or services in accordance with the contract. It also guarantees to pay any claims against the owner from contractors and suppliers who have not been paid for labor, material, and equipment incorporated into the project.

Laborer: A building worker without any specific skill. Specific trades have their own laborers, e.g., a bricklayer's hod carrier.

Laced valley: Interlocking pattern formed in the valley of a roof by overlapping shingles in alternating rows, making a basket-weave pattern.

Lacing: Generally horizontal members that connect together and reduce the unsupported length of compression members.

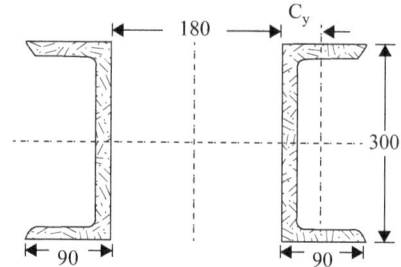

(a)

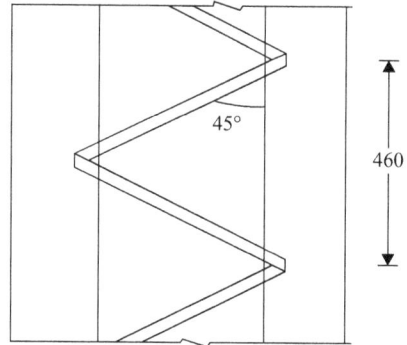

(b)

Lacing.

Lacquer: A fast-drying usually clear coating that is highly flammable and dries by solvent evaporation only. Can be reconstituted after drying by adding solvent.

FYI: The word lacquer is derived from the word lac that describes the secretions of the lac beetle. This insect, found mainly in Asia, deposits its secretions on branches of trees and this crop is later harvested. In its original state the resin developed by the insects, contains a red dye. This dye is separated from the resin by boiling in water. Then the residue resin, known as seed lac, is melted, strained, cooled, and flaked and then becomes shellac.

Lagging: With regard to piles, pieces of timber or other material that is attached to the sides of piles to increase resistance to penetration through soil.

Also, insulation that is used to prevent heat diffusion, as from a steam pipe.

Also, a wooden frame built especially to support the sides of an arch until the keystone is positioned.

Ladder stand: A mobile, fixed-size, self-supporting ladder consisting of a wide flat tread ladder in the form of stairs.

Laid up: Work done or being performed on a masonry wall installation.

Laitance: An accumulation of fine, powdery aggregate particles on fresh cement caused by the upward movement of water. Laitance indicates that too much water was used in the mix resulting in poor surface adhesion for a waterproofing layer.

Lally column: A steel column used as a support for girders, beams, joists, etc.

Lally columns.

Lamella: Shell construction in which the shell is formed by a lattice of interlacing members.

Also, the thin cementing layer between wood cells.

Lamellar tearing: A separation or crack in the base metal caused by through-thickness weld shrinkage strains of adjacent weld metal.

Typically, the cracks appear in solders of several passes in the joints angle in Ts and Ls; they are always associated with points high stress concentration.

It is generally agreed that three conditions must be satisfied for lamellar tearing to occur:

* **Transverse strain:** The shrinkage strains on welding must act in the through-thickness direction.
* **Weld orientation:** The fusion boundary will be roughly parallel to the plane of the inclusions.
* **Material susceptibility:** The plate must have poor ductility in the through-thickness direction.

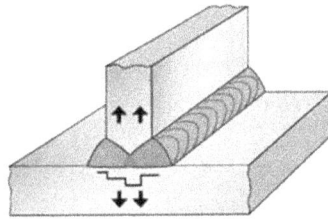

Lamellar tearing in T-butt weld.

Laminar flow: Laminar flow occurs when a fluid flows in parallel layers, with no disruption between the layers, i.e., without mixing. There are no cross currents, eddies or mixing perpendicular to the direction of flow. Flow that is not laminar is called turbulent flow.

Laminar flow is also known as streamline flow.

Laminar tear: A planar separation that develops in flanges of rolled shapes and within thick plates near certain large welds, as high weld shrinkage develops stresses across plate thickness.

Laminate: A product made by bonding together two or more layers (laminations) of material or materials.

Terms that begin with the word laminated:

* **Laminated glass:** Glass composed of two sheets of glass fused together with a sheet of transparent plastic between the sheets. It is a form of safety glass; when broken laminated glass will generally not leave the opening.

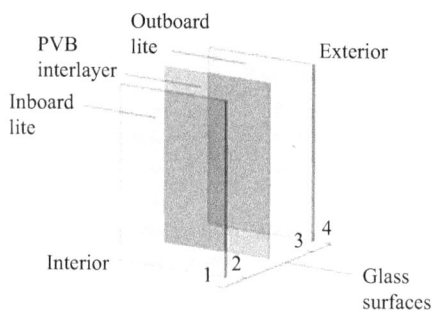

Laminated glass.

- **Laminated shingles:** Shingles that have extra layers or tabs, giving a shake-like appearance. See **Dimensional shingles** and **Architectural shingles**.
- **Laminated square edge countertop:** The edge on a countertop that is made by covering the square front corner with plastic laminate. See **Plastic laminate countertop**.
- **Laminated strand lumber (LSL):** A type of reconstituted timber made of separated strands glued together under pressure.
- **Laminated timbers:** An assembly made by bonding layers of veneer or lumber with an adhesive so that the grain of all laminations is essentially parallel.
- **Laminated veneer lumber (LVL), beam:** Laminated veneer lumber or micro-laminated beam, made from thin layers of wood (veneers) that are glued together. The veneer segments may run either perpendicular or parallel to the load and they have no arch or camber.
- **Laminated veneer lumber (LVL) panel:** A structural lumber manufactured from veneers that are laminated into a panel with the grain of all veneer running parallel. The resulting panel is generally manufactured in 3/4 to 1-1/2 inch thicknesses and ripped to common lumber widths of 1-1/2 to 11-1/2 inches, or wider.
- **Laminated veneer lumber core (LVLC):** A door manufactured by laminating veneer with all grain laid-up parallel. It can be manufactured by using various species of wood fiber in various thicknesses.

Laminating: The bonding together of two or more layers of materials.

Lamp: The actual bulb or encased gas source producing light in a fixture. Lamps use electricity to produce light in any of several ways: by heating a wire for incandescence, by exciting a gas that produces ultraviolet light from a luminescent material, by generating an arc that emits visible light and some ultraviolet light, or by inducing excitation of mercury through radio frequencies. Light-emitting diodes, packaged as traditional bulbs, are also considered to be lamps.

Lamp life: The useful operating life of a lamp.

Lamps: Lamps use electricity to produce light in any of several ways: by heating a wire for incandescence, by exciting a gas that produces ultraviolet light from a luminescent material, by generating an arc that emits visible light and some ultraviolet light, or by inducing excitation of mercury through radio frequencies. Light-emitting diodes, packaged as traditional bulbs, also fall under this definition.

Lancet: A narrow window with a sharp, pointed arch, commonly associated with gothic revival architecture.

Lancet windows.

Land sale-leaseback: The same concept as a sale-leaseback, but in this case only the land is sold and leased back using a ground lease.

Landfills: Waste disposal sites for solid waste from human activities.

Landing: A platform between flights of stairs or at the termination of a flight of stairs. Often used when stairs change direction. Many building codes set minimum dimensions for landings.

Landlord: The lessor or owner of the leased property.

Landlord-paid tenant improvements (LPTI): The total cost (outlay) of necessary tenant improvements paid by the landlord netted against any contribution made by the tenant.

Landscape area: The project site area less the building footprint, hardscape areas, water bodies, etc.

Landscape coefficient (K_L): A coefficient used to calculate the evapotranspiration rate; it takes into account the species factor, vegetation density factor, and microclimate factor of the area.

Landslide: Earthquake triggering land disturbance on a hillside where one land mass slides over the other.

Langelier's index: A calculated number that is used to predict whether or not water will precipitate, be in equilibrium with, or dissolve calcium carbonate. It is sometimes erroneously assumed that any water that tends to dissolve calcium carbonate is automatically corrosive.

Lanyard: A flexible line of rope, wire rope, or strap that generally has a connector at each end for connecting the body belt or body harness to a deceleration device, lifeline, or anchorage.

Lap: To cover the surface of one shingle or roll with another.

Also, with regard to veneer, a condition where the pieces of veneer are so misplaced that one piece overlaps the other and does not make a smooth joint.

Other terms beginning with the word lap:

- **Lap cement:** An asphalt-based roof cement formulated to adhere overlapping plies or asphalt roll roofing.
- **Lap seam:** A lap seam occurs where overlapping materials are seamed, sealed, or otherwise bonded.
- **Lap siding:** Horizontal siding that is installed by overlapping the top edge of each course with the bottom edge of the course directly above it.
- **Lap setting:** An adjustment on snips that determines how close the tips come together or cross over.

Laser/Laser level: A tool used in building construction that projects a light beam out on a level plane for use as a reference in aligning objects so they are level (e.g., grade stakes, ceiling tiles, and plumbing up walls and structure).

Latch: A beveled metal tongue operated by a spring-loaded knob or lever. The tongue's bevel lets you close the door and engage the locking mechanism, if any, without using a key. Contrasts with dead bolt.

Late Quaternary: The late Quaternary refers informally to the past 0.5–1.0 million years. Faults that have slipped during this time are sometimes considered active.

Geologic time scale					
EON	ERA	Period		EPOGH	Present
Phanerozoic	Cenozoic	Quaternary		Holocene	
					0.01
				Pleistocene	
					1.6
		Tertiary	Neogene	Pliocene	
					5.3
				Miocene	
					23.7
			Paleogene	Oligocene	
					36.6
				Eocene	
					57.8
				Paleocene	
					66.4
	Mesozoic	Cretaceous			
					144
		Jurassic			
					208
		Triassic			
					245
	Paleozoic	Permian			
					286
		Pennsylvanian	Carboniferous		
					320
		Mississippian			
					360
		Devonian			
					406
		Silurian			
					438
		Ordovlcian			
					505
		Cambrian			
					570
Precambrian	Proterozoic				
					2500
	Archean				
					3800
	Hadean				
					4550

Age in millions of years before present

Geologic time scale.

Terms that begin with the word lateral:

- **Lateral (electric, gas, telephone, sewer, and water):** The underground trench and related services (i.e., electric, gas, telephone, sewer, and water lines) that will be buried within the trench.
- **Lateral bracing:** Structural bracing against loads produced in the horizontal.
- **Lateral buckling:** Buckling of a structural member involving lateral deflection and twist. Also referred to as **lateral-torsional buckling**.
- **Lateral force coefficients:** Factors applied to the weight of a structure or its parts to determine lateral force for seismic structural design.
- **Lateral sewage line:** A sewage line that connects one sewage pipe with another.
- **Lateral spreads:** Landslides that form on gentle slopes as the result of liquefaction of a near-surface layer from ground shaking in an earthquake.

Latex: A colloidal dispersion of a polymer or elastomer in water that coagulates into a film upon evaporation of the water.

Latex-based paint: A general term used for water-based emulsion paints made with synthetic binders such as 100% acrylic, vinyl acrylic, terpolymer, or styrene acrylic. A stable emulsion of polymers and pigment in water.

Latewood: The portion of the annual growth ring that is formed after the earlywood formation had ended. It is usually denser and stronger mechanically than earlywood. Also called **Summerwood**.

Lath: A building material of narrow wood, metal, gypsum, or insulating board that is fastened to the frame of a building to act as a base for plaster, shingles, or tiles.

Lathe: A machine tool that rotates the workpiece on its axis to cut, sand, knurl, drill, and otherwise deform with tools that are applied to the workpiece. The objects that are created have are symmetrical about the axis of rotation.

Lattice: An open framework of criss-crossed wood or metal strips that form regular patterned spaces.

Lauan plywood underlayment: A plywood underlayment made from Lauan wood. See **Underlayment, particleboard underlayment, Plywood underlayment, Cement board underlayment, Gypsum-based underlayment**, and **Untempered hardboard underlayment**.

Laundry room pan: A pan that is placed under a washer to catch water should the washer overflow. The pan may catch and hold the water or channel the water into a drain.

Layout: Marking the actual field location of any part of a building excavation, assembly, structure, mechanical, plumbing, electrical, et al, and locating that part on and within the building under construction.

Layout board: A board that has been marked to show the distance between each of the trusses. It is used while trusses are being installed to ensure that they are positioned properly.

Lazy Susan: Type of specialty shelves that revolve.

Leaching: To make a soluble material (chemical or mineral) drain away from soil, ash, or similar material by the action of percolating liquid.

With regard to concrete, the process by which a liquid dissolves and removes the soluble components of a material. Leaching occurs from the interior of the concrete to the surface.

Leach field: A method used to treat/dispose of sewage in rural areas that are not accessible to a municipal sewer system. Sewage discharged into a leach field.

Lead: A metal, previously used as a pigment in paints. Use of lead was discontinued in the early 1950s by paint industry consensus standard, and banned by the Consumer Products Safety Commission in 1978 because of its toxicity. It has a somewhat sweet taste that small children are drawn to, especially when it is in the form of peeling paint.

Lead is an element, that is represented by the symbol Pb. Its atomic structure is permanently arranged and is not changed by chemical reactions, but lead can combine chemically with other atoms or molecules to make new compounds.

Also, lead is an electrical conductor.

Other terms beginning with the word lead:

- **Lead accredited laboratory:** A laboratory that has been evaluated and received accreditation through the EPA's National Lead Laboratory Accreditation Program (NLLAP) to perform lead measurement or analysis, usually over a specified period of time.
- **Lead inspection:** A surface-by-surface investigation to determine the presence of lead-based paint and the provision of a report explaining the results of the investigation.
- **Lead light:** A window with small panes of glass set in grooved rods of cast lead (or came). The glass can be clear, color, or stained. Also referred to as **lead glazing and stained glass**.
- **Lead-based pain:** Paint or other surface coatings that contain lead equal to or in excess of 1.0 mg/cm² or more than 0.5% by weight.
- **Lead-based paint activities:** In the case of target housing and child-occupied facilities, lead inspection, lead risk assessment, and lead abatement.
- **Lead-based paint hazard:** Any condition that causes exposure to lead from lead-contaminated dust, lead-contaminated soil, or lead-contaminated paint that is deteriorated or present in accessible surfaces, friction surfaces, or impact surfaces that would result in adverse human health effects as identified by the EPA pursuant to TSCA section 403.
- **Lead-based paint hazard control:** Activities to control and eliminate lead-based paint hazards, including interim controls, and lead abatement.
- **Lead-contaminated dust:** Surface dust in a residential dwellings or child-occupied facility that contains an area or mass concentration of lead at or in excess of levels identified by the EPA, pursuant to Hazard Standards for Lead in Paint, Dust and Soil in the Toxic Substances Control Act (TSCA), Section 403.
- **Lead-contaminated soil:** Bare soil on residential real property and on the property of a child-occupied facility that contains lead at or in excess of levels identified by the EPA pursuant to the Toxic Substances Control Act (TSCA), section 403.
- **Lead-based paint free certification:** A rental dwelling certified by a certified lead-based paint inspector to contain no lead at or above 1.0 mg/cm².
- **Lead-hazard free dwelling:** A dwelling that has been evaluated and contains no lead-based paint and has interior dust and exterior soil lead levels below the applicable federal standards.
- **Lead-hazard screen:** A limited lead risk assessment activity that involves limited paint and dust sampling as described in The Code of Federal Regulations of the United States of America (CFR), Sec. 745.227(c).
- **Lead-specific detergent:** A cleaning agent manufactured specifically for cleaning and removing leaded dust or other lead contamination.

Leader: A vertical drainage pipe for conveying storm water from roof or gutter drains to a building house storm drain, building house drain (combined), or other means of disposal. The leader includes the horizontal pipe to a single roof drain or gutter drain.

Leader head: Architectural appeal is the most common purpose for the use of leader heads. They add a finishing touch to a high-quality gutter system. These items are custom fabricated. Also known as **conductor heads**, **collector boxes**, and **scupper boxes**.

Leader heads.

Leader pipe: A conduit for carrying water from a gutter, scupper, drop outlet, or other drainage unit from roof to ground level. Also known as **Downspout**.

Leading edge: The edge of a floor, roof, or formwork for a floor or other walking/working surface (such as the deck) which changes location as additional floor, roof, decking, or formwork sections are placed, formed, or constructed. A leading edge is considered to be an "unprotected side and edge" during periods when it is not actively and continuously under construction.

Leading edge.

Leakage rate: The speed at which an appliance loses refrigerant, measured between refrigerant charges or over 12 months, whichever is shorter. The leakage rate is expressed in terms of the percentage of the appliance's full charge that would be lost over a 12-month period if the rate stabilized. (EPA Clean Air Act, Title VI, Rule 608).

Lean manufacturing: A business philosophy/strategy that focuses on eliminating waste. It is usually employed along with the concept of kaizen, or continuous improvement.

Lean-to: A structure depending upon another structure for support and having only one slope such as a shed.

Lease: A contract by which one party conveys land, property, services, etc., to another for a specified time, usually in return for a periodic payment.

A lease is a contractually binding agreement that grants a right to exclusive possession or use of property, usually in return for a periodic payment called rent.

Lease buyout: The process by which a landlord, tenant, or third party pays to extinguish the tenant's remaining lease obligation and rights under its existing lease agreement.

Leased fee: The landlord's ownership interest of a property that is under lease. Its value is based on the anticipated income from rent, and the reversionary property value upon lease expiration.

Least-squares fit: When plotting data points on a graph, the least-squares-fit is the line or curve that comes closest to going through all the points.

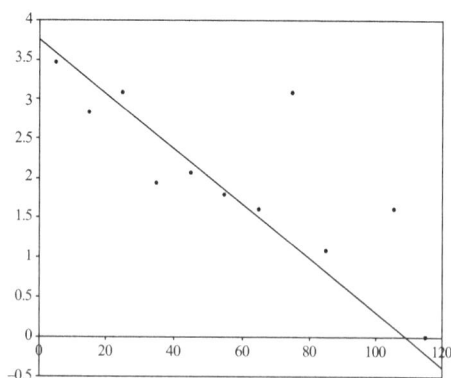

Plot showing least squares fit.

Least toxic chemical pesticide: Any pesticide product for which all active ingredients and known inert ingredients meet the least toxic Tier 3 hazard criteria under the City and County of San Francisco's hazard screening protocol. Least toxic also applies to any pesticide product, other than rodent bait, that is applied in a self-contained, enclosed bait station placed in an inaccessible location, or applied in a gel that is neither visible nor accessible.

Ledger: In scaffolding, the horizontal members running along the scaffold.

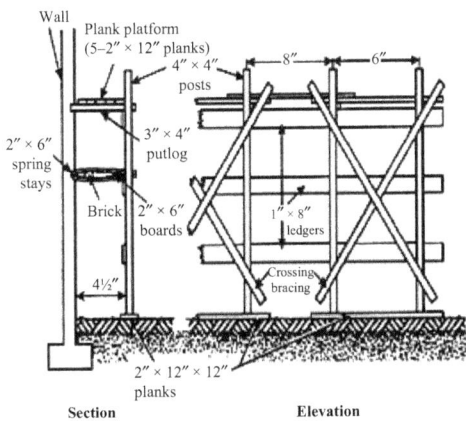

Ledgers on scaffolding.

Also, the wooden perimeter frame lumber member that bolts onto the face of a foundation wall or along the bottom of the side of a girder and supports the wood structural floor. Also referred to as **ledger strip**.

Ledger with joists attached.

LEED®Accredited Professionals (AP): Individuals who have successfully completed the LEED professional accreditation exam.

LEED® Certification (Leadership in Energy and Environmental Design): US Green Building Council's Leadership in Energy & Environmental Design. The US Green Building Council oversees the LEED® Rating System.

Leeward: The direction toward which the wind is blowing, which is opposite the side from which the wind blows. Opposite of windward.

Leeward and windward vis-a-vis wind direction.

Left hand thread: A screw thread that is screwed in by rotating counterclockwise.

Leftover paint: The following steps are recommended for dealing with leftover (post-consumer) paint:

- Purchase the correct amount of paint
- Store paint properly
- Use leftover paint
- Reuse (donate) or recycle
- As a last resort, dispose of paint properly.

Also referred to as **Post-consumer paint management**.

Leg: The flat projecting part of a structural angle.

Legionella: A waterborne bacterium that causes Legionnaire's disease. It grows in slow-moving or still warm water and can be found in plumbing, showerheads, and water storage tanks. Outbreaks of Legionella pneumonia have been attributed to evaporative condensers and cooling towers.

Lessee: The person in possession of a building under a lease from the building's owner. Also referred to as a **tenant**.

Lessor: The person who rents or leases a property to another.

Also known as a **landlord**.

Letter of intent: A notice from an owner to a contractor that states that a contract will be awarded to the contractor; some letters of intent require that certain events must occur or specific conditions be met prior to award. The letter can sometimes serve as a formal notice to proceed on the project.

Let-in brace: Nominal 1-inch-thick boards applied into notched studs diagonally. Also, "L" shaped, metal straps that are installed by the framer at the rough stage to give support to an exterior wall or wall corner.

Level: True horizontal.

Also, a surveying instrument used to establish a datum for the location of elevations and to plumb vertical surfaces.

Level cut: A true horizontal cut. The actual angle required for a level cut is determined by the slope of the member to be installed. A level cut is at right angles to the plumb cut.

Level payment mortgage: A mortgage with identical monthly payments over the life of the loan.

Level-type swage: (Swage is method of shaping a saw tooth to provide side clearance on both sides of each tooth.) Level-type swage is a device for widening the tips of saw teeth by drawing out the tooth point between a lever-actuated die and a fixed anvil.

Leveling: Finding levels during surveying, or providing levels for new construction.

Leveling plate: A steel plate used on top of a foundation on which a structural column can be placed.

Leveling screws: Any screw or bolt used to level a piece of equipment or structural column base plate.

Lever lock: A lever handle and lever arm operator available as an option on awning units.

Leverage: The exertion of force using a lever or an object used in the manner of a lever and the mechanical advantage gained in this way, e.g., using a crow-bar to move a large stone.

In business the ratio of a company's loan capital (debt) to the value of its common stock (equity).

Lewis: A device consisting of expanding wedges used for lifting heavy stone masonry.

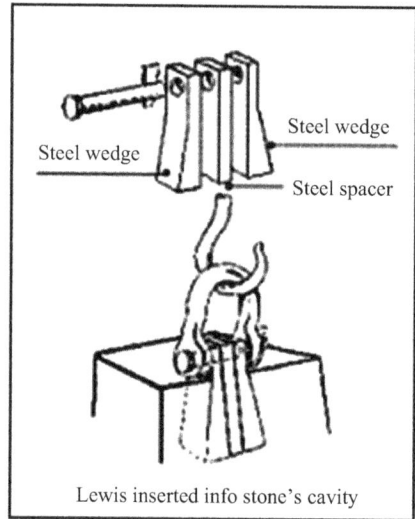

A Lewis device.

Liberty™: Self-adhering low slope roofing. Liberty™ systems are applied without torches, open flames, hot asphalt, or messy solvent-based adhesives.

License: A written document (usually) issued by a governmental agency authorizing a person to perform specific acts in or in connection with the construction or alteration of buildings, or the installation, alteration, and use and operation of service equipment therein.

Lien: An encumbrance that usually makes real or personal property the security for payment of a debt or discharge of an obligation. See **Mechanic's lien**.

Lien release: A written document from the contractor to the owner that releases the Lien, Mechanic's, or Material following its satisfaction.

Lien waiver: A written document from a contractor, subcontractor, material supplier or other construction professional, having lien rights against an owner's property, relinquishing all or part of those rights. Lien waivers are generally used for processing progress payments to prime or main or subcontractors as follows: **Conditional Lien Waiver, Unconditional Lien Waiver**, and **Final Lien Waiver**.

Life-cycle assessment (LCA): A technique to assess the environmental aspects and potential affects associated with a product, process, or service, by: (1) **compiling** an inventory of relevant energy and material inputs and environmental releases; (2) **evaluating** the potential environmental affects associated with identified inputs and releases, and; (3) **interpreting** the results to help make a more informed decision. Also referred to as life-cycle analysis.

Life-cycle costing (LCC): An accounting methodology to evaluate the economic performance of a product or system over its useful life; it considers the operating costs, maintenance expenses, and other economic factors.

Lifeline: Lifelines are structures that are important or critical for a community to function, such as roadways, pipelines, powerlines, sewers, communications, and port facilities.

Also, a component consisting of a flexible line connected vertically to an anchorage at one end (vertical lifeline), or connected horizontally to anchorages at both ends (horizontal lifeline), and that serves as a means for connecting other components of a personal fall arrest system to the anchorage.

Lift: The height a body or column of fluid is raised (sucked up); for example, from a reservoir to the pump inlet. Lift is sometimes used to express a negative pressure or vacuum. Lift is the opposite of head.

Also, in masonry terms, a movement of scaffolding to enable the next vertical section of wall to be laid up.

Also, with regard to roofing, the sprayed polyurethane foam that results from a pass. It usually is associated with a certain pass thickness and has a bottom layer, center mass, and top skin in its makeup.

Lift slab: A method of concrete construction in which floor and roof slabs are cast on or at ground level and lifted into position using jacks. See **Jacking operation**.

Light: Space in a window sash for a single pane of glass. Also, a pane of glass.

Light framing: The use of dimension lumber, trusses, and other small cross-section members to provide support and enclosure for a building.

Light-gauge steel stud: A cold formed steel stud made of sheet metal of 22 gauge to 16 gauge in thickness.

Light pollution: Waste light from building sites that produce glare, is directed upward to the sky, or is directed off the site. Waste light does not increase nighttime safety, utility, or security and needlessly consumes energy and natural resources. Beware of simple contract requirements that say "contractor to provide security lighting" or words to this effect, as light pollution may result in law suites from the local community.

Light reflectance: The percentage of light that is not absorbed by the surface of a material.

Lighting outlet: An outlet intended for the direct connection of a lampholder, a luminaire (lighting fixture), or a pendant cord terminating in a lampholder.

Lightning point: The terminus of a grounding system, usually the highest point on a building or structure and made of a solid conductor such as copper or aluminum.

Lignin: The second most abundant constituent of wood. Lignin is located principally in the secondary wall and the middle lamella, which is the thin cementing layer between wood cells. Chemically it is an irregular polymer of substituted propylphenol groups, and thus no simple chemical formula can be written for it.

Limestone: A white colored rock that is made up mostly of calcium carbonate. Rain water dissolves limestone; therefore, it is not good as a building material.

Limit stop: A faucet control unit used to adjust maximum water temperature.

Limit switch: A safety control that automatically shuts off a piece of equipment, e.g., a furnace, if it gets too hot.

Limited access zone: An area alongside a masonry wall that is under construction and clearly demarcated to limit access by employees. See **Controlled access zone**.

Line: An object or form whose actual or visual length greatly exceeds any actual width or depth it many have.

Also, a tube, pipe, electric cord, or hose that acts as a conductor.

Other terms beginning with the word line:

- **Line current:** Alternating electrical current of 110 volts or more.
- **Line loss:** The reduction in the quantity of natural gas flowing through a pipeline that results from leaks, venting, and other physical and operational circumstances on a pipeline system, including leaks, theft, or fuel used by compressors to maintain pressure necessary for transportation.

  Also, a voltage drop caused by resistance in wire during transmission of electrical power over distance.
- **Line of action of a force:** The infinite line defined by extending along the direction of the force from the point where the force acts.
- **Line of balance scheduling:** Some projects may include activities that are repetitive, e.g., a multi-story building with similar floors. These repetitive activities can be modeled in CPM, but they can be difficult to visualize. The Line-of Balance is a graphical technique that can be used in conjunction with CPM and may help you better visualize how work crews interact.

Linear: A structure is said to behave linearly when its deformation response is directly proportional to the loading (i.e., doubling the load doubles the displacement response). For a material, linear means that the stress is directly proportional to the strain.

Linear actuator: A device that converts hydraulic energy into linear motion, a cylinder or ram.

Linear elastic: A force–displacement relationship that is both linear and elastic. For a structure, this means the deformation is proportional to the loading, and deformations disappear on unloading. For a material, the concept is the same except strain substitutes for deformation, and stress substitutes for load.

Lineal foot: A 12-inch ruler is a linear foot. Some people use the term "linear foot," to distinguish between a square foot, or a cubic foot. For example, if the room is 10 × 10 (linear feet) it is 100 square feet.

Lineament: An extensive linear topographic feature or geophysical anomaly that usually reflects an underlying structure.

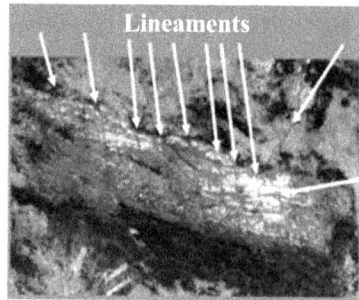

Liner/Liner insert: As pertains to range hoods, a specialized range hood that is designed specifically to fit inside of a wood hood or other enclosure.

Liner or insert liner for a range hood.

Lineside treatment: Treating feedwater carried out prior to the water being transferred to the water tanks.

Linoleum: Genuine linoleum, not to be confused with vinyl, was invented nearly 150 years ago and is still relevant today. Environmentally preferred linoleum is made from natural, raw materials. Linseed oil that comes from the flax plant is the primary ingredient. Other ingredients include wood or cork powder, resins, and ground limestone. Mineral pigments provide its colors. Sheet linoleum is available in many thicknesses. If the opportunity arises visit some old WW II Army barracks. The linoleum on the floor will probably be original and looking as good as new.

Linseed Oil: A drying oil made from the flax seed. It is used as a solvent in many oil- based paints. "Boiled" linseed oil can be used to protect wood from water damage. Sometimes used as a furniture polish.

Lintel: A horizontal structural member that supports the load over an opening such as a door or window.

Liquefaction: The transformation of a granular material (soil) from a solid state into a liquefied state as a consequence of increased pore-water pressure induced by vibration. Normally solid soil suddenly changes to a liquid state (usually sand or granular soil in proximity to water) due to vibration.

Liquid driers: A solution of soluble driers in organic solvents.

Liquidated damages: An amount stipulated in the contract that the parties agree is a reasonable estimation of the damages to one party in the event of a breach by the other party.

To collect, the liquidated damages provision must be deemed a reasonable forecast of the damages likely to actually result from the breach. Where these conditions are met, the amount that was established becomes a maximum limitation on the defaulting party's liability. If the provision does not meet these conditions, or if it otherwise appears that inclusion of it was motivated by a desire to deter a breach rather than by a good faith effort to estimate probable damages, the provision will be considered a "penalty" and will be unenforceable. Recoveries will then be limited to actual damages, if any.

List price: Generally accepted to be the same as suggested retail price; the price to the consumer.

Listed: Equipment, materials, or services included in a list published by an organization that is acceptable to the authority having jurisdiction and is concerned with evaluation of products or services, that maintains periodic inspection of production of listed equipment or materials or periodic evaluation of services, and whose listing states that the equipment, material, or services either meets appropriate designated standards or has been tested and found suitable for a specified purpose.

Note: Some owners may require equipment to be both listed and labeled.

Lite: A piece of glass. In windows and doors, lite refers to separately framed panes of glass (as well as designs simulating the look of separately framed pieces of glass). Sometimes spelled **light**.

Liter (l): Basic unit of volume in the metric system equal to 1,000 cubic centimetres (1.056 US quarts).

Lithology: Lithology is the description of rock composition (what it is made of) and texture.

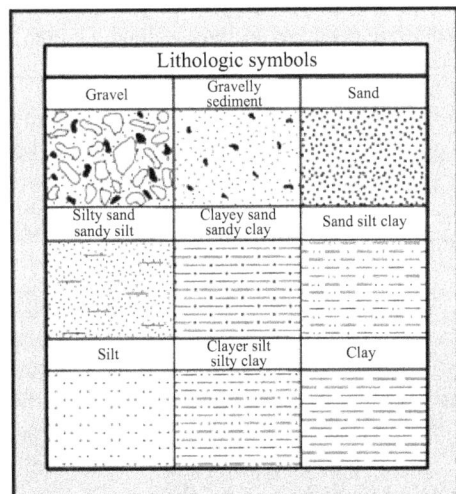

Lithologic symbols		
Gravel	Gravelly sediment	Sand
Silty sand sandy silt	Clayey sand sandy clay	Sand silt clay
Silt	Clayer silt silty clay	Clay

Lithopone: A white pigment of barium sulfate and zinc sulfide. Lithopone was once a primary substitute for lead carbonate or "white lead" pigments. It has been largely replaced by titanium dioxide.

Lithosphere: The solid outer crust of the earth including the crust and upper mantle.

Live load: All occupants, materials, equipment, constructions or other elements of weight supported in, on or by a building that will or are likely to be moved or relocated during the expected life of the building. These loads do not include dead load, wind load, snow load, or seismic load.

Live parts: As regards to electrical, energized conductive components.

Load: An external force.

Also, the term load is sometimes used to describe more general actions such as temperature differentials or movements such as foundation.

Load and resistance factor design (LRFD): A method of proportioning structural members such that no limit state is exceeded when all appropriate load combinations have been applied.

Load bearing wall: Load bearing walls include all exterior walls and any interior wall that is aligned above a support beam or girder. See **Non-load bearing wall**.

Load combination: The combination of loads that produce the worse loading condition in a structural member.

Load factor: In electrical engineering the load factor is defined as the average load divided by the peak load over a specified time period:

$$F_{Load} = \text{Average load} \div \text{Maximum load over a given time period.}$$

A high load factor means power usage is relatively constant. A low load factor means that at times a high demand is set, thereby imposing higher costs on the system; utility companies charge for both usage and demand.

The load factor is related to and sometimes confused with the demand factor.

$$F_{Demand} = \text{Maximum load in a given time period} \div \text{Maximum possible load}$$

The primary difference is that the denominator in the demand factor is fixed depending on the system. Because of this, the demand factor cannot be derived from the load profile but needs also the full load of the system.

Load table: A table of standard structural member designations that give the total safe uniformly distributed load-carrying capacities and live load-carring capacities of the members for different span lengths. Load tables also usually give the approximate weight per foot of each structural member.

Loader: Any of a variety of machines, wheel or track mounted, designed primarily to lift and load and stack materials.

Loading combinations: The systematic application of composite design forces or loading conditions used to determine the maximum stresses in structural members. For example, a designer may assume 100% dead load and 80% live load plus 60% wind load from the north, plus 90% snow load and a seismic design category III area as loading combinations.

Loading diagram: A diagram that shows all design loads and design criteria that a member is to be designed for. The loads include dead load, live load, snow drift, concentrated loads, moments, etc. The design criteria include: deflection requirements, load combinations, net uplift, one-third increase in allowable stress allowed or not, etc.

Loading ramp: A hinged, mechanically operated lifting device that is used for spanning gaps and adjusting heights between loading surfaces, or between loading surfaces and carriers.

Loan: The amount to be borrowed.

Loan to value ratio: The ratio of the loan amount to the property valuation and expressed as a percentage. For example, if a borrower is seeking a loan of $200,000 on a property worth $400,000 it has a 50% loan to value rate. The higher the loan to value, the greater the lender's perceived risk. Loans above normal lending LTV ratios may require additional security.

Local buckling: Metal members are relatively slender and usually consist of an assembly of thin plates. When the cross section of a (usually) steel shape is subjected to large compressive stresses, the thin plates that make up the cross section may buckle before the member's full strength is attained. When a cross-sectional element fails in buckling, then the member's capacity is reached. Consequently, local buckling becomes a limiting state for the strength of steel shapes subjected to compressive stress. Such buckling is sometimes referred to as overall buckling.

Lock: A fastening device in which a bolt is secured and can be operated by a key. Also referred to as **latches and catches**.

Lock block: A concealed block the same thickness as the door stile or core which is adjacent to the stile at a location corresponding to the lock location and into which a lock is fitted.

It is not be possible to fit a lock, latch or handle into a hollow door unless a lock block is installed; the manufacturer fits the block into the door during construction. After the door has been produced, it is stamped with "Lock" or "Lb," indicating which side of the door the block has been positioned. Without this stamp, you may end up fitting the hinges on the wrong side of the door and have nothing to fit the lock or handles to.

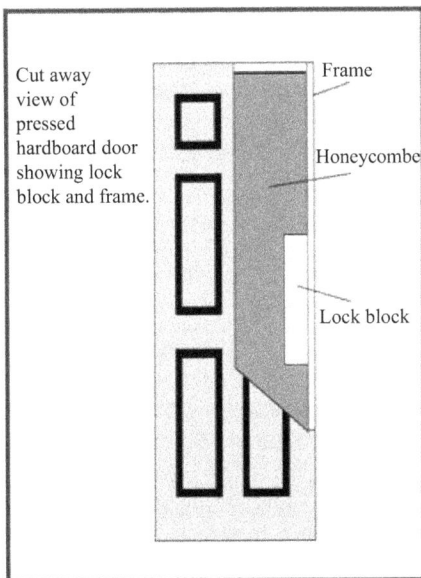

Cut away view of pressed hardboard door showing lock block and frame.

Frame

Honeycombe

Lock block

Lock stile: The vertical member (stile) of a casement sash that closes against the surrounding frame.

Locked fault: A locked fault is a fault that is not slipping because frictional resistance on the fault is greater than the shear stress across the fault (it is stuck). Such faults may store strain for extended periods that is eventually released in an earthquake when frictional resistance is overcome.

Tools whose names begin with locking:

• **Locking bar clamp:** An adjustable, vise-type clamp with a jaw that slides along a steel bar to extend clamping capabilities. The clamp can be locked onto a work piece, leaving both hands free for work.

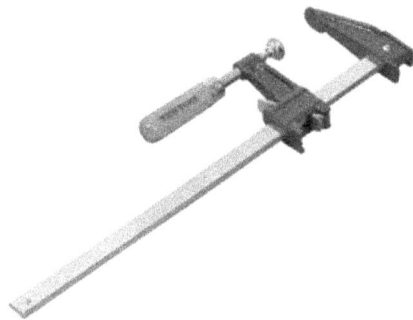

Locking bar clamp.

• **Locking C-clamp:** an adjustable, vise-type C-clamp that can be locked onto a work piece, leaving both hands free for work.

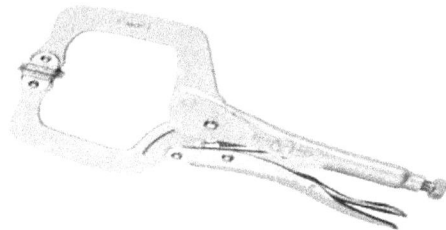

Locking C-clamp.

• **Locking chain clamp:** An adjustable, vise-type chain clamp that can be locked around an odd-shaped or circular work-piece, leaving both hands free for work.

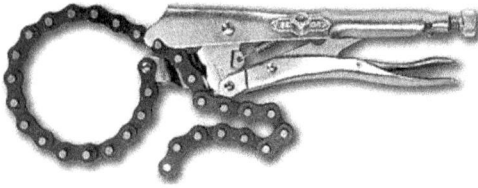

- **Locking hold-down clamp:** An adjustable, vise-type clamp designed to screw securely into a pre-drilled hole on a drill table or workbench, holding a work piece securely to the work surface.

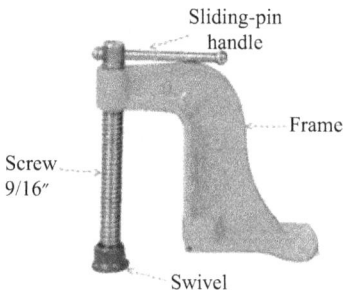

A 3-inch hold-down clamp.

- **Locking pinch-off tool:** An adjustable, vise-type tool that can be locked onto a work piece, specifically a piece of tubing to stop the flow of fluid or gas.

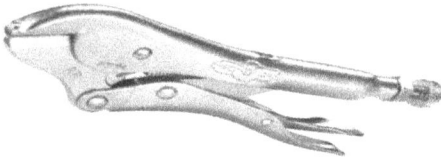

Locking pinch-off tool.

- **Locking pliers:** An adjustable, vise-type pliers that can be locked onto a workpiece, leaving both hands free for work. This tool can be used as pliers, a pipe wrench, an adjustable wrench, wire cutters, a ratchet, or a clamp. Locking pliers are available in various sizes and shapes.

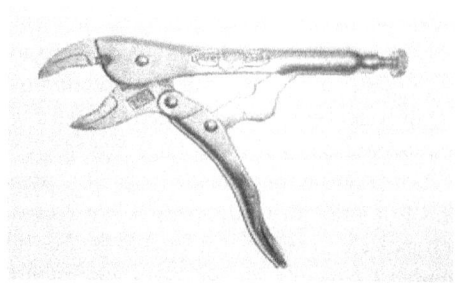

Locking pliers.

- **Locking wrench:** A versatile tool. The jaws of a locking wrench can be locked in a holding position that exerts pressure up to one ton. A locking wrench can also be used as a hand vise, holding clamp, pipe wrench, and hand vise pliers.

Locking wrench.

Lockout: To lock a switch in the "off" position by means of a padlock, or to lock electrified equipment behind a locked door, to which only qualified persons have the key.

Lockout procedures should be established and enforced, always. There should also be lockout procedures established and enforced for confined spaces, always.

Lockset: A complete door lock system comprised of the lock mechanism together with knobs, keys, plates, strikes, and other accessories.

Log: A segment sawed or split from a felled tree, such as, but not limited to, a section, bolt, or tree length.

Log rules: In the United States over 100 log rules have been developed, using a variety of methods. Some were based upon the lumber tallies of individual mills, others were developed by diagramming the cross-section of boards in the ends of logs, and others were developed using mathematical formulas. In general, log rules account for the taper that exists in all logs, saw kerf (or the loss of wood as sawdust), and a fixed procedure for removing wood on the outside of the logs for slabs. The Doyle, Scribner, and International log rules are widely used, but not exclusively used rules, in the United States.

- **The Doyle Log Rule:** This rule was developed around 1825. It is based on a mathematical formula and is widely used throughout the southern United States. This rule allows for a saw kerf of 5/16 inch and a slabbing allowance of 4 inches that is about twice the normal amount. Because of this, the Doyle Rule is somewhat inconsistent. It underestimates small logs and overestimates large logs.
- **The Scribner Log Rule:** This rule was developed around 1846. It is a diagram rule, created by drawing the cross-sections of 1-inch boards within circles representing the end view of logs. A space of 1/4 inch was left between the boards to account for saw kerf. The Scribner Rule does not have an allowance for log taper and typically under-estimates logs, particularly if the log length is long. The Scribner Decimal C is a different form of the Scribner Rule; it rounds the volumes to the nearest 10 board feet.
- **The International 1/4-Inch Log Rule:** This rule was developed in 1906 and is based on a reasonably accurate mathematical formula. The rule allows for a 1/4-inch saw kerf and a fixed taper allowance of 1/2 inch per 4 feet of log length. Deductions are also allowed for shrinkage of boards and a slab thickness that varies with the log diameter. Overall, the International 1/4-inch Log Rule is the most consistent and is often used as a basis of comparison for log rules.

Logged: The process of recording a receipt of a construction document or the entry and exit of personnel into a construction site.

Long logs: Logs over 20 feet long, commonly 32 to 40 feet long. Also known as long wood.

Long (ship) bit: An auger bit ranging in size from 12" to 30".

Longitudinal direction: Oriented in the direction of the long axis of an object.

Longitudinal wave: See **P wave**.

Longspan designation: A standard way of communicating the longspan joist safe uniformly distributed load-carrying capacities for a given clear span such as 18LH06 or 36LH10 where the first number is the nominal joist depth at midspan and the last number is the section number. See **Joist designation** and **Joist girder designation**.

Longspan joist: A structural load-carrying member with an open web system that supports floors and roofs utilizing hot-rolled or cold-formed steel and is designed as a simple span member. These carry higher loads than a regular joist.

Lookout: A short wood bracket or cantilever that supports an overhang portion of a roof.

Loop pile: Carpet pile in which fibers are looped and both ends are attached to the carpet backing. See **Pile** and **Cut pile**.

Loose: As used here, something that is not fixed in place, not secure, unsecured, unattached; detached, unfastened, untied; wobbly, unsteady, or movable.

Terms beginning with loose:

- **Loose angle strut:** A single or double angle either welded or bolted at the first bottom chord panel point and extended to brace another member such as a beam, joist girder, frame, or wall.
- **Loose collar:** The flanged collar that is fixed against the circular saw by attachment to the arbor by means of a nut.
- **Loose knots:** See **Knots, loose**.
- **Loose lintels:** Usually steel angles used over window and door openings to support masonry work above bearing on masonry and not connected to the structure of the building.
- **Loose side:** In knife-cut veneer, that side of the sheet that was in contact with the knife as the veneer was being cut, and containing cutting checks (lathe checks)

because of the bending of the wood at the knife edge.

- **Loose-laid membranes:** Roof membranes that are not attached to the substrate except at the perimeter of the roof and at penetrations. Typically, loose-laid membranes are held in place with ballast, such as water-worn stone, gravel, pavers, etc.

Loss estimation and loss reduction: Estimation and reduction of loss or damage that will result from earthquakes.

Lost formwork: Formwork used in the placement of concrete or backfill that is designed to be buried or incorporated into the structure and not removed.

Lot: A portion or parcel of land considered as a unit. A **zoning lot**.

The line that divides one land unit from another, or from a street or other public space is referred to as a lot line.

Louver: A vented opening into the home that has a series of horizontal slats and arranged to permit ventilation but to exclude rain, snow, light, insects, or other living creatures.

Love wave: A type of seismic surface having only horizontal motion transverse to the direction of propagation. (Love waves were named after named after A. E. H. Love.)

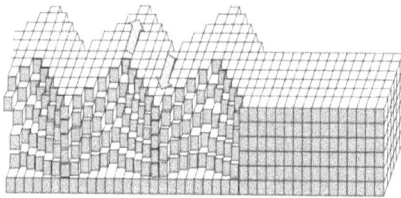

Love wave.

Terms beginning with low:

- **Low consumption toilet:** A class of toilet designed to flush using 1.6 gallons or less of water. Also known as **water-saving toilets**.

- **Low-emissivity (Low E) glass:** A membrane that is placed between double panes of glass to filter the light coming through the window. It transfers more heat through the glass in the winter and blocks heat in the summer.

- **Low-emissivity II glass (Low E II):** A high-performance Low-E glass, providing the best winter U-value and warmest center glass. It offers significant improvement in reducing solar heat gain coefficient values, providing customers one of the coolest summer glass temperatures of all Low E products. Additionally, ultraviolet light transmission is greatly reduced. The Low E II coated glass products are specifically designed for insulating glass units normally as a second surface coating. See **Low E** and **Pyrolitic definitions**.

- **Low-emitting vehicles:** Vehicles classified as zero-emission vehicles (ZEVs) by the California Air Resources Board. See Summary of **Referenced Standard**, above, for more information.

- **Low-impedance:** Low resistance to A/C current.

- **Low pressure decorative laminate:** A decorative surface paper that is saturated with reactive resins. During hot press lamination, the resin flows into the surface of the substrate, creating a hard crosslinked thermosetting permanent bond and permanently changing the characteristics of both the paper and the board.

- **Low rise:** Usually defined as a structure that is less than 75 feet in height.

- **Low slopes:** Roof pitches less than 4 inches per foot (4/12) 4/12 are considered low sloped roofs. Special installation practices must be used on roofs sloped 2/12–4/12. Shingles cannot be installed at slopes less than 2/12. See **Slopes**.

- **Low temperature flexibility:** The ability of a membrane or other material to remain flexible (resist cracking when flexed), after being cooled to a low temperature.

- **Low voltage wiring:** Wiring commonly used for television antennas, doorbells, thermostats, intercoms, and some specialty lighting systems.

- **Low-profile water closet:** A water closet with a short tank that cannot usually be detached from the bowl. See **Water closet and Turbo toilet**.

Lower levels: Those areas or surfaces to which an employee can fall. Such areas or surfaces include, but are not limited to, ground levels, floors, platforms, ramps, runways, excavations, pits, tanks, material, water, equipment, structures, or portions thereof.

Lower unit: With regard to cabinets, a cabinet unit that is designed to sit on the floor. Also called a base unit. See **Vanity cabinet, Upper unit**, and **Full height cabinet**.

Lumber: Timber sawn into rough planks or otherwise partly prepared. When lumber leaves the saw and planning mill it is not further manufactured than by sawing, resawing, passing lengthwise through a standard planing machine, crosscutting to length, and matching.

Other terms beginning with lumber:

- **Lumber dimensions**
  - **Actual size:** The dimensions obtained when an individual piece of lumber is measured with a caliper and tape.
  - **Manufactured size:** The dimensions for a given state of manufacture that are provided in product specifications. Examples are rough-green, surfaced-dry, and so forth. The manufactured size stated in the American Lumber Standards for a surfaced-dry 2 × 4 is 1.5 × 3.5 inches.
  - **Nominal size:** The size in name only. The commercial name by which lumber is usually known and sold on the market, e.g., a 2 × 4. It is the basis on which lumber volume in board feet is calculated.
- **Lumber, matched:** Lumber that is edge dressed and shaped to make a close tongued-and-grooved joint at the edges or ends when laid edge to edge or end to end.

- **Lumber, rough:** Lumber that has not been dressed (surfaced) but which has been sawed, edged, and trimmed.
- **Lumber, structural:** Lumber that is intended for use where allowable properties are required. The grading of structural lumber is based on the strength or stiffness of the piece as related to anticipated uses.
- **Lumber, surfaced:** Lumber that has been run through a planer.
- **Lumber, timbers:** Lumber that is nominally 5 inches or more in its least dimension. Timbers are generally used as beams, stringers, posts, caps, sills, girders, purlins, etc.
- **Lumber-core plywood:** Plywood that has thin sheets of veneer glued to a core of narrow boards. Lumber-core plywood differs from regular plywood that is made up of successive layers of alternating grain veneer.

Lumens: A unit of measurement of the amount of brightness that comes from a light source; a lumen is a unit of luminous flux equal to the light emitted in a unit solid angle by a uniform point source of one candle intensity. Lumens define "luminous flux," that is energy within the range of frequencies we perceive as light. For example, a wax candle generates 13-lumens; a 100-watt bulb generates 1,200-lumens.

Luminaire: A complete lighting unit consisting of a housing; lamp(s); light controlling elements; brightness controlling element; lamp holder(s); auxiliary equipment, such as ballast or transformer, if required; and a connection to a power supply.

Lump sum contract: A written contract between the owner and contractor in which the owner agrees to pay the contractor a specified sum of money for completing the work without requiring a cost breakdown. See also **Stipulated sum agreement**.

Lump sum fee: A fixed dollar amount that includes all costs of services including overhead and profit.

Lunette: A crescent-shaped window framed by moldings or an arch.

M shapes: A hot rolled shape called a **miscellaneous shape** with symbol M that cannot be identified as W, HP, or S Shapes.

Macerator: A device installed in a drain line between a toilet and the soil stack to reduce solids to liquid form.

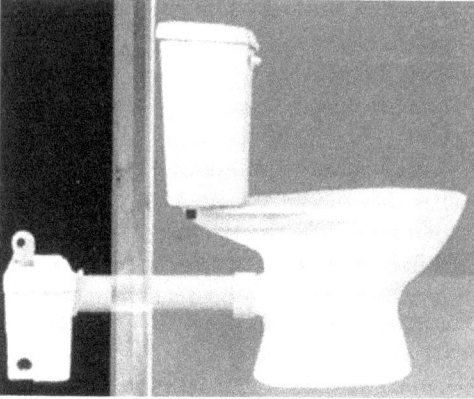

A macerating toilet.

Machine isolators: Calibrated mountings with springs used to attenuate vibration generated by machines. For seismic locations they are modified in order to absorb lateral movement and to keep the machine or equipment upright. These devices are available commercially.

Machine texture: Any texture that is applied to drywall using a machine.

Macrozones: Large zones of earthquake activity such as zones designated by the International Building Code map.

Macro-economy: A term generally used in reference to matters of economy or economic factors and forces portrayed or operating at the macro-level (as opposed to micro-level), used synonymously with national economy.

Magma: Liquid or molten rock material, it is called lava when it reaches the earth's surface.

Magnetic anomaly: A feature in the magnetic field produced by the distribution of magnetized material in the earth.

Magnetic polarity reversal: A magnetic polarity reversal is a change of the earth's magnetic field to the opposite polarity. This has occurred at irregular intervals during geologic time. Polarity reversals can be preserved in sequences of magnetized rocks and compared with standard polarity-change time scales to estimate geologic ages of the rocks. Rocks created along the oceanic spreading ridges commonly preserve this pattern of polarity reversals as they cool, and this pattern can be used to determine the rate of ocean ridge spreading. The reversal patterns recorded in the rocks are termed as **Sea-floor magnetic lineaments**.

Magnetite: The mineral form of black iron oxide, Fe_3O_4, that often occurs with magnesium, zinc, and manganese and is an important ore of iron. Magnetite forms a layer of gray to black colored material on internal boiler surfaces giving added protection against corrosions, caustic embrittlement, and fouling.

Magnification factor: An increase in lateral forces at a specific site for a specific factor.

Terms beginning with magnitude:

- **Magnitude:** A scalar value having physical units.
- **Magnitude, earthquake:** A measure of earthquake size that describes the amount of energy released. See **Magnitude, moment**.
- **Magnitude, body-wave (m_b):** A measure of the of the magnitude of an earthquake determined from P waves.
- **Magnitude, local (M_L):** A measure of the magnitude of a local earthquake originally defined by Charles Richter using the maximum amplitude recorded by horizontal-component seismographs.
- **Magnitude, moment (M):** The magnitude of an earthquake determined from seismic moment.

The **moment magnitude scale** is a logarithmic scale of 1 to 10 (a successor to the **Richter scale**) that enables seismologists to compare the energy released by different earthquakes on the basis of the area of the geological fault that ruptured in the quake.

- **Magnitude, surface-wave** (M_S): The magnitude of an earthquake determined from surface waves.

Magnum: A Marvin trade name for heavily constructed window products that are designed for applications where a heavy duty product is necessary. Under this trade name are: **Magnum double hung** that is a heavy duty double hung product made with larger than standard parts; **Magnum hopper** that is a heavy duty window that is designed to tilt into the room for ventilation purposes; and, **Magnum tilt-turn** that has hardware that allows the sash to either be tilted into the room for ventilation or swung into the room for egress or cleaning. Other companies are now competing with Marvin for the heavy-duty market.

Main: With regard to the water and drain system, the primary artery of supply of these systems, in which all the branches connect. The principal artery of the venting system to which vent branches may be connected is referred to as the **main vent** and also the **vent stack**.

In electrical work, the primary conduit for current is called the **main feeder**. A set of large switches or breakers that allow electricity to the structure to be turned off without removing the meter is referred to as the **main disconnect**. On newer structures the main disconnect is located on the meter base.

Major axis: The axis of a structural member possessing the largest section modulus and radius of gyration, thus having the greatest flexural and axial compressive strength.

Makeup water: Water fed into the system to replace what is lost through evaporation, drift, blowdown, and other causes.

Male fitting: A fitting that is inserted into another fitting.

Mall: An enclosed or roofed area that is used as a pedestrian circulation space. Usually a mall connects no more than three stories or portions of stories of a building or buildings. A mall may house a single or multiple tenants.

Malleable fittings: Fittings that are made of cast iron, a metal that is soft and pliable.

Manhole: A hole through which one may go especially to gain access to an underground or enclosed structure. Manholes are usually round to prevent the cover from falling into the opening. Also spelled **man hole**.

Regarding boilers, an oval opening in the boiler plate allowing access for inspection and washing out. Normally the man hole is large enough for a person to enter the boiler. There are special safety procedures necessary for a person to enter a man hole. See **Hand hole.**

Manifest: The document that lists the contents of a truck, ship, plane, or train.

Manifold: A pipe or chamber branching into several openings or ports.

A manifold installation is one in which a number of branches are connected. For example, parallel water heaters that are connected for large hot water demand applications.

Mansard roof: A roof style with four sides similar to a HIP ROOF but each side is divided into an upper and lower section, the lower section having a steeper slope than the upper section. Often, the center of the Mansard roof consists of a flat roof.

Mansard roofs had a long history in France, but became popular in Paris during the Second Empire of Louis Napoleon (Napoleon III). Legend has it that in Paris, buildings were taxed on the basis of floors (storys) "under the roof." Since the Mansard style had one full floor within, and not under the eaves of the sharply sloping roof, that top floor was not taxed. Obviously the French tax collectors of the time are not like today's Assessor's Office inspectors.

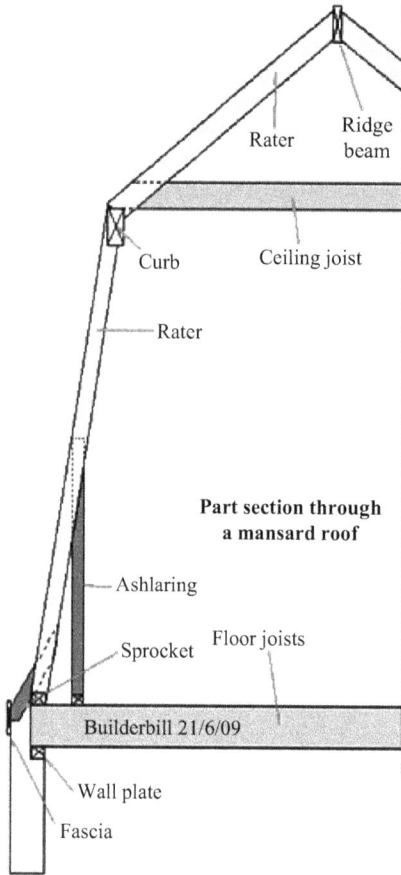

Mansard roof cross-section.

(labels in figure: Ridge beam, Rater, Curb, Ceiling joist, Rater, Part section through a mansard roof, Ashlaring, Sprocket, Floor joists, Builderbill 21/6/09, Wall plate, Fascia)

Mantel The shelf above a fireplace opening. Also used in referring to the decorative trim around a fireplace opening.

Also, the main bulk of the earth between the crust and the core.

Also, a fragile mesh cover that is fixed around a gas jet, kerosene wick, etc., to give an incandescent light when heated.

Manual control: A control actuated by the operator, e.g., a foot pedal, lever, and starter button.

Manual override: A way of manually actuating automatically-controlled equipment. For example, an emergency generator usually has a manual override in the event the automatic starter fails.

Manufactured housing: Housing units partially or completely built in a factory.

Manufactured wood: A wood product such as a truss, beam, gluelam, microlam, or joist that is manufactured out of smaller wood pieces and glued or mechanically fastened to form a larger piece. See **Oriented strand board**, **Engineered wood products**, and **Glue laminating (Glulam).**

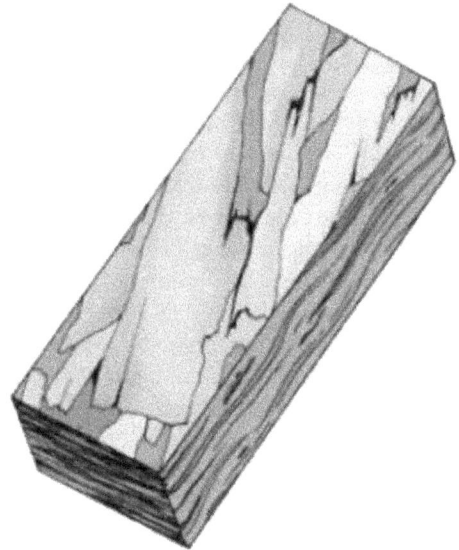

Manufactured wood.

Manufacturer's specifications: The written installation and/or maintenance instructions that are developed by the manufacturer of a product and that may have to be followed in order to maintain the product warrantee.

Mapp gas: A colorless, flammable gas made by combining liquefied petroleum gas with Methylacetylene-Propadiene. It is a stable, non-toxic fuel used in brazing and soldering.

Mapping GIS software: Computer-mapping programs that perform any of a wide variety of map-making tasks (for both on-screen and file-oriented use).

Marine borer: Any mollusc or crustacean that lives usually in warm seas and destroys wood by boring into and eating it. The gribble and shipworm are the best known since they penetrate any wood in favorable water.

Damage by marine-boring organisms to wood structures in salt or brackish waters is a worldwide problem, except in heavily polluted waters. Evidence of attack is sometimes found in rivers even above the region of brackishness. The rapidity of attack depends upon local conditions and the kinds of borers present. Along the Pacific, Gulf, and South Atlantic Coasts of the United States, attack is rapid. Untreated pilings may be completely destroyed in a year or less. Along the New England States coast the rate of attack is slower due to cold water temperatures, but is still sufficiently rapid to require protection of wood where long life is required.

Marine paint: Coating specially designed for immersion in water and exposure to marine atmosphere. See **Anti-fouling paint**.

Marine Stewardship Council's Blue Eco-Label: Certification that applies to products that meet the MSC principles and criteria for sustainable fishing, including sustainable harvest of the target stock, acceptable impact of the fishery on the ecosystem, effectiveness of the fisher management system (including all relevant biological, technological, economic, social, environmental, and commercial aspects), and compliance with relevant local and national local laws and standards and international understandings and agreements.

Mark: An identification number or method of relating to the erector which part, e.g., structural member, of the building goes at what location when being erected.

Marking crayon: A non-toxic, weatherproof marking tool for use on oily, slick, wet, cold, or dry surfaces.

Marquee sign: A sign that is placed flat against the front or side facia of a marquee.

Masking: The process of covering part of a surface that you do not want to paint during the present application. Masking involves placing tape on materials adjacent to the surface being painted to keep them clean. Typically masking is placed on trim that will be painted or stained a different color than the surface to which it is attached.

Mason: A mechanic who places and finishes concrete.

Masonry: Stone, brick, concrete, hollow-tile, concrete block, or other similar building units or materials. Normally bonded together with mortar to form a wall. The masonry units may be manufactured products or they may be natural building stones.

Masonry includes nonbearing walls (walls that carry their own mass and is some cases wind and seismic forces), bearing walls (walls that carry additional gravity loads above their own and in some cases in-plane and out-of-plane forces), beams, arches, and shells. Masony is frequently used compositely with metals, concrete, and wood.

Masonry is a compound material. In its simplest forms it is an assemblage of dry masonry units or masonry units and mortar. In complex assemblages different types of masonry units are used along with different types of mortar. They may include steel reinforcing bars, grout, tie rods, embedded metals and wood, metal ties, anchors, and joint reinforcement. Some masonry assemblages include air space, as in a cavity wall.

Masonry opening: An area in a masonry wall left open for windows or a door.

Masonry primer: An asphalt-based primer that is used to prepare masonry surfaces for bonding with other asphalt products.

Mass: A constant quantity or aggregate of matter. Mass is a property of all matter. It is measured in, for example, grams. Mass is independent of gravity, unlike weight that depends on gravity.

Also, a property of an object measured by the degree that it resists acceleration.

Terms beginning with mass:

- **Mass concrete:** Unreinforced concrete, as often used in foundations or other applications where the added strength of reinforcement is not required.
- **Mass determination:** The mass per unit area of a specimen calculated by dividing the total mass, determined after a specified

procedure of oven-drying and cooling of a unit, by the average projected area of the two faces of the unit as normally laid in a wall.

• **Mass transit:** Movement of large groups of persons in a single vehicle, such as a bus or train car.

Massed produced cabinets: Milled cabinets that are built in large quantities and in standard sizes. They are sold in high volume through retailers. See **Built-in cabinets, Milled cabinets**, and **Custom cabinets**.

Master: Regarding hardware, the keying of locks that will allow a single key to operate most locks. A grand master will operate all locks in a keying sequence.

Mastic: A pasty material used as a cement (as for setting tile) or a protective coating (as for thermal insulation or waterproofing). See **Asphalt plastic roofing cement**.

Mat: A thin layer of woven, non-woven, or knitted fiber that serves as reinforcement to the material or membrane.

Matching edge band (ME): Regarding doors, an edge band that is the same species or laminate pattern as the face veneer.

Material: Something needed for an activity, e.g., paint remover.

Also, facts, information, or ideas for use in creating a design.

Also, cloth or fabric.

Also, regarding historic preservation, the physical elements that were combined or deposited to form a property. Historic material or historic fabric is that from a historically significant period, as opposed to material used to maintain or restore a property following its historic period(s).

Material balance: A relationship, often diagrammed, that shows how materials are allocated and used.

Material Safety Data Sheets (MSDS): MSDS contain product information on chemicals, chemical compounds, and chemical mixtures; e.g., paints and custodial supplies among many others. MSDSs can also include instructions for safe handling, storage, and disposal of products. Sales clerks should make MSDSs available to retail customers.

Maul: A heavy long handled hammer used for splitting a piece of wood along its grain. One side is shaped like a sledge hammer and the other is a broad head axe shape. Often called a **splitting maul** and **woodsplitting maul**.

Maul or splitting maul or woodsplitting maul.

Maximum credible earthquake (MCE): An earthquake that is about 50% higher than the design base earthquake (DBE).

Maximum intended load: With regard to scaffolding, the total load of all persons, equipment, tools, materials, transmitted loads, and other loads reasonably anticipated to be applied to a scaffold or scaffold component at any one time.

Maxwell diagram: A graphical method of determining stresses in a truss by combining force polygons of all the joints into one stress diagram.

MC shapes: A hot rolled shape called a **miscellaneous channel** with the symbol MC. See **M shapes**.

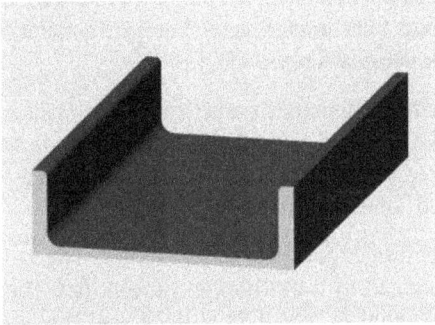

MC shape.

Mean: A measure of central tendency (for a distribution of values) defined as the average value of a variable in a sample and calculated by adding together all the values observed in a data set and dividing by the number of values observed.

Measured drawings: Drawings that depict existing conditions or other relevant features of structures, landscapes, or objects. Measured drawings are usually produced in ink on archivally stable material, such as polyester film and preserved.

Measuring tape: A basic tool used to measure the length of materials. A measuring tape has marks on it that make it possible to read measurements with accuracy of up to 1/16th of an inch or more. Measuring tapes often have special marks at each 16" interval for easy location of 16-inch-on-center framing members.

Terms beginning with mechanical:

- **Mechanical controls:** Dials, slides, switches, knobs, buttons, and so forth used to operate ranges, ovens cook-tops and other equipment.

- **Mechanical damage:** Damage to a roof by means of items puncturing or otherwise unnecessarily penetrating the roof system or any of its components. Screws or nails stuck in the roof and heel marks along base flashings are examples of mechanical damage.

- **Mechanical engineer:** An engineer who applies the principles of mechanics and energy to the design of machines and devices.

- **Mechanical fasteners:** Devices such as screws, plates, battens, nails, or other materials that are used to secure roofing materials.

- **Mechanical trowels:** A piece of concrete finishing equipment that has three or four steel trowels attached to an internal combustion engine and produces a concrete finish ready for hand finishing.

- **Mechanical ventilation:** Ventilation provided by machine-powered equipment such as motor-driven fans and blowers, but not by devices such as wind-driven turbine ventilators and mechanically operated windows.

- **Mechanical window:** A term for a product, usually vinyl, in which the corners are assembled using screws or other fastening mechanisms, as opposed to a welded corner construction. Also referred to as a **Mechanically fastened window**.

Mechanically-fastened membranes: Generally used to describe roofing membranes that have been attached at defined intervals to the substrate. Mechanical fastening may be performed with various fasteners and/or other mechanical devices, such as plates or battens.

Mechanically-fastened membranes.

Mechanic's lien: A lien on real property, created by statue in many years, in favor of persons supplying labor or materials for a building or structure, for the value of labor or materials supplied by them. In some jurisdictions, a mechanics lien also exists for the value of professional services. Clear title to the property cannot be obtained until the claim for the labor, materials, or professional services is settled. Timely filing is essential to support the encumbrance, and prescribed filing dates vary by jurisdiction.

Property owners should make sure that their general contractor and construction managers pay their employees or sub-contractors to avoid a mechanic's lien, since the owner could be forced to pay their debts even though the owner has already paid the contractor/CM. If the worker or supplier does not sue to enforce the mechanic's lien, he/she may still sue for the debt.

Mechanicals: The wiring, plumbing, and heating and cooling systems in a building.

Also, the components with moving parts such as furnaces, plumbing fixtures, etc.

Mechanized parking garage equipment: Special devices in mechanical parking garages that operate in either stationary or horizontal moving hoistways, that are exclusively for automobiles. Normally, people are not stationed on any level other than the receiving level; during the parking process automobiles are moved on and off the elevator directly into parking spaces or cubicles by means of a power-driven transfer device.

Median: The middle value of a data set (or sample) when the values are arranged in order (by size ranking, in ascending or descending order). Note that for an odd number of values in an ordered data set, the median is identified as the value which divides the data set into two data sets of equal size on each side of the median or middle value. For an even number of values arranged in order, the median is found by simply calculating the value midway between the two middle values. Note that the position of the median value of an ordered data set containing n observations may be found by using the formula: position of the median = $n/2 + 1/2$.

Medium density fiberboard (MDF): The generic name for a panel or core manufactured from lignocellulosic fibers combined with a synthetic resin or other suitable binder and bonded together under heat and pressure in a hot press by a process in which the added binder creates the entire bond.

Medium density fiberboard core (MDFC): Wood fiber and/or agri-fiber-based materials that comply with ANSI A208.2.

Medium density overlay (MDO): Typically MDO is kraft paper saturated with resin and cured under high heat and pressure to make a hard, smooth, paintable surface.

Meeting edges: Two adjacent door edges not separated by a mullion or transom bar. These are found in pair, Dutch door, and door and transom applications.

Meeting notes: A written report consisting of a project number, project name, meeting date and time, meeting place, meeting subject, a list of persons attending, and a list of actions taken and/or discussed during the meeting. Generally, this report is distributed to all persons attending the meeting and any other person having an interest in the meeting. Meeting notes are kept until the contract is closed.

Meeting rail: The part of a window where two sets of panes meet and overlap.

Mega: SI Prefix for 10^6 or 1,000,000.

Member: A structural element such as a beam, column, girder, and brace. Also, a structural part of a window, such as a rail, stile, or lintel. The term is also used for other single elements of construction.

Member release: An idealization to model how members are attached to "each other." It designates whether forces and moments at the ends of a member are considered fixed to or released from the member's point of attachment.

Membrane: A solid sheet of waterproof material that covers an entire roof area.

Membrane pressed panel: Insert panel produced by molding to profile a wood or composite core (usually MDF or particleboard) then pressing veneer to the core using a flexible pressing surface. Also referred to as **Bladder pressed panel.**

Mende-process board: Made in a continuous ribbon from wood particles. Thermosetting resins bond the particles. Thickness ranges from 1/32 - 1/4 inch.

Mercalli Scale: See **Intensity**.

Messenger cable: On an overhead electrical drop, the messenger cable is one of the three intertwined cables in the drop wire that contains the neutral lead and carries the weight of the other leads.

Metal building system: A building system consisting of a group of coordinated components that have been designed for a certain loading. These components are mass produced and assembled in various combinations with other structural materials to produce a building.

Metal-enclosed power switchgear: A switchgear assembly completely enclosed on all sides and top with sheet metal (except for ventilating openings and inspection windows) containing primary power circuit switching, interrupting devices, or both, with buses and connections. The assembly may include control and auxiliary devices. Access to the interior of the enclosure is provided by doors, removable covers, or both.

Metal flashing: Accessory components fabricated from sheet metal and used to weatherproof terminating roof covering edges. Frequently used as through-wall flashing, cap flashing (coping), counterflashing, stepflashing, etc. See **Flashing**.

Metal lath: Sheets of metal that are slit to form openings within the lath. Used as a plaster base for walls and ceilings and as reinforcing over other forms of plaster base.

Metal roof coatings: The list that follows may not a complete listing of metal roof coatings:

- **Kynar 500 and Hylar 5000:** Kynar 500 and Hylar 5000 are trade names for polyvinylidene (PVDF) paint finishes that provide very strong longevity and durability including fade and chalk resistance that leads the coatings industry. Kynar 500 is produced by Arkema Chemicals, and Hylar 5000 is produced by Solvay Solexis.
- **Stone coated:** Metal roofing that is made from zinc or aluminum coated steel that is then coated with the same granules as composition shingles. It is attractive but faces similar problems of streaking, ganule loss, and organic growth as traditional composition shingles.
- **Siliconized polyester:** A solvent-based system with polyester resin. Silicone additives are used to increase resin stability and coating flexibility. Standard polyester finishes are commonly used on agricultural metal roofs where price is of greater concern than performance.
- **Super polyester:** A siliconized polyester with fade resistant pigmentation. The pigments enhance the performance of traditional polyesters, but the coatings are still prone to chalking as the resin breaks down over time.
- **Plastisol:** A coating that is traditionally used in the siding industry. It is composed of PVC particles embedded in a plasticizer that provides some flexibility and durability. It is **not recommended** as a roofing application in the United States.

Metal stirrup: Used to connect a wood column or post to a concrete part. It holds the wood member securely while preventing the long grains of the wood from directly contacting the moist concrete. This keeps the moisture that collects on the concrete from being pulled into the wood by capillary action.

Metal stud: A structural steel member used for framing walls just as a regular wooden one.

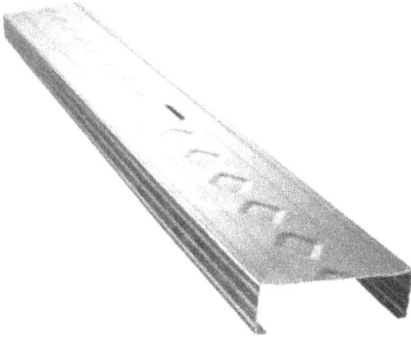

Typical 25-gauge galvanized steel wall framing stud, with dimensions 3-5/8 inches × 10 feet.

Metallic lathers: Workers who install steel reinforcing rods, welded wire fabric, pretensioning cables, and lath in concrete and plaster work.

Metalling: Broken stone or cinders that are used in the construction or repair of roads or railways. Metalling originally referred to the process of creating a gravel roadway; road metal was used extensively in the construction of roads by the Roman Empire. Later "Road Metal" became the name of stone chippings mixed with tar to form the road surfacing material tarmac. A road of such material is called a "metalled road" in Britain, a "paved road" in Canada and the United States, and a "sealed road" in Australia and New Zealand.

Road metal.

Metamorphic rock: Metamorphic rocks arise from the transformation of existing rock types (sedimentary rock, igneous rock, or another older metamorphic rock.), in a process called metamorphism. The original rock (protolith) is subjected to heat (temperatures greater than 150–200°C) and pressure (1500 bars, a bar being about equal to the atmospheric pressure on Earth at sea level) causing both physical and/or chemical change.

Meter: The fundamental unit of length in the metric system, equal to 100 centimeters or ~ 39.37 inches.

Also, a device that measures and records the quantity, degree, or rate of something, especially the amount of electricity, gas, or water used. Meters provide the types of data needed for performance measurement and subsequent improvements.

A mounting plate for an electric watt-hour meter is a **meter base**. A meter base consists of a grounded backplate holding mounting lugs for the electric meter. One set of lugs connects to the electric utility mains, the other set to the customer's wiring. The lugs are arranged so that electricity must pass through the meter to get from the utility mains to the customer premises. The meter base is set in a weather-tight metal enclosure that is usually sealed to prevent meter tampering.

An adapter that connects a water meter to the water supply line is a **meter tailpiece.**

Angle meter tail piece.

A **meter tile** is a cylindrically shaped casing that forms the pit to hold a water meter. It is usually made of plastic.

Meter tile.

Method statement: A document that details the way a work task or process is to be completed. The method statement usually outlines the hazards involved and include a step by step guide on how to do the job safely. The method statement should also detail the control measures that will be used to ensure safety. Method statements may be required by neighboring owners when potentially hazardous work is proposed and by planning authorities to ensure that a proposal is buildable.

Methymercury: Any of various toxic compounds of mercury containing the complex CH_3Hg-; it often occurs in pollutants and bioaccumulates in living organisms, especially in higher levels of the food chain.

Mews: A row or street of houses or apartments that have been converted from stables or built to look like former stables. Mews properties are often separated from their main house and converted to sought-after dwellings.

Mews.

Mezzanine: An intermediate floor between the floor and ceiling of any space. Codes vary, but the New York City Building Code states that "when the total gross floor area of all mezzanines occurring in any story exceeds thirty-three and one-third percent of the gross floor area of that story such mezzanine shall be considered as a separate story."

Microclimate factor (kmc): A coefficient used in calculating the landscape coefficient; it adjusts the evapotranspiration rate (the sum of evaporation and planttranspiration from the Earth's land surface to atmosphere) to reflect the climate of the immediate area.

Microirrigation: Involves irrigation systems with small sprinklers and microjets or drippers designed to apply small volumes of water. The sprinklers and microjets are installed within a few centimeters of the ground; drippers are laid on or below grade.

Microlam: A manufactured structural wood beam. It is constructed of pressure and adhesive bonded wood strands of wood. They have a higher strength rating than solid sawn lumber.

Micron: One-millionth of a meter or approximately 0.00004 inch.

Micron rating: The size of the particles a filter will remove.

Microzonation: Seismic zoning, generally by use of maps, for land areas smaller than regions shown in typical seismic code maps, but larger than individual building sites.

Migration: The absorption of oil or vehicle (defined as a substance that facilitates the use of a drug, pigment, or other material mixed with it.) from a compound into an adjacent porous surface.

Mil: A unit of measure. One mil is equal to 0.001 inches. 25.400 Microns is often used to indicate the thickness of a roofing membrane. Mils are also a measurement of thickness of paint.

Milar: Plastic, transparent copies of a blueprint. Also spelled **mylar**.

Mild steel: Structural steelwork and reinforcement generally come in two qualities: mild steel and high-tensile steel. Mild steel's name comes from the fact it has less carbon than high-tensiler steel. It is softer and less ductile than higher carbon steels; it bends a long way instead of breaking. Mild steel is used in nails, some types of wire, staplers, staples, railings, and most common metal products.

Mile (ml): A unit of linear measurement on land, equivalent to 5,280 feet (1,760 yd) or 1.6 kilometers.

Milestone: An informational marker that is used to delineate important events; milestones mark the progress of a project. Sometimes milestones are contractual.

Milestone schedule: A schedule of milestones spanning from the start of construction to occupancy. The schedule is often used as the main measure of progress to keep the project on schedule. It is usually summary in nature.

Mill finish: The surface finish found on aluminum when it is extruded at the mill.

Mill order: The actual final purchase order for the mill or manufacturer based on quantities derived from the final steel shop drawings. This order replaces or confirms the advanced bill.

Mill test report: A report of a heat of steel (a heat number is an identification number that is stamped on a material plate after it is removed from the ladle and rolled at a steel mill) that indicates the customer's order number, grade of steel, number and dimensions of pieces shipped, and the chemical compositional makeup of hot rolled structural steel members. It also indicates physical properties, such as yield strength, tensile strength, elongation, impact, and ultimate strength.

Milled: A surface that has been accurately sawed or finished to a true plane.

Milled cabinets: Cabinets built by a manufacturer in a cabinet mill. See **Built-in cabinets, Custom cabinets**, and **Mass produced cabinets**.

Milli: SI prefix for 10^{-3} or 0.001 or 1/1000. Thus, a unit of measurement equaling one thousandth (1/1000) of an ampere is a **milliampere**, and a unit of length equal to one thousandth of a meter (0.03937 inches) is a **millimeter (mm)**.

Millwork: Generally all building materials made of finished wood and manufactured in millwork plants including all doors, window and door frames, blinds, mantels, panelwork, stairway components (ballusters, rail, etc.), moldings, and interior trim. Millwork does not include flooring, ceiling, or siding.

Mineral: A solid inorganic substance of natural occurrence.

Terms beginning with mineral:

Mineral fibers: There are many mineral fibers, e.g., asbestos is a mineral fiber that is no longer used because of its adverse affects on health. A common type of mineral fiber insulation is called rock wool.

Mineral granules: See **Granules**.

Mineral spirits: Paint thinner. A solvent distilled from petroleum.

Mineral stain: Olive and greenish-black streaks believed to designate areas of abnormal concentration of mineral matter; common in hard maple, hickory, and basswood. Also called **mineral streak**.

Mineral stain in hard maple.

Mineral stabilizers: Finely ground limestone, slate, traprock, or other inert materials that are added to asphalt coatings for durability and increased resistance to fire and weathering.

Mineral-surfaced roofing: Roofing materials whose surface or top layer consists of mineral granules. See **Mineral surfaced sheet**.

Mineral-surfaced sheet: A roofing sheet that is coated on one or both sides with asphalt and surfaced with mineral granules. See **Mineral surfaced roofing**.

Minimum efficiency reporting value (MERV): A filter rating established by the American Society of Heating, Refrigerating, and Air Conditioning Engineers (ASHRAE 52.2–1999, Method of Testing General Ventilation Air Cleaning Devices for Removal Efficiency by Particle Size). The MERV efficiency categories range from 1 (very low efficiency) to 16 (very high).

Minispread or mini-"widespread faucet": A faucet that looks similar to an 8-inch spread faucet. The handles are separate from the spout, i.e., not mounted on an escutcheon and have a 4" spread. The minispread is an option when the look of a widespread is desired, but space is limited.

A minispread faucet.

Minor axis: The axis of a structural member possessing the smallest section modulus and radius of gyration, thus having the least flexural and axial compressive strength.

Minor change in work: A written order by the architect to make a change that does not involve adjustment in the contract sum or extension of the contract time and is not inconsistent with the intent of the contract documents.

Mission coupling: A neoprene flex coupling, connecting PVC to PVC or clay to PVC.

Miter cut: A single cut made at an angle to the member length. See **Bevel cut**.

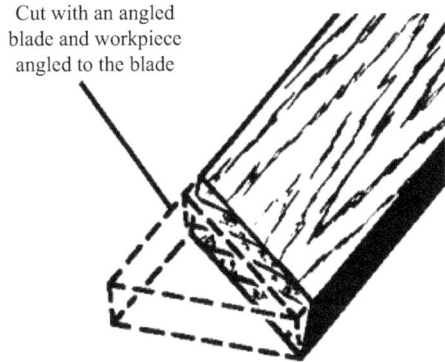

Cut with an angled blade and workpiece angled to the blade

Miter cut.

Miter joint: The joint of two pieces at an angle that bisects the joining angle. For example, the miter joint at the side and head casing at a door opening is made at a 45° angle.

Mitered panel: An insert panel with a solid lumber edge banded around the core then veneered and profiled. Also called a **rim banded panel**.

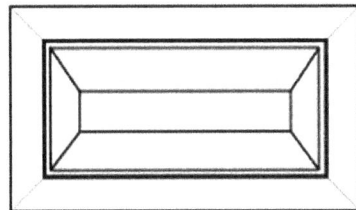

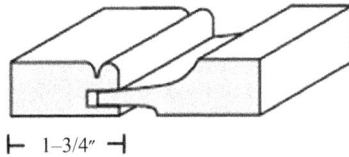

├─ 1–3/4″ ─┤

Mitered panel.

Mix: The proportions of the ingredients of concrete, mortar, etc.

Mixed-mode ventilation: A combination of natural ventilation and mechanical ventilation that allows the building to be ventilated mechanically or naturally, and at times both simultaneously.

Mixing valve: A valve that mixes hot and cold water to achieve a specified delivery temperature.

Mjölir: Thor's hammer: not for sale. Also spelled **Mjolnir**, **Mjollnir**, **Mjölner**, or **Mjølner**.

Mock Tudor: An architectural style that is popular in suburban development; this style is often copied poorly.

Mock Tudor.

Modal analysis: The determination of seismic design forces based upon the theoretical response of a structure in its several modes of vibration to excitation.

Mode: Any of the distinct kinds or patterns of vibration of an oscillating system, e.g., the shape of an earthquake vibration curve.

Also, in statistics, the value that occurs most frequently in a given set of data.

Model codes: A group of codes and standards accepted by more than one of the Building Code regulatory agencies such as Building Officials and Code Administrators International, Inc. (BOCA) and International Conference of Building Officials (ICBO).

Modified bitumen roof system: A single membrane roof system made from either asphalt or coal tar pitch with added plasticizers. Installation methods for modified bitumen roof systems include both hot mop and torch down.

Modified Mercalli: See **Intensity**.

Modillion: An ornamental bracket found under a cornice, similar in appearance to dentil, only larger. See **Dentil**.

Modillion.

Modular housing: Housing that has its major components assembled in a factory and is then shipped to the building site where it is joined with other components to form the finished structure.

Modulate: To control within an infinite range between 0% and 100% as opposed to on/off control.

Modulus of elasticity: The mathematical description of an object or substance's tendency to be deformed elastically, i.e., nonpermanently, when force is applied to it. The elastic modulus of an object is defined as the slope of its stress-strain curve in the elastic deformation range.

Said differently, it is the ratio, within the elastic limit of a material, of stress to corresponding strain under a given loading condition. In the case of wood, an ortotropic material with three mutually perpendicular axis, its properties, e.g., strength and stiffness, along its grain and in each of the two perpendicular directions are different. Design moduli of elasticity are based on bending tests of lumber and are usually given for the longitudinal direction only.

Modulus of rupture: The ultimate strength that is determined in a flexure or torsion test. In a flexure test, modulus of rupture in bending is the maximum fiber stress at failure. In a torsion test, modulus of rupture in torsion is the maximum shear stress in the extreme fiber of a circular member at failure. Alternate terms are flexural strength and torsional strength.

Moenstone®: A blend of ceramic-like quartz (70%) and acrylic polymer (30%) that is impervious to chips or water spots.

Moenstone® granite double bowl sink.

Mohorovicic (moho) discontinuity: The boundary between the *crust* and upper *mantle* usually identified by an increase in the velocity of propagation of seismic waves.

Moisture: Water or other liquid diffused in a small quantity as vapor, within a solid, or condensed on a surface.

Terms beginning with moisture:

- **Moisture barrier:** A material that retards the passage of water vapor from one space to another. A polyethylene sheet is commonly used as a vapor retarder. Also referred to as a vapor barrier.

- **Moisture content:** The amount of moisture in a material determined under prescribed conditions and expressed as a percentage of the mass of the moist specimen, i.e., the original mass comprising the dry substance plus any moisture that is present.

  Moisture content is used in a wide range of scientific and technical areas. It can be given on a volumetric or mass (gravimetric) basis.

  The moisture content of wood changes with changes in the conditions under which it is used. To give best service, the wood should be installed at a moisture content close to the midpoint between the high and low values it will usually attain in use. Wood products shrink as they dry and swell as they absorb moisture, either liquid or vapor, from the atmosphere. Unless these changes in dimension are kept to a minimum, they may result in unsatisfactory service of wood products and structures.

  The moisture content of concrete must be viewed from the context of total water content of the fresh concrete mixture and the available moisture content of the hardened concrete. The total water content of a fresh concrete mixture is a function of the total cementitious materials and water cement ratio (w/cm). Moisture content is an important factor affecting concrete volume changes.

  Moisture content is also referred to as **water content**.

- **Moisture distribution:** When referring to wood, moisture distribution is the variation in moisture content within a board and also the variation in moisture content between boards.

- **Moisture expansion:** An increase in dimension or bulk volume that is caused by reaction with water or vapor. This reaction may occur over time and is usually expedited by exposure to water or water vapor at elevated temperatures and pressures.

- **Moisture gradient:** When referring to wood, the difference in moisture content between areas of a board. Usually refers to the moisture content difference between the surface and core of a board.
- **Moisture relief ven:** A vent installed through the roofing membrane to relieve moisture vapor pressure that has been trapped within the roofing system.
- **Moisture scan:** The use of a mechanical device (capitance, infrared, or nuclear) to detect the presence of moisture within a roof assembly. See **Non-destructive testing**.

Molybdenum: A soft white metallic element that looks like silver. It is used in ferrous alloys for high-speed cutting steels, die steels, and in structural steels. Molydenite is the chief ore of molybdenum.

Molding: A material such as wood, plastic, or stone strip having an engraved, decorative surface.

(**Molding:** The British spelling of mold and molding; molding is used throughout this book.)

There are numerous moldings:

- **Aaron's rod:** A molding that is used as a building ornament. It is either straight or rounded molding decorated with an entwined snake, and sometimes leaves, vines, and/or scrolls.
- **Astragal:** A semi-circular molding that is attached to one of a pair of to cover the air gap where the doors meet; used especially on fire doors.
- **Baguette:** A thin, half-round molding, that is smaller than an astragal. Baguettes are sometimes carved, and enriched with foliages, pearls, ribbands, laurels, etc. When enriched with decorations it was also called a **chapelet**.
- **Bandelet:** Any small band or flat molding that crowns a Doric architrave (a main beam resting across the tops of columns, specifically the lower third entablature (a horizontal, continuous lintel on a classical building supported by columns or a wall, comprising the architrave, frieze, and cornice.)). It is also called a **tenia**.
- **Baseboard:** Baseboards are used to conceal the junction of an interior wall and floor, to protect the wall from impacts and to add decorative features. There are many baseboard profiles. Also called **base molding** or **skirting board**. See **Shoe molding**.
- **Baton:** see **Torus**.
- **Batten:** A symmetrical molding that is placed across a joint where two parallel panels or boards meet. Also called or **board and batten**.
- **Bead molding:** A narrow, half-round convex molding that when repeated forms a small semicylindrical molding or ornamentation (reeding).
- **Beading/bead:** A molding in the form of a row of half spherical beads that are larger than pearling. Other forms of this category of molding include **bead and leaf, bead and reel**, and **bead and spindle**.
- **Beak:** A small fillet molding left on the edge of a larmier (a flat, jutting part of a cornice; the eaves of a house), that forms a canal, and makes a kind of pendant. See also **Chin-beak**.

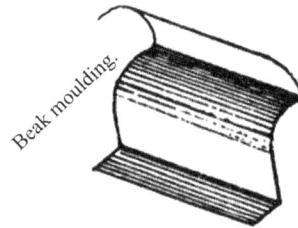

Beak molding.

- **Bed molding:** A narrow molding that is used at the junction of a wall and ceiling. Bed moldings can be either sprung or plain.
- **Bolection:** A molding that is raised, projecting from the face frame. It is located at

the intersection of the different surface levels between the frame and inset panel on a door or wood panel. It will sometimes have a rabbet (a step-shaped recess cut along the edge or in the face of a piece of wood, typically forming a match to the edge or tongue of another piece) at the back, the depth of the difference in levels, so that it can lay over the front of both the face frame and the inset panel and therefore can give more space to nail the molding to the frame, leaving the inset panel free to expand.

Cartouche.

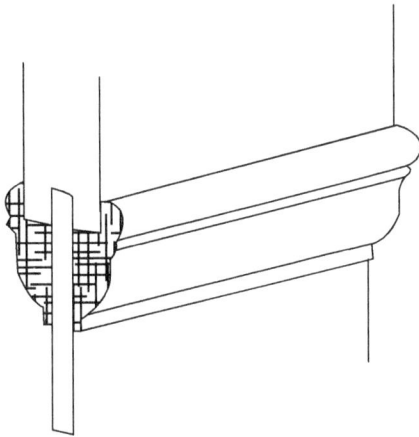

Bolection molding.

- **Cable molding or ropework:** A convex molding carved to look like a twisted rope or cord. It is used as a decorative moldings of the Romanesque style in England, France, and Spain.
- **Cabled fluting or cable:** A convex circular molding sunk in the concave fluting of a classic column, and rising about one-third of the height of the shaft.
- **Casing:** The final trim or finished frame around the top, and both sides of a door or window opening.
- **Cartouche:** A carved tablet, shield, emblem, or drawing representing a scroll with rolled-up ends. It is used ornamentally to be inscripted or have a coat of arms enscripted on it. As regards heraldry, an **escutcheon.**

- **Cavetto:** A concave, quarter-round molding. Sometimes employed in the place of the cymatium (a waved molding that tops an entablature (a horizontal, continuous lintel on a classical building supported by columns or a wall, comprising the architrave, frieze, and cornice)) of a cornice. It forms the crowning feature of the Egyptian temples, and took the place of the cymatium in many of the Etruscan temples.
- **Chair rail:** A horizontal molding that is placed part way up a wall to protect the surface from chair-backs, and used simply as decoration.
- **Chamfer:** A beveled edge connecting two adjacent surfaces.
- **Chin-beak:** A concave quarter-round molding.
- **Corner guard:** Corner guards are used to protect the edges of walls at an outside corner, or to cover a joint on an inside corner.
- **Cove molding/Coving:** A concave-profile molding that is used at the junction of an interior wall and ceiling. Crown molding A wide, sprung molding that is used at the junction of an interior wall and ceiling.
- **Crown molding:** A general term for any molding placed at the top or "crowning" of an architectural element.
- **Cyma:** A molding of double curvature, combining the convex ovolo and concave cavetto. When the concave part is

uppermost, it is called a cyma recta but if the convex portion is at the top, it is called a Cyma reversa. The crowning molding at the entablature is of the cyma form, it is called a **cymatium**. See **Cavetto**.

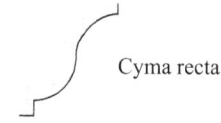
Cyma recta

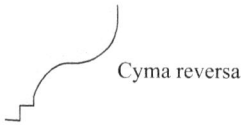
Cyma reversa

Escutcheon.

- **Dentils:** Small blocks spaced evenly along the bottom edge of the cornice.
- **Drip cap:** This molding is placed over a door or window opening to prevent water from flowing under the siding or across the glass.
- **Echinus:** Similar to the ovolo molding and found beneath the abacus (the flat slab on top of a capital, supporting the architrave (a main beam resting across the tops of columns, specifically the lower third entablature)) of the Doric capital or decorated with the egg-and-dart pattern below the Ionic capital.
- **Egg-and-dart:** One of the most widely used classical moldings with egg shapes alternating with V-shapes and known from Ancient Greek temples. See **Egg and dart**.

Egg and Dart at the top of an Ionic Capital.

- **Escutcheon:** An ornamental or protective plate, as for a keyhole. In heraldry a shield or shield-shaped emblem bearing a coat of arms.

- **Fillet:** A small, flat band separating two surfaces, or between the flutes of a column.
- **Fluting:** Vertical, half-round grooves cut into the surface of a column in regular intervals, each separated by a flat astragal. This ornament was used for all but the Tuscan order.

Spiral fluted columns in apamea, Syria.

- **Godroon/Gadroon:** An ornamental band with the appearance of beading or reeding. It is said to be derived from raised work on linen. In France tit has been applied to varieties of the bead and reel, in which the bead is often carved with ornament. In England the term is used by auctioneers to describe the raised convex decorations under the bowl of stone or terracotta vases. The godroons radiate from the vertical support of the vase and rise half-way up the bowl.

- **Guilloche:** Interlocking curved bands in a repeating pattern often forming circles enriched with rosettes.
- **Molding inlay:** Profiled wood trim pieces that surround the perimeter of door panels or glazing, but do not protrude above the surface of the surrounding stiles and rails.

Molding inlay.

- **Molding overlay:** Profiled wood trim pieces that surround the perimeter of panels or glazing and protrude above the surface of the surrounding stiles and rails.

Molding overlay.

- **Keel molding:** A keel molding has a sharp edge resembling in cross-section the keel of a ship.
- **Ogee:** An S-shaped line of molding; a double continuous S-shaped curve.

- BA simple, convex quarter-round molding that can also be enriched with the egg-and-dart or other pattern.
- **Panel molding:** A decorative pattern, originally used to trim out raised panel wall constructional. It is most useful fabricated as a frame, surrounding attractive wall covering for a paneled effect on walls.

Panel molding corner.

- **Picture rail:** A functional molding installed 7–9 feet above the floor from which framed pictures and paintings are hung using picture wire and picture rail hooks. Picture rails are frequently found in older homes with plaster walls, as hammering in nails to hang pictures from would cause damage to the plaster.
- **Rosette:** A circular, floral decorative element found in Mesopotamian design and early Greek stele. Rosettes have been a part of revival styles in architecture since the Renaissance.
- **Scotia:** A concave molding with a lower edge projecting beyond the top and therefore used at the base of columns as a transition between two torus moldings with different diameters.
- **Screen molding:** A panel molding that covers the seam where screening is fastened to the screen frame. Also used as a shelf edge.
- **Shingle molding:** A molding that is installed at the top edge of the fascia board and provides enhanced detail. It butts

against the bottom row of shingles, hence its name. Also known as **rake molding**.

- **Shoe molding:** These moldings are often used at the bottom of the baseboard to cover a small gap or uneven edge between the flooring and the baseboard. Also called **base Shoe molding, Toe molding** and **Quarter-round**. See **Baseboard** (molding).

Shoe molding.

- **Skirting:** Molding around the base of a wall.

Skirting.

- **Torus:** A convex, semi-circular molding, larger than an astragal, often at the base of a column, that may be decorated with leaves or plaiting.

- **Trim molding:** A general term used for moldings that add detail or cover up gaps.
- **Wainscot cap:** Sometimes called a dado cap, this trims out the upper edge or top of a wainscot. Also called a **Wainscot molding**.

Mole: A large solid structure on a shore that serves as a pier, breakwater, or causeway. Also, a harbor formed or protected by such a structure. See **Pier** and **Jetty**.

Mole run: A meandering buckle or ridging in a roof membrane that is not associated with insulation or deck joints.

Moment: The resultant of a system of forces causing rotation without translation. A moment can be expressed as a couple.

Other terms beginning with moment:

- **Moment connection:** A diagram that represents graphically the moment at every point along the length of a member.
- **Moment of inertia (I):** Moment of inertia has two distinct but related meanings: (1) it is a property of an object relating to the magnitude of the moment required to rotate the object and overcome its inertia, and; (2) it is the property of a two-dimensional cross-section shape with respect to an axis, usually an axis through the centroid of the shape.
- **Moment magnitude:** As regards earthquakes, the measure of total energy released by an earthquake. It is based on the area of the fault that ruptured in the quake. It is calculated in part by multiplying the area of the fault's rupture surface by the distance the earth moves along the fault.
- **Moment magnitude scale (MMS and denoted as M_w or M):** This scale is used by seismologists to measure the magnitude of earthquakes. It is based on the seismic moment of the earthquake. The scale was developed in the 1970s to succeed the 1930s-era Richter magnitude scale and it is now the scale that is used to estimate magnitudes for all modern large earthquakes by the United States Geological Survey. Even though the formulae are different, the new scale retains the familiar continuum

of magnitude values defined by the Richter scale. See **Richter magnitude scale.**

- **Moment plate:** A welded steel plate that is used to develop a rigid connection to the supporting member so that moment transfer can occur.
- **Moment release:** See **Pin connection.**
- **Moment resisting-connection:** See **fixed connection.**
- **Moment tensor:** A mathematical representation of the movement on a fault during an earthquake, comprising of nine generalized couples, or nine sets of two vectors. The tensor depends on the source strength and fault orientation.

Monitor: A screen on which an electronics system displays information.

Monitor cover: The cast iron lid that fits over a flange casting that is mounted on meter tile in a water meter pit. It consists of an outer lid, an inner lid, and the flange casting in one or two pieces.

Mono truss: A truss that has only one slope so that its outline is a triangle. Generally installed to rest on an exterior wall and on an inside bearing wall or they bear on the vertical member at the high end of the truss. Mono trusses may be used to form the outside portion of a Mansard roof.

Monoblock faucet: A single-handle faucet, usually a lever.

Monoblock faucet.

Monolithic: Used to describe something without seams; formed from a single material.

Monolithic pour: A nonstop material pour. An example of monolithic pour is when the footings, foundation, and floor slab are all formed and then poured at the same time. Another example is a seamless roof formed from or composed of a single material.

Monolithic dome: A dome composed of a series of arches, joined together with a series of horizontal rings called parallels.

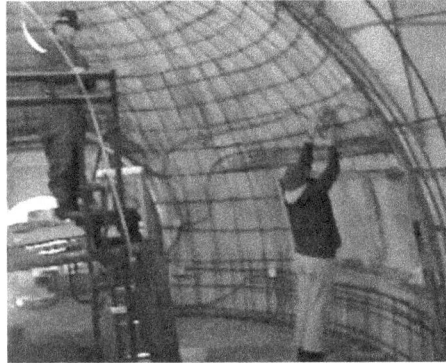

A monolithic dome.

Monomer: A substance composed of low molecular weight molecules capable of reacting with like or unlike molecules to form a polymer.

Monopost: An adjustable metal column used to support a beam or bearing point.

Monorail: Usually a single rail support for a material handling system.

Monument: A permanent surveying mark placed in the ground or onto bedrock indicating elevation, and location of that point as a reference.

Mop-and-flop: A roofers' term where the back side of a roofing material is mopped, then the piece is turned over and set in place.

Mopping: The application of hot bitumen with a roofer's hand mop or mechanical applicator to the substrate or to the felts of a bituminous membrane. Methods of mopping include: **solid mopping**, a continuous mopping of the surface; **spot mopping**, a mopping pattern in which hot bitumen is applied in roughly circular areas, leaving a grid of unmopped, perpendicular bands on the roof; **sprinkle mopping**, a random mopping pattern in which heated bitumen beads are strewn onto the substrate with a brush or mop; and, **strip mopping**, a mopping pattern in which hot bitumen is applied in parallel bands.

Mortar: A binder for masonry. The traditional product was lime mortar. Nowadays mortars are comprised primarily of cement mixed with sand, with the addition of lime or plasticizer added to make them workable or "buttery." Sometimes referred to as **grout**.

Mortar bed: A type of isolation membrane that is made by spreading a layer of mortar, usually between 1/2" to 1-1/2" thick, over the substrate. After the mortar bed cures, the tiles are attached. See **Thinset tiles, Isolation membrane**, and **Cement board**.

Mortgage: A loan that is usually secured by the building and land for which the loan is given. A **mortgagee** is the lender who makes the mortgage loan.

Other terms beginning with mortgage:

• **Mortgage broker:** A broker who represents numerous lenders and helps consumers find affordable mortgages.
• **Mortgage company:** A company that borrows money from a bank, lends it to consumers to buy homes, then sells the loans to investors.
• **Mortgage deed:** A legal document establishing a loan on property.
• **Mortgage loan:** A contract in which the borrower's property is pledged as collateral. It is repaid in installments. The mortgagor (buyer) promises to repay principal and interest, keep the home insured, pay all taxes and keep the property in good condition.
• **Mortgage origination fee:** A charge for work involved in preparing and servicing a mortgage application.

Mortise: A slot cut into a board, plank, or timber, usually edgewise, to receive the tenon (or tongue) of another board, plank, or timber to form a joint; a mortice and tenon joint.

Mortise and tenon: See **Blind mortise and tenon**.

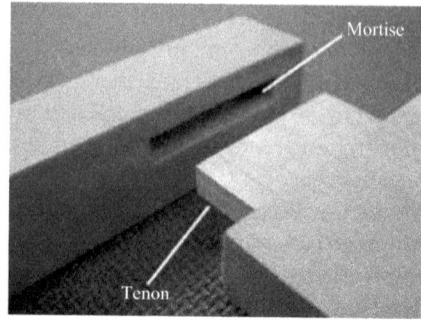

Mortise and tenon.

Mortise chisel: This chisel has a straight, non-beveled edge, unlike a standard chisel. It is designed to take a heavy pounding and for prying.

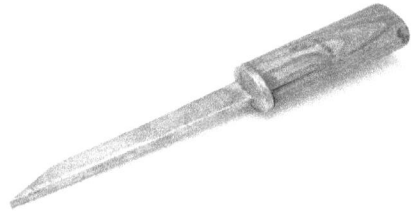

Mortise chisel.

Mortise lock: A lock fitting a rectangular-shaped cavity in the edge of a door.

Mortise lock.

Motor: A machine, especially one powered by electricity, internal combustion, or hydraulic fluid power that supplies motive power for some other device with moving parts. It usually provides rotary mechanical motion

Motor control center: An assembly of one or more enclosed sections having a common power bus and principally containing motor control units.

Movable bridge: A bridge in which the deck moves to clear a navigation channel. There are various types of movable bridges: swing bridges, drawbridges, bascule bridges, and lift bridges.

Moving walks: A passenger carrying device on which passengers stand or walk and in which the passenger carrying surface remains parallel to its direction of motion and is uniniterrupted. They can be horizontal moving walks or inclined moving walks. Moving walks are common in large airports.

MT shapes: A hot rolled structural tee shape with symbol MT that are cut from nonstandard I-shapes.

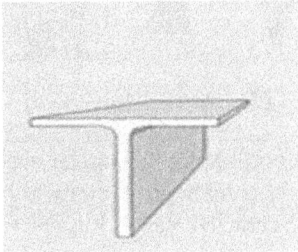

MT 12 × 38.

Mud: A common name for mortar used by bricklayers.

Also, solid particles contained within feedwater. These generally collect at the low points of the boiler where they form a sticky mud like substance.

Other terms beginning with mud:

- **Mud cracking:** Surface cracking of a material that looks similar to dried, cracked mud.
- **Mud flow:** The mass movement of material finer than sand, lubricated with large amounts of water.
- **Mud hole:** An oval opening in the boiler plate allowing access for inspection and washing out.

- **Mud slab:** A nonstructural thin slab of concrete placed directly on the ground for purposes of providing a hard dry surface to work upon.
- **Mud room:** A small room or entranceway in a house where muddy overshoes and wet garments can be removed before entering other rooms of the home.
- **Mudsill:** The bottom horizontal member of an exterior wall frame that rests on top of a foundation. Also known as a **Sill plate** or **Sole plate**.

Müller-Breslau principle: The Müller-Breslau principal states that the height of an influence line (an influence line is a graph of a response function of a structure as a function of the position of a unit load moving across the structure) for the reaction, or for the axial force, shear, or moment at any point in the structure is proportional to the deflected shape of that structure, when the structure is subjected to a unit distortion in the direction of that reaction, axial force, shear, or moment.

Mulling: The act of attaching two or more window or door units together. The joint is then finished with a mullion center cap or mull trim.

Mullion: A vertical divider in the frame between windows, doors, or other openings.

Multiple coats: More than one layer of coating applied to a substrate

Multiple ply membrane: A roof system with more than one layer. Multiple ply membrane roof systems are also called built-up roofs and hot tar roofs. They are usually made from roll roofing materials that are bonded together with asphalt. A three-ply roof has a base sheet, ply sheet, and cap sheet. A five-ply roof has a base sheet, three ply sheets, and a cap sheet. Hot tar (asphalt) is used to bond the plies and make the roof watertight.

Multipoint lock: A locking system, operated with one handle, that secures a window or door at two or more locking points.

Multiport valve: A rotary type backwash valve. One valve can replace up to 6 regular gate valves.

Multi-lite sash: A sash divided into many lites.

Muntin: Profile or molding, either vertical or horizontal, used to separate glass in a sash into multiple lites. Generally refers to components used to construct divided lite grids or grilles simulating a divided lite look. Also referred to as a **bar**.

Diagonal muntins.

Muntin grilles: Wood, plastic, or metal grilles.

Muriatic acid: Commonly used as a brick cleaner after masonry work is completed.

Mushroom: The unacceptable occurrence when the top of a caisson concrete pier spreads out and hardens to become wider than the foundation wall thickness.

Mylar: A trade name for a clear, durable plastic sheet used for covering an inside storm panel or for removable, roll-up glazing over an entire window frame.

Nail: A small metal spike with a broadened flat head. Typically nails are driven into wood with a hammer to join things together or to serve as a peg or hook.

Also, a medieval measure of length for cloth, equal to 2 1/4 inches.

Other terms beginning with nail, nailer, and nailing:

- **Nail guide line:** A painted line on laminated shingles, to aid in the proper placement of fasteners.
- **Nail gun:** See stud gun and pneumatic nailer.
- **Nail inspection:** An inspection made by a municipal building inspector after the drywall material is hung with nails and screws. This inspection is conducted prior to taping.

- **Nail puller:** This tool has a V-notch that slips under the nail head and a long handle to provide extra leverage to pull it up.
- **Nail sets:** This tool is used to sink a nail head below the work surface. Tips vary: point, flat or a cup. Cushioned grip versions protect hands, allowing users to focus on the work and not their knuckles.
- **Nail-holding hammer:** A magnet that is set into the head grips an iron or steel nail. To start nailing the magnetic slot on top of the hammer head accepts 3D to 6D nails and finishing nails.

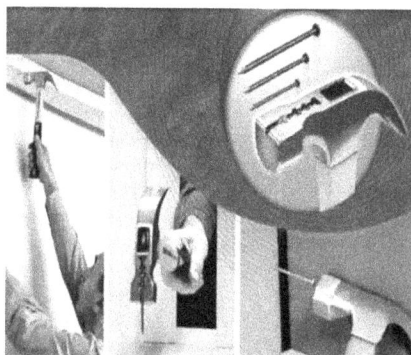

Nail-holding hammer.

Another nail-holding hammer, the **Cheney hammer** is solid with a pretty typical and functional claw. However, it is the nail holding method that sets this hammer apart. The mechanism just above the claw was patented by Arthur E. Taylor in 1925 (granted in 1927). It uses a pair of spring loaded ball bearings that hold the nail securely, but release it easily once the nail is initially sunk into the wood.

Cheney hammer.

- **Nail-pop:** When a nail is not fully driven, it sits up off the roof deck.

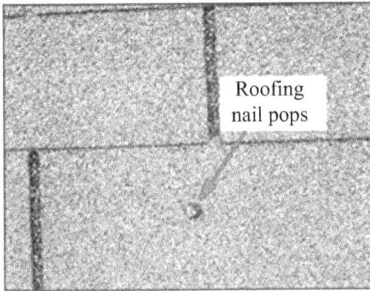

Nail-pop.

- **Nailer:** A piece or pieces of dimensional lumber and/or plywood secured to the structural deck or walls that provide a receiving medium for the fasteners used to attach membrane or flashing.

  Generally, nailers are the same thickness as the adjacent insulation. They may be treated with a non-oil-borne preservative, and they should be of sufficient width to fully support the horizontal flashing flange of a metal flashing (where used). Also referred to as **blocking**.

- **Nailing:** The application of nails. With regard to roofing, nailing may be exposed, i.e., the nail heads are exposed to the weather or concealed nailing, i.e., the nail heads are concealed from the weather by an overlapping material.

- **Nailing fin:** An accessory component or integral extension of a window or patio door frame that generally overlaps the conventional stud construction and through which nails are driven to secure the frame in place.

- **Nailing pattern:** Refers to a specific method or pattern at which nails are applied. For instance, a nailing pattern for base sheets on plywood roof decks can be "Nine and Eighteen." This means one row of nails on the outside edge of the sheet set at nine inches (9") on center, and two rows in the center of the sheet, each set at eighteen inches (18") on center.

  Besides roofing, building codes usually specify nailing patterns/schedules for all aspects of construction.

Nanometer: One billionth of a meter.

Naphtha: A petroleum distillate used mostly by professionals (as opposed to do-it-yourself painters) for cleanup and to thin solvent-based coatings. A volatile organic compound (see **VOC**).

National Earthquake Hazards Reduction Program: The NEHRP leads the federal government's efforts to reduce the fatalities, injuries, and property losses caused by earthquakes. Congress established NEHRP in 1977, directing that four federal agencies coordinate their complementary activities to implement and maintain the program. These agencies are FEMA, the National Institute of Standards and Technology, the National Science Foundation, and the US Geological Survey.

National historic landmark: A district, site, building, structure, or object of national historical significance, designated by the Secretary of the Interior under authority of the Historic Sites Act of 1935 and entered in the National Register of Historic Places.

National Register of Historic Places: The comprehensive list of districts, sites, buildings, structures, and objects of national, regional, state, and local significance in American history, architecture, archeology, engineering, and culture kept by the NPS under authority of the National Historic Preservation Act of 1966.

Native vegetation and adapted vegetation: Plants that are indigenous to a locality or plants that are adapted to the local climate and are not considered invasive species or noxious weeds. Native vegetation and adapted vegetation require limited irrigation following planting, do not require active maintenance such as mowing, and provide habitat value.

Natural: Existing in or caused by nature; something that is not made or caused by humankind.

A word used to describe door color and matching veneers that contain any amount of sapwood and/or heartwood, i.e., natural birch, maple, ash.

Other terms beginning with natural:

- **Natural areas:** Native or adapted vegetation or other ecologically appropriate features.
- **Natural break:** A naturally occurring transition in a material. For example, on walls natural breaks occur at corners or where one material such as painted walls intersects another type of material such as wallpaper.
- **Natural convection:** A heat transfer process involving motion in a fluid (such as air) caused by the difference in density of the fluid and the action of gravity. This is an important part of heat transfer from the glass surface to room air. See **Forced convection**.
- **Natural finish:** A transparent finish that does not seriously alter the original color or grain of the natural wood. Natural finishes are usually provided by sealers, oils, varnishes, water repellent preservatives, and other similar materials.
- **Natural frequency:** The constant frequency of a vibrating system in the state of natural oscillation. Also referred to as fundamental frequency.
- **Natural gas:** A colorless, odorless fuel derived from the earth. Natural consists primarily of Methane (CH4). Odors are added to aid in leak detection.
- **Natural resins:** Resins from trees, plants, fish, and insects. Examples: damars, copals.
- **Natural ventilation:** Ventilation provided by thermal, wind, or diffusion effects through doors, windows, or other intentional openings in the building.

Nearside: For joists and joist girders, when looking at the member with the tagged end to the right, it is the side that you see first and is closest to you (generally, a metal tag wired to each joist shows the job number, mark number, and plant location. The bundle tags show standard erection warnings, plant location, job number, list number, part quantity, part number, paint color, date, and time built).

Needle rasps: Tiny rasps that are used for model-making or detailed carving and fitting. These rasps come in varied shapes: rectangular, tapered, triangular, round, half-round, and square. **Needle files** are available for metalwork.

Negative-pressure smoking rooms: Rooms with mechanical airflow devices (such as exhaust fans) to lower air pressure below that of surrounding areas. The negative pressure causes air to flow from surrounding areas into the space to provide ventilation.

Neoprene: A synthetic rubber material that can be molded or extruded into shapes. Also called **Polychloroprene**.

Nest of saws: This tool is actually four saws in one. The tool comes with three interchangeable blades for sawing around the home and yard. It features a hardwood pistol-grip handle. A quick-change wingnut and bolt hold the steel blades in place.

Nest of saws.

Nesting: Installing a second layer of shingles aligning courses with the original roof to avoid shingle cupping; the edge of the new shingle is butted against the bottom edge of the existing shingle tab.

Nesting of shingles.

Net free vent area: The area, measured in square inches, open to unrestricted air flow and commonly used as a yardstick to measure relative vent performance.

Net lease: A lease in which the tenant pays, in addition to rent, all operating expenses such as real estate taxes, insurance premiums, and maintenance costs. See **Gross lease**.

Net operating income (NOI): The potential rental income plus other income, less vacancy, credit losses, and operating expenses.

Net present worth/Net present value: The difference between the present value of cash inflows and the present value of cash outflows. NPV is used in capital budgeting to analyze the profitability of an investment or project.

Neutral: In chemistry neither acid nor alkaline, i.e., having a pH of about 7.

Other terms beginning with neutral:

• **Neutral axis:** The surface in a structural member where the stresses change from compression to tension. The neutral axis represents zero strain and therefore zero stress. The neutral axis is perpendicular to the line of applied force.

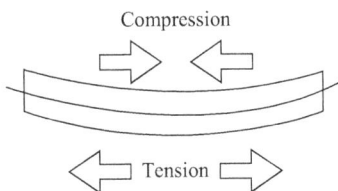

The line is the neutral axis.

• **Neutral color:** A color or shade such as light gray or beige.
• **Neutral conductor:** One of three electrical conductors provided to a residential structure. It is the intermediate conductor in a three-wire electrical system. The neutral conductor is usually grounded or maintained at zero potential. It is connected with one of the current carrying leads to provide 120 volt power.

• **Neutral pressure:** There are pressure differences within buildings brought about by wind and stack effects. Wind creates a positive pressure on the windward side of a building. Warmer air rises and pushes out at the top, thereby creating a suction that pulls in cooler air at the bottom.

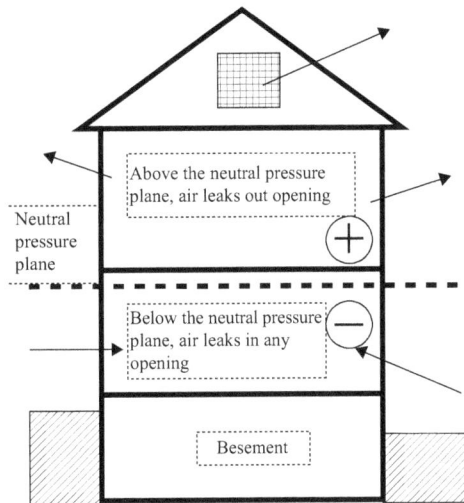

Neutral pressure plane.

Fire door tests are likely to be conducted under positive pressure. The tests are performed with the neutral plane at 40" above the floor, with negative pressure below 40" and positive pressure above. As a result door preps and hardware located below the 40" mark need only meet neutral pressure standards, but everything above 40" must meet positive pressure standards.

• **Neutral wire:** Usually color-coded white, the neutral wire carries electricity from an outlet back to the service panel.

Unlike a neutral conductor the neutral wire is intended to carry current, it is the return path for current supplied by the hot wire.

Because there is current in the neutral wire, is it not safe to handle. A break in the neutral wire would cause the side attached to the load to become hot, and depending on the resistance of the wire and the current it carries, there could be a significant voltage on the wire even in normal operation.

However, the neutral conductor (ground wire) can be attached to the enclosure of the device. It carries no current, and even if disconnected it will still be safe as long as there isn't a short circuit to the hot wire.

Should there be a short between the enclosure and the hot wire, and the ground wire is intact, the ground wire will carry enough current to blow the fuse or circuit breaker. Without the ground wire this fault would make the device dangerous to the touch. See **Neutral conductor, Hot wire,** and **Ground**.

Newel post: The large starting post to which the end of a stair guard railing or balustrade is fastened.

Newton: The SI unit of measure for force.

Newton (N): A unit of force equivalent to the force that produces an acceleration of one meter per second on a mass of one kilogram.

Night seal: To temporarily seal the edge of a roof membrane in order to protect it from moisture entry. Also referred to as night tie-off and water cut-off.

Nineteen-inch selvage: A prepared roofing sheet with a 17" (430 mm) granule surfaced exposure and a nongranule-surfaced 19" (485 mm) selvage edge. This material is sometimes referred to as **SIS, double-coverage or according to ASTM Standard D 371 89, Standard Specification for Asphalt Roll Roofing (Organic Felt) Surfaced with Mineral Granules, Wide Selvage.** See **Splitsheet**.

Ninety-degree corner clamp: Meant for accurate 90° joints, this tool allows for gluing and nailing while the pieces are secure. Some 90° corner clamps can be screwed to a bench top.

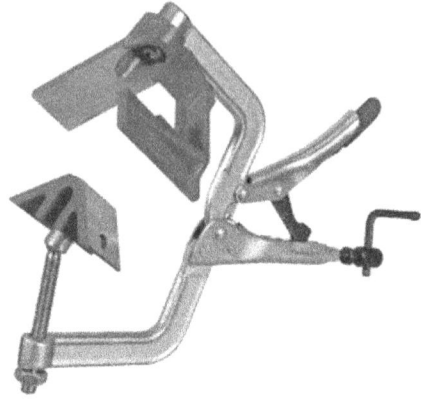

Ninety-degree corner clamp.

Ninety-pound: Granule-surfaced or fiber glass or organic felt roll roofing that has a mass of ~90 pounds per 100 square feet.

Nitrile: Nitrile gaskets are resistant to solvents, oils, and greases and are used under these conditions.

Nitroglycerin: An explosive compound made from a mixture of glycerol and concentrated nitric and sulfuric acids, and an important ingredient of most forms of dynamite. See **Dynamite**.

No-cutout shingle: A shingle made of one solid strip of material.

Node: A point at which lines or pathways intersect or branch; a central or connecting point.

Noggin: A short length of timber fixed crossways between joists, studs or rafters; also the infill between the studs of a traditional timber-framed building.

A noggin.

Also, the brickwork or other infill between the studs of traditional timber-framed construction.

Also a noggin.

Noise Reduction Coefficient (NRC): The average sound absorption coefficient measured at four frequencies: 250, 500, 1,000 and 2,000 Hz expressed to the nearest integral multiple of 0.05. Rates the ability of a ceiling or wall panel or other construction to absorb sound. NRC is the fraction of sound energy, averaged over all angles of direction and from low to high sound frequencies, that is absorbed and not reflected.

Nominal size: The approximate size and or dimension of a material or building module.

As applied to timber or lumber, the size by which it is known and sold in the market; nominal size often differs from the actual size. For example, a 2×4 actually measures 1-1/2 by 3-1/2 inches.

Terms beginning with non:

- **Nonautomatic:** With regard to electrical equipment, an action that requires personal intervention for its control. As applied to an electric controller, nonautomatic control does not necessarily imply a manual controller, but only that personal intervention is necessary.
- **Nonbearing wall:** A wall supporting no load other than its own weight. Also spelled **non-bearing wall**. Also called a **non-load-bearing wall**. See **Load bearing wall**.

- **Noncombustible:** Generally, a construction material is considered to be noncombustible if it will not ignite and burn when subjected to fire. However, if the material liberates flammable gas when heated to any temperature up to 1380°F for five minutes is not considered noncombustible. Materials are also not considered to be noncombustible if the effects of age, fabrication or erection techniques, moisture, or other interior or exterior atmospheric conditions cause them to not meet the criteria stated above.

  Note: This definition should be vetted against local codes; codes are usually very specific about what is deemed to be noncombustible.
- **Noncompact section:** A steel section that does not qualify as a compact section and the width-thickness ratios of its compression elements do not exceed the values designated in the AISC Manual.
- **Nonlinear load:** With regard to electrical equipment, a load where the wave shape of the steady-state current does not follow the wave shape of the applied voltage. Electronic equipment, electronic/electric-discharge lighting, adjustable-speed drive systems and similar equipment may be nonlinear loads.
- **Nonoccupied spaces:** Generally, rooms used by maintenance personnel and not open to occupants. Examples include janitorial closets, cleaning supply storage, and equipment rooms. Local building and zoning codes may define occupied, nonoccupied, and nonregularly occupied spaces and the like. These local definitions need to be read and understood as the net size and economic viability of a project may rest on them.
- **Nonpotable water:** Water that does not meet EPA's drinking water standards, and therefore, is not suitable for human consumption.
- **Nonregularly occupied spaces:** Corridors, hallways, lobbies, break rooms, copy rooms, office supply closets, kitchens,

restrooms, and stairwells. Having identifying these spaces my come in handy when arguing over the allowable size of a building with the Department of Building and Safety, or equivalent.

- **Nonrigid structure:** A structure that cannot maintain its shape and may undergo large displacements and would collapse under its own weight when not supported externally.
- **Nonstructural components:** Building components that are not intended primarily for the structural support and bracing of the building.
- **Nonvolatile:** The portion of a coating left after the solvent evaporates; sometimes called the solids content.
- **Nonwoven:** Random arrangement of the reinforcement fibers of a scrim sheet or mat.
- **Non-bonded door core:** Stiles and rails (edge bands) are not glued to the core prior to face materials.
- **Non-breathing membrane:** A membrane that does not permit water vapor or air to permeate it.
- **Non-CFC foam:** Insulation that minimizes the use of Chlorofluorocarbons.
- **Non-destructive testing (NDT):** A method to evaluate the disposition, strength, or composition of materials without damaging the object under test.

  The most common NDT methods are:
  - Acoustic Emission Testing (AET)
  - Acoustic Resonance Testing (ART)
  - Electromagnetic Testing (ET)
  - Infrared Testing (IRT)
  - Leak Testing (LT)
  - Magnetic Particle Testing (MT)
  - Dye Penetrant Testing (PT)
  - Radiographic Testing (RT)
  - Ultrasonic Testing (UT)
  - Visual Testing (VI - Visual Inspection)

NDT does not permanently alter the article being inspected and is therefore a valuable technique that can save both money and time in product evaluation, troubleshooting, and research. Also spelled **nondestructive testing**. Also referred to as **nondestructive examination (NDE), nondestructive inspection (NDI)**, and **nondestructive evaluation (NDE)**.

With regard to roofing, NDT is used to evaluate moisture content in roofing assemblies. The three common test methods are **electrical capacitance, infrared thermography**, and **nuclear back-scatter**.

- **Non-ferrous:** Not containing iron.
- **Non-flammable Liquid having no measurable flash point.**
- **Non-friable:** A material that, when dry, cannot be crumbled, pulverized or reduced to powder by hand pressure.

  Friable and non-friable are important considerations when assessing asbestos. Asbestos-containing materials are categorized as friable or nonfriable in order to show how easily they may release asbestos fibres when disturbed.

  A material that is friable is one that can be crumbled, pulverized or powdered by hand pressure. If a friable asbestos-containing material is damaged or disturbed it presents an inhalation risk because asbestos fibers are more easily released into the air. Examples of friable materials include sprayed fireproofing on structural steelwork and thermal insulation on pipes.

  A non-friable asbestos product is one in which the asbestos fibers are bound or locked into the product matrix so that the fibers are not readily released. Examples of nonfriable asbestos products include vinyl asbestos floor tiles, acoustic ceiling tiles, and asbestos cement products.

- **Non-keyed cylinder:** A handle without a keyed cylinder. The door cannot be locked from the exterior.
- **Non-potable:** Not suitable for drinking.
- **Non-structural part:** The part of a building that is not essential for supporting a load or for keeping the structure intact.
- **Non-traffic bearing:** With regard to roofing, a membrane system that requires some form of protection barrier and wearing surface.

- **Non-volatile content:** The portion of a material that will not evaporate.
- **Non-veneer panel:** Any wood based panel that does not contain veneer and caries ana APA span rating, such as wafer board or oriented strand board.
- **Non-vulcanized material:** A material that retains its thermoplastic properties throughout its service life.

Neo angle base: A shower base designed to allow the shower to fit into a corner using minimal floor space while maintaining an elegant look.

Neo angle shower stall base.

Nipple: A short length of pipe installed between couplings or other fittings.

Normal strain: Strain measures the intensity of deformation along an axis. Normal strain is usually denoted by ε. Average normal strain between two points is calculated as

$$\varepsilon = \Delta L/L,$$

where L is the original distance between the points, and ΔL is the change in that distance. Normal strain is often simply called strain.

Normal stress: Stress acting perpendicular to an imaginary plane cutting through an object. Normal stress has two senses: compression and tension. Normal stress is often simply called stress.

Normal user control: A switch or other device that activates a system or component and is provided for use by an occupant of a building.

Nosing: The projecting edge of a molding or drip or the front edge of a stair tread.

Not restricted: Allowed, unlimited.

Notch: An indentation or incision on an edge or surface, e.g., a crosswise groove at the end of a board.

Note: A formal document showing the existence of a debt and stating the terms of repayment.

Notes: Instructions placed on plans by the architect or engineer. They are an integral part of the contract documents and must be followed.

Notice of award: A letter from an owner to a contractor stating that a contract has been awarded to the contractor and a contract will be forthcoming. It may also function as a notice to proceed. See **Letter of intent**.

Notice to proceed: A notice from an owner directing a contractor to begin work on a contract, subject to specific stated conditions.

Nozzle: The part of a heating system that sprays the fuel of fuel-air mixture into the combustion chamber.

No-hub connector: A connector for no-hub iron pipe. The connector consists of a rubber sleeve and a stainless steel band secured by hose clamps. A variation, a neoprene sleeve with two adjustable steel bands, is used for connecting dissimilar materials, as when connecting new plastic pipe to an existing cast-iron drainpipe.

No-hub connector.

Number one common oak grade: The grade of oak strip flooring that may have bright spots of sap, pinworm holes, machine defects, streaky or inconsistent color, grain variations and a few knots. See **Select and better oak grade** and **Number two common oak grade**.

Number two common oak grade: The grade of oak strip flooring that may have pronounced bright spots of sap, pinworm holes, machine defects, streaky or inconsistent color, grain variations, and knots. Some defects found in number two common oak grade are so severe that the installer will want to cut out some bad spots or even discard some severely flawed strips of wood. See **Select and better oak grade** and **Number two common oak grade**.

Nylon carpet: A carpet made from man-made nylon fibers. See **wool carpet, Berber carpet, indoor-outdoor carpet, sculptured carpet,** and **shag carpet**.

O-ring: A round rubber washer used to create a watertight seal.

Oakum: A loose hemp or jute fiber that is impregnated with tar or pitch. Oakum is used to caulk large seams or for packing plumbing pipe joints.

Obscure glass: Any textured glass (frosted, etched, fluted, ground, etc.) used for privacy, light diffusion, or decorative effects. Also called **visionproof glass**.

Obsolescence: In reference to the inadequacy, disuse, outdated, or nonfunctionality of facilities, infrastructure, products, or production technologies due to effects of time, changing market conditions, or decay (a factor considered in depreciation to cover the decline in value of fixed assets due to the invention and adoption of new production technologies, or changing consumer demand).

Occupancy group: A category establishing the intended use for a building. The act of occupying the building for its intended use by the owner.

Occupancy phase: A stipulated length of time following the construction phase, during which contractors are bonded to ensure that materials, equipment, and workmanship meet the requirements of their contracts, and that supplier- and manufacturer-provided warranties and guarantees remain in force.

Occupancy schedule: A schedule of the activities and events required to effect occupancy or the use of a facility for its intended purpose. It is used to determine if construction progress will meet the occupancy date.

Oceanic spreading ridge: An oceanic spreading ridge is the fracture zone along the ocean bottom where molten mantle material comes to the surface, thus creating new crust. This fracture can be seen beneath the ocean as a line of ridges that form as molten rock reaches the ocean bottom and solidifies.

Oceanic trench: An oceanic trench is a linear depression of the sea floor caused by the subduction of one plate under another.

Offset ridge: When the ridge of a joist that has the top chord pitched two ways is not in the center of the member or bay.

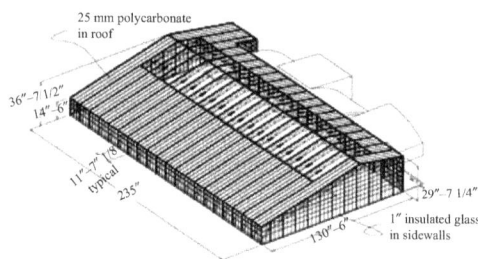

Building with an offset ridge.

Also, in geology, a ridge that consists of resistant sedimentary rock that has been made discontinuous as a result of faulting.

Off-site salvaged materials: Materials recovered from an off-site source and reused.

Off-ratio foam: SPF where the 1 to 1 ratio of the A and B components has been compromised thereby resulting in a lower quality material.

Office building: While not universal the following terms normally apply:

- **Low-rise:** Fewer than seven stories high above ground level.
- **Mid-rise:** Between seven and twenty-five stories above ground level.
- **High-rise:** Higher than twenty-five stories above ground level.

Offset: To move or dimension something off center for purposes of installation, access, and or clarification.

Also, a tubular component that permits the offsetting of a drainage run in the same basic direction.

Offset snips: A cutting tool that has offset handles to keep hands above work. Specifically designed for long, inside cuts.

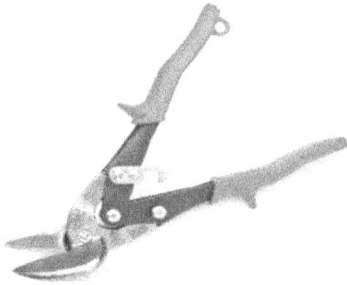

Offset snips.

Offset wrench: A tool that is used to tighten or loosen bolts and nuts in difficult to reach locations.

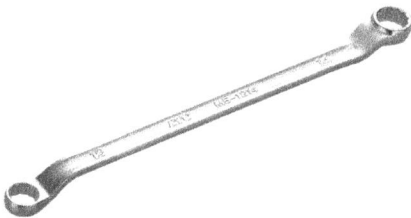

Offset wrench.

Another offset wrench.

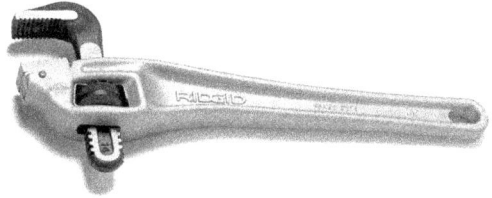

And, an offset pipe wrench.

Ogee: An S-shaped line of molding; a double continuous S-shaped curve. See **Moldings**.

Ogee molding.

Also, a pointed arch with a curve near the apex.

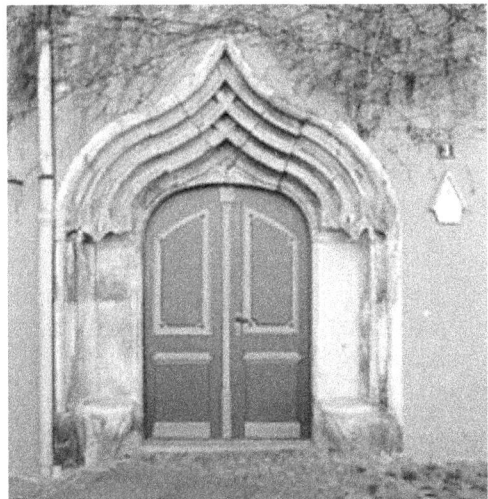

Ogee arch.

Ohm: The unit by which resistance to electrical current is measured. An ohmmeter measures electrical resistance in ohms.

From Ohm's Law: (current = voltage/resistance, or in other words, current = voltage/ohms), a mathematical expression of the relationship between these three elements.

Oil buffer: One type of buffer (for elevators with speeds of more than 200 feet per minute), that uses a combination of oil and spring to cushion the elevator. It is located in the elevator pit.

Oil paint: A paint that contains drying oil, oil varnish, or oil-modified resin as the film-forming ingredient. The term is commonly and incorrectly used to refer to any paint soluble by organic solvents.

Oleoresin: A natural plant product that contains oil and resins. Turpentine is an example.

On the flat: A measurement of distance horizontally on a plan, no slopes involved.

One-call: An agency that will mark the location of all utility lines. This service is not available in all areas and the name of the agency varies from state to state.

On-demand water heaters: Water heaters that heat water only when it is needed and then apply only the amount of heating required to satisfy the user's immediate needs. Many utility companies offer sizable rebates for installing on-demand water heaters. Also called **instantaneous hot water heaters** and **tankless hot water heaters**.

One-dimensional calculations: Items that are simply counted or measured in just one direction. "Each" and "Lineal feet" are examples of one-dimensional units of measure.

One-piece toilet: A toilet with both the tank and bowl as a single vitreous china fixture. One-piece toilets usually have a lower profile than two-piece toilets.

On-site salvaged materials: Materials recovered from and reused at the same building site. Do not assume all materials that are in good condition can be reused. Check first with local codes.

On-site supervision: Personnel who are assigned to the site with supervisory responsibilities.

One-step distributor: An industry term for a wholesale company that buys building products from a manufacturer and sells them to builders, contractors, and homeowners. A wholesaler that buys building products from the manufacturer and sells them to lumberyards and home centers that in turn sell to builders, contractors, and homeowners is referred to as a **two-step distributor**.

One-wide (1W): A term used to describe one frame with single or multiple sash or panels.

One-quarter bond: A course of brick where the vertical joint between bricks is one fourth of the way across the length of the brick in the course below it.

One-quarter bond.

One-third bond course of brick where the vertical joint between bricks is one third of the way across the length of the brick in the course below it.

One-third bond.

One-quarter turn: A handrail piece that is used to turn a 90° corner. See **Top rail, Volute, Goose neck**, and **Balustrade**.

Ongoing commissioning: A continuous process that methodically identifies and corrects system problems to maintain optimal building performance; the process includes regular measurement and comparative analysis of building energy data over time.

Ongoing consumables: Goods with a low cost per unit and are regularly used and replaced in the course of business. Examples include paper, toner cartridges, binders, batteries, and desk accessories.

Oolite: The small round particles that make up a sedimentary rock. On mass they look just like fish eggs.

Opacity: The quality or state of a material that makes it impervious to the rays of light.

Terms beginning with open:

• **Open circuit:** With regard to electrical wiring, a discontinuous circuit through which no current can flow.

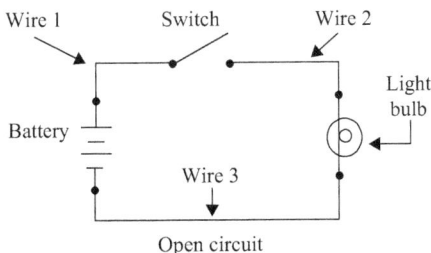

Open circuit

With regard to a hydraulic circuit, the pump works continuously at low pressure. A valve can redirect the flow to an actuator. Because the pump is working continuously the pressure increases as the resistance increases. It's called open because the valve, when in the neutral position, provides a return path to the tank (open center).

In a closed circuit, full pressure is applied to all valves at all times. The pump will adjust its flow rate, pumping very little oil until the operator moves the valve. It's called closed because the valve in neutral does not provide a return path to the tank.

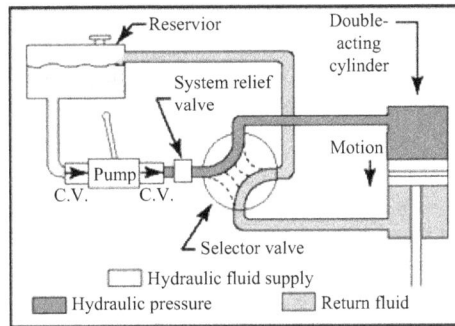

Open hydraulic circuit.

• **Open competitive selection:** When clients advertise impending bids so that anyone interested in the project who meets qualifications may submit a bid. All qualified bidders are given access to the prescribed project information; open bid project information is not private. Government agencies are generally required to use open competitive selection. Also referred to as **open bid**.

• **Open glass wall:** A glass wall that is a combination of door-leafs, either connected or not connected. An open glass wall is used as a design element to achieve a degree of enclosure against weather or for security and expansive openness. There are several open glass wall systems that can be divided into two systems: folding and individual panel; the folding systems has panels that are hinged together and form a continuous train, whereas the individual panel system is composed of separate panels.

- **Open grain:** A wood grain where the annual growth rings are pronounced and there is an obvious difference between the pore size of springwood and summerwood. Oak and ash are examples of open-grained wood.

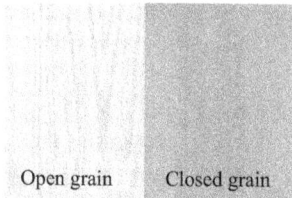

Open grain   Closed grain

Open and closed grain wood.

- **Open hole inspection:** When an engineer (or municipal inspector) inspects the open excavation and examines the earth to determine the type of foundation (caisson, footer, wall on ground, etc.,) that should be installed in the hole.
- **Open sides and ends, per OSHA:** The edges of a platform that are more than 14 inches (36 cm) away horizontally from a sturdy, continuous, vertical surface (such as a building wall) or a sturdy, continuous horizontal surface (such as a floor), or a point of access. Exception: For plastering and lathing operations the horizontal threshold distance is 18 inches (46 cm).
- **Open specification:** A bid specification written in such a way as to allow multiple numbers of products for the item being required.
- **Open time:** The time after an adhesive has been applied and permitted to cure when the two surfaces can be bonded.
- **Open valley:** A valley where both sides of the roof are trimmed back from the centerline to expose the valley flashing material beneath.
- **Open web joist:** A fabricated structural steel truss usually made of steel angle sections and or bar members.
- **Open-grid pavement:** For LEED purposes, a pavement that is less than 50% impervious and contains vegetation in the open cells. Open-grid pavement consists of a thin, open-graded asphalt mix layered

over the top of a coarse stone aggregate; water passes through the asphalt surface and is stored in the aggregate, from which it slowly percolates into the soil.

Opencast: The method of mining near the surface, by cutting into it from above rather than digging underground.

Opening per OSHA: A gap or void 30 inches (76 cm) or more high and 18 inches (48 cm) or more wide, in a wall or partition, through which employees can fall to a lower level.

Operator: A crank-operated device for opening and closing casement or jalousie windows.

Operating expenses: Cash outlays necessary to operate and maintain a property. Examples of operating expenses include real estate taxes, property insurance, property management and maintenance expenses, utilities, and legal or accounting expenses. Operating expenses do not include capital expenditures, debt service, or cost recovery.

Opportunity cost: The cost of an alternative that must be forgone in order to pursue a certain action. Put another way, the benefits you could have received by taking an alternative action.

Also, the difference in return between a chosen investment and one that is necessarily passed up. For example, suppose you invest in a stock and it returns only 2% over the year. In placing your money in the stock, you gave up the opportunity of another investment, say, a risk-free US Bond yielding 6%. In this situation, your opportunity costs are 4% (6% − 2%).

Orange peel texture: A finish applied to drywall with a machine that splatters mud onto the walls leaving a bumpy texture that is similar to the pattern on an orange peel.

Orbital sander: This sander is used for finishing work; the square pad can sand inside corners. It must be used with care to avoid swirl marks, especially when sanding dense hardwoods. It accepts sheet sandpaper.

Orbital weld: A circumferential, full fusion weld used to join together two lengths of tubing. It is a Gas tungsten arc (GTAW) welding process. Usually to join coiled lengths.

Order of operations: The mathematic rules that specify the order in which mathematic operations must be accomplished to produce the correct result. The order of operations is as follows:

- Operations inside parenthesis.
- Squares and square roots.
- Multiplication and division.
- Addition and subtraction.

Orders: Columns influenced by the Greeks and Romans are placed into specific orders, such as Doric, Ionic, Corinthian, Composite, and Tuscan.

Organic: Being or composed of hydrocarbons or their derivatives originating from plant or animal matter.

Organic felt: An asphalt roofing base material manufactured from cellulose fibers.

Organic shingles: Composition shingles made with an organic mat.

Oriel: A type of bay window that protrudes from building, but does not touch the ground.

Oriel windows in San Francisco.

Orientation: The placement of a room, window, or building with respect to sun, wind, earth, access, or view.

Oriented strand board (OSB): A manufactured 4'× 8' wood panel made out of 1–2" wood chips and glue. Often used as a substitute for plywood. Used especially as decking.

Orifice: A restriction, the length of which is small in respect to its cross-sectional dimensions, e.g., a pipe or tube.

Original basis: The total amount paid for a property, including equity capital and the amount of debt incurred.

Oscillation: To move or swing back and forth at a regular speed.

Also, in physics, to vary in magnitude or position in a regular manner around a central point.

Also, in electricity, current or voltage is said to oscillate when it varies in magnitude in a regular manner.

Outdoor air: The ambient air that enters a building either through a ventilation system (with intentional openings) for natural ventilation or by infiltration (ASHRAE 62.1 – 2004).

Overspray: The loss of spray particles (from coatings, SPF, etc.) in the air.

Owner: (1) An entity that possesses the exclusive right to hold, use, benefit-from, enjoy, convey, transfer, and otherwise dispose of an asset or property; (2) a person or entity who awards a contract for a project and undertakes to pay the contractor, also called contract owner; (3) an employee or executive who has the principle responsibility for a process, program, or project; and (4) a person directly employed by the organization holding title to the project building and recognized by law as having rights, responsibilities, and ultimate control over the project building.

Outcrop: The area where a particular rock body reaches the surface.

Outdoor air: The ambient air that enters a building either through a ventilation system or by infiltration (ASHRAE 62.1-2004).

Outfeed: The side of a power tool where the board exits.

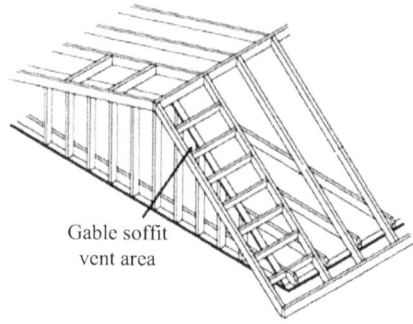

Outrigger extension to a rafter.

Out of phases: The state where a structure in motion is not at the same frequency as the ground motion.

Also, when equipment in a building is at a different frequency from the structure.

Outside casing: That portion of the window frame that is exposed to the outdoors. Also referred to as **outside facing, outside trim**, and **exterior casing**.

Outside corner: The point at which two walls form an external angle, one you usually can walk around.

Outside glazing: Glazing that is installed from the outside.

Outstanding leg: The leg of a structural angle that is projecting toward or away from you when viewing.

Outswing French door: A French door with panels that swing to the outside. One, two, three, or four panel units available as stationary or operating.

Oven-dry wood: Wood dried to a relatively constant weight in a ventilated oven at 102–105°C.

Oven-dry weight: The weight obtained by drying wood in an oven at 102°F (plus or minus 3°F) until there is no more weight loss.

Terms beginning with over or that use over as a preface:

- **Over speed governor:** A mechanical speed control mechanism. Normally it will be a centrifugal device used to stop and hold the movement of an elevator by

Table saw rear outfeed support.

Outlet: With regard to electrical wiring, a point on the wiring system at which current is supplied to equipment lighting, etc.

Outlet: The opening through which the water exits the pump. Also called the **discharge**.

Outrigger: A projecting member run out from a main structure to provide additional stability or to support something.

Also, a structural member that is usually perpendicular to a joist or beam and attaches to it. It then bears on a beam or wall and cantilevers across.

Outrigger from a beam.

Also, an extension of a rafter beyond the wall line. Usually a smaller member nailed to a larger rafter to form a cornice or roof overhang.

initiating the activation of the safety gear as well as the cutting of power to the drive motor and brake.

- **Overburden:** Unconsolidated materials overlying rock.
- **Overcurrent:** Any current in excess of the rated current of equipment or the ampacity of a conductor. It may result from overload, short circuit, or ground fault.
- **Overdriven:** The term used for fasteners driven through roofing material with too much force, breaking the material.
- **Overdry:** Lumber dried to the point of having too low of a moisture content.
- **Overexposed:** Installing shingle courses higher than their intended exposure.
- **Overflow:** In washbasins there is usually a hole near the rim of the bowl that drains into the waste outlet pipe The overflow by-passes the stopper mechanism in the case of a blocked drain.

    In toilets there is usually a vertical tube inside the toilet tank that directs water into the bowl in case the ballcock malfunctions. It is usually part of the flush valve. It prevents potential water damage caused by a tank overflow. On most toilets, the overflow tube also has a refill tube flowing into it that directs water from the ballcock through the overflow tube to the bowl, after a siphon break.

- **Overhand bricklaying:** The process of laying bricks and masonry units such that the surface of the wall to be jointed is on the opposite side of the wall from the mason, requiring the mason to lean over the wall to complete the work. Related work includes mason tending and electrical installation incorporated into the brick wall during the overhand bricklaying process.
- **Overhang:** An outward projecting eave-soffit area of a roof; the part of the roof that hangs out or over the outside wall. See also **Cornice**.
- **Overhead:** The upper portion of the elevator hoistway. Measured from the top landing level to the bottom of the machine room floor slab (or the hoistway roof for machine room less elevators or basement drives). Codes establish the required

overhead. Overhead may rise above the roof line thereby affecting a building's aesthetics.

- **Overhead costs:** Costs that cannot be identified with or charged to projects or units of production except on some more or less arbitrary allocation basis. These are costs that are incurred in the home office.
- **Overhead door:** A door commonly found on garages, mounted in a track or frame enabling it to move above the opening when in the open position.
- **Overlap:** To cover and extend beyond something, of which there are many examples, for example:
    - With regard to **kitchen range hoods**, sizing the range hood to be wider than the cooking equipment. Whenever possible, the range hood should overlap the cooking surface by 3-inches on each end. For island hoods, this recommendation should be considered mandatory.
    - With regard to **siding**, a joint that is made by placing the edge of a piece of siding over a previously installed piece of siding.
    - A waterproof seam created by placing the edge of one long side of a metal roofing panel over the long edge of the panel adjacent to it would be called an **overlapping seam**.
- **Overlay:** A thin layer of paper, plastic, film, metal foil, or other material bonded to one or both faces of panel products, or to lumber, to provide a protective or decorative face, or a base for painting.
- **Overlay door:** A door that is on the outside of the frame and, when closed, the door hides the frame from the view.

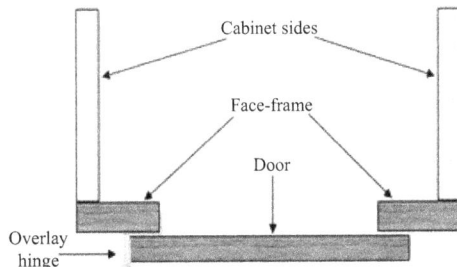

Overlay door.

- **Overlaying shingles:** The process of laying a new layer of shingles on top of an old layer of shingles. Overlaying shingles can make a roof more susceptible to hail damage. Weight is also an issue of concern. Most building codes do not allow more that two layers of shingles.

- **Overload:** With regard to electrical systems, operation of equipment in excess of normal, full-load rating, or of a conductor in excess of rated ampacity that, when it persists for a sufficient length of time, would cause damage or dangerous overheating. A fault, such as a short circuit or ground fault, is not an overload.

- **Overrun:** To move or extend over or beyond as with a cost overrun.

  Also, to rotate faster than another part of a machine as an overrunning clutch

- **Overspray:** Fine particles of paint that are carried in the air from the paint sprayer. These particles may then land on other surfaces that are not intended to be painted.

- **Overstressing:** Subjecting materials to stresses that are in excess of their allowable stresses.

Owner's representative: A representative of the owner of a construction project. The owner's representative ensures that the design meets the owner's needs for both function and aesthetic appeal. During construction the representative makes sure that bids are correctly solicited, change orders are analyzed and approved, funds are properly managed, and, generally, makes certain that the owner receives full value.

Owner-architect agreement: A written form of contract between architect and client for professional architectural services.

Owner-builder: An owner who takes on the responsibilities of the general contractor to build a specific project.

Owner-construction agreement: A contract between an owner and a contractor for a construction project.

Oxidation: The interaction between oxygen molecules and all the different substances they may contact, from metal to living tissue. More precisely oxidation may be defined as the loss of at least one electron when two or more substances interact. Those substances may or may not include oxygen.

Oxidation can be destructive, as with the rusting of steel and iron. However, sometimes oxidation is not such a bad thing, as in the formation of super-durable anodized aluminum.

We commonly use the words oxidation and rust interchangeably, but not all materials that interact with oxygen molecules actually disintegrate into rust. In the case of iron, the oxygen creates a slow burning process that results in the brittle brown substance known as rust. When oxidation occurs in copper the result is a greenish coating called copper oxide. The metal itself is not weakened by oxidation, but the surface develops a patina after exposure to air and water.

The opposite of oxidation is reduction, the addition of at least one electron when substances come into contact with each other.

OX, OXO, OXXO: Designations of the arrangement of sliding and fixed panels in sliding glass doors and windows; X indicates a sliding panel; O indicates a fixed panel; see also XO, XOO, XOOX.

P-delta effect: The secondary effect of column axial loads and lateral deflection on the moments in structural members.

P-trap: A plumbing part, shaped like the letter P. A P trap holds water that traps sewer gases in the line and prevents them from entering the structure. See **Traps.**

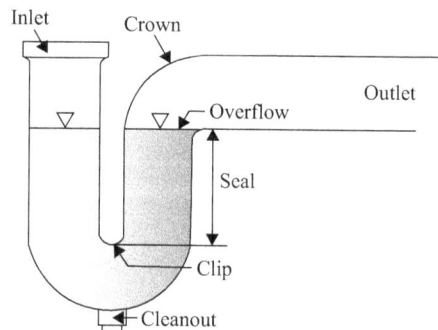

P trap.

P-wave: A seismic wave that involves particle motion in the direction of propagation. It is the fastest traveling wave generated by an earthquake and therefore the first to arrive at any point. See **Waves**.

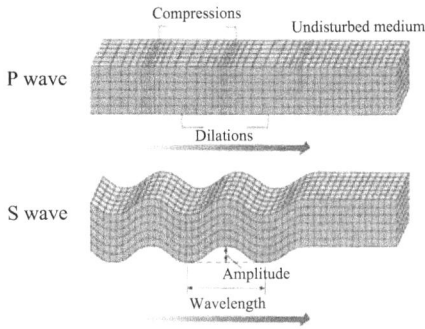

Showing the difference between a P wave and S wave.

Packing: Fibrous material that is used on faucets to prevent leaks.

Packing nut: The nut that holds the stem of a faucet in position and holds the packing material.

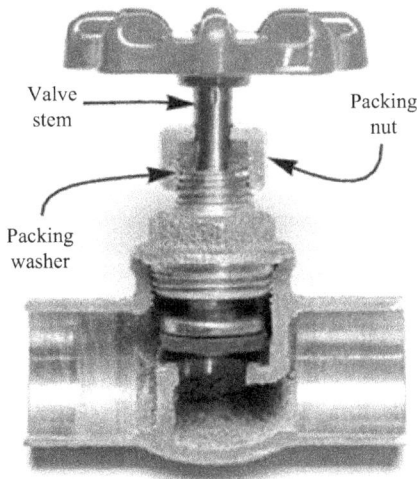

In line shut off valve showing a packing nut.

Pad out: To shim or add strips of wood to a wall or ceiling in order that the finished ceiling/wall will appear correct. Also referred to as **pack out**.

Padding: A material installed under carpet to add foot comfort, isolate sound, and to prolong carpet life.

Padstone: A block of concrete or stone used to distribute the weight of a beam or joist upon which it rests.

A padstone.

Paint: A combination of pigments with suitable thinners or oils to provide decorative and protective coatings. Paints are generally classified into two categories: water-based and solvent-based. Solvent-based paints use a carrier consisting of volatile organic compounds (VOCs) that can cause smog, ozone pollution and indoor air quality problems. Newer formulations contain more paint solids or environmentally friendly solvents.

Liquid paint is made up of three main components: pigment, solvent, and binder (resin). Additives may be used in small amounts in comparison to the main ingredients. They include: flatting agents (pigments added to reduce gloss such as Talc, Zinc sterate, and silica), rheology modifiers (used primarily in a latex paint system is to control flow properties), wetting agents (to reduce the surface tension of a liquid in which it is dissolved), and curing agents (a hardener or activator added to synthetic resin to develop the proper film forming properties).

Other terms beginning with paint:

• **Paint grade material:** Trim material that has flaws or joints that will be hidden if the material is painted.

- **Paint remover:** A chemical that softens old paint or varnish and permits it to be easily scraped off. Also called **stripper**.
- **Paint sprayer:** A high pressure paint pump that sprays paint through a sprayer tip or nozzle onto the surface. There are air powered and airless paint sprayers. See **Airless spraying**.
- **Paint thinner:** See **Mineral spirits**.

Paleoseismicity: Refers to earthquakes that have been recorded geologically. Most of them are not known from human descriptions or seismograms. Geologic records of past earthquakes can include faulted layers of sediment and rock, injections of liquefied sand, landslides, abruptly raised or lowered shorelines, and tsunami deposits.

Paleoseismology: The study of geologically recent earthquakes.

Palladian window: A large, arch-top window flanked by smaller windows on each side.

Palladian window.

Pallets: Wooden platforms used for storing and shipping material. Forklifts and hand trucks are used to move these wooden platforms around.

Pan: The bottom flat part of a roofing panel that is between the ribs of the panel.

Installing pan roofing.

Also, the concave piece of **pan and cover tile** whose rounded surface touches the top side of the roof substrate. Also called **pan tile**.

Pan and cover tile.

Pan flashing: A sheet metal flashing that covers an equipment platform and is designed to counter flash the base flashings surrounding the platform.

Pane: A sheet of glass for glazing a window. After installation, the pane is referred to as a **lite**.

Panel: A thin flat piece of wood, plywood, or similar material, framed by stiles and rails as in a door (or cabinet door), or fitted into grooves of thicker material with molded edges for decorative wall treatment.

Also, the distance between two adjacent panel points (the point where one or more web members intersect the top or bottom chords of a joist or joist girder.

Other terms beginning with panel:

- **Panel, flat:** A door panel in which the perimeter does not contain a machined profile (panel raise). Constructed with veneer on the face and a composite core for a stained finish, or medium-density fiberboard (MDF) for a painted finish.

- **Panel products:** A general descriptor that includes products such as: plywood, waferboard and oriented strandboard, hardboard, insulation board, and composites that use these materials.
- **Panel, raised:** A door panel whose faces are raised above the perimeter and whose edges are shaped to fit into grooves in the stiles, rails, and mullions. These panels are typically bladder pressed or rim banded for a stained finish or MDF for a painted finish.
- **Panel saw:** Any type of sawing machine that cuts sheets into sized parts. Panel saws are used by cabinet shops to easily cut plywood and melamine sheets into cabinet components. They are also used by sign shops to cut sheets of aluminum, plastic and wood for their sign blanks.

Panel saw.

- **Panel stock:** Stock panels that are 4' wide and 8' long.

Panelboard: A single panel or group of panel units designed for assembly in the form of a single panel, including buses and automatic overcurrent devices, and equipped with or without switches for the control of light, heat, or power circuits; designed to be placed in a cabinet or cutout box placed in or against a wall, partition, or other support; and accessible only from the front.

Paneling: Wood used to cover the entire expanse of a wall, from top to bottom.

Panning: In replacement window work, the outside aluminum trim that can extend around the perimeter of the window opening. Panning is used to cover up the old window material.

Paper, building: A general term for papers, felts, and similar sheet materials used in buildings without reference to their properties or uses. Generally comes in long rolls.

Paper burns: Tiny fibers that are raised in the drywall surface when sanded excessively.

Parallel chord: A type of joist or joist girder that has its top and bottom chords parallel to each other. The member can be sloped and still have parallel chords.

Parallel strand lumber (PSL): A composite wood product where veneers are clipped into 1" strips and pressed together with all grain parallel into a large billet (a thick piece of wood). Billets are cut into different sizes for high strength beam and column applications. Also referred to as **Parallam**.

Parallam mantle.

Parallelogram: A 4-sided shape in which the diagonals drawn from opposite corners are not equal and opposite sides are parallel to each other.

Parapet wall: Often found around flat roof systems, a parapet wall is a low wall that rises above the roof deck. Parapet walls are to prevent people from falling and to prevent roof fires from entering upper floor windows and vice-versa.

Pargetting or pargeting Parallel strand lumber (PSL): A composite wood product where veneers are clipped into 1" strips and pressed together with all grain parallel into a large billet (a thick piece of wood). Billets are cut into different sizes for high strength beam and column applications. Also referred to as Parallam.

Parallam mantle.

Paring chisel: This tool allows the user to make light finishing cuts with the blade flat on the stock, even when working in the middle of a wide board.

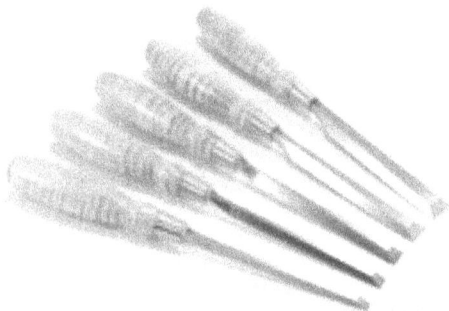

Paring Chisels.

Parquet flooring: A wood floor that consists of small pieces of wood that are arranged into a specific design. Parquet flooring is generally made from pre-manufactured interlocking blocks. See **Plank flooring, Strip flooring**, and **Plug-and-plank flooring.**

Parquetry: Geometrically patterned wood inlay.

Part number: See **Mark** and **Piece mark**.

Part stop: A strip of wood with weatherstripping attached that prevents air and water infiltration. Part stops are commonly found at the head jamb of a double hung unit.

Partially amortized mortgage loan: The payments do not repay the loan over its term. Therefore, a lump sum (balloon) is required to repay the loan.

Partially restrained: A type of connection that displays a moment rotation behavior that cannot be described as either pinned or fixed.

Partially shielded: An exterior light fixture that is shielded so that the lower edge of the shield is at or below the centerline of the lamp to minimize light emitted above the horizontal plane.

Participation mortage: A type of mortgage that allows the lender to share in part of the income or resale proceeds. The lender participates in the income of the mortgaged property beyond a fixed return, or receives a yield on the loan in addition to the straight interest rate.

Particle board: A plywood substitute made of course sawdust that is mixed with resin and pressed into sheets. Used for closet shelving, floor underlayment, stair treads, etc.

Nonstructural panels that require smooth faces are configured with small particles on the outside and coarser particles on the interior (core). Panels designed for a structural application may have flakes aligned in orthogonal directions in various layers to mimic the structure of plywood. Three-and five-layer constructions are most common.

Types of particle board include:

- **Extruded:** Particle board that is made by ramming binder-coated particles into a heated die. The binder then cures to form a rigid mass as the material is moved through the die.
- **Mat-formed:** Particles that have been previously coated with a binding agent are formed into a mat having substantially the same length and width as the finished panel. This mat is then pressed in a heated flat-platen press to cure the binding agent.
- **Mende-process:** Particle board that is made in a continuous ribbon from wood particles with thermosetting resins used to bond the particles; thickness ranges from 1/32 to 1/4 inch.
- **Multilayer:** Particle board that is made of wood particles that are classified or sized differently. The particles are placed into the preprocessed panel shape to produce a panel with specific properties.

Parting stop, strip or bead: A small wood piece used in the side and head jambs of double hung windows to separate the upper sash from the lower sash.

Partition wall: A wall that subdivides spaces within any story of a building or room.

Party wall: A fire division on an interior lot line common to two adjoining buildings.

Buildings joined with a party wall.

Pascal: A SI unit of pressure or stress equal to one Newton per square meter.

Pass: In roofing, the term that is used to describe the application of one layer of Spray Polyurethane Foam (SPF). The speed of a pass will determine foam thickness.

Pass line: The distinct line formed between two passes of Spray Polyurethane Foam (SPF). This line is the top skin of the bottom pass of the SPF.

Passage hardware: Interior door hardware that does not lock. See **Privacy hardware**.

Patera: A round or oval raised surface design.

Paterae.

Patina: A surface coloring, usually brown or green, produced by the oxidation of bronze, copper, or other metal. Patinas occur naturally and are also produced artificially for decorative effect.

Patina on a copper dome.

Patio: An open court.

Pattern layer: With regard to vinyl flooring, one of three layers of material typically found in vinyl floor covering. The pattern layer is the inner foam layer that is sandwiched between the **backing** and the **wear** layer. See **Backing layer** and **Wear layer**.

Pattern rafter: A rafter used as a guide for making all other common rafters.

Patterned glass: One or both surfaces of glass with a rolled design; used for privacy and light diffusion.

Patternmaker's rasp: While similar to regular rasps that have individual teeth for fast cutting, these tools have their teeth staggered and they are smaller to give a finer finished surface in wood. They also have cut edges for working in corners.

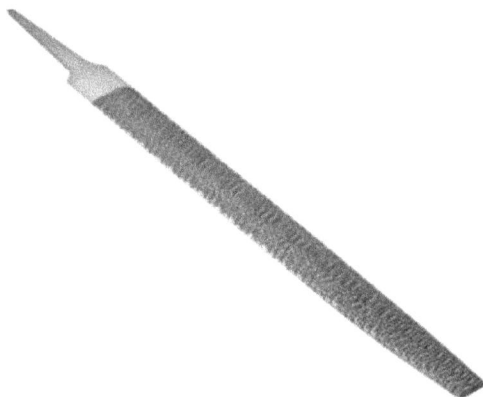

Paternmakers' rasp.

Parking footprint: The area of the site that is occupied by the parking structure.

Parking tier: A general level of parking.

Partially shielded: An exterior light fixture that is shielded so that the lower edge of the shield is at or below the centerline of the lamp to minimize light emitted above the horizontal plane.

Paver/paving: Materials, commonly masonry, laid down to make a firm, even surface. Any tile or masonry unit that can be used as a surface upon which one may drive or walk.

Payment bond: Security from a surety company to the owner, on behalf of a prime or subcontractor. The payment bond guarantees payment to all persons providing labor, materials, equipment, or services in accordance with the contract.

Payment schedule: A pre-agreed upon schedule of payments to a contractor usually based upon the amount of work completed. Such a schedule may include a deposit prior to the start of work. There may also be a temporary "retainer" (5–10% of the total cost of the job) at the end of the contract for correcting any small items that have not been completed or repaired.

PB manifold system: A system that distributes hot and cold water to individual plumbing fixtures from a single panel of valves.

Pea gravel: Aggregate or stones roughly the size of peas.

Pea shingle: Shingle consisting of rounded stones that pass through a 10 mm grid.

Pea shingle.

Peak: Highest part of the roof where the roof planes meet, also called the ridge. See **Apex**.

Peak hour demand: The time when the largest demand for services, e.g., electricity and hot water, is needed.

Peak load period: The period of the day in which the service system has the greatest demand.

Pedestal: A metal box installed at various locations along utility easements that contain electrical, telephone, or cable television switches and connections.

Pedestal sink: A free standing washbasin that is supported vertically by a column from the floor. The pedestal is designed to partially conceal pipe workings and usually comes with a way to attach the back to the wall. Also referred to as **pedestal lavatory**.

Pedestal sink.

Pediment: An ornamental detail placed over a door, portico, or window, often found in a triangular shape.

Pediment.

Pedogenic: Pertaining to processes that add, transfer, transform, or remove soil constituents.

Penalty clause: A provision in a contract that provides for a reduction in the amount otherwise payable under a contract to a contractor as a penalty for failure to meet deadlines or for failure of the project to meet contract specifications.

Pending and potential change order: An unexecuted change order or items that have been identified to have a cost or time impact on the project and that may result in a change order.

Pendentive: A curved triangle of vaulting formed by the intersection of a dome with its supporting arches.

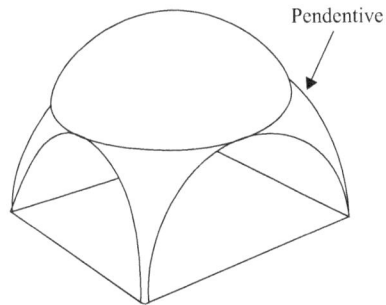

A dome supported on pendentives.

Penetrating finish: A finish that sinks into the substrate, as opposed to settling on the surface.

Penetration: With regard to roofing, any object passing through a roof. Also, the consistency (hardness) of a bituminous material expressed as the distance, in tenths of a millimeter (0.1 mm), that a standard needle penetrates vertically into a sample of material under specified conditions of loading, time, and temperature.

Penny: As applied to nails, it originally indicated the price per hundred. The term now serves as a measure of nail length and is abbreviated by the letter "d." Normally, 16d (16 "penny") nails are used for framing. See **Common nails**.

penny size	length (inches)	length (nearest mm)
2d	1	25
3d	1¼	32
4d	1½	38
6d	2	51
7d	2¼	57
8d	2½	65
9d	2¾	70
10d	3	76
12d	3¼	83
16d	3½	89
20d	4	102
30d	4½	115
40d	5	127
50d	5½	140
60d	6	152

Penny nail lengths.

Pentachlorophenol (penta): Pentachlorophenol (PCP) was one of the most widely used biocides in the US prior to regulatory actions that cancelled and restrict certain non-wood preservative uses of it in 1987. Now its commercial uses include: utility poles, fences, shingles, walkways, building components, piers, docks and porches, and flooring and laminated beams. Additionally, there are also some agricultural uses that are sometimes referred to as **outdoor residential**, i.e., wood protection treatment to buildings/products, and fencerows/hedgerows.

Penthouse: A part of the superstructure above the roof.

Percent of completion method: An accounting method that recognizes revenues and gross profit each period based upon the progress of construction, i.e., the percentage of completion.

Percent elongation: In tensile testing, the increase in the gauge length of a specimen

measured at or after fracture of the specimen within the gauge length. Usually expressed as a percentage of the original gauge length.

Also, the maximum amount that a material can be lengthened or stretched before breaking; expressed as a percentage of the original length of material tested. See **Yield strength** and **Elastic limit**.

Percentage humidity: The weight of water vapor in air divided by weight of vapor contained in saturated air, expressed as a percentage.

Percentage lease: A lease in which the rent amount is based on a percentage of gross sales (monthly or annually) made by the tenant.

Performance bond: Some owners require a performance bond from the winning bidder as a guarantee that the bidder, now contractor, will complete the work in accordance with the contract.

Perforated pipe: Pipe (PVC) that has been perforated for underground drainage collection.

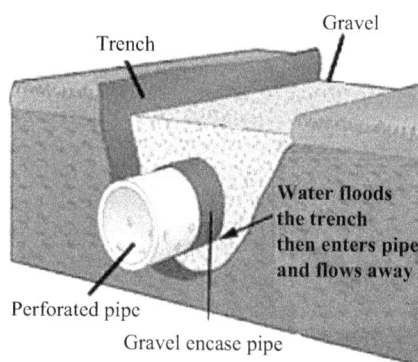

Gravel encase pipe

Perforated pipe (sometimes hose) is also used to discharge water through small, multiple, closely spaced orifices, or nozzles that are placed in a segment of its circumference for irrigation purposes.

Perlite: An aggregate used in lightweight insulating concrete and in preformed perlitic insulation boards, formed by heating and expanding siliceous volcanic glass.

Permeability: The capacity of a porous material to conduct or transmit fluids. Also, the amount of a fluid moving through a barrier in a unit time, unit area, and unit pressure gradient not normalized for, but directly related to, thickness.

Perculation: The general movement of liquid through soil.

Percolation test (perc. test): A test to determine how well water passes through soil. Commonly used to determine the feasibility of installing a leech field type sewer system and for sizing such a leech field.

Performance based design: A concept of designing a project for optimum performance within a given life cycle. By definition the building program is to include the careful analysis of all physical, economical, environmental, aesthetic, and sociological factors, including natural and man-made hazard mitigation, all to agreed upon levels.

Performance bond: An amount of money (usually 10% of the total price of a job) that a contractor must put on deposit with a governmental agency as an insurance policy that guarantees the contractors' proper and timely completion of a project or job.

Pergola: A garden structure with an open wooden-framed roof, often latticed, supported by regularly spaced posts or columns. The structure is often covered by climbing plants such as vines or roses to shade a walk or passageway.

Perimeter: The measurement around the outside of an object. The perimeter of a rectangle or triangle is the sum of the lengths of its sides. On a circle the perimeter is also referred to as the circumference, and is calculated by multiplying pi (3.14159) and the diameter.

Perimeter backing: A type of backing used on some types of specialty vinyl floor covering and is installed with adhesive placed only around the perimeter of the room. See **Backing layer**.

Perimeter drain: Usually 3–4" perforated plastic pipe that goes around the perimeter (either inside or outside) of a foundation wall (before backfill) and collects and diverts ground water away from the foundation. In urban areas it is generally "daylighted" through a curb and onto the street. In more rural areas the drain may be directed into a sump pit from which the water seeps into the ground. Sometimes a sump pump may be placed in the pit to discharge any accumulation of water.

Period: The elapsed time in seconds of a single cycle of oscillation. The inverse of **frequency**.

Period of significance: The span of time that it takes for a property to attained the significance for which it meets the National Register criteria.

Perlite: A natural volcanic glass having distinctive concentric cracks and a relatively high water content. Perlite in a fluffy heat expanded form is used in lightweight insulating concrete, fire-resistant rigid insulation board (R = 2.78 per inch) and potting soil.

Perm: A unit of water vapor transmission, defined as one grain of water vapor per square foot per hour per inch of mercury (Hg) pressure difference (1 inch of mercury = 0.491 psi).

Permeability: A measure of the ease with which water penetrates a material.

Permit: A governmental authorization to perform a building process as in:

Some of the more common permits that contractors must obtain:

- **Building permit:** Authorization to build or modify a structure.
- **Demolition permit:** Authorization to tear down and remove an existing structure.
- **Electrical permit:** A separate permit required for most electrical work.
- **Grading permit:** Authorization to change the contour of the land.
- **Plumbing permit:** A separate permit required for new plumbing and larger modifications of existing plumbing systems.
- **Septic permit:** A health department authorization to build or modify a septic system.

• **Zoning\Use permit:** Authorization to use a property for a specific use e.g., a garage, a single family residence etc.

And, there are other permits that localities may require. It is not unusual for each permit to be independently issued. Therefore, as a contractor you may have to spend considerable time going from one office to another to obtain all the permits necessary to begin a project.

Permissible exposure limit (PEL): The maximum worker exposure to lead under the OSHA lead in construction standard. No employee may be exposed to lead at airborne concentrations greater than 50 µg/m³ averaged over an eight-hour period.

Personal fall arrest system: A system used to stop an employee in a fall from a working level. It consists of an anchorage, connectors, a body harness, and may include a lanyard, deceleration device, lifeline, or suitable combinations of these. Using a body belt only for fall arrest is prohibited.

Personal fall arrest system.

Perviousness: The permeability of an area. A concrete driveway would be considered impervious while a lawn would be deemed pervious.

Also, the percentage of area covered by a paving system that is open and allows moisture to soak into the earth below the paving system. See **Percolation test/Perc. test.**

Petcock: A small faucet for draining liquids or relieving air pressure.

Petrology: The study of rocks, their origin and from what they are made.

Phase: A stage in a periodic motion measured with respect to a reference and expressed in angular measure.

In AC power systems, load current is drawn from a voltage source that typically takes the form of a sine wave. Ideally, the current drawn by the loads in the system is also a sine wave. With a simple, resistive load such as a light bulb, the current sine wave is always aligned with the voltage sine wave. This is called single-phase. A single-phase power system normally uses three wires, called hot, neutral, and ground, and the voltage is typically 120/240. Most home and office outlets operate in this manner. With some loads, such as motors, and in high voltage systems, the current sine wave is purposely delayed and lags behind the voltage sine wave. The amount of this lag is expressed in degrees and is called a phase difference. A common example is three-phase power, where the system has three "hot" wires, each 120° out of phase with each other.

Phased application: With regard to roofing, the installation of separate roof system or waterproofing system component(s) during two or more separate time intervals. Application of surfacings at different time intervals are typically not considered phased application. See **Surfacing**.

Phenol: A poisonous and carcinogenic chemical often found in municipal water supplies. Also known as carbolic acid, phenol is a benzene derivative.

Phillips head: A trademark used for a screw with a head having two intersecting perpendicular slots (+) and for a screwdriver with a tip shaped to fit into these slots.

Photoelectric cell: A device that generates an electric current or voltage dependent on the degree of illumination. Also called a **photocell**.

Photosynthesis: A process that plants use to synthesize nutrients from water and minerals using sunlight.

Photovoltaic or solar energy: Electricity from photovoltaic cells that convert the energy in sunlight into electricity.

Physical disability/Physical handicap: A person is considered physically disabled if one or more of the following conditions exist: (1) impairment requiring use of a wheelchair; or (2) impairment causing difficulty or insecurity in walking or climbing stairs or requiring the use of braces, crutches or other artificial supports; or impairment caused by amputation, arthritis, spastic condition or pulmonary, cardiac or other ills rendering the individual semi-ambulatory; or (3) total or partial impairment of hearing or sight causing insecurity or likelihood of exposure to danger in public places; or (4) impairment due to conditions of aging and incoordination.

Note: Do not assume that this definition is complete and accurate in all places. If disabilities are at issue check your local laws before proceeding with you design.

Physical Vapor Deposition (PVD): A process that protects a faucet's finish, making it tarnish-, scratch-, and corrosion-resistant. PVD technology integrates the faucets finish (such as polished brass or satin) with the cast-brass base, thereby making the PVD finish very durable.

Physiographic: The character and distribution of landforms.

Pi (Greek alphabet π): A mathematical constant used in many calculations involving circles. PI is equal to 3.141596.

Pickling: Immersing something, e.g., a pipe, into an acid bath for removal of scale, oil, dirt, etc.

Picogram: One trillionth of a gram.

Picograms per lumen-hour: A measure of the amount of mercury in a lamp per unit of light delivered over its useful life. If you are responsible for purchasing mercury-containing lamps seek to buy lamps that contain no more than 70 picograms of mercury per lumen-hour.

Picture framing: With regard to roofing, a square or rectangular pattern of buckles or ridges in a roof covering, generally coinciding with insulation or deck joints; generally, a function of movement of the substrate.

Picture molding: A molding used to support hooks for picture hanging. Applied around a room's circumference near the ceiling line.

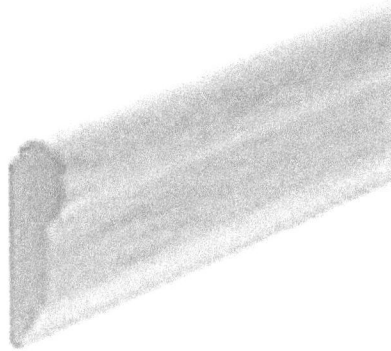

Picture molding

Picture rail hook (fits over picture molding).

Picture window: A large, non-operating window. It is usually longer than it is wide to provide a panoramic view.

Piece mark: An identification number that distinguishes one piece of steel or assembly from another. Frequently, piece marks follow a code that tell the field worker the exact area, level and location of the piece of steel or assembly. See **Mark and Part number**.

Pier: A column, usually rectangular in horizontal cross section, used to support other structural members. See **Caisson, Mole**, and **Jetty**.

Also, a structure leading out from the shore into a body of water; a platform supported on pillars or girders. Large piers are used as entertainment areas, typically incorporating arcades and places to eat.

Also, a breakwater or mole.

Pier and grade beam structure: A structure utilizing piers and grade beams. It is built upon footings that have been constructed at a lower level to rest on bedrock or stable soil. This type of construction is commonly used on steep hillsides.

Piezo igniter: A special feature on gas water heater. During installation, or if the pilot flame is extinguished for any reason, the Piezo igniter allows the pilot flame to be re-lit without matches with just the push of a button located outside the water heater.

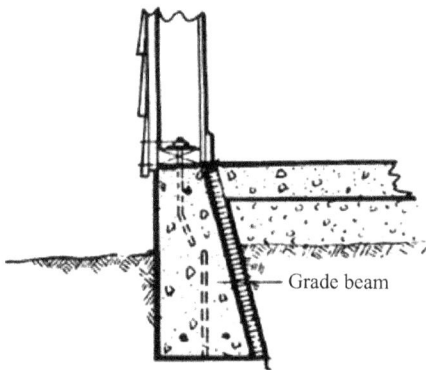

Grade beam

Pigtails: The electric cord that the electrician provides and installs on an appliance such as a garbage disposal, dishwasher, or range hood.

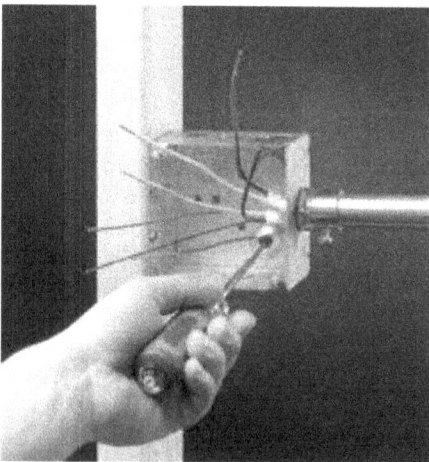

Pigtails.

Pigment: Insoluble, finely ground materials that gives paint its properties of color and hide. Titanium dioxide is the most important pigment used to provide hiding in paint. Other pigments include anatase titanium, barium metaborate, barium sulphate, burnt sienna, burnt umber, carbon black, China clay, chromium oxide, iron oxide, lead carbonate, strontium chromate, Tuscan red, zinc oxide, zinc phosphate, and zinc sulfide.

Pigment provides color, opacity to UV light which ensures a longer life for the coating and they also increase the porosity and increase the hardness of the paint.

Pigments are comprised of three parts: Primary, Secondary, and Colorant:

- **Primary** is commonly composed of titanium dioxide. The amount greatly impacts the hiding capabilities of the paint and UV protection.
- **Secondary** are pigments such as calcium carbonate, mica, silicas, talc, and have little to no impact on hiding. They are used as fillers to help control viscosity, leveling, and sheen.
- **Colorants** are tinting liquids dispersed to bring the paint to the final color.

Pilaster: An upright architectural member that is rectangular in plan and is structurally a pier but is architecturally treated as a column. Pilasters provide support for vertical roof loads or lateral loads on the wall. Pilasters usually project a third of their width or less from the wall.

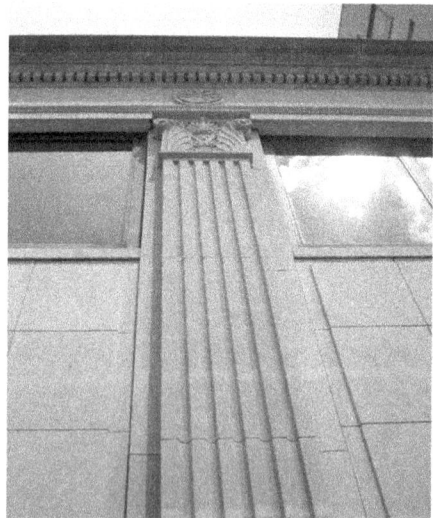

A decorative pilaster.

Pile: A long slender column, usually of timber, steel, or reinforced concrete, that is driven into the ground by a **pile** driver to carry a vertical load. Piles may also be cast-in-place in which a **piling rig** would be used to drill the hole.

Normally a reinforced concrete structure called a **pile cap** is constructed above the pile(s) to transfer loads into the pile(s).

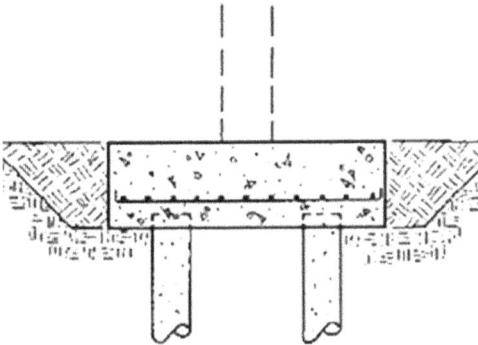

Pile cap.

With regard to carpeting, fibers that have been attached to a carpet backing. See **Loop pile** and **Cut pile**.

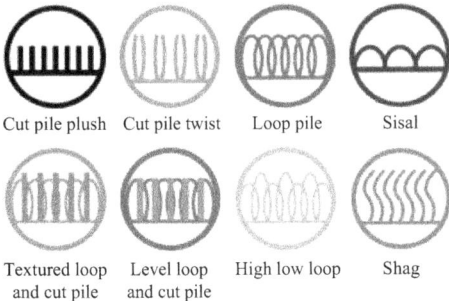

| Cut pile plush | Cut pile twist | Loop pile | Sisal |

| Textured loop and cut pile | Level loop and cut pile | High low loop | Shag |

Various carpet piles.

Pillar: A column used for supporting parts of a building or home.

Pilot hole: A small-diameter, pre-drilled hole that guides a nail or screw.

Pilot light: A small, continuous flame (in a hot water heater, boiler, or furnace) that ignites gas or oil burners when needed. Pilot lights have been replaced by electric starters on new appliances.

Pin knot: A knot smaller that one-half inch in diameter.

Pin support: In two dimensions, a pin support restrains two translation degrees of freedom but does not restrain rotation. Since the rotation degree of freedom is unrestrained at a pin connection, it transfers no moment. When considering reaction forces, a pin support is usually considered to have two force components: one each about the x and y axes respectively. Also referred to as a **pin connection**.

Pincer: This tool is ssentially a large nail puller and cutter. It is designed for removing brads and small nails.

Pinch dogs: Perfect for edge gluing, pinch dogs pull joints tightly without the need for clamping. The dog is hammered into the butt ends of the work.

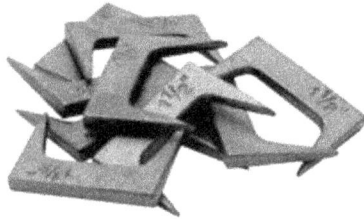

Pinch Dogs.

Pin: External (male) threaded end of pipe.

Pinhole: A tiny hole in a coating, film, foil, membrane, or laminate.

Pinnacle: Projecting or ornamental cap on the high point of a roof.

Pintle: A pin or bolt, especially one on which something turns, as the gudgeon of a hinge.

Pipe: A hollow cylinder of metal used for the conveyance of water or gas or used as a structural column which comes in sizes of standard, extra strong and double-extra strong.

Terms beginning with pipe:

• **Pipe boot:** A prefabricated flashing piece used to flash around circular pipe penetrations. See **Collar** and **Pitch pocket**.

Pipe boot (Cut off top to fit pipe.)

• **Pipe bridge:** A structural system where two joists are used to carry loads such as piping or ducts. The two joists have to have diagonal bridging and their top and bottom chords have to be laced together with structural members to provide stability for the whole structure.

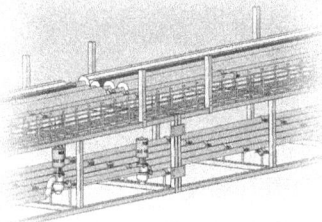

Pipe bridge.

• **Pipe clamp fixtures:** Long reaching fixtures that are attached to a pipe. When positioned they are tightened by turning a handle. They come in various configurations.

Pipe clamp.

• **Pipe dope:** Compound placed on threads that helps to seal threaded pipe joints.
• **Pipe flashing:** A flashing that is placed around any pipe that penetrates the roof. A gasket-like sleeve fits around the pipe and its base slides under the upper shingle and over the lower shingle.
• **Pipe insulation:** A preformed rigid insulation for piping.
• **Pipe plugs:** Plugs used to close off a pipe temporarily for the purpose of pressure testing.

Pit: That portion of the hoistway extending from the sill level of the lowest landing to the floor at the bottom of the hoistway.

Pitch: The inclined slope of a roof or the ratio of the total rise to the total width of a house, i.e., a 6-foot rise and 24-foot width is a one-fourth pitch roof. Roof slope is expressed in the inches of rise, per foot of horizontal run.

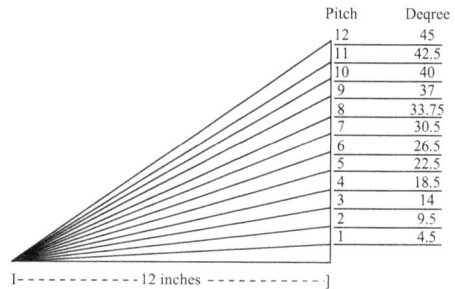

Pitch	Degree
12	45
11	42.5
10	40
9	37
8	33.75
7	30.5
6	26.5
5	22.5
4	18.5
3	14
2	9.5
1	4.5

I- - - - - - - - - - - 12 inches - - - - - - - - - - -]

Various roof pitches.

Also, an accumulation of resin in the wood.

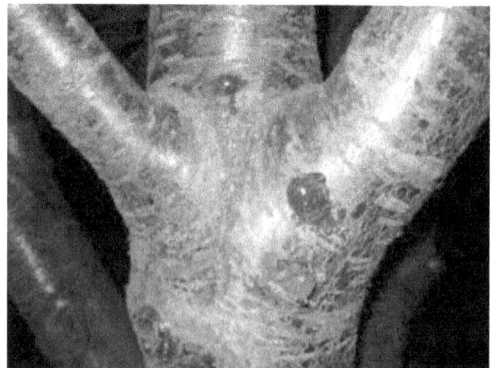

Cherry tree pitch.

Also, the nominal distance between two adjacent th readroots or crests.

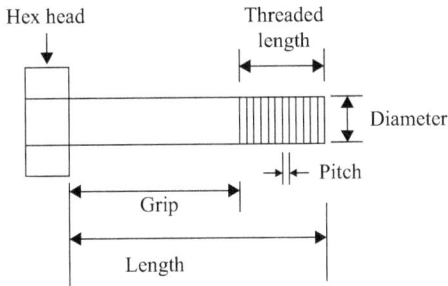

A bolt, showing pitch, thread length, etc.

Pitch pocket: An opening between growth rings that usually contains or has contained resin or bark or both.

Pitch pockets.

Also, a flanged, open bottomed enclosure made of sheet metal or other material, placed around a penetration through the roof, filled with grout and bituminous or polymeric sealants to seal the area around the penetration. Also referred to as **pitch pan**. See **Collar** and **Pipe boot**.

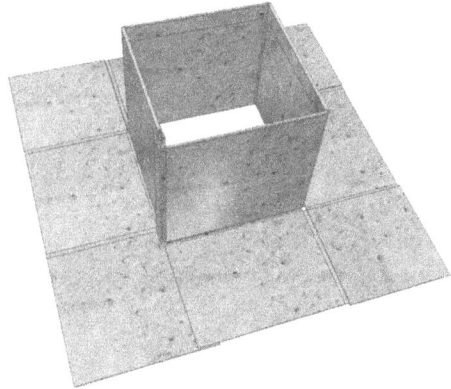

Pitch pocket.

Pith: The small cylinder of primary tissue of a tree stem around which the annual rings form; the center of a tree.

Pittsburgh lock seam: A method of interlocking two separate pieces of metal.

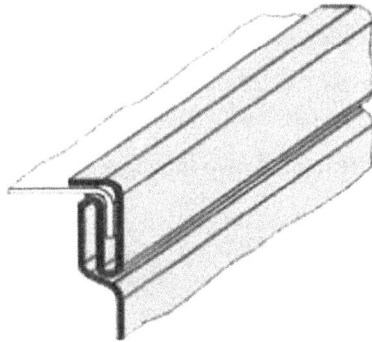

Pittsburgh lock seam.

Pivot window: A unit with a sash that swings open or shut by revolving on pivots at either side of the sash or at top and bottom.

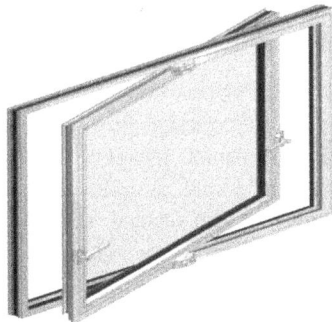

Vertical pivot window.

Placing plan: See **Erection plan** and **Framing plan.**

Plain die stock: A type of hand tool used for turning a die.

Plain sawing: Plain sawing or **slash sawing** is the most common method of sawing logs into lumber. This is really a matter of slicing the log into the required sizes.

Plain sliced: Veneer sliced parallel to the pith of the log and approximately tangent to the growth rings to achieve flat cut veneer. Plain sliced veneer can be cut using either a horizontal or vertical slicing machine or by the half-round method using a rotary lathe. Also known as **flat cut.**

Plan: A set of pictures of whatever project is being worked on. A plan provides instruction for the construction of the project depicted.

Plan north: The north arrow symbol on a contract drawing.

Plan view: A drawing of a project with the view from overhead, looking down.

Plane frame: A two-dimensional structural framework.

Plank flooring: Wood floor containing wood strips that are over 3-1/4" wide. See **Strip flooring, Parquet flooring,** and **Plug-and-plank flooring.**

Plank matched: A face containing specially selected and assembled dissimilar (in color, grain, and width) veneer strips of the same species, and sometimes grooved at the joints between strips, to simulate lumber planking. Plank matched wood provides a casual and rustic effect.

Plaster: Made from concrete, water and aggregate, as are concrete, grout and mortar. It is mixed using sand as its aggregate, using no course aggregate such as gravel. It produces a hard, concrete-like surface. Plaster is often used as a finish coat on the exterior of a block wall.

Plasterboard: Two sheets of cardboard with a gypsum plaster filling. Usually nailed or screwed to studs, joists, or rafters as a carrier for a plaster skim finish. Plasterboards with chamfered edges can be jointed so that they act as a finish without being skimmed with plaster.

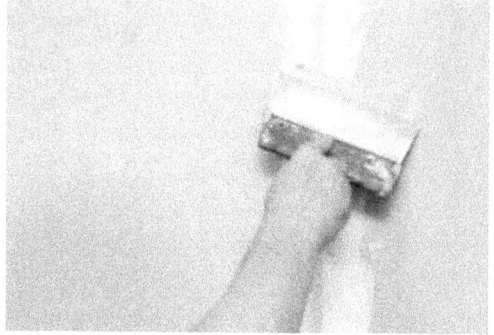

Plasterboard.

Plastic: A synthetic material that is made from a wide range of organic polymers such as polyethylene, PVC, nylon, etc. Plastic can be molded into shape while soft and then set into a rigid or slightly elastic form. The material is commonly stiffer than rubber. See **Ductile** and **Inelastic.**

Other terms beginning with plastic:

• **Plastic cement:** A roofing industry generic term used to describe Type I asphalt roof cement that is a trowelable mixture of solvent-based bitumen, mineral stabilizers, other fibers and/or fillers. Generally, intended for use on relatively low slopes, not vertical surfaces. See **Asphalt roof cement** and **Flashing cement.**

• **Plastic design:** Plastic analysis procedures that are based on the considerations of equilibrium, yield mechanism, and plastic strength conditions. A goal of plastic analysis and design is to utilize the reserve strength of members and connections beyond the elastic limit due to the redistribution of internal forces. Therefore, the analysis focuses on the internal forces at the limit level when the yield mechanism forms. Plastic analysis plays an important role in seismic design where structures are expected to form mechanisms during strong ground motions.

- **Plastic film:** A flexible sheet made by the extrusion of thermoplastic resins.
- **Plastic laminate:** A thin sheet of synthetic material used as a surfacing. For example a **plastic laminate countertop** uses plastic laminate veneer that is glued over supporting material such as plywood or medium density fiberboard (MDF). In practice, many people refer to plastic laminate by the brand name **Formica**. See **Solid surface countertop, Solid plastic countertop, Wood block countertop, Cultured countertop, Cultured marble countertop, Stone countertop**, and **Tile countertop**.
- **Plastic-tipped hammer:** This steel-core hammer has two replaceable plastic faces, one hard, and one soft. The hammer can strike various materials without marring, including wood, metals, plastic, and stone.

Plasticiser: A material, frequently solvent-like, incorporated in a plastic or a rubber to increase its ease of workability, flexibility, or extensibility. Also spelled plasticizer.

Plasticity: As used here, plasticity refers to the softness of concrete, mortar, or soil or to how easily it is to mold and shape. Concrete that has a high plasticity flows easily and is easy to work with. When concrete loses its plasticity it becomes hard and can no longer be worked.

Plat: A map or chart of an area showing boundaries of lots and other parcels of property.

Plate: Normally a 2 × 4 or 2 × 6 that lays horizontally within a framed structure, e.g., a **sill plate** or **sole plate** is the bottom horizontal member of a wall or building to which vertical members are attached, and a **top plate** is the horizontal member of a frame wall that supports ceiling joists, rafters, or other members.

Sill plate.

Also, a thin, flat piece of metal of uniform thickness that is usually over 8 inches to 48 inches in width.

Plate girder: A built-up structural beam. Typically plate girders are large beams that are built-up by welding plates together.

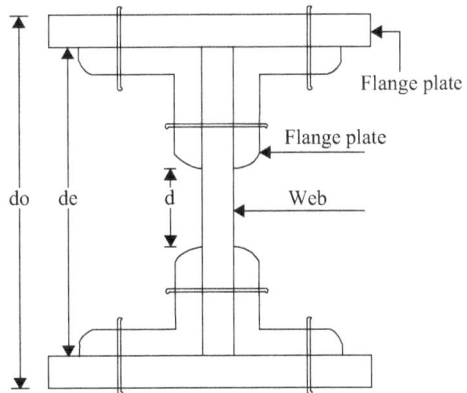

Plate glass: Flat glass produced by grinding and polishing to create parallel plane surfaces affording excellent vision. Although the term is still used commonly, most window glass is now produced using the float process. See **Float glass**.

A mirror that is made from high quality glass is a **plate glass mirror**. Plate glass mirrors can be up to 1-1/4" thick. They can be glued directly to a wall, held in a frame that is hung on a wall, or held in place with plastic clips.

Plate joiner: A tool that is designed to quickly and accurately cut matching slots in pieces that are to be joined with a small carbide-tipped blade.

Plate, tectonic: A large unit of the earth's lithosphere that moves relative to other plates and the interior of the earth.

The theory and study of plate formation, movement, interaction, and destruction, and also the theory that explains seismicity, volcanism, mountain building and paleomagnetic evidence in terms of plate motions is referred to as **plate tectonics**.

Platform: A landing in a flight of stairs, usually between floors.

Also, the entire floor assembly of an elevator on which passengers stand or the load is carried.

Also, a work surface elevated above lower levels. Platforms can be constructed using individual wood planks, fabricated planks, fabricated decks, and fabricated platforms.

Platform frame: Light timber construction in which the exterior walls and bearing walls consist of studs that are interrupted at floors by the entire thickness of the floor construction.

Pleasing match: A door face containing components that provide a pleasing overall appearance. The grain of the various components need not be matched at the joints. Sharp color contrasts at the joints of the components are not permitted. Pleasing match could logically apply to all aspects of construction.

Pleistocene: The Pleistocene is the time period between about 10,000 years before present and about 1,650,000 years before present. As a descriptive term applied to rocks or faults, it marks the period of rock formation or the time of most recent fault slip, respectively. Faults of Pleistocene age may be considered active though their activity rates are commonly lower than younger faults.

Plenum: The main hot-air supply duct leading from a furnace. In suspension ceiling construction, the space between the suspended ceiling and the main structure above.

Pliability: The material property of being flexible or moldable.

Pliers: This hand tool is used to hold objects firmly. Generally, pliers consist of a pair of metalfirst-class levers joined at a fulcrum that is positioned to create short jaws on one side of the fulcrum and longer handles on the other side, thereby creating a mechanical advantage for the users grip. Pliers come in many sizes and shapes to meet specific needs.

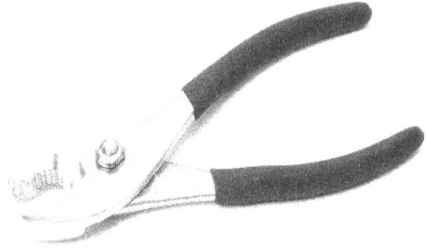

Flat-Nose Pliers.

Plinth block: A decorative wood block placed between the vertical casing and the top casing of a unit to provide an elegant interior casing profile.

Bullseye plinth block.

Plot plan: An overhead view plan that shows the location of the home on the lot. Includes all easements, property lines, set backs, and legal descriptions of the home. Most building departments require a plot plan whenever you apply for a building permit. See **Plan view**.

Plough: To cut a lengthwise groove in a board or plank. An exterior handrail normally has a ploughed groove for hand gripping purposes. Sometimes spelled **plow**.

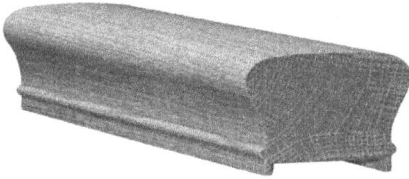

Ploughed (plowed) handrail.

Plumber's putty: Pliable putty used to seal joints between drain pieces and fixture surfaces.

Plug: A rod, plate, or angle welded between a two web member or between a top or bottom chord panel to tie them together. Plugs are usually located at the middle of the member. See **Tie** or **Filler**.

Also, a device to which the conductors of a cord are attached, which is used to connect to the conductors permanently attached to a receptacle.

Plug-and-plank flooring: Wood flooring that is installed by installing fasteners through the tops of boards. Holes for fasteners are predrilled along with a countersink hole. After the fasteners are tightened into place, the holes are filled with plugs that are usually sanded level with the finish floor. See **Plank flooring, Strip flooring** and **Parquet flooring**.

Plug weld: A weld in a slot in a piece of steel that overlaps another piece. A principle use for a plug weld is to transmit shear in a lap joint. See **Slot weld** and **Puddle weld**.

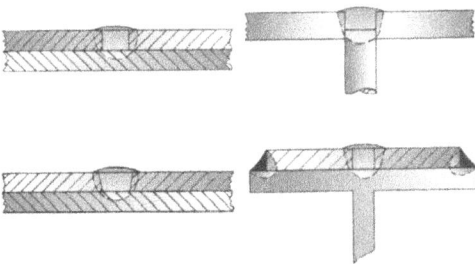

Various plug welds.

Plum: A large stone or piece of solid concrete used as a filler in mass concrete.

Plumb: Exactly vertical and perpendicular. A lead weight attached to a string to determine plumb is a **plumb bob**.

Plumb cut: True vertical cut when a member is installed. The actual angle required for a plumb cut is determined by the slope of the member. A plumb cut is at a right angle to the level cut.

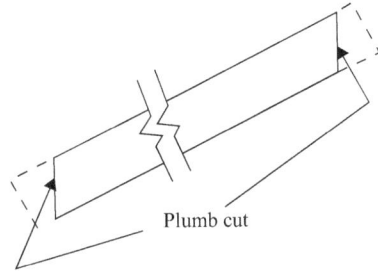

Plumb cut

Plumbing: The system of pipes, tanks, fittings, and other apparatus required for the water supply, heating, and sanitation in a building. Plumbing is usually distinguished from water supply and feces sewage systems, in that a plumbing system serves one building, while water and sewage systems serve a group of buildings.

Also, the work of installing and maintaining such a system.

Other terms beginning with plumbing:

- **Plumbing boots:** Metal saddles used to strengthen a bearing wall/vertical stud(s) where a plumbing drain line has been cut through and installed.
- **Plumbing fixtures and fittings:** Receptacles, devices, or appliances that are either permanently or temporarily connected to the building's water distribution system and receive liquid or liquid-borne wastes and discharge wastewater, liquid-borne waste materials, or sewage either directly or indirectly to the drainage system of the premises. This includes water closets, urinals, lavatories, sinks, showers, and drinking fountains.
- **Plumbing ground:** The plumbing drain and waste lines that are installed beneath a basement floor.

- **Plumbing jacks:** Sleeves that fit around drain and waste vent pipes at, and are nailed to, the roof sheeting.
- **Plumbing rough:** Work performed by the plumbing contractor after the Rough Heat is installed. This work includes installing all drain and waste lines, copper water lines, bath tubs, shower pans, and gas piping to furnaces and fireplaces.
- **Plumbing stack:** A plumbing vent pipe that penetrates the roof.
- **Plumbing tree:** Prefabricated set of drain waste, vent, and supply lines.
- **Plumbing trim:** Work performed by the plumbing contractor to get the home ready for a final plumbing inspection. Includes installing all toilets (water closets), hot water heaters, sinks, connecting all gas pipes to appliances, disposal, dishwasher, and all plumbing items.
- **Plumbing waste line:** Pipe used to collect and drain sewage waste.

Plunger: A rubber suction cup ~6" in diameter attached to a wooden dowel handle used to free drain clogs. Commonly known as a **plumber's helper**.

Plutonic: Igneous rocks that have formed from magma at a great depth in the earth's crust.

Ply: A term to denote the number of layers of roofing felt, veneer in plywood, or layers in built-up materials, in any finished piece of such material.

Plycap: Covers and beautifies plywood's rough sandwich edge in installation where it is exposed to view.

Plywood: A panel (normally 4'× 8') of wood made of three or more layers of veneer, compressed and joined with glue, and usually laid with the grain of adjoining plies at right angles to give the sheet strength.

Terms associated with plywood:

- **Plywood, cold-pressed:** An interior-type plywood manufactured in a press without external applications of heat.
- **Plywood, exterior:** A general term for plywood bonded with a type of adhesive that by systematic tests and service records has proved highly resistant to weather, microorganisms, steam, dry heat, and cold, hot, and boiling water.
- **Plywood, interior:** A general term for plywood manufactured for indoor use or construction subjected to only temporary moisture. The adhesive used may be interior, intermediate, or exterior.
- **Plywood, marine:** Plywood panels manufactured with the same glueline durability requirements as other exterior-type panels but with more restrictive veneer quality requirements.
- **Plywood, molded:** Plywood that is glued to the desired shape either between curved forms or more commonly by fluid pressure applied with flexible bags or blankets (bag molding) or other means.
- **Plywood, postformed:** The product formed when flat plywood is reshaped into a curved configuration by steaming or plasticizing agents.
- **Plywood underlayment:** An underlayment made from plywood that has been specially manufactured for use as a vinyl floor underlayment. See **Underlayment, Particleboard underlayment, Lauan plywood underlayment, Cement board underlayment, Gypsum-based underlayment**, and **Untempered hardboard underlayment**.

Pneumatic: To move or convey an object by air or gas under pressure.

Many tools (besides those listed) are pneumatically powered:

- **Pneumatic drill:** A tool that combines a hammer directly with a chisel. Commonly known as a jackhammer.
- **Pneumatic nailer:** A tool that uses compressed air to drive a nail, usually into wood, but can drive special nails into hard surfaces like concrete. It can drive nails up to 3-1/2 inches long.
- **Pneumatic stapler:** This stapler can drive crown-style staples up to 1/2-inch wide and two inches long. Smaller models are available for installing carpeting, roofing felt, floor underlayment and insulation.

Pocket door: A sliding door which rolls on a track and opens into a cavity in the wall.

Pocket window: A window unit designed for replacement applications that is installed into the existing window frame after removal of the sash, balance hardware and parting stops. Also called an insert window, these units allow existing interior and exterior trim to be maintained.

Point load: A point where a bearing/structural weight is concentrated and transferred to the foundation.

Point of origin: The point from which the damage began. For example, in a fire the point of origin is where the fire started.

Pointing: The application of mortar or high strength grout to a joint or hole in masonry or concrete.

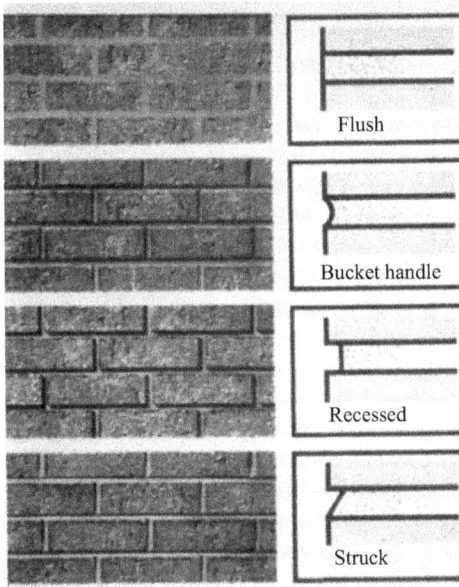

Various types of pointing.

When re-pointing the process includes raking out all disintergrated mortar and then applying new mortar.

Poise: A centimeter-gram-second unit of dynamic viscosity that is equal to one dyne-second per square centimeter. Absolute viscosity is measured using the poise as the basic measurement unit; this method utilizes a partial vacuum to induce flow in the viscometer.

Poisson distribution: A Poisson distribution is a probability distribution that characterizes discrete events occurring independently of one another in time.

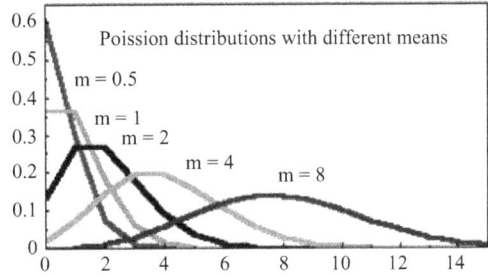

Poisson distribution.

Poisson's ratio: Defined as the ratio of the unit lateral strain to the unit longitudinal strain. It is constant for a material within the elastic range. For structural steel, the value is usually taken as 0.3. It gradually increases beyond the proportional limit, approaching 0.5.

Polar moment of inertia (J): The sum of any two moments of inertia about axes at right angles to each other. It is taken about an axis that is perpendicular to the plane of the other two axes.

Polarity: The correct flow of electricity that is achieved when the hot and neutral wires of the power supply circuits are connected to the corresponding hot and neutral wires of an appliance or outlet.

Poly under top tier (PUTT): A way to give protection to unitized lumber; a piece of plastic is inserted between the next-to-top and top tiers.

Polybutylene (PB): A form of plastic resin that was used extensively in the manufacture of water supply piping from 1978 until 1995; it was popular because of its low cost and ease of installation.

While scientific evidence is scarce, it is believed that oxidants in the public water supplies, such as chlorine, react with the polybutylene piping and acetal fittings. This caused them to scale and flake and become brittle. Micro-fractures resulted, and the basic structural integrity of the system was reduced to the point that failure occured. Other factors, such as improper installation may also have affected system performance.

Polycarbonate: A plastic material used for glazing.

Polychromatic: Having various or changing colors.

Polyester fiber: A synthetic fiber usually formed by extrusion. Scrims made of polyester fiber are used for fabric reinforcement.

Polygon: A plane figure with at least three straight sides and angles, and typically five or more.

Polymer: A substance, the molecules of which consist of one or more structural units repeated any number of times; vinyl resins are examples of true polymers.

Polymerization: The interlocking of molecules by chemical reaction to produce very large molecules. The process of making plastics and plastic-based resins.

Polypropylene: A tough, lightweight plastic made by the polymerization of high-purity propylene gas.

Polyurethane: A paint and varnish resin that forms a protective coating on wood. Sold under the names Varathane, Urethane, and Durathane.

Polyvinyl acetate (PVA): Water-based primer commonly used on drywall. When used on wood, PVA can cause raised grains.

Polyvinyl butyral (PVB): Plastic material used as the interlayer in the construction of some types of laminated glass.

Polyvinyl chloride (PVC): A synthetic resin used in the binders of coatings. Tends to discolor under exposure to ultraviolet radiation. Commonly called "vinyl." Also spelled **polyvinylchloride**.

Ponding: The excessive accumulation of water at low-lying areas, as on a roof.

Pooching: A term sometimes used to describe the effect of the area immediately surrounding a tapped hole being raised up as a result of the tension from the stud. Tapped holes are often bored out for the first couple of threads to eliminate this problem.

Here, a stud is a fastener (not a 2 × 4) that is threaded at both ends with an unthreaded shank in between. One end is secured into a tapped hole, the other is used with a nut.

Double thread stud.

Pop rivet: A relatively small headed pin with an expandable head for joining relatively light gauge metal.

Popcorn texture: Finish applied to drywall ceilings that contains large clumps of texturing material similar to popcorn or cottage cheese. Also called acoustic texture or cottage cheese texture.

Poppet: A mushroom-shaped valve with a flat end piece that is lifted in and out of an opening by an axial rod. Also referred to as a **poppet valve**.

Popouts: The breaking away of small portions of a concrete surface due to internal pressure usually leaving shallow, conical depressions. Popouts are typically caused by reinforcing steel corrosion, cement-aggregate reactions, or from internal water freezing.

Pop-up drain: A type of drain assembly for lavatory and bath. When a lavatory lift rod or bath overflow plate lever is lifted, the pop-up drain closes so the lavatory or tub retains water.

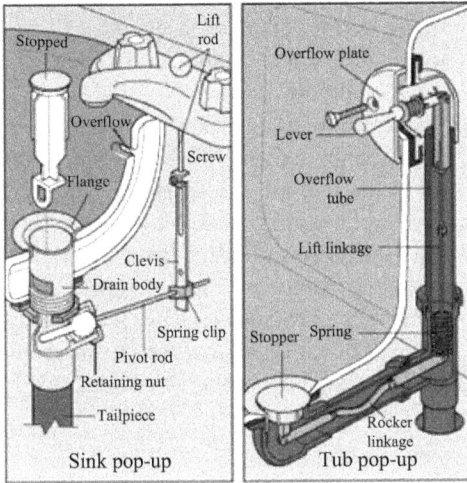

Pop-up sink/tub drain.

Porch: A covered area attached to a house at an entrance.

Porosity: The ratio of the volume of all the pores in a material to the volume of the whole. It is a measure of how much of a material is open space.

Porous concrete pipe: Concrete pipe that is porous allowing for water to enter for underground drainage collection.

Port: An opening in a burner head through which gas or an air-gas mixture is discharged for ignition.
   Also, a harbor.

Portable heaters: Compact wheeled portable heaters are typically used in residential construction. They range in size from 25,000 to around 200,000 BTUs and vary in cost depending on size, features, and quality.

Portal frame: A rigid frame structure that is designed to resist longitudinal loads where diagonal bracing is not permitted. It has rigidity and stability in its plane.

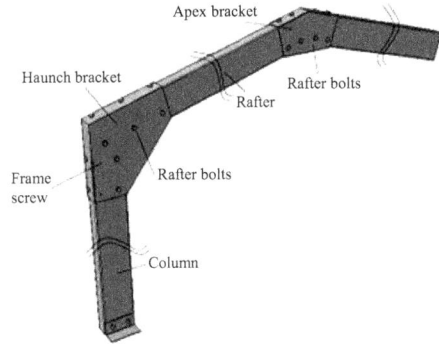

Portal frame.

Porte cochere: A covered, drive-through structure that extends from the side of a home, providing shelter for people getting in and out of vehicles.

Portico.

Portland cement: Cement made by heating clay and crushed limestone into a brick and then grinding to a pulverized powder state.

Portland cement: A hydraulic cement used almost universally for making concrete and other cement based products. So-called because concrete made with it resembles limestone from the Isle of Portland.

Positioning device system: A body belt or body harness system rigged to allow an employee to be supported on an elevated vertical surface, such as a wall, and work with both hands free while leaning.

Positive displacement pump: A positive displacement pump moves fluid by trapping a fixed amount of the fluid and then forcing that trapped volume into the discharge pipe.

Some positive displacement pumps work using an expanding cavity on the suction side and a decreasing cavity on the discharge side. Liquid flows into the pump as the cavity on the suction side expands and the liquid flows out of the discharge as the cavity collapses. The volume is constant given each cycle of operation.

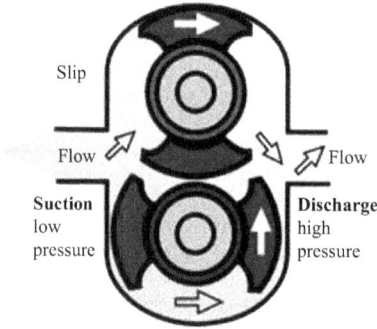

Positive displacement pump

In theory positive displacement pumps produce the same flow at a given speed (RPM) no matter what the discharge pressure. However, because there is always a slight increase in internal leakage as the pressure increases, a truly constant flow rate cannot be achieved.

The following types of pumps (not a complete list) use the principle of positive displacement:

• **Plunger pump:** A reciprocating plunger pushes the fluid through one or two open valves, closed by suction on the way back.

• **Progressive cavity pump:** A type of positive displacement pump, it transfers fluid by means of the progress, through the pump, of a sequence of small, fixed shape, discrete cavities, as its rotor is turned. Also known as a **progressing cavity pump**, **eccentric screw pump**, and **cavity pump**.

• **Piston pump:** A type of positive displacement pump where the high-pressure seal reciprocates with the piston. They are used to move liquids and compress gases.

• **Diaphragm pump:** A pump that uses a combination of the reciprocating action of a rubber, thermoplastic or teflon diaphragm and suitable non-return check valves to pump a fluid. Also called a membrane pump.

• **Screw pump:** A screw pump uses one or several screws to move fluids or solids along the screw(s) axis. In its simplest form (the Archimedes' screw pump), a single screw rotates in a cylindrical cavity, thereby moving the material along the screw's spindle.

• **Hand pump:** A manually operated pump. Hand pumps are widely used in every country in the world for a variety of industrial, marine, irrigation, and leisure activities. There are many different types of hand pump available, mainly operating on a piston, diaphragm or rotary vane principle with a check valve on the entry and exit ports to the chamber operating in opposing directions.

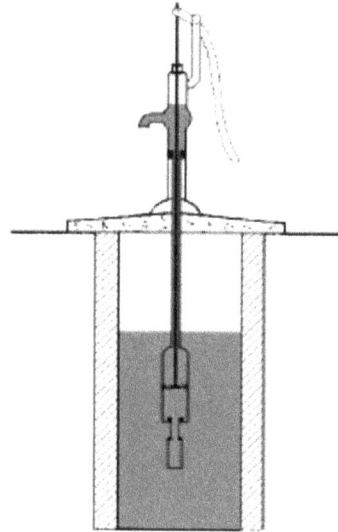

A simple hand pump.

Positive drainage: With regard to roofing, the drainage condition in which consideration has been made during design for all loading deflections of the deck, and additional roof slope has been provided to ensure drainage of the roof area within 48 h of rainfall, during ambient drying conditions.

While not always using the words positive drainage most building codes address the subject.

Post: A vertical framing member usually designed to carry a beam. Often a 4" × 4", a

6" × 6", or a metal pipe with a flat plate on top and bottom.

Post formed countertop: Type of plastic laminate countertop that includes an integral rolled backsplash and a rolled front edge. Post-formed countertops are fabricated in a shop. See **Plastic laminate countertop** and **Flat-laid countertop**.

Post stressed concrete: Concrete strengthened with steel wires that are stressed after the concrete has cured. Also referred to as post-tensioned. See **Prestressed concrete**.

Post-and-beam: A basic building method that uses just a few large posts and beams to support an entire structure.

Postconsumer content: The percentage of material in a product that is recycled from consumer waste. LEED projects may specify postconsumer content such as paper, paperboard, and fibrous wastes that are collected from municipal waste streams (**postconsumer fiber**) and materials recycled from consumer waste (**postconsumer material**).

Pot life: The period of time during which a material with multiple ingredients can be applied or administered after being mixed together. Applies especially to epoxies.

Potable water: Water suitable for drinking that meets or exceeds EPA drinking water standards; it is supplied from wells or municipal water systems.

Potential energy: The energy stored in a raised object (e.g., the weights in a grandfather clock). Potential energy equals mgh, where m is mass, g is the acceleration of gravity, and h is the vertical distance from a reference location. It is called potential energy because the energy can be regained when the object is lowered. This type of potential energy is sometimes called gravitational potential energy in order to distinguish it from elastic potential energy. See **Elastic energy**.

Potentiometer: A three-terminal resistor with a sliding contact. It forms an adjustable voltage divider. If only two terminals are used, one end and the wiper, it acts as a rheostat.

Potentiometers are commonly used to control electrical devices such as volume controls on audio equipment and as an adjustable resistor used to vary the sensing distance of the touch-less faucet. When operated by a mechanism potentiometers can be used as position transducers, for example, in a joystick.

Informally a potentiometer is referred to as a pot.

Pound: A unit of weight equal to 16 ounces (0.45 kilograms). Abbreviated as **LB** or #.

Pour: The common term for the continuous placement of concrete in a form.

Pour stop: The construction of a wood or metal stop for the control placement of concrete usually at the edge of a slab. In composite deck construction this is a cold formed steel angle that becomes part of the floor slab as a lost form at its edge.

Pourable sealer: A type of sealant often supplied in two parts, and used at difficult-to-flash roof penetrations, typically in conjunction with pitch-pockets to form a seal.

Poured-in insulation: Loose insulation that comes in bags and is poured into place. Most commonly used to fill cavities in masonry walls.

Powder actuated: A fastening method that uses a powdered charge to imbed the fastener into the member.

Powder-post damage: Small holes filled with dry, crumbled wood due to the work of beetles (mostly Lyctus and Anobiid) in seasoned and unseasoned wood.

The damage produced by these beetles may be small round holes (1/16" to 1/12" in diameter). The first sign of an infestation is usually a small amount of powder, hence the name powder-post damage, called **frass**. The frass appears in and around a piece of wood that has been attacked by the beetle larvae. If the powder is a little gritty it is a sign of anobiid beetle damage. If the frass is soft and not gritty it is a sign of lyctid beetle activity.

Power: Work per unit of time. Measured in horsepower (hp) or watts (W).

Terms beginning with power:

- **Power cut:** When the power does not work in an area; sometimes called a **power outage or blackout**.
- **Power outlet:** An enclosed assembly that may include receptacles, circuit breakers, fuseholders, fused switches, buses, and watt-hour meter mounting means; intended to supply and control power to mobile homes, recreational vehicles, park trailers, or boats or to serve as a means for distributing power required to operate mobile or temporarily installed equipment.
- **Power pack:** Portable battery operated unit used for power outages and outdoor activities. Some power packs may include jumper cables, a built-in air compressor and an emergency light. Others power pack units may contain a pump, reservoir, relief valve, and directional control.
- **Power pole auger bit:** A specialty auger bit that is designed for drilling creosoted poles and heavy construction timber. Also called a **utility pole auger bit**.
- **Power roller:** Paint roller with a pump unit attached to bring the paint onto the roller so it can be spread evenly on the surface.
- **Power vent:** A vent that includes a fan to speed up air flow. Often installed on roofs.

Pozzolan: A siliceous volcanic ash that is used to produce hydraulic cement. When combined with calcium hydroxide, pozzolan exhibits cementitious properties. Also, any of various artificially produced substances resembling pozzuolana ash.

Pre: A prefix that means in front of, before, prior to, etc. For example:

- **Precast concrete:** Concrete components made in a factory or yard and transported to the site.
- **Prefabricate:** To manufacture or construct parts or sections of structural assemblies beforehand that are ready for quick assembly and erection at a jobsite.
- **Prefabricated housing:** Housing units partially or completely built in a factory.
- **Preferred parking:** Parking that is made available to particular users. Preferred parking includes designated spaces close to the building, designated covered spaces, discounted parking passes, and guaranteed passes in a lottery system.
- **Preliminary drawings:** The drawings that precede the final approved drawings. Usually these drawings are stamped or titled "PRELIMINARY," or "PRELIMINARY/NOT FOR CONSTRUCTION," and the "PRELIMINARY" is removed from the drawings upon being reviewed and approved by the architect and/or owner and/or code official.
- **Preliminary lien notice:** A written notice that is given to the owner by subcontractors, suppliers, and other contributors to the project. The notice states that if bills are not paid a mechanic's lien may be placed against the property, even though the owner has paid the prime contractor.

  The notice may explain how the owner can protect himself against this consequence by requiring the prime contractor to furnish a release by the person giving the owner notice before making payment to the prime contractor or any other method or device that is appropriate under the circumstances. Some states (known as "lien notice states") require lien notices to be filed and some do not. Failure to file a lien notice in a lien-notice state usually results in forfeiture of one's lien rights. See **Mechanic's lien**.
- **Prepared backfill:** A backfill material that has been designed for its intended use by specification.
- **Preservation:** The act or process of applying measures to sustain the existing form, integrity, and material of a [historic] structure, landscape or object. Work generally focuses upon the ongoing preservation maintenance and repair of historic materials and features, rather than extensive replacement and new work.

  Action to mitigate wear and deterioration of a historic property without altering its historic character by protecting its condition, repairing when its condition warrants with the least degree of intervention is referred to as **preservation maintenance**.

Preservation maintenance includes: daily, monthly, and annual maintenance activities; larger capital projects; and, identifying, caring for, and maintaining historic. See **Preventive maintenance.**

- **Preservative:** Any pesticide that will prevent the action of wood-destroying fungi, insect borers, and similar destructive agents when the wood has been properly coated or impregnated with it. Normally an arsenic derivative. Chromated Copper Arsenate (CCA) is an example.
- **Prestressed concrete:** Concrete strengthened with steel wires that are stressed before the concrete is poured. See Post stressed concrete.
- **Preventive maintenance:** Routinely scheduled equipment inspection, cleaning, and repair conducted to detect and prevent equipment failure and keep materials and systems in working order.
- **Pre-consumer content:** Formerly known as post-industrial content, is the percentage of material in a product that is recycled from manufacturing waste, e.g., sawdust. Excluded are materials such as rework, regrind, or scrap generated in a process and capable of being reclaimed within the same process that generated it.
- **Pre-hanger:** A company that buys doors, framing, hardware, glass lites and other components, and prepares (or pre-hangs) the unit for installation.
- **Pre-hung door:** A door unit that comes with the jamb assembled, door stop in place, and the slab connected to the jamb with hinges.
- **Pre-pasted wallpaper:** Wallpaper with adhesive that has been applied to the paper then dehydrated. Pre-pasted wallpaper can be attached by dipping it into water and then placing it on the wall. However, in practice, many installers will apply a thin layer of paste.
- **Pre-qualification:** A screening process of prospective bidders. The owner gathers background information from potential contractors or construction managers for selection purposes. Qualifying considerations usually include such items

as competence, integrity, dependability, responsiveness, bonding rate, bonding capacity, work on hand, similar project experience, etc.

- **Pre-tinning:** Coating a metal with solder or tin alloy, prior to soldering or brazing it.

But not all words beginning with pre mean in front of, before, prior to, etc., for example:

- **Precision end trimmed (PET):** Lumber trimmed smooth on both ends and varying no more than 1/16" nor more than 20% of the pieces. PET may be a condition of sale.
- **Premises wiring:** Interior and exterior wiring, including power, lighting, control, and signal circuit wiring together with all their associated hardware, fittings, and wiring devices, both permanently and temporarily installed, that estends from the service point or source of power, such as a battery, a solar photovoltaic system, or a generator, transformer, or converter windings, to the outlet(s). Such wiring does not include wiring internal to appliances, luminaires (fixtures), motors, controllers, motor control centers, and similar equipment.
- **Premium:** The amount payable on a loan.
- **Present value (PV):** The sum of all future benefits or costs accruing to the owner of an asset when such benefits or costs are discounted to the present by an appropriate discount rate.

   A comparison technique that compares the present values of the cash flows for any two alternatives is referred to as the **present value method**. The best user alternative is based on the lower present value amount.

- **Press brake:** A machine used in cold-forming sheet metal or strips of metal into desired profiles.

Pressure: Force per unit area. Usually measured in pounds per square inch (PSI) or kilopascals (kPa). A kilopascal is equal to 1000 newtons per square meter or 0.0102 kg/sq cm (0.145 lb/sq in).

   Pressure is a similar idea to stress, the force intensity at a point, except that pressure means something acting on the surface of an object rather than within the material of the object. When discussing the pressure within a fluid, the meaning is equivalent to stress.

Terms beginning with pressure:

- **Pressure-assist toilet:** Toilets that remove waste by using forced air and water (that produces a brief whooshing sound), and empties the bowl quickly.
- **Pressure balance valve:** On shower faucets, the pressure balance valve maintains water temperature by preventing pressure changes that can occur when someone turns on the water somewhere else in the house.
- **Pressure drop:** The difference in pressure between any two points of a system, usually a pipe or tube, or a component.

  Pressure drop is the result of frictional forces on the fluid as it flows through the pipe or tube. Fluid velocity and fluid viscosity are important determinants of pressure drop. As flow velocities and fluid viscosities increase a larger pressure drop occurs across a section of pipe or a valve or elbow. Lower velocity will result in a lower pressure drop.

  Liquid and gas always flow in the direction of lowest pressure. The pressure drop increases proportionally to the frictional shear forces within the piping network. A piping network that has a high degree of surface roughness, many pipe fittings and joints, changes in diameter, turns, etc., will affect the pressure drop.
- **Pressure line:** The line that carries fluid from a pump outlet to a pressurized port of an actuator. (A type of motor for moving or controlling a mechanism or system.)
- **Pressure override:** The full-flow pressure minus the cracking pressure. (Cracking pressure is the minimum upstream pressure at which the valve will operate. Typically the check valve is designed for and can therefore be specified for a specific cracking pressure.) The pressure override is a measure of the increase in pressure over the cracking pressure when additional flow passes through the valve after it cracks.
- **Pressure reducing valve:** A valve that limits the maximum pressure at its outlet regardless of the inlet pressure.

- **Pressure relief valve (PRV):** A device mounted on a hot water heater or boiler that is designed to release any high steam pressure in the tank to prevent tank explosions.
- **Pressure switch:** An electric switch operated by fluid pressure.

  Also, switches that read the available pressure in air and hydraulic lines. These switches are often used as a safety device to prevent equipment from operating when there is not enough air pressure or hydraulic fluid pressure.
- **Pressure tank:** Used with a water pump to provide pressure in plumbing lines when the pump is not running. The pressure tank contains water and air. As water is pumped into the pressure tank the air compresses. Then, when a faucet is opened the air pressure in the tank pushes the water through the lines. When pressure in the tank drops below a preset limit the pump turns on and runs until the required pressure is reached.
- **Pressure test:** A test of plumbing piping for leaks.
- **Pressure tubing:** Tubing used to conduct fluids under pressure or at elevated temperatures or both, and produced to stricter tolerances than pipe.
- **Pressure-treated lumber:** Lumber that has been saturated with a preservative.
- **Pressurized irrigation water:** Water that is not fit for human consumption but is provided to a structure for use as irrigation water for plants and lawns. See also Graywater.

Primary air: Air that is mixed with gas before the gas leaves a burner port to burn. The ideal burning condition generally is 10 cubic feet of air per one cubic foot of gas. The opening(s) through which primary air is admitted into a burner is the primary air inlet.

Primary members: This is the main load carrying members of a structure such as a beam or joist girder.

Prime contractor: Any contractor having a contract directly with the owner.

Prime sash: The balanced or moving sash of a window unit.

Prime window: A primary window, as opposed to a storm or combination unit added on.

Primer: The first, base coat of paint when a paint job consists of two or more coats. A first coating formulated to seal raw surfaces and holding succeeding finish coats.

Also, with regard to roofing, a thin, liquid-applied solvent-based bitumen that may be applied to a surface to improve the adhesion of subsequent applications of bitumen.

Also, a material that is sometimes used in the process of seaming single-ply membranes to prepare the surfaces and increase the strength (in shear and peel) of the field splice.

Priming: With regard to boilers, priming is the carryover of varying amounts of droplets of water in the steam (foam and mist) that lowers the energy efficiency of the steam and leads to the deposit of salt crystals on the super heaters and in the turbines. Priming may be caused by improper construction of boiler, excessive ratings, or sudden fluctuations in steam demand. Priming is sometimes aggravated by impurities in the boiler-water.

Principal: The original amount of the loan, the capital.

Principle of superposition: The principle of superposition states that the resultant is the algebraic sum of the effects when applied separately.

Prismatic beam: A structural beam with uniform cross section.

Privacy hardware: Interior door hardware that locks. See Passage hardware.

Private/private use: Plumbing fixtures in residences and apartments, private bathrooms in hotels and hospitals, and restrooms in commercial establishments; these fixtures are intended for the use of a family or an individual. See **Public/Public use**.

Pro dealer: Building product dealers and/or distributors that cater to professional customers, e.g., home builders and remodeling contractors.

Process water: Water used for industrial processes and building systems, such as cooling towers, boilers, and chillers.

Professional engineer: A designation reserved, usually by law, for a person or organization professional qualified and duly licensed to perform such engineering services as structural, mechanical, electrical, sanitary, civil, etc.

Professional liability insurance: Insurance provided for design professionals and construction managers that protect the owner against the financial results and liability of negligent acts by the insured. Usually referred to as Errors and Omissions (E&O) insurance.

Professional services: Services that are provided by a professional, in the legal sense of the word, or by an individual or firm whose competence can be measured against an established standard of care.

Program schedule: A schedule that spans from the start of design to occupancy. The schedule includes the milestones that control the progress of the project from start to finish. See **Milestones**.

Progress meeting: A meeting dedicated to project progress during the construction phase.

Progress paymen: Partial payments on a contractor's contract amount Payments are made periodically by the owner for work accomplished by the contractor to date as determined by calculating the difference between the completed work and materials stored and a predetermined schedule of values or unit costs. See Estimate cost-to-complete and Estimated final cost.

Progress schedule: A diagram, usually a line diagram, that shows the proposed and actual starting and completion times for the respective project activities.

Progressive collapse: A process by which the collapse of part of a building leads to further collapse; a "house of cards" collapse.

Project: The overall scope of work necessary to complete a specific construction job.

Terms beginning with project:

- **Project budget:** The target cost of the project as established by the owner with the concurrence of the project team. The project budget usually includes the cost of construction other line-item costs such as land, legal fees, interest, design fees, CM fees, etc. that the owner wishes to have included in the budget.

- **Project building:** The real property, including an occupied and operational building(s) and its associated grounds that is registered for and actively pursuing LEED certification.

- **Project completion graph (S-curve):** While there are variations, S-graphs commonly show an estimated cumulative spending profile and an actual spending profile as work progresses. The graph is based on a direct relationship of cash flow to the percentage of project completion. For a typical single-phase construction project, the first and last stages of construction take up 15–20% each of the construction schedule. The remaining interim period, representing 60–70% of the schedule is generally a more-or-less straight line, that connects the ramping up and down sections of the schedule.

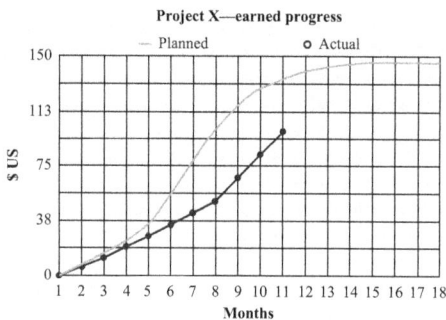

Project X—earned progress

— Planned   o Actual

Typical project completion S-curve.

- **Project contingency graph:** A graph that projects the use of hard cost contingency through project completion. Each month the committed hard cost contingency is plotted as a percentage of total hard cost contingency against the project schedule. The projected hard cost contingency line provides a baseline for analysis. When planning a construction project it is common to assume that up to 40% of the hard

cost contingency will be committed during the first 30% of the project schedule and that the remaining 60% will be a straight-line projection to the end of the project; the 40% allows for the high-risk period during project startup.

- **Project cost:** All costs for a specific project including costs for land, professionals, construction, furnishings, fixtures, equipment, financing, and any other project related costs.

- **Project directory:** A written list of all people connected with a specific project. The list usually includes a classification or description of the person (i.e., Owner, Architect, Attorney, General Contractor, Civil Engineer, Structural Engineer, etc.); name, address, e-mail, telephone and FAX numbers opposite their respective classifications or description. Emergency and after-hour telephone numbers should also be included. It is normal to control the directory on need to know basis.

- **Project manager:** A person or construction firm in direct charge and coordination of all contractors involved with supplying, fabricating, and building a project. Also referred to as a **construction manager**.

- **Project manual:** An organized book that sets forth the bidding requirements, conditions of the contract and the technical work specifications for a specific project. The manual documents and augments the drawings. The project manual contains the technical specifications, general conditions, supplementary and special conditions, the form of contract, addenda, change orders, bidding information, and proposal forms as appropriate.

- **Project meeting:** A meeting that includes supervisors from the contractor's home office and the on-site project team. It is dedicated to job-site progress and progress payments. Sub-contractors are usually included when appropriate.

- **Project site:** Synonym for construction site.

- **Project team:** The team usually consists of the architect/engineer, construction manager or general contractor, and the owner or owner's representative. Depending on the size of the project the team may also include several levels of management/ supervision, and the designated leaders of prime and subcontractors.

Projection: A thing or part that extends outward beyond a prevailing line or surface. For example, with regard to a kitchen range hood, he depth of the range hood, front to rear. The range hood should be deep enough to come out to the front edge of the front burner.

Also, a prediction or an estimate of something in the future, based on present data or trends; a plan for an anticipated course of action.

Projected window: A window in which the sash opens on hinges or pivots. Refers to casements, awnings, and hoppers.

Propellant: The gas used to expel materials from aerosol containers.

Property manager: A person directly employed by the organization who oversees operations, maintenance, and upkeep of the project building on behalf of the owner or serves as the primary liaison between the owner and project building tenants.

Property survey: A survey to determine the boundaries of your property. Some banks require a property survey prior to granting a mortgage.

Property type: A grouping of individual properties based on a set of shared physical or associative characteristics.

Proportion: The relationship between one part of an object or composition and another part and to the whole, or between one element and another.

Proportional limit: The highest stress at which stress is directly proportional to strain. It is the highest stress at which the curve in a stress-strain diagram is a straight line. Proportional limit is equal to elastic limit for many metals. See Elastic limit and Yield strength.

Proportioner: A pumping unit comprised of two positive displacement pumps that is designed to dispense two components at a precise ratio.

Proportions of aggregate: The respective proportions of the fine and course aggregate in concrete by mass.

Proprietary specification: A specification that ames the products and materials by manufacturer's name, model number, or part number.

Protected Harvest certification: Certification standards that reflect the unique growing requirements and environmental considerations of each crop and bioregion. Each crop- and region-specific standard is divided into the following three parts: production, toxicity, and chain-of-custody.

Protected membrane roof (PMR): An insulated and ballasted roofing assembly, in which the insulation and ballast are applied on top of the membrane. Sometimes referred to as an inverted roof assembly.

Protected pattern: A pattern left on a surface that was exposed to heat and smoke. Items on the surface, such as appliances and dishes, protect the areas they rest on from heat and smoke. When the surface is cleaned a pattern remains because the exposed areas discolor more than do the areas that were covered.

Protection: With regard to property, action that is taken to safeguard a property by defending or guarding it from further deterioration, loss, or attack or shielding it from danger or injury. Protection is especially important for historic properties.

Protective system: A method of protecting employees from cave-ins, from material that could fall or roll from an excavation face or into an excavation, or from the collapse of adjacent structures. Protective systems include support systems, sloping and benching systems, shield systems, and other systems that provide the necessary protection.

Provenance: The history of physical custody of an object or collection, and its origin.

Proximity switch: A photoelectric switch that is triggered as an object passes nearby without any physical contact.

A proximity switch emits an electromagnetic field or a beam of electromagnetic radiation that looks for changes in the field or return signal from its target. Different targets, e.g., plastic and metal, require different sensors.

Also referred to as a proximity sensor.

Pruning saw: This tool is used for trimming trees and bushes. It has a curved blade that folds back into a curved hardwood handle.

Pry bar: This tool has a curved blade to fit behind nailed wood such as moldings or between two sections that need to be separated. A long handle provides leverage to pry things apart.

Psychrometric chart: A chart that shows dry bulb and wet bulb temperatures used to determine the relative humidity of air and the dew point temperature. Other engineering data referring to moisture in air are also shown.

A psychrometer is an instrument that uses the difference in readings between two thermometers, one having a wet bulb and the other having a dry bulb, to measure the moisture content or relative humidity of air.

Public/Public use: With regard to the design of plumbing fixtures in buildings or structures. If the classification for public or private use is unclear, project teams are usually required to default to public-use flow rates.

Puddle weld: The weld used in fastening the composite steel deck to the structural steel frame.

Pugging: Traditional infill between timber floor joists intended to enhance the acoustic insulation of the floor. It may occupy the whole depth of the floor or only part of it. Materials used include sand, mortar, concrete, straw, and sea shells.

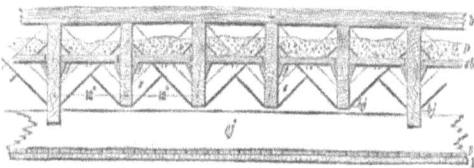

Pugging.

Pull: A handle used on drawers and cabinet doors.

Pullout strength: The strength determined/ by measuring the maximum force needed to pull an insert from hardened concrete. For example, anchor bolt pullout strength is the force required to pull a single bolt out of its foundation. The separation can occur between the epoxy grout and the concrete foundation or it can occur between the anchor bolt and the epoxy grout itself.

Pull-out spray: A single-control kitchen faucet that offers a retractable hose and spray-head to be used for food preparation and cleaning.

Pulse system: A furnace that produces heat through multiple explosions of gas.

Pultrusion: The process used to produce fiberglass composite profiles or components for the production of windows and doors. The term also is used generally to refer to the composite profiles or lineals cut and processed to make window and door components.

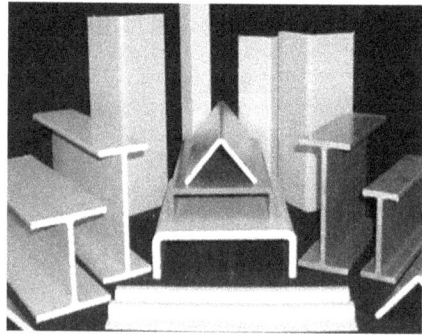

Pultrusion products.

Pump: A mechanical device that uses suction or pressure to raise or move liquids, compress gases, or force air into inflatable objects. A pump converts mechanical force and motion into hydraulic fluid power.

Pump mix: Special concrete that will be used in a concrete pump. Generally, the mix has smaller rock aggregate than regular mix.

Punch list: A list of discrepancies that is prepared by the owner or owner's representative for corrected the contractor to correct prior to acceptance.

Punch out: To inspect and make a punch list.

Puncture resistance: The extent to which a material is able to withstand the action of a sharp object without perforation.

Pure CM: A contractual form of the construction management system exclusively performed in an agency relationship between the construction manager and owner. (CM without risk).

Purged line: A plumbing line in which the faucet has been opened and allowed to run for a specified length of time, usually 1–5 min.

Purlin: A horizontal member that spans across adjacent rafters or beams, commonly installed to provide a fastening surface for the roofing material. A purlin that is heavier than a common purlin is called a **principal purlin**.

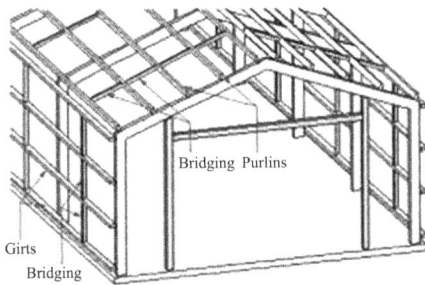

Purlins, bridging, and girts.

Putlog or Putlock: A horizontal scaffold member one end of which is built into the wall. Putlog scaffolds are seldom used because they can be dangerous, and because the hole in the wall has to be repaired when the scaffold is taken down.

Putlog scaffold.

Putty: A type of dough used in sealing glass in the sash, filling small holes and crevices in wood, and for similar purposes.

Pyrolytic glass: A glass product that is coated, usually to provide low-emissivity or solar-control benefits, during the manufacturing process at the molten glass stage. Commonly referred to as a hard coat, this type of coating offers a surface that is generally as durable as an ordinary glass surface, and therefore requires no special handling and does not need to be used in an insulating glass unit. The other type of glass coating is a sputter-coat that is applied in a secondary process. Sometimes referred to as a soft-coat, these types of coatings generally require some additional care in handling and fabrication and must be used within an insulating glass unit.

Pythagorean theorem: A mathematic relationship between sides of right triangles specifying that the length of the hypotenuse (longest side) squared equals the sum of the lengths of the other two sides squared. It is often expressed as: "$a^2 + b^2 = c^2$." The theorem can be used to solve for any side of a right triangle where the lengths of the other two sides are known. To solve for side "a," use the formula "$a^2 = c^2 - b^2$." To solve for side "b," use the formula "$b^2 = c^2 - a^2$."

Q-scheduling: Q Scheduling is a new technique that is gaining in popularity within the construction industry. It is a scheduling technique that reveals a relation between the sequence of doing a job and the cost to be incurred. The Q-schedule is similar to the Line-of-Balance but it allows for a varying volume of repetitive activities at different segments or locations of the construction project.

Quad outlet: An electrical outlet with four ports.

Quadrant: A quarter of a circle. The name is also used for various things of this shape, e.g., molding, a corner kerbstone, and a historic navigational instrument.

Qualified person, per OSHA: One who, by possession of a recognized degree, certificate, or professional standing, or who by extensive knowledge, training, and experience, has successfully demonstrated his ability to solve or resolve problems relating to the subject matter, the work, or the project.

Quality assurance (QA): (The terms "quality assurance" and "quality control" (see below) are often used interchangeably to refer to ways of ensuring the quality of a service or product. The terms, however, have different meanings.)

Quality assurance is planned and systematic activities that are implemented to ensure that all requirements for a product or service are met. Having some fun with words, one ensures, i.e., makes certain that something shall occur, so that one can assure, i.e., tell someone positively or confidently that it has been accomplished to the utmost.

Quality control (QC): The observation techniques and activities used to fulfill requirements for quality.

Quality engineering: That part of the quality assurance procedure where the required level of quality is accurately inserted into the construction documents by the A/E.

Quantity survey: An inventory of all materials that go into the construction project.

Quarry tile: A man-made or machine-made clay tile used to finish a floor or wall. Generally 6" × 6" × 1/4" thick .

Quarter round: A small trim molding that has the cross section of a quarter circle. See **Moldings**.

Quartered: Veneer produced by cutting in a radial direction to the pith to achieve a straight (vertical) grain pattern. In some species, principally red oak and white oak, ray fleck is produced. Also referred to as quarter-sliced and quarter cut.

Quaternary: The Quaternary is the geologic time period comprising about the last 1.65 million years.

Quartersawing: A method of cutting boards from a log by sawing from the bark side of the log toward the center axis of the log. The method is called quartersawing because logs are usually split into quarters before they are sawn. Quartersawn boards have a consistent grain that runs at a 45–90° angle to the face of the board. See **Plain sawing**.

Quartz: A natural crystalline form of silica. Quartz sand is used as an abrasive, a building stone, in cement manufacture, in glass and porcelain, and in foundry moulds.

Quartzite: The metamorphic equivalent of a quartz sandstone, that has recrystallised into closely fitting granules.

Queen post truss: A truss with two posts directly supporting the purlins. See **King post truss**.

R-factor or value: A measure of insulation. A measure of a material's resistance to the passage of heat. The higher the R value, the more insulating "power" it has. For example, typical new home's walls are usually insulated with 4" of batt insulation with an R value of R-13, and a ceiling insulation of R-30.

Rabbet: A rectangular longitudinal groove cut in the corner edge of a board or plank.

A siding joint that is made by thinning the edge of two boards to about half their width and then overlapping the two thinned edges is called a **rabbet joint**.

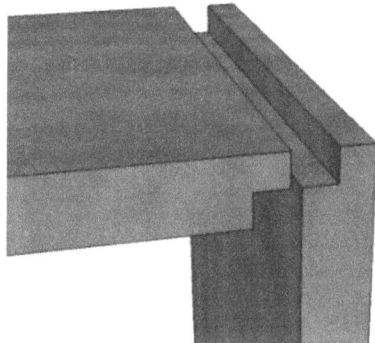

Double rabbet joint.

Rabbet plane: This tool has a wide blade that is ideal for trimming narrow dadoes and rabbets.

Rabbet plane.

Raceway: An enclosed channel of metal or nonmetallic materials designed expressly for holding wires, cables, or busbars, with additional functions as permitted in the NEC. Raceways include, but are not limited to, rigid metal conduit, liquidtight flexible conduit, electrical nonmetallic tubing, electrical metallic tubing, under floor raceways, cellular concrete floor raceways, cellular metal floor raceways, surface raceways, wireways, and busways.

Racking: A method of installing shingles in a straight up the roof manner. Racking is generally considered to be an improper installation, although some manufacturers list racking as an acceptable, but not preferred, method of installation for certain types of their shingles.

Racking of shingles.

Also, racking is the distortion of a rectangular shape to a skewed parallelogram.

Radial: Radiating from or converging to a common center.

In the case of a tree a radial section would be coincident with a radius from the axis of the tree or log to the circumference, i.e., from its pith to bark.

Terms beginning with radial:

- **Radial arm saw:** A circular saw that runs on an overhead track. The track mechanism swings in relation to the table to make miter cuts.

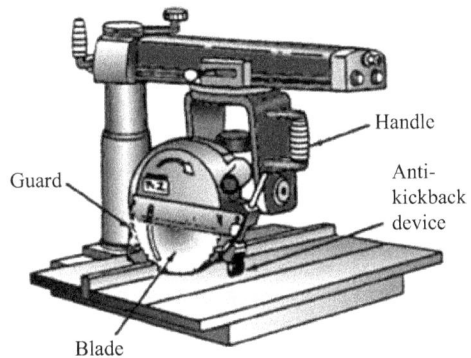

Radial arm saw.

- **Radial clearance:** The angular clearance on the sides of a saw tooth or saw blade.
- **Radial drill press:** A drill press with the head mounted on a tube that is laterally and vertically adjustable, allowing greater throat clearance.

Radiant heating: A method of heating, usually consisting of a forced hot water system with pipes placed in the floor, wall, or ceiling. Also, **electrically heated panels**.

Radiation: Energy transmitted from a heat source to the air around it. Radiators actually depend more on convection than radiation.

Radiator: Heating device that transfers heat from water or steam running inside of it to the air and objects around it.

Radiometric: Radiometric pertains to the measurement of geologic time by the analysis of certain radioisotopes in rocks and their known rates of decay.

Radius: A line extending from the center of a circle to the outside edge. It is equal to one-half the diameter of the circle.

Radius of gyration (k or r): The distance from the neutral axis of a section to an imaginary point at which the whole area of the section could be concentrated and still have the same moment of inertia.

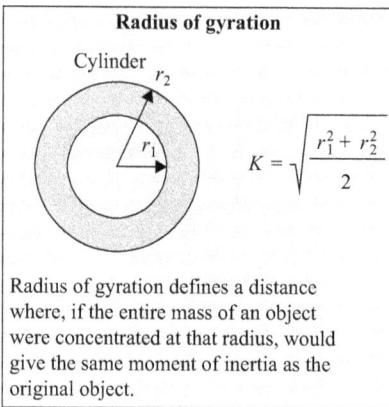

Radius of gyration

Cylinder

$$K = \sqrt{\frac{r_1^2 + r_2^2}{2}}$$

Radius of gyration defines a distance where, if the entire mass of an object were concentrated at that radius, would give the same moment of inertia as the original object.

Radius of gyration.

Radius plane: A plane used to round or chamfer the edges of a board

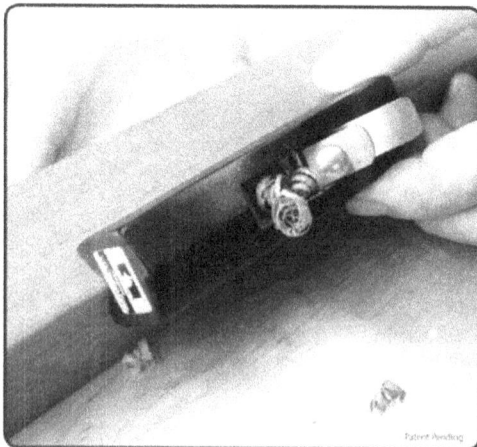

Radius plane.

Radon: A naturally occurring, heavier than air, radioactive gas common in many parts of the country. Radon gas exposure is associated with lung cancer. Mitigation measures may involve crawl space and basement venting and various forms of vapor barriers.

Radon system: A ventilation system beneath the floor of a basement and/or structural wood floor and designed to fan exhaust radon gas to the outside of the home.

Rafter: The main beam supporting a roof system or a sloping roof framing member.

When constructing with lumber, rafters are usually 2 × 10's and 2 × 12's. The rafters of a flat roof are sometimes called roof joists.

There are numerous terms associated with rafters:

• **Rafter, common:** In wood-frame construction, one of a number of slanting structural members (extending from the ridgeboard down to the eaves) that support the roof; these members are usually of the same size and evenly spaced along the length of the roof ridge.

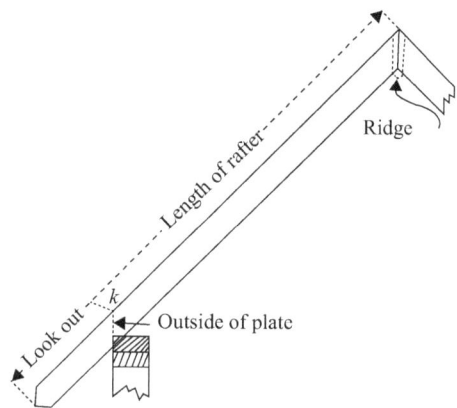

Common rafter.

• **Rafter, cripple:** A rafter that extends from a hip rafter to a valley rafter. These are the most complex rafters and are cut on both ends at 45° with these cuts facing in opposite directions to each other.

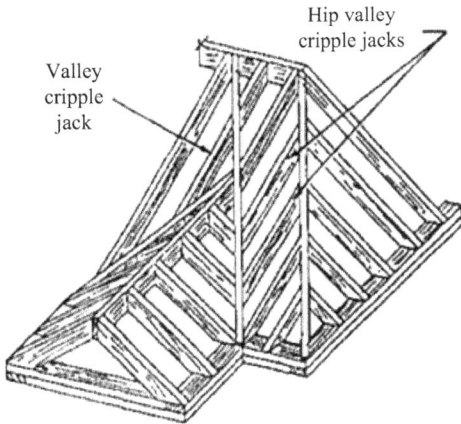

Valley and hip cripple jack rafters.

- **Rafter, hip:** A rafter that forms the intersection of an external roof angle.
- **Rafter, jack:** A rafter having less than the full length of the roof slope, as one meeting a hip or a valley. A jack rafter spans the distance from the wall plate to a hip, or from a valley to a ridge.
- **Rafter rise:** The vertical distance a rafter will span from where the rafters top edge is in line (plumb) with the outside of the wall to the roof peak.
- **Rafter, run of:** The horizontal distance a rafter will span. It is measured from the outside of the wall to the center of the ridge (one half the building width).

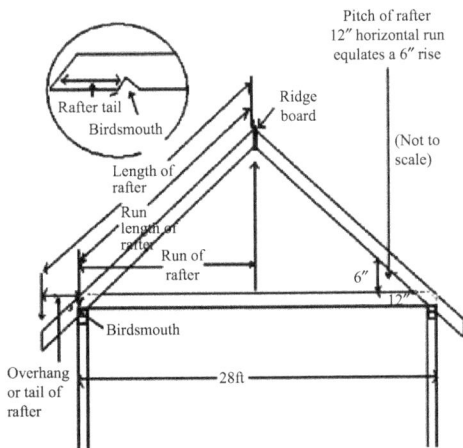

Run of rafter. Note difference between run of rafter and run length of rafter.

- **Rafter, valley:** A rafter that forms the intersection of an internal roof angle. The valley rafter is often made of double 2-inch-thick members.

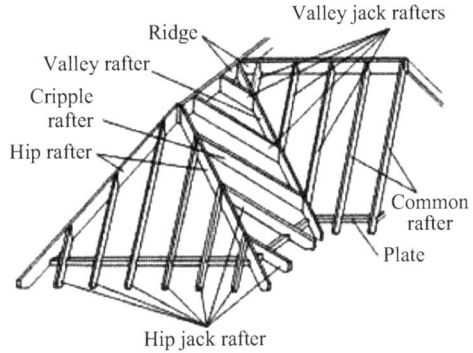

Hip, cripple, valley, valley jack and common rafters.

Rail: A horizontal member of the framework of a window sash or door; the cross members of panel doors or of a sash. See Stile.

Also, a wall or open balustrade placed at the edge of a staircase, walkway bridge, or elevated surface to prevent people from falling off.

Also, any relatively lightweight horizontal element, especially those found in fences (split rail).

Also, a horizontal member on the cabinet face frame. See **Stile**.

Also, a bar or series of bars that are typically fixed on upright supports to serve as part of a fence or barrier or used to hang things on.

Terms beginning with rail:

- **Rail, bottom:** With regard a door, the bottom rail of a stile and rail door. Sometimes called **horizontal edge**.
- **Rail, intermediate:** With regard to a door, a rail, other than the top and bottom rail, that separates panels or panels from glazing materials in a combination door. Also referred to as **cross rail**.
- **Rail, lock:** With regard to a door, an intermediate rail located at approximately adjacent to the lock.

Nomenclature of door and frame

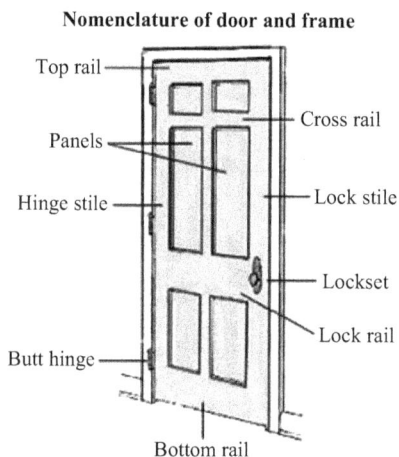

- Top rail
- Cross rail
- Panels
- Hinge stile
- Lock stile
- Lockset
- Lock rail
- Butt hinge
- Bottom rail

- **Rail, top:** With regard to a door, the uppermost rail of a stile and rail door. Sometimes called **horizontal edge**.
- **Rail fittings:** (Excuse me for being out of alphabetical order.) Fittings that are used to make hand rails. They have setscrews, rather than threads that hold pipes. Rail fittings are available in various sizes and configurations to suit a number of applications. Rail fittings are not used for plumbing or conduit.

Railroad tie: Black, tar and preservative impregnated, 6" × 8" and 6-8' long wooden timber that was used to hold railroad track in place. Often used as a member of a retaining wall.

Rails: With regard to elevators, normally steel T-sections with machined guiding surfaces installed vertically in a hoistway to guide and direct the course of travel.

Rainforest Alliance certification: A certification that is awarded to farms that protect wildlife by planting trees, control erosion, limit agrochemicals, protect native vegetation, hire local workers, and pay fair wages.

Rainproof: Constructed, protected, or treated so as to prevent rain from interfering with the successful operation of an apparatus under specified test conditions.

Raintight: Constructed or protected so that exposure to a beating rain will not result in the entrance of water under specified test conditions.

Raised panel door: A door made from panels, usually framed and held in place by stiles

and rails; often used in a frame and panel cabinet door. The edges of the panel are shaped to a thin edge so the panel will fit into the slots in the surrounding frame. Simulated raised panel doors are frequently made from pressed-wood fibers. See **Frame and panel cabinet door, Slab cabinet door**, and **Flat panel**.

Making a raised panel door.

Rake: The inclined edge at the end of a sloping roof plane. For example, the roof edge at the top of the gable end wall is a rake. The vertical face of the sloping end of a roof eave is called the **rake fascia**.

Rake

Also, a tool that is used to remove snow from roofs.

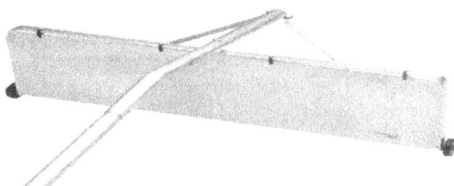

Roof rake that is used for removing snow.

Also, the angle at which the leading edge of the teeth are cut on a saw blade. The rake angle determines the cutting characteristics of a saw blade. It is a measurement of the angle of the tooth face in relation to an imaginary radial line drawn from the exact center of the saw blade through the very tip of the saw tooth. Positive rake teeth have a forward tilt while negative rake teeth have a backward tilt. Blades with a high positive rake are aggressive and fast-cutting. Blades with a lesser rake take smaller bites and are easier to control for fine finishes and precise cuts.

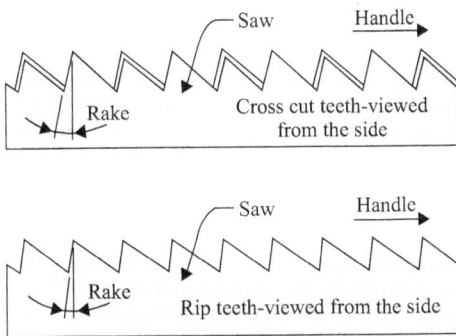

Rake on a saw.

Rake siding: The practice of installing lap siding diagonally.

Rake-starter: See Bleeder strip.

Raked: A mortar joint between courses of brick in which the mason uses a tool called a rake to remove the excess mortar to create a uniform depth.

Raker: A diagonal brace for supporting sheeting of an excavation.

Ram: The falling weight of a pile-driving machine.

Also, a hydraulic water-raising or lifting machine; the piston of a hydraulic press; and the plunger of a force pump (A pump with a solid piston and valves used to raise a liquid or expel it under pressure.).

Ranch: A single story, one level home.

Random lengths (RL): Lumber of various lengths, usually in even two-foot increments. Lumber offered as random length will contain a variety of lengths which can vary greatly between manufacturers and species. A random length loading is presumed to contain a fair representation of the lengths being produced by a specific manufacturer.

Random matched: A face containing veneer strips of the same species that are selected and assembled without regard to color or grain, resulting in variations, contrasts and patterns of color and grain. Pleasing appearance is not required. Also referred to as **mismatched**. See **Veneer**.

Random-orbit sander: This sander spins like a disc sander but moves in a circular orbit, like an orbital sander, chewing through wood grain without leaving cross-grain scratches. It can sand in any direction.

Random-tab shingles: Shingles on which tabs vary in size and exposure.

Random-tab shingles.

Rapidly renewable materials: Agricultural products, both fiber and animal, that take 10 years or less to grow or raise and can be harvested in an ongoing and sustainable fashion.

Rasp: A coarse file that is used to smooth and shape wood.

Rate of return: The percentage return on each dollar invested. Also known as yield.

Rated load: The manufacturer's specified maximum load to be lifted by a hoist or to be applied to a scaffold or scaffold component.

Rating: The stated operating limit of a piece of equipment, expressed in a unit of measure such as volts or watts.

Raw water: Water supplied to the plant before any treatment.

Rayleigh wave: A seismic surface wave involving elliptical motion in a vertical plane oriented in the direction of propagation of the wave.

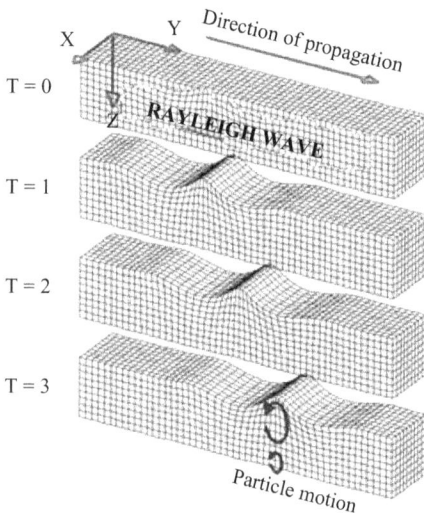

Rays: A ribbon-like aggregate of wood cells that extend radially across the grain, radiating out from the pith to the bark. See **Fleck**.

Reaction: A force exerted by a support on an object: sometimes called support reaction. Using this definition, a reaction is an external force.

Reaction wood: Wood that forms in a tree in response to leaning. In softwoods it is called **compression wood** because it forms on the underside of the stem or limb. In hardwoods it is called tension wood because it forms on the upper side.

This wood is often very dense, hard, and brittle. In hardwood trees, reaction wood forms on the upper side of the lean and is called tension wood.

Woolly surfaces and excessive longitudinal shrinkage are often symptoms of tension wood.

The dense hard wood is less likely to accept an even stain, it is prone to failure under load and tends to crack and split easily when nailed and screwed, it is difficult to carve and shape and, changes in moisture affect it in a more significant way than other wood: try to avoid it.

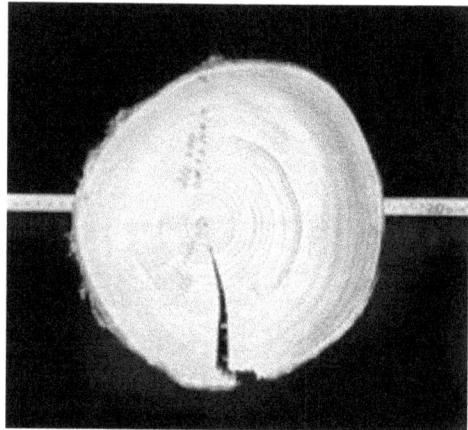

Reaction (compression) wood in a softwood tree.

Reactive colloidal silica: An engineered dispersion of nano-sized silica particles in an aqueous suspension (no metal salts).

Readily accessible: A space that can be reached, entered, or viewed without difficulty, moving obstructions, or requiring any action that may harm people or property.

Ready mixed concrete: Concrete mixed at a plant or in trucks en route to a job and delivered ready for placement.

Ready-set tile: A type of small mosaic tile that comes pre-fastened to a mat. Ready-set tiles are much faster to install than individual mosaic tile.

Real estate investment trust (REIT): An investment vehicle in which investors purchase certificates of ownership in a trust that invests the money in real property and then distributes any profits to the investors. A REIT is not subject to corporate income tax as long as it complies with the tax requirements for a REIT, but shareholders must include their share of the REIT's income in their personal tax returns.

Reamer: A tool used to shape or enlarge holes or bores.

Rebond pad: A urethane foam carpet pad made by gluing small pieces of foam together. See **Synthetic felt pad, Waffle type sponge rubber pad**, and **High density urethane foam pad.**

Receptacle: A typical household will have many 120-volt receptacles for plugging in lamps and appliances and 240-volt receptacles for the range, clothes dryer, air conditioners, etc. A single receptacle is a single contact device with no other contact device on the same yoke. A multiple receptacle is two or more contact devices on the same yoke.

An outlet where one or more receptacles are installed is called a **receptacle outlet.**

Recess-mount medicine cabinet: A medicine cabinet that is recessed into the wall, usually between two studs. See Surface-mount medicine cabinet.

Recessed pointing: Flat pointing set back from the surface of the bricks. See **Pointing**.

Recessed tub: A tub that has a decorative finish on one side only and is surrounded by three walls. The apron, or skirting, may not be included and must be ordered separately.

Recessed tub.

Reciprocating saw: A tool that will chew through most any cutting task. This saw can cut through wood, metal or plastics easily, making it. It is ideal for demolition tasks, cutting rough openings, and cutouts for plumbing and heating ducts. Also called a **ricip saw.**

Reciprocating saw.

Reciprocation: Back-and-forth straight line motion or oscillation.

Recirculated air: Air removed from a space and reused as supply air, delivered by mechanical or natural ventilation.

Reclaimed water: Wastewater that has been treated and purified for reuse.

Recommissioning: Applies to buildings that were previously commissioned as part of new construction or retrocommissioning. Recommissioning involves periodic conducting of the original commissioning tests from the commissioning or retrocommissioning process to ensure that the original results are maintained over time.

Reconnaissance: A preliminary survey or research. A contractor would "recon" a project prior to estimating and bidding.

With regard to cultural preservation, a reconnaissance study is synthesis of cultural resource information describing the kinds of cultural resources in a study area and summarizing their significance; sometimes called a cultural resource overview.

Reconstruction: Rebuilding (new construction) the form, features, and detailing of a non-surviving historic structure or landscape, or any part thereof, for the purpose of replicating its appearance at a specific time and in its historic location.

Record drawings: A set of contract document drawings, marked up as construction proceeds to reflect changes made during the construction process. The drawings should show the exact location, geometry, and dimensions of all elements of the constructed project as installed. At some point these record drawings should be made into **as-built drawings**. See **As-built drawings**.

Recording fee: A charge for recording the transfer of a property, paid to a city, county, or other appropriate branch of government.

Recoverable heat: Heat energy that is actually recovered to do useful work, e.g., the heating of domestic hot water with the waste heat from an air conditioning system.

Recovered fiber: Includes both postconsumer fiber and waste fiber from the manufacturing process.

Recovery capacity: The amount of water in gallons per hour raised 100°F at a given thermal efficiency and BTU per hour input.

Rectangular hollow section: A structural steel component in the shape of a steel tube with a rectangular cross section.

Recurrence interval: The average time between specific events at a particular site.

Recycling: The collection, reprocessing, marketing, and use of materials that were diverted or recovered from the solid waste stream.

Red/Brown: When referring to color and matching, veneers containing all heartwood, ranging in color from light to dark. See **Veneer**.

Red label shingles: Shingles made from high-grade wood with some slight sapwood and very little flat grain. Most residential structures use red or blue label shingles.

Red lined prints: Blueprints that reflect changes and that are marked with red pencil.

Reducer: A fitting with different size openings at either end and used to go from a larger to a smaller pipe.

Reduction of area: The decrease in the area of fracture section of a tensile test specimen as a percentage of the original area.

Redundants: The reactions that are not necessary for static equilibrium.

Reflection: The throwing back by a body or surface of light, heat, or sound without absorbing it, e.g., the reflection of light.

With regard to earthquakes, seismic energy that has been returned (reflected) from an interface of materials of different elastic properties.

Reflective glass: Window glass that is coated to reflect radiation striking the surface of the glass.

Reflective insulation: Sheet material with one or both faces covered with aluminum foil.

Reflectivity: See **Light reflectance**.

Reflex gauge glass: A form of gauge glass considered to be best practice for steam boilers. A flat glass on the outer side and prisms on the water side giving a very clear indication of the water level.

Reflex flat gauge glass.

Refraction: Light, radio waves, etc., being deflected in passing obliquely through the interface between one medium and another or through a medium of varying density. It is a result of the wave traveling at different speeds at different points along the wave front. For example, seismic energy that has been deflected by passing from one material to another with different elastic properties.

Refrigerants: The working fluids of refrigeration cycles that absorb heat from a reservoir at low temperatures and reject heat at higher temperatures.

Refrigerant-gas: A substance that remains a gas at low temperatures and pressure and can be used to transfer heat. Freon is an example and is used in air conditioning systems. Many refrigerant gasses are harmful to the environment and have been banned. Check carefully before purchasing AC equipment.

Refusal: The point at which the desired bearing condition is reached in soil or rock.

Regionally harvested (or extracted) and processed materials: Materials that come from within a 500-mile radius of the project site. Use of these materials will earn LEED points for a project.

Register: A grill placed over a heating duct or cold air return.

Registered Professional Engineer: A person who is registered as a professional engineer in the state where the work is to be performed. However, a professional engineer registered in any state is deemed to be a registered professional engineer within the meaning of this standard when approving designs for "manufactured protective systems" or "tabulated data" to be used in interstate commerce.

Reglaze: To replace a broken window.

Reglet: A preformed slot in concrete or masonry for the eventual installation of flashing or membrane.

   Also, a plastic or wood molding placed in a concrete or masonry opening to provide a uniform groove for a spline-type gasket to hold window glass.

Regression analysis: A regression analysis is a statistical technique applied to data to determine, for predictive purposes, the degree of correlation of a dependent variable with one or more independent variables; in other words, to determine if there is a strong or weak cause and effect relationship between to things. See **Least squares**.

Regular building occupant: A worker who either has a permanent office or workstation in the project building or typically spends 10 hours per week or more in the project building. For a residential building, this includes all persons who live in the building.

Regularly occupied spaces: In commercial buildings, these are areas where people sit or stand as they work; in residential applications these spaces are living and family rooms.

Regulatory requirements: In reference to land use, they are restrictions or guidelines on development or use of land, properties, or facilities as defined in accordance with design standards, building construction requirements, land use plans, occupancy codes, and zoning classifications as determined by the controlling/governing municipal or county agencies.

Rehabilitation: Changing the use of a historic structure through repair, alterations, and additions while preserving those portions or features, that convey its historical, cultural and architectural values.

Reimbursable expenses: Amounts expended for or on account of the project that, in accordance with the terms of the appropriate agreement, are to be reimbursed by the owner.

   These expenses may cover costs for services that could not or intentionally were not quantified at the time the fee arrangement was made.

Reinforced concrete: Concrete reinforced with steel bars.

Reinforced membrane: A roofing or waterproofing membrane that has been strengthened by the addition or incorporation of one or more reinforcing materials, including woven or nonwoven glass fibers, polyester mats or scrims, nylon, or polyethylene sheeting.

Reinforcing: The process of strengthening a member with some additional piece of material. An additional member added to a structural member to provide additional strength is referred to as reinforcement.

Reinforcing bar: Ribbed steel bars installed in foundation concrete walls, footers, and poured in place concrete structures designed to strengthen concrete. They are bent into special shapes according to the engineer's bending schedule. Ironworkers usually place rebar. Reinforcing bars come in various thickness' and strength grades. Also referred to as **reinforcing steel, reinforcement steel, rerod, deformed bars, reo and reo bar**, but in your author's experience they are universally referred to as **rebar.**

Rebars.

Relamp: The act of reinstalling light bulbs, lamps, and various luminaries.

Relative humidity: The ratio of partial density of water vapor in the air to the saturation density of water vapor at the same temperature and the same total pressure.

Relative rigidity: The comparative stiffness of interconnected structural members in view of relative distribution of the horizontal force. (Only identical stiffness of interconnected members can share the total load equally.)

Relaxation: A decrease in load or stress of a member under a sustained constant deformation.

Also, the loss of clamping force in a bolt that occurs typically without any nut rotation occurring. Commonly occurs as a result of embedment but can also be due to gasket creep, metal creep (at elevated temperatures), differential thermal expansion, and stress relaxation.

Release film: A thin strip of plastic attached to the underside of composition shingles to prevent the shingles from sticking together during shipment. When installed, the release film lines up with the sealant strip on the face of the shingle in the course below; it does not have to be removed for application. When heated by the sun asphalt penetrates the release film and bonds to the sealant strip of the shingles in the underlying course. Also called **release tape**.

Release of lien: A written action properly executed by an individual or firm supplying labor, materials or professional services on a project that releases his mechanic's lien against the project property.

Release powder: Powder spread on the concrete surface before a concrete stamp is used. Prevents the concrete stamp from sticking to the concrete and adds a second color to the concrete.

Relief valve: A device designed to open if it detects excess temperature or pressure.

Remote: A remote electrical, gas, or water meter digital readouts that are installed near the front of the home in order for utility companies to easily read the home owners usage of the service.

Remote-control circuit: Any electrical circuit that controls any other circuit through a relay or an equivalent device.

Render: Cement-based wall plaster.

Also, a first coat of plaster applied to a brick or stone surface.

Renewable energy: Energy from sources that are not depleted when used. This includes energy from the sun, wind, and small hydro-power. Ways to capture energy from the sun include photovoltaic, thermal solar energy systems, and bioenergy. One issue with bioenergy is the amount of fossil fuel energy used to produce it.

Renewable energy certificates (RECs): Tradable environmental commodities representing proof that a unit of electricity was generated from a renewable energy resource; RECs are sold separately from the electricity itself and thus allow the purchase of green power by a user of conventionally generated electricity.

Repairability: The process of determining whether an item is repairable or should be replaced. If an item can be made to look and function the same as it did before the damaging event at a cost that is less than replacement cost it should be repaired. If not, it should be replaced.

Repair: Work to restore damaged or worn-out property to a normal operating level. As a basic distinction, repairs are curative and maintenance is preventive. Both repairs and maintenance are usually considered expenses, although major repairs may fall under the category of capital expences; this is something to be discussed with your accountant.

Repair clamp: A sleeve type clamp used to repair a split pipe, held in place by bolts.

Pipe repair clamp.

Replacement value: The cost of an item purchased from a retail vendor, reflecting its age and condition.

Replication: Constructing a building so that it is an exact replica or imitation of an historic architectural style or period.

Representative number: The quantity of components chosen for close inspection by an owner or building inspector. This is usually, but not always, one per room for multiple similar interior components such as windows, and electric receptacles; one on each side of the building or home for multiple similar exterior components.

Reproduction: The construction or fabrication of an accurate copy of an object.

Reroofing: The process of re-covering, or tearing-off and replacing an existing roof system.

Resawing: Cutting timber along the grain (lengthwise), parallel to the wide face, to reduce larger sections into smaller sections or veneers. Resawing changes the thickness of the lumber but not its width.

Reshoring: The construction operation in which shoring equipment (also called reshores or reshoring equipment) is placed, as the original forms and shores are removed, to support partially cured concrete and construction loads.

Resident architect: An architect who is permanently assigned to a job site. The resident architect oversees the construction work for the purpose of protecting the owner's interests during construction.

Resident engineer: An individual permanently assigned to a job site for the purpose of representing the owner's interests during the construction phase, i.e., owner's representative.

Residential interior design: Concerned with the planning and/or specifying of interior materials and products used in private residences.

Residual: The residual is the difference between the measured and predicted values of some quantity.

Residual stress: Pre-induced stresses within a structural member due to uneven cooling of the shape after hot-rolling.

Resin: A sticky flammable organic substance, insoluble in water, that is exuded by most notably fir and pine trees. There are also solid and liquid synthetic organic polymers that are used as the basis of plastics, adhesives, varnishes, or other products.

Resins act as a binder and give coatings physical properties such as hardness and durability.

Examples of resins are acrylic, alkyd, copal ester, epoxy, polyurethane, polyvinyl chloride, silicone.

Resistance: The capacity of a structure or structural member to resist the effects of loads or forces imposed on it.

Also, anything that impedes the flow of electricity, particularly in direct (DC) current. Electrical resistance is measured in ohms.

Also, a measure of a body's ability to prevent heat from flowing through it, equal to the difference between the temperatures of opposite faces of the body divided by the rate of heat flow. Also known as heat resistance. Used widely in the window industry.

Also, relating to concrete, cycles of freezing and thawing may adversely affect concrete elements. Since the 1930s, air entrainment has been used to enhance the freeze-thaw resistance of portland cement concrete exposed to an external environment. The typical deterioration of concrete exposed to freeze-thaw conditions is random cracking, surface scaling and joint deterioration due to D-cracking. The first two are primarily due to lack of adequately entrained air in the concrete mass or the surface layer, respectively, and the latter phenomenon is primarily related to non-durable aggregate.

When water in the concrete whether from precipitation or contact with moist sub grade freezes, it expands. This movement of water generates pressures that, when in excess of the tensile strength of concrete or mortar layer at a surface, causes cracking and scaling. Concrete has to be critically saturated (>91%), which is generally true for concrete surfaces. Entrained

air bubbles are microscopic in size (0.01 inches or less), evenly distributed in the paste fraction, and take on water during the freezing cycle to relieve pressure buildup. Generally, an air entrainment of 4–8% and, more importantly, an air bubble spacing factor of less than 0.01 inch provides satisfactory freeze-thaw performance under most conditions.

Resisting force: See **Internal force.**

Resonance: Induced oscillations of maximum amplitude produced in a physical spectrum when applied oscillatory motion and the natural oscillatory frequency of the system are the same. When the site and building periods coincide, the buildings resonate with the ground. Then the amplitude of building vibration gradually approaches infinity by time, resulting in structural failure. The ground may vibrate at a period of 0.5–1.0 s. Structures may vibrate at a period of 0.1–6 s depending on the type of structure.

Examples:

- 1 story structure = 0.1 s.
- Up to 4 story structure = 0.5 s.
- 10–20 story structure = 1–2 s.
- Water tank structure = 2.5–6 s.
- Large suspension bridge = 6 s.

Resorcinol glue: An adhesive made from resorcinol resin and formaldehyde. Resorcinol glue is waterproof and is impervious to salt and fresh water, temperature extremes, weather, solvents, oils, grease, mild acids, or alkali. It is highly resistant to degradation by molds, fungi, bacteria, and insects.

Response spectrum: With regard to earthquakes, the maximum response (generally acceleration) of a site plotted against increasing periods.

A response spectrum is a plot of the peak or steady-state response (displacement, velocity or acceleration) of a series of oscillators of varying natural frequency, that are forced into motion by the same base vibration or shock. The resulting plot is used in assessing the peak response of buildings to earthquakes.

Restoration: The act or process of accurately depicting the form, features, and character of a historic structure, landscape, or object as it appeared at a particular period of time by means of the removal of features from other periods in its history and reconstruction of missing features from the restoration period.

Restriction: A reduced cross-sectional area in a line or passage that produces a pressure drop.

Retain: To keep secure and intact.

Retaining wall: A structure that holds back a slope and prevents erosion. Many building codes require retaining walls over four feet in height to be designed by a licensed engineer.

Resultant: The resultant of a system of forces is a single force or moment whose magnitude, direction, and location make it statically equivalent to the system of forces.

Retainer: A fee paid in advance to an architect, attorney, or engineer (or anyone), in order to secure or keep their services when required.

Retaining wall: A wall designed to resist the lateral displacement of soil, water, or any other type of material.

Retention: A percentage withheld from a contractor's payment until the project is completed or until an agreed upon time after the work is complete.

Retrofit: Any change to an existing facility, such as the addition or removal of equipment or a required adjustment, connection, or disconnection of equipment.

Retro-sizing: Refers to units that are sized for replacement purposes.

Return air: Air removed from conditioned spaces and either recirculated in the building or exhausted to the outside.

Return line: A line used to carry exhaust fluid from the actuator back to sump.

Return period of earthquakes: The time period in years in which probability is 63% that an earthquake of a certain magnitude will recur. See **Recurrence interval**.

Reuse: A method of returning materials to active use in the same or a related capacity thereby extending the lifetime of the materials. Examples of ongoing consumables that can be reused include binders, staplers, and other desk accessories, whether they are reused on-site or donated to other facilities.

Reverse board and batten: Vertical siding in which narrow boards called battens are installed first with gaps between them. Wider boards are then installed over the gaps.

Reverse casehardening: A final stress-and-set condition in dry lumber whereby the outer fibers are under tensile stress and the inner fibers are under compressive stress. This condition is not reversible.

As opposed to **casehardening** that is a condition of stress-and-set in dry wood whereby the outer fibers are under compressive stress and the inner fibers under tensile stress. The stresses persist after the lumber is dry and cause warp if the wood is remachined after drying.

Reverse osmosis: A method of producing pure water; a solvent passes through a semi permeable membrane in a direction opposite to that for natural osmosis when it is subjected to a hydrostatic pressure greater than the osmotic pressure.

Reverse painting: A technique where paint is applied to the back side of the surface (typically glass) and viewed through the front. This process requires the painting to be done in reverse order; what appears closest to the viewer, as a detail or highlight must be painted first rather than last. Any lettering must likewise be painted in the mirror image so it will appear right facing when viewed from the front.

Reverse trap water closet: A water closet having a siphonic trapway at the rear of the bowl and an integral flushing rim and jet.

Reverse trap

Reversing valve: A four-way directional valve used to reverse a double-acting cylinder or reversible motor. In a heat pump a reversing valve reverses the direction of refrigerant flow. This changes the heat pump refrigeration cycle from cooling to heating or vice versa allowing a facility to be heated and cooled by a single piece of equipment, by the same means, and with the same hardware.

Re-cover: The addition of a new roof membrane or steep-slope roof covering over a major portion of an existing roof assembly without removal of the existing roofing.

Rheostat: An electrical instrument used to control a current by varying the resistance.

Rib: A fabricated fold or bend in a sheet of deck that projects up from a horizontal plane.

Ribbon: In carpentry, a (normally) 1×4 board let into the studs horizontally to support the ceiling or second-floor joists. Also called a girt.

Ribbon window: A continuous band of windows.

Office ribbon window.

Richter magnitude scale: A measure of earthquake size that describes the amount of energy released. The measure is determined by taking the common logarithm (base 10) of the largest ground motion observed during the arrival of a P-wave or seismic surface wave and applying a standard correction for distance to the epicenter. (Each unit of the Richter scale represents a 10 times increase in wave amplitude. This corresponds to an~ 31.6 times increase of energy discharge for each unit on the Richter scale.)

The scale was replaced by the **moment magnitude scale (MMS).** The numerical values of earthquakes that were adequately measured by the Richter scale are approximately the same. See **Moment magnitude scale(MMS).**

Ridge: As used here, a ridge is the line or edge formed where the two sloping sides of a roof meet at the top.

Other terms beginning with ridge:

- **Ridge board:** The board placed on the ridge of the roof onto which the upper ends of other rafters are fastened.
- **Ridge cap:** A material or covering applied over the ridge of a roof.
- **Ridge course:** The last or top course of roofing materials, such as tile, roll roofing, shingles, etc., that covers the ridge and overlaps the intersecting field roofing.
- **Ridge shingles:** Shingles that are used to cover the ridge board.

Ridge shingles that also serve as a ridge cap. They are often used to cove a ridge vent.

- **Ridge tile:** A curved tile that covers the ridge on a pitched roof.
- **Ridge vent:** A vent that is placed along the ridge of the roof. It allows ventilation of the roof by raising the level of the ridge slightly leaving room for air flow. A filtration fabric placed in the side vents allows air to move through while preventing insects from entering. (Not to be confused with rigid vent that is hard plastic ridge vent material.)

Ridge vent.

Rifflers: These tools are for shaping and smoothing details that other files can't handle. One end is fine, the other end is coarse.

Rifflerfile cut, flat taper.

Rift: A fault trough formed in a divergence zone or in other areas in tension. See **Graben**.

Rift cut: Refers only to oak vaneer. It is a veneer that is produced by cutting at a slight right angle to the radial to produce a parallel grain pattern and quartered appearance without excessive ray fleck.

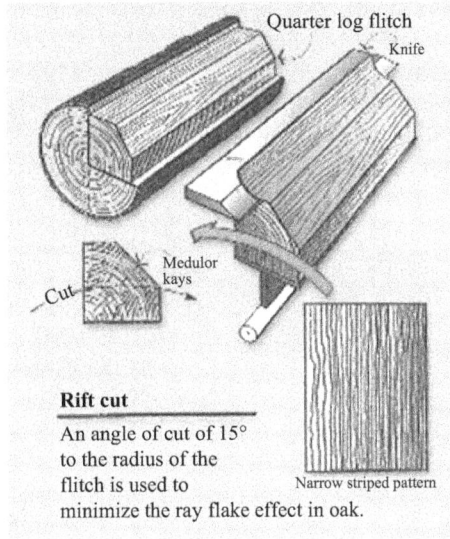

Quarter log flitch

Knife

Medulor kays

Cut

Rift cut

An angle of cut of 15° to the radius of the flitch is used to minimize the ray flake effect in oak.

Narrow striped pattern

Riftsawn: Wood that has been cut so that growth rings are at an angle of 30-60° to the board face.

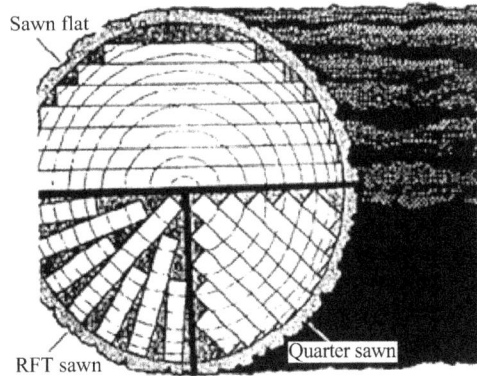

Sawn flat

RFT sawn

Quarter sawn

Riftsawn, quartersawn, and sawn flat (plainsawn).

Rigger: A person who erects and maintains scaffolding, lifting tackle, cranes, etc.

Right-lateral fault: See **Fault, right-lateral**.

Rigid: An idealized concept that means something that does not deform under loading. In fact, all objects deform under loading, but in modeling it can be useful to idealize very stiff objects as rigid.

Other terms beginning with rigid:

- **Rigid connection:** A connection where moment is transferred from one member to another. See also Fixed-end support and Fixed connection.

• **Rigid frame:** A structural framing system that consists of members joined together with moment or rigid connections that maintain their original angular relationship under load without the need for bracing in its plane. See **Frame** and **Stability**. Also called a **rigid structure**.

• **Rigid pipe construction:** In whirlpool construction, pipes that carry water into the whirlpool system may be flexible or rigid. Flexible pipes are more likely to allow water to puddle, resulting in bacteria build-up; rigid pipes are more likely to keep the pipes free of standing water.

Rigidity: Relative stiffness of a structure or element. In numerical terms, equal to the reciprocal of displacement caused by a unit force. See Shear modulus.

Right triangle: A three sided shape which includes a 90° angle between two of its sides.

Right-angle drill: This tool allows users to reach around corners or into confined spaces to bore holes or drive screws. It is excellent for smaller jobs such as drilling pilot holes or driving screws inside a cabinet.

Right-angle drill.

Rim joist: A joist that runs around the perimeter of the floor joists and home.

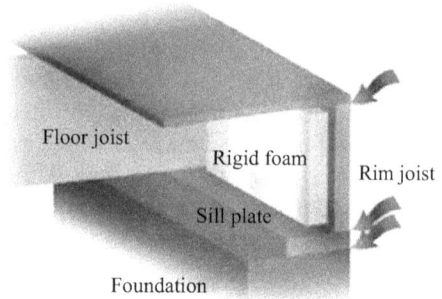

Rim speed: The speed of a saw blade at the extreme periphery when rotated.

Rimless sink: A sink with edges that overlap the hole in the countertop. Rimless sinks are usually made of heavy materials such as cast iron.

Rimless sink.

Rimmed sink: A sink with a rim that attaches to the edge of the sink and to the countertop.

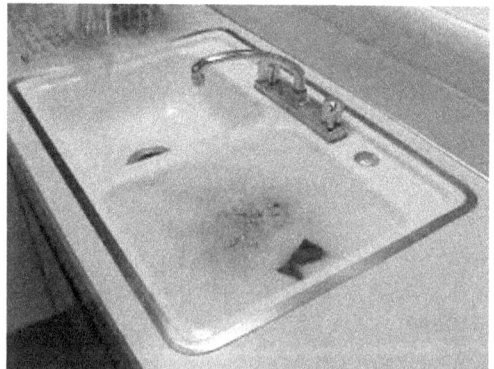

Rimmed sink.

Ring porous: Hardwood that shows a distinct zone between early and late wood, such as oak and ash.

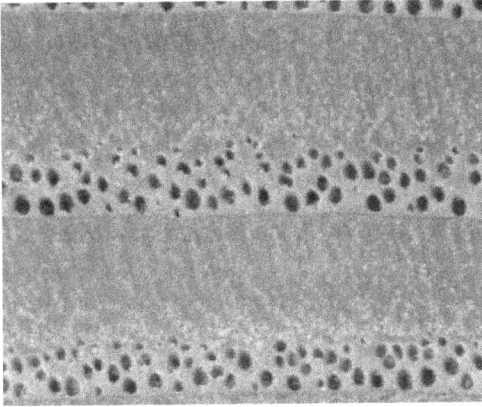

Ring porous.

Rip cut: A cut that is made parallel to the direction of a piece of wood's grain.

Rip hammer: This (22-oz.) hammer has a straight ripping claw. They come with fiberglass, wood and solid steel handles and are used for rough carpentry work.

The rip hammer differs from a claw hammer in that it is much flatter and is used to rip apart the wood that has been nailed together.

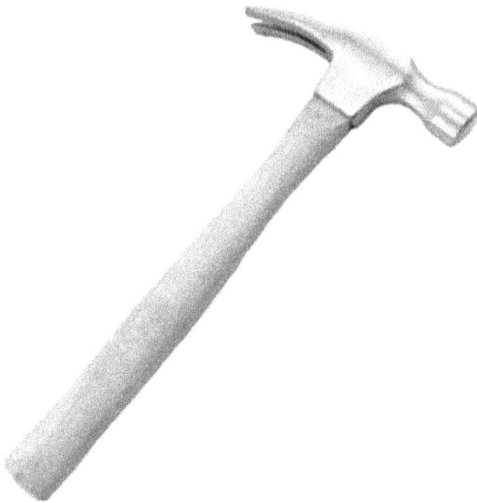

Rip hammer.

Rip saw: A rip saw is designed for making rip cuts. The cutting edge of each tooth has a flat front edge and it is not angled forward or backward. This design allows each tooth to act like a chisel (rather than being knife-like, as with a crosscut saw). This prevents the saw from following grain lines that could curve the path of the saw.

Rise: The vertical distance from the eaves line to the ridge.

Also, the vertical distance from stair tread to stair tread.

Also, any difference in elevation; the term is commonly used for almost all verticle distances.

Riser: Each of the vertical boards closing the spaces between the treads of stairways.

Also, a vertical run of plumbing, electric wiring, and or HVAC ducts.

Riser and panel: The exterior vertical pipe (riser) and metal electric box (panel) the tradespeople provide and installs at the rough electric and plumbing stage.

Rising damp: Water soaking up through the walls of the building.

Rising damp.

Risk evaluation, risk reduction, and risk management: All relate to exposure of life and property to earthquake hazards.

Rivet: Before large pre-formed structural steel members became available they were made up by joining steel plates with steel rivets. Rivets were also used to connect structural steelwork. Rivets have been mostly replaced by readymade sections and with bolting and welding that are faster and safer for connections. The presence of rivets in an existing structure can help to date it; rivets started to be replaced with bolts in the 1950s.

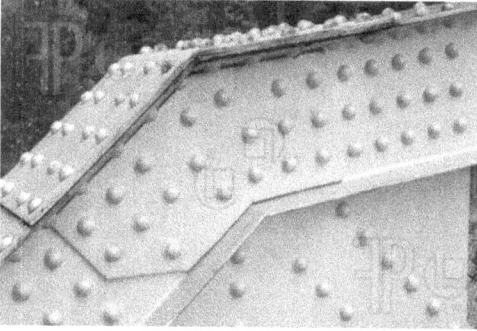

Detail of steel bridge that was constructed using rivets.

Road base: An aggregate mixture of sand and stone.

Road surface: The durable surface material that is intended to carry vehicular or foot traffic. In the past cobblestones and granite setts (a broadly rectangular quarried stone) were extensively used, but these have for the most part been replaced by asphalt or concrete. Most road surfaces are marked to guide traffic. Today, permeable paving methods are beginning to be used for low-impact roadways and walkways.

Roadbuilding: Roadbuilding typically involves the construction of roads, approaches, railroads, and curbs and gutters. The term usually includes resurfacing and repairing of roadways, but it does not include the routine maintenance of the roadways.

Rock avalanche: A large mass of rock, sliding or flowing very rapidly under the force of gravity.

Rock fall: Large block(s) of rock falling under the force of gravity.

Rodding: Tamping technique that involves consolidating concrete by the use of a push stick or rod.

Rock 1, 2, 3: When referring to drywall, this means to install drywall to the walls and ceilings (with nails and screws), and before taping is performed.

Rod: A smooth solid round bar, e.g., a one used for the web system of a bar joist (Bar joists form a lightweight, long-span system used as floor supports and built-up roofing supports.).

Also, a unit of length equal to 5½ yards, 16½ feet or 1/320th of a statute mile. Since the adoption of the international yard in 1959, it has been equivalent to exactly 5.0292 meters. Also called a **perch** or **pole**.

Rodded: The process of distributing the concrete mix in a test cylinder or slump cylinder. The concrete is "rodded" using a steel rod in an up and down movement.

Roll good: A general term applied to rolls of roofing felt, ply sheet, etc., that are typically furnished in rolls.

Roll/Rolling: To press, spread, or level with a roller, e.g., a paint roller.

Also, to move along a surface by rotation without sliding, e.g., roll a cart that is carrying materials; to move on wheels.

To install the floor joists or trusses in their correct place. To roll the floor means to install the floor joists.

Other terms beginning with roll rolled and roller:

• **Roll roofing:** A roofing material produced in rolls, made by saturating organic mat with asphalt or coal-tar pitch and embedding mineral granules on the surface exposed to the weather. Rolls are usually 36-inch wide with and 108 square feet of material. Weights are generally 45–90 pounds per roll.

• **Roll tiles:** Tile shingles that use caps and pans that form a series of peaks and valleys on the finished roof. Roll tiles include barrel and S tiles.

• **Roll type standing seam:** A standing seam roof in which the panels are placed next to each other with standing edges touching. The edges are then mechanically crimped to fasten and seal the seam.

- **Rolled aluminum:** A term that describes aluminum profiles for screen and energy panel surrounds that are fabricated by the use of a roller or series of rollers to produce a desired profile.
- **Rolled steel joist (RSJ):** A steel beam, especially one with a cross section in the form of a letter H or I.
- **Rolled wall covering:** Any wall covering that is provided in rolls. Examples include: fabric, vinyl, and paper.
- **Roller:** A cylinder that rotates around a longitudinal axis. Rollers are commonly used in various machines and devices to move, flatten, or spread something.
- **Roller guides:** Guide shoes that use rollers that rotate on guide rails rather than sliding on the rails.
- **Roller hanger:** A pipe hanger similar to a clevis hanger except that the yoke bolts to a roller rod instead of a metal strap. This roller rod supports the pipe and permits horizontal movement.

Roller support.

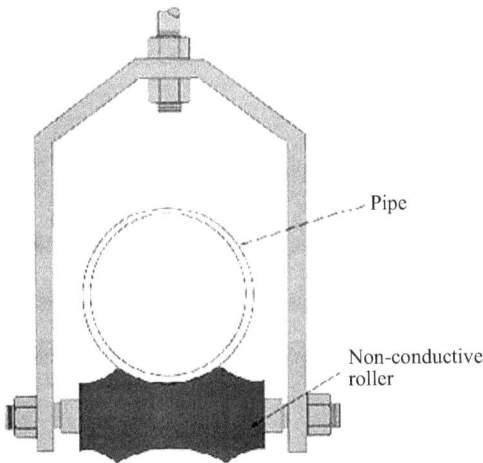

Roller hanger.

- **Roller support:** This type of support has two degrees of freedom: it can freely rotate about its axis or displace in one direction in the plane. Only one reactive force exists at a roller that acts perpendicular to the path of the displacement and its line of action passes through the center of the roller.

Rollover protective structure (ROPS): Vehicle structures such as roll-bars, frames, roll-protective cabs, etc., designed to prevent the vehicle operator from being crushed as a result of a rollover.

Roman tub filler: Similar to a basin sink faucet, but with a larger valve and higher flow levels used for filling baths and whirlpools more quickly.

Romex®: A named brand of nonmetallic sheathed electrical cable that is used for indoor wiring. The word Romex® is often used incorrectly in a generic sense to refer to any type of non-metallic sheathed electrical cable.

Roof: The top inner surface of a covered area or space.

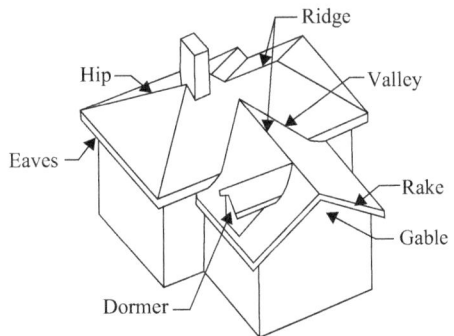

Common roof terms.

Terms beginning with roof, roofer, and roofing.

- **Roof assembly:** An assembly of interacting roof components: the roof deck, vapor retarder (if present), insulation, and roof covering.
- **Roof cement:** See Asphalt roof cement or Coal tar roof cement. Also called roofer's cement.
- **Roof covering:** The exterior roof cover or skin of the roof assembly, consisting of membrane, panels, sheets, shingles, tiles, etc.
- **Roof curb:** A raised frame used to mount mechanical units such as air conditioning units and exhaust fans, skylights, etc.
- **Roof deck:** Surface of the sheathing placed over the roof framing.
- **Roof diaphragm:** The entire roof system including rafters or trusses, bracing, sheathing, rough fascia, ridge boards, fasteners, and so forth. All elements of the roof system work together to form a diaphragm that resists wind and other forces and secure the top of exterior walls.
- **Roof drain:** Used with a roof membrane system, it fastens into the roof deck and carries water into a drainpipe. It is usually covered with a strainer that filters out leaves and other debris that may clog the drainpipe. You should consider other options before installing a roof drain as they do clog up especially in cold climates.
- **Roof jack:** Sleeves that fit around the black plumbing waste vent pipes at, and are nailed to, the roof sheeting.
- **Roof joist:** The rafters of a flat roof. Lumber used to support the roof sheeting and roof loads. Generally, 2 × 10's and 2 × 12's are used.
- **Roof louvers:** Rooftop rectangular shaped roof vents. Also referred to as box vents, mushroom vents, airhawks, soldier vents.
- **Roof overhang:** A roof extension beyond the exterior wall of a building.
- **Roof plane:** A roofing area defined by having four separate edges. One side of a gable, hip, or mansard roof.

- **Roof seamer:** A mechanical device used to crimp metal roof panels and make the seams watertight.

    Also, a machine used to weld membrane laps of PVC (Thermoplastic) roofing material.
- **Roof sheathing:** The wood panels or sheet material fastened to the roof rafters or trusses on which the shingle or other roof covering is laid. Also called **roof sheeting**.
- **Roof slope:** The angle that a roof surface makes with the horizontal, expressed as a ratio of the units of vertical rise to the units of horizontal length (sometimes referred to as run). For English units of measurement, when dimensions are given in inches, slope may be expressed as a ratio of rise to run, such as 4:12, or as a percent.
- **Roof system:** This includes the roof framing, sheathing, trusses, and roofing material. It is a structural part because it helps hold the bearing walls in place, resisting forces that attempt to move the walls such as wind and earthquakes.
- **Roof valley:** The "V" created where two sloping roofs meet.
- **Roof window:** An operable unit similar to a skylight placed in the sloping surface of a roof.
- **Roofer:** The contractor who supplies and installs roofing materials.
- **Roofing:** The materials that are applied to a roof to make it watertight.
- **Roofing work:** The hoisting, storage, application, and removal of roofing materials and equipment, including related insulation, sheet metal, and vapor barrier work, but not including the construction of the roof deck.
- **Roofing felt:** Asphalt-saturated organic mat that is produced in rolls. Used as shingle or siding underlayment, or anywhere a moisture-resistant barrier is needed. Also called tar paper or organic felt.
- **Roofing tape:** A asphalt-saturated tape used with asphalt cement for flashing and patching asphalt roofing.

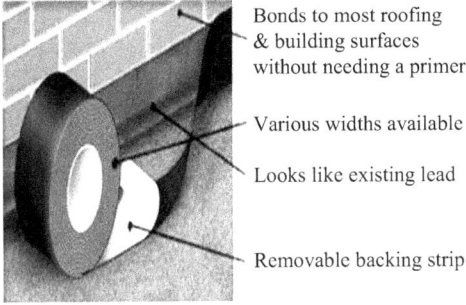

Bonds to most roofing
& building surfaces
without needing a primer

Various widths available

Looks like existing lead

Removable backing strip

Roofing tape.

Rope: A construction of twisted fibers or wire to form a inter-twisted strong cord. **Wire rope** is an assembly of multi-wire twisted around a fiber or steel core.

Rope grab: A deceleration device which travels on a lifeline and automatically, by friction, engages the lifeline and locks so as to arrest the fall of an employee. A rope grab usually employs the principle of inertial locking, cam/level locking, or both.

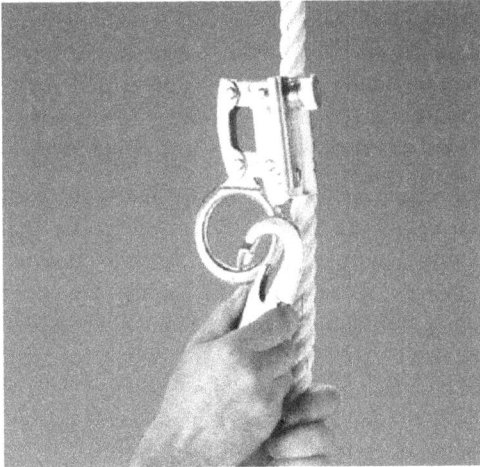

Rope grab.

Rose: A circular cover plate attached to the stile directly behind a knob or door handle. May be plain or have a decorative design embossed into the cover.

Rosette: A circular or oval decorative wood piece used at the termination of a stair rail into a wall. See **Balustrade**.

Rosin: A natural resin obtained from living pine trees or from dead tree stumps and knots.

Rosin paper (specifically Rosin-sized sheathing paper): A non-asphaltic paper used as a sheathing paper or slip sheet in some roof systems.

Rotary actuator: A device for converting hydraulic energy into rotary motion; a hydraulic motor.

Rotary cut: Veneer produced by centering the entire log in a lathe and turning it against a broad cutting knife. See **Veneer.**

Rotation: Motion of an object where the path of every point is a circle or circular arc. A rotation is defined by a point and vector which determine the axis of rotation. The direction of the vector is the direction of the axis and the magnitude of the vector is the angle of rotation.

Roto-gear: A term used to describe the steel drive worm, gears and crank device used for opening awnings and windows

Awning window operator (roto-gear).

Rough: As used here, generally work or the shape of something that is in a rough, preliminary fashion.

Other terms beginning with rough and roughing:

• **Rough arch:** A brick arch in which the bricks are rectangular and the arch shape is formed by means of the mortar joints being wedge-shaped. See **Axed arch.**

Rough arch.

- **Rough cut:** Irregular shaped areas of generally uneven corrugation on the surface of veneer. See **Veneer**.
- **Rough electrical/Rough-in electrical:** Any electrical device or part that will be hidden by, or embedded in, the finish wall. See Rough plumbing/Rough-in plumbing.
- **Rough fascia:** A horizontal member that is fastened to the vertical edge of the rafter tail or truss and later covered by the fascia. Also commonly referred to as the sub-fascia.
- **Rough level:** The initial process of placing wet concrete at the approximate desired level. The final level is applied later, when the surface is finished with the desired texture, after the concrete has lost some, but not all of its plasticity.
- **Rough heating:** Work performed by the heating contractor after the stairs and interior walls are built. This includes installing all ductwork and flue pipes. Sometimes the furnace and fireplaces are installed at this stage of construction. Also called heat rough.
- **Rough opening:** The horizontal and vertical measurement of a window, door, or access portal that replaces studs or joists where a window, door, stairway, skylight, etc., are placed and before drywall or siding is installed.
- **Rough plumbing/Rough-in plumbing:** Generally roughing in means to bore holes through the studs for the pipes, install and connect pipes, but does not include connecting fixtures or any end elements. See Rough electrical/Rough-in electrical.

- **Rough sill:** The framing member at the bottom of a rough opening for a window. It is attached to the cripple studs below the rough opening.
- **Roughing-in:** The initial stage of a plumbing, electrical, heating, carpentry, and/or other project, when all components that won't be seen after the second finishing phase are assembled. Completion of the rough-in usually means all parts that penetrate through the wall, floor, and roof sheathing are in place. The rough-in work for these four trades is reviewed in the four-way inspection before the walls or ceilings are covered. See also Heating rough, Plumbing rough, and Electrical rough.

Round: A 360° round molding, most often used as a closet pole. See Molding.

Round front toilet: Once the standard-size toilet, these fixtures measure about 2-inches shorter than their elongated counterparts. See Elongated toilet.

Round-top: One of several terms used for a variety of window units with one or more curved frame members, often used over another window or door opening. Also referred to as arch-tops, circle-tops, and circle-heads.

Rounding: With regard to decimals, the process of rounding, either up or down, the numbers after the decimal point to the desired precision, i.e., if the third digit after the decimal point is five or higher and the desired precision is 1/100th, you should increase the second digit after the decimal point by one; if the third digit after the decimal point is four or lower, leave the second digit as is.

Rounding also refers to process of **rounding dimensions** to the nearest unit of desired precision, i.e., a measurement ending with a fraction less than 3/8 inch and a desired precision of 1/2 inch should be rounded down to the next lower inch.

Rounding to a unit is a term that is used when ordering materials so that a fractional portion is equal to a multiple of the amount contained in the smallest package in which the material may be purchased.

Round file: A standard file for cleaning up or enlarging holes and shaping tight internal curves. It works on metal or wood.

Router: A power tool that is used to cut holes or openings into wood panels without the need to start at an edge. A router is also used to make decorative pattern cuts in wood.

Routers may be fitted with one of a myriad of bits, allowing it to do such things as cutting cabinet joints (dado, rabbet, etc.,), trim plastic laminate, shape decorative edges, mill moldings, and carve signs and plaques.

Routine maintenance: Maintenance that usually consists of service activities such as tightening, adjusting, oiling, pruning, etc. See Cyclical maintenance.

Rowlock: A course of brick laid on edge with their ends exposed.

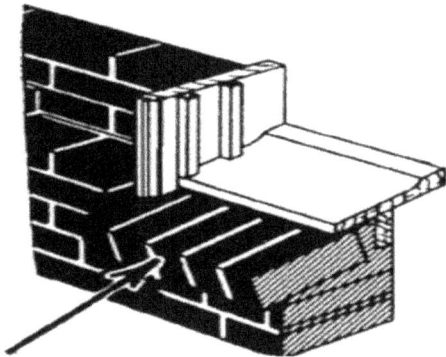

Brick rowlock.

Rubber and plastic mallets: These mallets are used to strike blows without damaging the surface. Useful for assembling furniture parts, setting dowel pins, metalwork, closing paint cans, etc.

Rubberized asphalt membrane: A shingle underlayment that adheres to the roof deck and seals around shingle nails driven through it during installation. Also referred to as bituthene, ice shield, or storm shield, it is placed on the roof where ice damming may occur to prevent water that may pass through the shingles from damaging the structure.

Rumford fireplace: A fireplace specially constructed to maximize heat output and minimizes smoke problems.

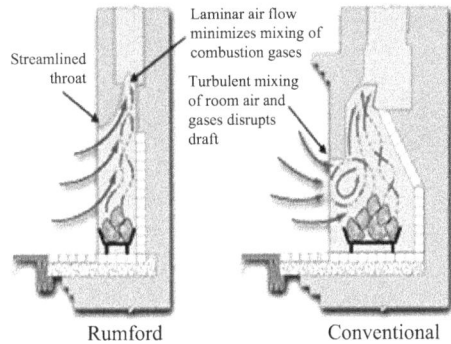

Streamlined throat

Laminar air flow minimizes mixing of combustion gases

Turbulent mixing of room air and gases disrupts draft

Rumford   Conventional

Rumford fireplace.

Run: With regard to a roof, the horizontal distance from the eaves to a point directly under the ridge; one-half the span.

With regard to stairs, the horizontal distance of a stair tread from the nose to the riser.

Rung: A rod or bar that forms the step in a ladder. Rungs attach to the two side rails of the ladder.

Also, an element of the **programmable logic controllers (PLC) ladder logic program**. PLC is a programming language that represents a program by a graphical diagram based on the circuit diagrams of relay logic hardware.

Running bond: Brick bond consisting of stretchers with head joints at the midpoint of the coursing below.

Running bond

Running match: See **Veneer.** Also referred to as **book match**.

Running trap: An in-line trap mounted in a horizontal drainpipe, where the inlet and outlet are parallel.

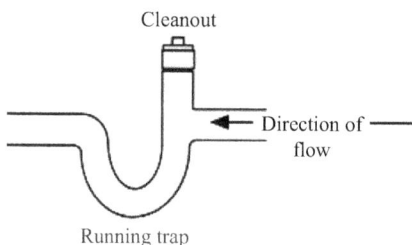

Running trap

Runout: The deviation from flatness of a circular saw near its periphery when rotated. Also referred to as wobble or warp.

Rupture front: The instantaneous boundary between the slipping and locked parts of a fault during an earthquake. Rupture in one direction on the fault is referred to as unilateral. Rupture may radiate outward in a circular manner or it may radiate toward the two ends of the fault from an interior point, behavior referred to as bilateral.

Rupture velocity: The speed at which a fault rupture propagates along a fault.

Ruptured grain: A break or breaks in the grain or between springwood and summerwood caused or aggravated by excessive pressure on the wood by seasoning, manufacturing, or natural processes. Ruptured grain appears as a single or series of distinct separations in the wood such as when springwood is crushed leaving the summerwood to separate in one or more growth increments.

Rust blush: The early stage of rust indicated by an orange or reddish color.

Rustic: Lacking excessive refinement, having a rough surface or finish.

S shapes: A hot rolled shape called an American Standard Beam with symbol S.

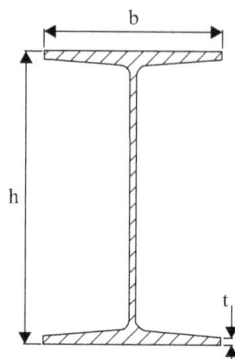

S beams come in many sizes. As an example, a S20×75 has the following dimensions: depth (h) 20 inches, flange width (b) 6-3/8 inches, flange thickness (t) 9/16 inches. The beam weighs 75 pounds per linear foot.

S-tile: A tile with a serpentine "S" shape. Also commonly referred to as Spanish tile.

S-wave: A seismic body wave involving shear motion transverse to the direction of propagation of the wave. See **Waves.**

Sabin: A unit of sound absorption, equal to one square foot (929 square centimeters) of a perfectly absorptive surface.

Sabre saw: This saw gives the user the ability to follow either curved or straight lines. Depending on the blade, it can cut metal, plastic, brick, etc. Once an opening is made its reciprocating blade can be used to cut enclosed holes.

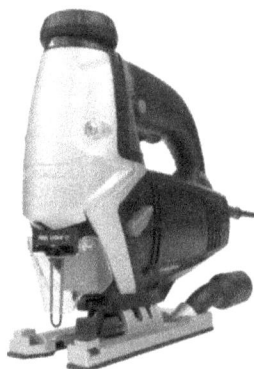

Sabre saw.

Saddle: A small second roof built behind the back side of a fireplace chimney to divert water around the chimney.

Saddle.

The part of the doorsill that spans the sill to floor construction joint and or is used as a stop for the bottom of the door. In interior doors the saddle is a closure piece from the floor to the door bottom. Also called a **sill** or **threshold**.

Door saddles.

Also, the shape of a roof.

Saddle roof.

Saddle angle: The angle connection or seat on the end of a header or frame which bears from the side on the top chord of a joist. This angle should be designed to carry the reaction of the header or frame to the center of the joist and must rest on and weld to both top chord angles.

Saddle valve: A connection that is used to tap into existing water supply lines. The Saddle T is clamped onto the pipe. When the valve is opened a drill-bit like point pierces the pipe and allows water into the Saddle T and the pipe connected to it. Also called a **saddle T, saddle tap**, and **flo tap**.

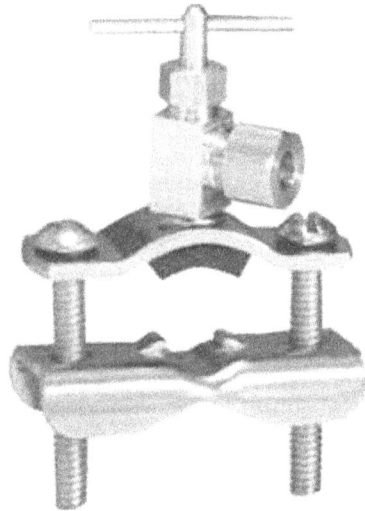

Saddle valve.

Sack mix: The amount of Portland cement in a cubic yard of concrete mix.

Safe area: Generally, an interior or exterior space that serves as a means of egress by providing a transitional area from a place where people gather. A safe area usually also serves as a normal means of entry and exit. If you are designing a building, check your local code for an exact definition.

Safe Drinking Water Act: An amendment to the Public Health Service Act that was passed in 1976 to protect public health by establishing uniform drinking water standards for the nation. In 1986 SDWA Amendments were passed that mandated the EPA to establish standards for 83 drinking water contaminants by 1992 and identify an additional 25 contaminants for regulation every 3 years thereafter.

Safety: A large clamp that anchors an elevator car to the building to keep the elevator from falling.

Terms beginning with safety:

- **Safety glass:** Laminated or wired glass that will hold and maintain the shards of glass in place when broken.
- **Safety harness:** The harness worn by Ironworkers, steel erectors, and others working where they may fall, thereby securing them to the structural frame with a safety line. OSHA requires this of all workers who erect structural steel and or are working on any structural frame prior to the installation of floor plates.
- **Safety report:** A report that is prepared following a regularly scheduled project safety inspection of the specific project. This should be on file in the event OSHA inspects the project.
- **Safety shutoff valve:** A device on a gas appliance that shuts off the gas supply to prevent a hazardous situation. A flame safety shutoff operates when the actuating flame becomes extinguished. In active earthquake areas houses are required to have such a valve that is activated with a shake.
- **Safety-monitoring system:** A safety system in which a competent person is responsible for recognizing and warning employees of fall hazards.

Safing: The installation of fire resistant material into a space, hole, opening, etc., through a fire rated assembly to insure the integrity of that assembly.

Sag: Undesirable excessive flow in material after application to a surface.

Sag pipe: A section of a sewer line that is placed deeper in the ground than normal in order to pass under utility piping, waterways, rail lines, highways, or other obstacles. The sewer line is raised again after passing under the obstacle.

Sag rod: A tension member used to limit the deflection of a girt or purlin in the direction of the weak axis.

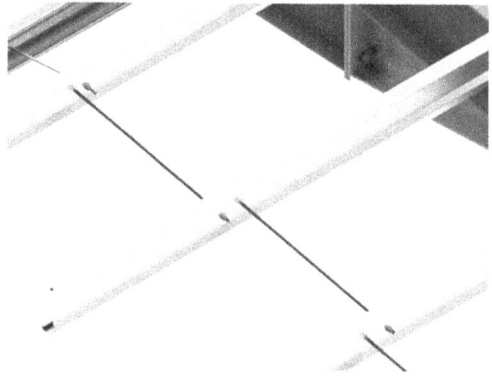

Sag rods.

Sailor course: A course of brick with each brick set vertically with the face, the long-wide side, of the brick exposed. See **Soldier course**.

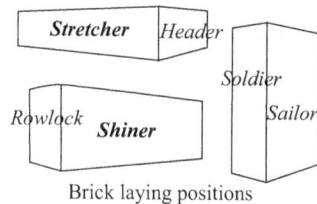

Brick laying positions

Sales contract: A contract between a buyer and seller that should explain:

- What the purchase includes?
- What guarantees there are?
- When the buyer can move in?
- What the closing costs are?
- What recourse the parties have if the contract is not fulfilled or if the buyer cannot get a mortgage commitment at the agreed upon time?

Sampling: The process of selecting units (e.g., masonry units, lumber, and people) from a population of interest so that by studying the sample we may fairly generalize our results back to the population from which they were chosen.

Sand: A general term given to small particles of various minerals, but especially to mica, feldspar, and quartz. It is the product of weathered rocks. Sands has many uses, e.g., as an abrasive and in concrete and mortar.

Terms beginning with sand and with sand as a prefix:

- **Sand boil:** Sand and water ejected from the ground as the result of liquefaction at shallow depths in an earthquake.
- **Sand float finish:** Lime that is mixed with sand, resulting in a textured finish on a wall.
- **Sand lime brick:** A kind of calcium silicate brick.
- **Sandblasting:** A method of etching the surface of material by spraying it with compressed air and sand.
- **Sanding, chatter, dust, burns:** The degree of defects allowed in sanding of the face.
- **Sandstone:** A common rock consisting mostly of quartz grains that are cementer together. Quartz is used in the manufacture of silica bricks that are used to line furnaces and as a building stone. Unfortunately, it does not weather well and should be used judiciously.

Sanitary fitting: A fitting that joins the assorted pipes in a drain, waste, and vent system, designed to allow solid material to pass through without clogging.

Sanitary sewer: A sewer system designed for the collection of waste water from the bathroom, kitchen, and laundry drains, and is usually not designed to handle storm water.

Sap: The fluid that circulates through a tree carrying the chemical food that enables the tree to grow. Sap is rich in minerals and nutrients.

Sap stain: A discoloration of sapwood that is caused by certain molds and fungi. Sap stain does not soften or cause disintegration of the wood.

Sapwood: Wood that is found near the surface of the tree, between the bark and the heartwood. Sapwood is lighter in color and less resistant to decay than heartwood.

Sash: A single light frame containing one or more lights of glass. The frame that holds the glass in a window, often the movable part of the window.

Also, a device, usually operated by a spring and designed to hold a single hung window vent up and in place.

Terms beginning with sash:

- **Sash balance:** A device for counter-balancing a sash of a double-hung window to hold it in the up position. Also referred to as spiral balance.
- **Sash cord:** Rope or chain in double-hung windows that attaches the sash to the counter balance.
- **Sash gang:** A frame in which one or several straight blades are clamped in a reciprocating frame.
- **Sash lift:** Protruding or recessed handle on the inside bottom rail of the lower sash on a double- or single-hung window.
- **Sash limiter:** A devise that limits how far a window may open. Limiters are available for many types of windows.
- **Sash lock:** A locking device that holds a window shut, such as a lock at the check rails of a double hung unit. Larger units utilize two locks.
- **Sash opening (SO):** The opening between wood frame members for both height and width (disregarding any jamb hardware tracks).
- **Sash retainer plate:** A nylon retainer plate used on double hung windows to secure the bottom sash.
- **Sash stiffener:** A reinforcement, usually inserted into a sash profile prior to assembly.
- **Sash stop:** A molding that covers the joint between window sash and the jamb.
- **Sash weights:** Concealed cast-iron weights used to counterbalance the sash of older double-hung windows.
- **Sash width:** The horizontal measurement across the face of a sash.

- **Sash window:** A traditional type of window that opens by sliding up and down. The frame is called a box-frame, because the side members are hollow wooden boxes in which the counterweights slide up and down. Unfortunately painters often caused them to stick, and the sash-cords had to be replaced frequently. Modern versions use springs instead of weights.

Repairing an old sash window.

Saturant: Asphalt that is used to impregnate an organic felt base material.

Saturated felt: There are numerous definitions. Generally, a felt that is impregnated with tar or asphalt.

Saturation coefficient: The ratio of the weight of water absorbed by a masonry unit during immersion in cold water to weight absorbed during immersion in boiling water; an indication of the probable resistance of brick to freezing and thawing.

Saw eye: The hole in the center of a circular saw blade, so it can be fitted on the arbor.

Saw guide: A supporting device that restrains the saw from deviating off line. It generally uses metal holders with babbit faces precision machined for accurate tolerances. Also referred to simply as a "guide."

Saw kerf: The groove cut by a circular saw blade.

Also, the width of the saw tooth at its outermost widest point.

Saw kerf counter flashing: A specially shaped counter flashing that is pressed into a saw kerf cut in masonry to prevent water penetration.

Sawing deviations: The deviations from targeted sawn sizes that are caused by the saws.

Scab on: A member fastened or welded to another member for reinforcement.

Scaffold: A temporary structure used to support people and material in the construction or repair of buildings and other large structures. Scaffolding is usually a modular system of metalpipes or tubes, although it can be constructed of lumber and other materials; there are numerous proprietary scaffolding systems. In Asia bamboo is commonly used for scaffolding.

Scaffold board: Timber boards used to make walkways on a scaffold.

Scagliola: A material developed in the 17th century in Northern Italy to duplicate marble. It is made from colored plaster and isinglass with inset marble chips. It can be polished to give a gloss finish.

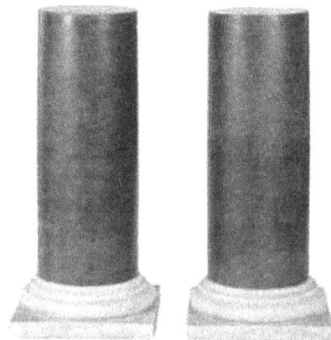

A pair of Italian Scagliola and plaster columns.

Scaling: Scaling is a disintegration process where local flaking or peeling of the surface portion of hardened concrete or mortar occurs.

Also, the forming of a solid crust of material that generally adheres to internal boiler surfaces causing localized overheating and circulation problems. Scale can be made up of many minerals with forms of calcium and silica being very common.

Also, logs are scaled (measured) to estimate the amount of lumber that can be obtained from them. Once logs have been processed into lumber they are again scaled to quantify the actual volumes produced.

Scanner: An optical or laser/camera measuring device. Scanners have a transmitter head and a receiver head that permit the electronics system to discern the shape and dimensions an object.

Scantling: The cross-sectional dimensions of a length of timber.

Also, the principal dimensions of a shaped stone.

Also, a piece of timber of a specific size.

Scalar: A mathematical entity that has a numeric value but no direction (in contrast to a vector).

Scallop depth: The depth of the arc created by the planer head on the finished surface of lumber. A planer's rotating cutterhead knives create scallops on the planed board. The depth of the scallop is dependent on the feed speed of the wood, the rotational speed of the planer cutterhead, and the number of cutterhead knives.

Scarf: A traditional woodworking joint for extending the length of a timber.

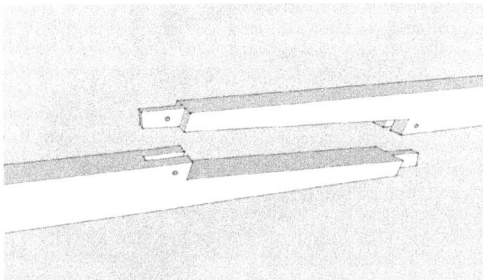

One of many types of scarf.

Scarfed: Shaped by grinding.

Scarp: A cliff, escarpment, or steep slope of some extent formed by a fault or a cliff or steep slope along the margin of a plateau, mesa, or terrace.

Schedule: As used here, a written or printed catalog or list of charges, items, prices, etc., arranged or organized in alphabetical, chronological, magnitudinal, or any other classification or order. For example: numbers that are assigned to different wall thicknesses of pipe is a pipe schedule; a table on the blueprints that lists the sizes, quantities, and locations of the windows, doors, and mirrors is a windows, doors, hardware, mirrors, schedule; and, the breakdown of a contract's cost into sub-items and sub-costs in a way that they can be evaluated for contractor progress payments is the **schedule of values**.

Schematic design phase: The schematic design phase of design involves preliminary design decisions for plans and specifications. See Design development phase.

Schoolmarm: A tree stem that branches into two or more trunks or tops.

Scissor joist: A non-standard type of joist where both the top chord and bottom chord are double pitched and parallel with each other.

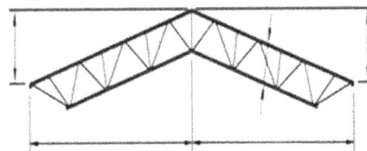

Scissor joist.

Scissor truss: A truss where the bottom chord is not horizontal. It is used where a sloped ceiling is desired in the inside of the building. The slope of the bottom chord is always less than the slope of the top chord.

Scissor truss.

Scoring: The process of cutting grooves into the face of panels thereby creating a different geometric visual with decorative and sometimes acoustical benefit.

Scrap out: The removal of all drywall material and debris after the drywall installation has been completed **(hung out)**.

Scratch coat: The first coat of plaster, which is scratched to form a bond for a second coat.

Screed: With regard to plaster, a small strip of wood, usually the thickness of the plaster coat, used as a guide for plastering.

With regard to concrete, a temporary rail, installed at a specific level, to enable concrete to be finished at the correct level.

Also sand and cement, mixed fairly dry and laid on a (usually concrete) floor and screeded and trowelled to make a smooth surface.

Screeding: With regard to concrete, to level off concrete to the correct elevation during a concrete pour; bringing the surface of concrete to the final, desired look, and finish by removing any excess or unwanted material.

Screen: A movable device, especially a framed construction such as a room divider or a decorative panel that is designed to divide, conceal, or protect.

Also, a coarse sieve that is used for sifting out fine particles, as of sand, gravel, or coal.

Also, a window or door insertion of framed wire or plastic mesh that is used to keep out insects and permit air flow. Usually called a **screen window** or **screen door**.

Also, the close-mesh woven screening material of metal, plastic, or fiberglass that is inserted into a window screen, to block the entry of insects but permit light, air, and vision through the screen. Also referred to as **wire cloth**.

Screen molding: See **Molding**.

Screen OM (outside measurement): The width and the height of a screen including wood or metal surrounds.

Screw: A threaded fastener.

Screw extractor: A device used to remove a broken screw, bolt, or stud from a threaded hole.

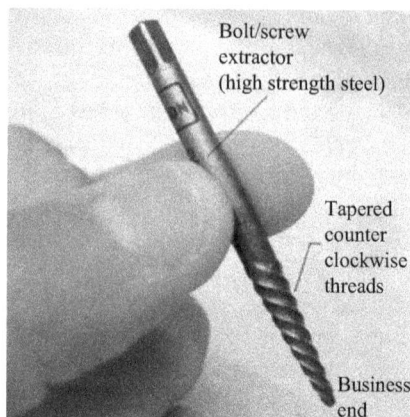

Bolt/screw extractor (high strength steel)

Tapered counter clockwise threads

Business end

Screw extractor.

Scribing: Cutting and fitting woodwork to an irregular surface; a scribe piece is an oversized piece of plastic laminate or wood that can be trimmed in the field to follow any minor irregularities of the wall.

Scrim: A woven, nonwoven, or knitted fabric, composed of continuous strands of material used for reinforcing or strengthening membranes. Scrim may be incorporated into a membrane by the laminating or coating process.

Scroll saw: This is a freehand curve-cutting machine. It uses fine-tooth blades that can cut intricate patterns. It gives smooth edges in thinner stock. The saw operates with little vibration, provides for quick blade changes and has easy-to-operate clamps. It is a stationary, i.e., not portable, tool.

Scroll saw.

A scroll saw is sometimes referred to as a jig saw, but a jig saw is different, as the picture below shows.

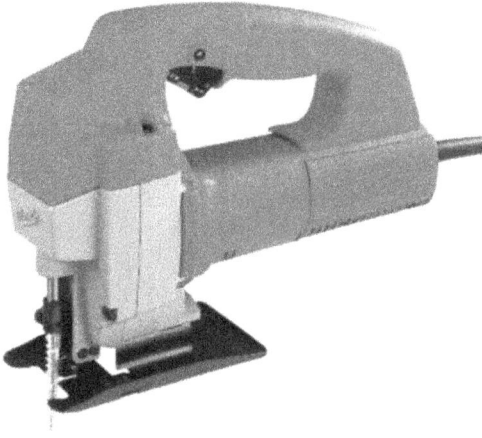

Jig saw.

Scope of work: The division of work to be performed under a contract or subcontract in the completion of a project, typically broken out into specific tasks with deadlines.

Sculptured carpet: A design of carpeting that is characterized by a mixture of high-pile and low-pile fibers arranged according to a specific configuration.

Scupper: An opening for drainage in a wall, curb, or parapet, usually connected to the downspout.

Scuttle: A unit that provides access to the roof from the interior of the building. See also Hatch.

Seal tab: See Sealant strip.

Sealable equipment: Equipment enclosed in a case or cabinet that is provided with a means of sealing or locking so the live parts cannot be made accessible without opening the enclosure. The equipment may or may not be operable without opening the enclosure.

Sealant: A compressible plastic material used to seal any opening or junction of two parts, such as between the glass and a metal sash, commonly made of silicone, butyl tape, or polysulfide. Sealant is used to weatherproof many types of construction joints where moderate movement is expected. The material comes in various grades: pourable, self-leveling, non-sag, gun grade, and cured or uncured tapes.

Sealant strip: Strips of asphalt placed on the face of the shingle where they will be covered by shingles in the course above. When the shingle is warmed by the sun, the sealant strip adheres to the shingle above, thus creating a tight, wind-resistant connection.

Sealed double glass: Two panes separated by a sealed space. See **Insulating glass**.

Sealer: A finishing material, either clear or pigmented, that is usually applied directly over raw wood for the purpose of sealing the wood surface. Sealers are designed to prevent absorption of finish coats into porous surfaces and to prevent bleeding.

Sealers are also applied to substrate to protect them from being stained by something that may drop or spill on it.

Seam: A joint formed by mating two separate sections of material. Seams may be made or sealed in a variety of ways, including adhesive bonding, hot-air welding, solvent welding, using adhesive tape, sealant, etc.

Seam strength: The force or stress required to separate or rupture a seam in the membrane material.

Seamless flooring: A mixture of a resinous matrix, fillers, and decorative materials applied in a liquid or vicious form that cures to a hard, seamless surface.

Seamless gutter: Aluminum gutter is often called seamless gutter because each straight section is made without seams.

Seasoning: Drying and removing moisture from green wood. Timber is considered seasoned when it has been dried so that the maximum moisture content anywhere in the piece does not exceed 15%.

Seat cut: The horizontal cut portion of a rafter which sets on a cap plate.

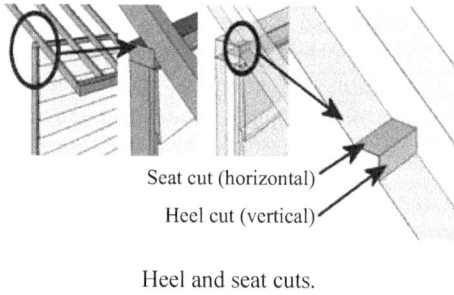

Seat cut (horizontal)
Heel cut (vertical)

Heel and seat cuts.

Seat depth: The out-to-out depth of the end bearing shoe or seat of a joist or joist girder which is the distance from the top of the top chord to the bottom of the bearing seat angle or plate.

Secant piles: Contiguous piles where each pile cuts into the one before, to make a more-or-less waterproof retaining-wall.

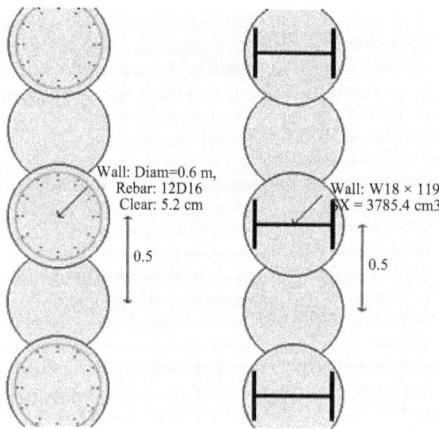

Wall: Diam=0.6 m,
Rebar: 12D16
Clear: 5.2 cm

0.5

Wall: W18 × 119
SX = 3785.4 cm3

0.5

Secant pile walls.

Second fix: Work that takes place after plastering, for example, fixing light switches, skirtings. See **First fix**.

Section: One of the five basic views found on a plan. A section is a view of the building as though it had been sliced through vertically and opened up so you could see what is inside. It may be thought of as a view showing a dissection of the building.

Section modulus: A property of a cross sectional shape that depends on shape and orientation. Section modulus is usually denoted S, and $S = I/c$, where I = moment of inertia about an axis through the centroid, and c is the distance from the centroid to the extreme edge of the section.

Sediment: Dregs, lees, precipitate, deposit, grounds; residue, remains; silt, alluvium; technical residuum. For example, the substance that settles on the bottom of a water tank.

Sedimentary rocks: Types of rock that are formed by the deposition of material on the Earth's surface. Sedimentation is the collective name for processes that cause mineral and/or organic particles (detritus) to settle and accumulate or minerals to precipitate from a solution. Before being deposited, sediment was formed by weathering and erosion and then transported to the place of deposition by water, wind, ice, mass movement, or glaciers which are called agents of denudation. See **Sedimentation**.

Sedimentation: The addition of soil particles to land and water bodies by natural and human-related activities. Sedimentation often decreases water quality and can accelerate the aging process of lakes, rivers, and streams.

Segmentation: The breaking up of a fault along its length into several smaller faults. This can happen as a result of other faults crossing it, topography changes, or bends in the strike of the faults. Segmentation can limit the length of faulting in a single earthquake to some fraction of the total fault length, thus also limiting the size of the earthquake.

Segregation: The separation of the components of a concrete mix during placement due to improper handling and or vibration. See **Moment magnitude scale (MMS)**.

Seiche: A standing wave on the surface of water in an enclosed or semi-enclosed basin (lake, bay, or harbor). A seiche is usually caused by strong winds and/or changes in atmospheric pressure. The seiche can continue, in a pendulum fashion, even after the cessation of the originating force.

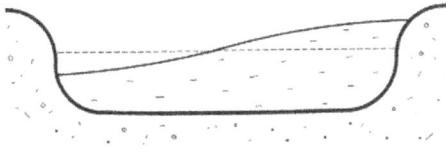

Seiche.

Seismic: Pertaining to earthquake activities.

Other terms beginning with seismic and the prefix seism:

- **Seismic gap:** A section of a fault that has produced earthquakes in the past but is now quiet. For some seismic gaps no earthquakes have been observed historically, but it is believed that the fault segment is capable of producing earthquakes on some other basis, such as plate-motion information or strain measurements.
- **Seismic hazard:** See **Earthquake hazard**.
- **Seismic hazard analysis:** An orderly approach to quantifying the extent to which a site or a region is exposed to earthquake hazards.
- **Seismic load:** Assumed lateral forces acting in any horizontal direction that produce stresses or deformations in a structural member due to the dynamic action of an earthquake.
- **Seismic moment:** A quantity used by earthquake seismologists to measure the size of an earthquake. Seismic moment is the basis of the moment magnitude scale that is often used to compare the size of different earthquakes and is especially useful for comparing the sizes of especially large (great) earthquakes. See **Moment magnitude scale (MMS)**.
- **Seismic reflection or refraction line:** A seismic refraction or seismic reflection line is a set of seismographs usually lined up along the earth's surface to record seismic waves generated by an explosion for the purpose of recording reflections and refractions of these waves from velocity discontinuities within the earth. The data collected can be used to infer the internal structure of the earth.
- **Seismic risk:** See **Earthquake risk**.

- **Seismic wave:** An elastic wave generated in the earth by an earthquake or explosion.
- **Seismic zone:** Areas defined on a map within which seismic design requirements are constant.
- **Seismic zonation:** Geographic delineation of areas having different potential for hazardous effects of earthquakes.
- **Seismicity:** The worldwide or local distribution of earthquakes in space and time; a general term for the number of earthquakes in a unit of time, or for relative earthquake activity.
- **Seismograph:** An instrument for amplifying and recording the motions of the earth caused by seismic waves.
- **Seismology:** The study of earthquakes, earthquake sources, and the propagation of seismic waves.
- **Seismometer:** The sensor that detects the seismic wave energy and transform it into an electric voltage.
- **Seismotectonic zone or province:** A geographic area characterized by similar geology and earthquake characteristics.

Select and better oak grade: A grade of oak strip flooring in which at least 50% of the wood is clear of defects except for a few small bright spots of sap. The other 50% may have pinworm holes, machine defects, and no more than one small tight knot for every three lineal feet of wood. See number one common oak grade and number two common oak grade.

Select structural: The highest grade of structural joists and planks. This grade is applied to lumber of high quality in terms of appearance, strength, and stiffness.

Select tight knot (STK): A grade term frequently used for cedar lumber. Lumber designated STK is selected from mill run for the tight knots in each piece, as differentiated from lumber which may contain loose knots or knotholes.

Self: As used here, something having its own persona.

Terms beginning with self:

- **Self-adhering membrane:** A membrane that can adhere to a substrate and to itself at overlaps without the use of an additional adhesive. The undersurface of a self-adhering membrane is protected by a release paper or film that prevents the membrane from bonding to itself during shipping and handling.

- **Self-cleaning glass:** Glass that is treated with a special coating. Currently, commercially available products feature a coating that uses the sun's UV rays to break down organic dirt through what is called a photocatalytic effect. The coating also provides a hydrophilic effect, which reduces the surface tension of water to cause it to sheet down the surface easily and wash away dirt.

- **Self-loosening fasteners:** Sometimes threaded fasteners come loose without human intervention. This loosening can be due to creep, embedding, stress relaxation or the fastener self-rotating (which is often called vibration loosening).

  Creep, embedding, and stress relaxation will generally not completely loosen a fastener; these loosening mechanisms occur without the nut rotating relative to the bolt.

  Self -loosening sometimes refers to the nut rotating relative to the bolt without human intervention. It is possible for fasteners to self rotate under the action of transverse joint movement. It can completely loosen a tightened fastener with the nut becoming detached from the bolt.

- **Self-drilling screw:** A fastener that drills and taps its own hole during application.

- **Self-retracting lifeline:** A deceleration device containing a drum-wound line which can be slowly extracted from, or retracted onto, the drum under slight tension during normal employee movement, and which, after onset of a fall, automatically locks the drum and arrests the fall. Also called a **self-retracting lanyard**.

- **Self-rimming sink:** A sink with no metal ring that has a built-in lip of the same material which supports it in the vanity top or kitchen counter. Also referred to as surface mounted sink. See **Rimless sink and rimmed sink**.

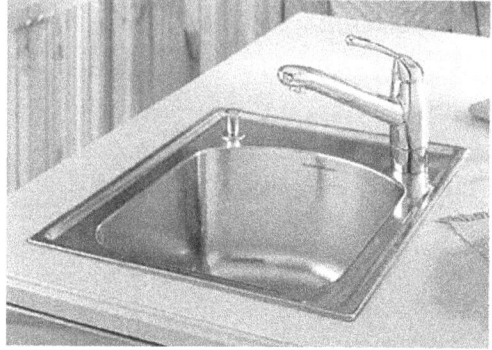

Self-rimming sink.

- **Self-sealing shingles:** Shingles containing factory-applied strips or spots of self-sealing adhesive. After installation, heat and sun will activate sealant to seal the shingles to each other. The factory applied adhesive is called a self-sealing strip or a spot.

- **Self-tapping screw:** A type of screw, commonly used with light gauge metal, that has a drill-bit style tip that forms its own hole in the metal. Also referred to as **self-drilling screw**.

- **Self-vulcanizing membrane:** Membrane that is initially thermoplastic in nature but that cures after installation.

Selvage: An edge or edging that differs from the main part of a fabric, granule-surfaced roll roofing or cap sheet, or other material.

Also, a specially defined edge of the material lined for demarcation that is designed for some special purpose, such as overlapping or seaming.

Selvage edge: An edge designed for certain sheet good materials, e.g., mineral-surfaced sheets. With mineral surfaced sheets, the surfacing is omitted over a portion of the longitudinal edge of the sheet (e.g., mineral surface cap sheet) in order to obtain better adhesion of the overlapping sheet; usually two, four, or nineteen inches in width.

Semi-gloss paint or enamel: A paint or enamel made so that its coating, when dry, has some luster but is not very glossy. Semi-gloss paints are formulated to give this result (usually 35–70° on a 60° meter). Bathrooms and kitchens are often painted with semi-gloss paint.

Separately derived system: A premises wiring system whose power is derived from a battery, from a solar photovoltaic system, or from a generator, transformer, or converter windings, and that has no direct electrical connection, including a solidly connected grounded circuit conductor, to supply conductors originating in another system.

Septic system: An on-site waste water treatment system. It usually has a septic tank that promotes the biological digestion of the waste, and a drain field that is designed to let the left over liquid soak into the ground.

Sequence: The order of a series of operations or movements. For example a detailed system-level documentation for each base building system that defines which operational states are desired under which conditions: running vs. idle systems; full-load or part-load operation; staging or cycling of compressors, fans, or pumps; proper valve positions; desired system water temperatures; target duct static air pressures depending on other variables (e.g., outside air temperatures, room air temperatures, and/or relative humidity); and any reset schedules or occupancy schedules **is referred to as the sequence of operations**.

Sequencing valve: A pressure operated valve that at its setting diverts flow to a secondary line while holding a predetermined minimum pressure in the primary line.

Serpentine seam: A carpet seam that is made by seaming two pieces of carpet that have been cut in meandering curved or S-shaped patterns. Serpentine seams are more time consuming and difficult to install than straight seams, but are believed to make the seam less visible. See **Straight seam**.

Service: As used here, generally fulfilling a need either by equipment or with people.

Other terms beginning with service:

- **Service cable:** Service conductors made up in the form of a cable.
- **Service conductors:** The conductors from the service point to the disconnecting means.
- **Service drop:** The overhead service conductors from the last pole or other aerial support to and including the splices, if any, connecting to the service-entrance conductors at the building or other structure.
- **Service, electrical:** The conductors and equipment for delivering electric energy from the serving utility to the wiring system of the premises served.
- **Service entrance conductors, overhead system:** The service conductors between the terminal of the service equipment and a point usually outside the building, clear of building walls, where joined by tap or splice to the service.
- **Service entrance conductors, underground system:** The service conductors between the terminals of the service equipment and the point of connection to the service lateral. See **Service lateral**.
- **Service entrance panel:** The main power cabinet where electricity enters a home wiring system.
- **Service equipment:** The main control gear at the service entrance, such as circuit breakers, switches, and fuses.
- **Service lateral:** The underground service conductors between the street main, including any risers at a pole or other structure or from transformers, and the first point of connection to the service-entrance conductors in a terminal box or meter or other enclosure, inside or outside the building wall. Where there is no terminal box, meter, or other enclosure, the point of connection is considered to be the point of entrance of the service conductors into the building.
- **Service point:** The point of connection between the facilities of the serving utility and premises wiring.

- **Service sink:** A deep fixed basin that is supplied with hot and cold water. It is used for rinsing of mops, disposal of cleaning water, or washing clothes, and other household items.
- **Service slab:** A 4-inch or so high slab of concrete placed over the floor slab as a pad for the installation of a piece of mechanical equipment.
- **Service tee:** A tee fitting with male threads on one run opening and female threads on the other two.

Servo mechanism: An automatic device that uses error-sensing negative feedback to correct the performance of a mechanism. Also called a servo.

The term applies to systems where the feedback or error-correction signals help control mechanical position, speed or other parameters. For example, an automotive power window that requires you to hold the control while the window moves is not a servomechanism because there is no automatic feedback that controls position; you do this by observation. By contrast the car's cruise control uses closed loop feedback, which classifies it as a servomechanism.

Servo valve: An electrically operated valve that controls how hydraulic fluid is transported to an actuator. An electrical signal is sent to the servo valve that opens it. This allows hydraulic fluid to flow into the cylinder. The voltage level and direction of the electrical signal determine the speed and direction of the device's movement.

Servo valves and servo-proportional valves change an analog or digital input signal into a smooth set of movements in a hydraulic cylinder. They provide precise control of position, velocity, pressure, and force with good post movement damping characteristics. Also referred to as an **Electro-hydraulic servo valve** or **EHSV**.

Set: The curing of concrete. Also, the act of setting a nail by recessing it beyond the face of the material it is nailed into.

Set screw: A threaded fastener that is typically used to hold a sleeve, collar or gear on a shaft. It is a threaded member that normally does not have a head. Unlike most other threaded fasteners it is basically a compression device normally used to generate axial thrust. Various socket types are provided to allow the set screw to be rotated. These types include hexagon socket, fluted socket, screwdriver slot, and square head.

Various point designs are available (the part of the set screw that rotates against the shaft being secured) and include:

- **Flat:** It causes little damage to the shaft and is used when frequent adjustment is required.
- **Oval:** A rounded end that is typically used when frequent adjustment is required. The oval end prevents/reduces indentation.
- **Cone:** A pointed end that generates the highest torsional holding power. It is typically used for a permanent connection.
- **Cup:** A hollowed end, it is the most commonly used point style. Used when the digging in of the point is not undesirable.
- **Dog:** A flat end with the threads stopping short of the end with the end fitting into a hole.

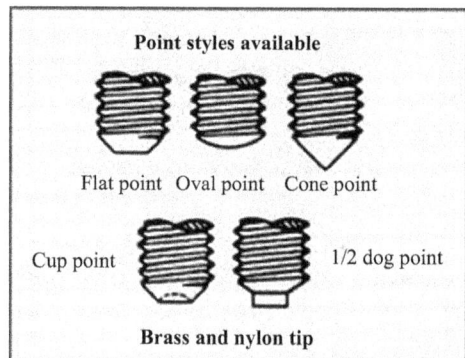

Point styles available

Flat point   Oval point   Cone point

Cup point   1/2 dog point

Brass and nylon tip

Setback: The distance from the property line to the foundation of the structure. Minimum set backs are established by local government to maintain desired appearance standards by keeping structures from being built to close to the edge of the property.

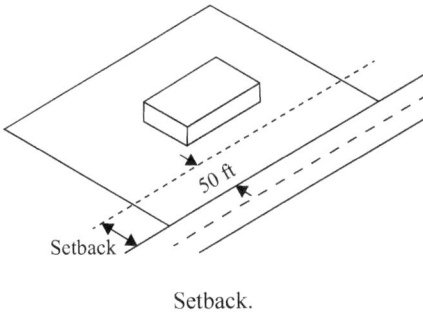

Setback.

In cities with tall buildings setbacks are also established to allow sunlight in.

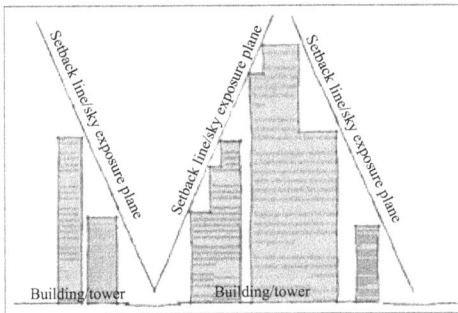

Setbacks.

Also, the distance from the outside edge of an angle or other member to the edge of a gusset plate or angle welded near the end.

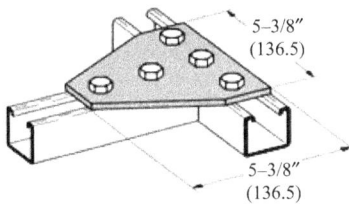

Setback = 5-3/8 inches.

Also, a current running opposite to the main flow of water, i.e., an **eddy**.

Setback thermostat: A thermostat with a clock that can be programmed to come on and off at various temperatures and at different times of the day/week.

Setpoints: Normal ranges for building systems and indoor environmental quality, outside which action is taken.

Settlement: The small downward movement of foundations when the weight of the building comes onto them. There are many possible causes for settlement: compression of the soil due to insufficient pre-construction compaction, poorly designed foundations, movement caused by freeze-thaw cycles, etc.

Setting: The place or type of surroundings where something is positioned or where an event takes place. For example, the physical environment of a historic property; the character of the place in which the property played its historical role.

Also, a speed, height, or temperature at which a machine or device can be adjusted to operate.

Setting block: In glazing, a small synthetic block of material used to support the glass at its base bearing surface.

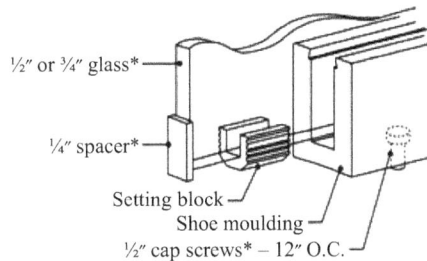

Setting block.

Sewage: The waste matter that passes through sewers.

Sewage ejector: A pump used to "lift" wastewater to a gravity sanitary sewer line. Used in basements and other locations that are situated below the level of the side sewer.

Sewer: An artificial conduit that is usually underground. Sewers carry off (usually) urban waste water and refuse.

Other terms beginning with sewer:

- **Sewer lateral:** The portion of the sanitary sewer that connects the interior waste water lines to the main sewer lines.
- **Sewer stub:** A short section of pipe connected to the main sewer line or septic

tank and extending toward the home. It is designed so the drain line coming from the home can easily be connected to it. The footings on a full basement home should be positioned so that the stub is lower than the bottom of the footings. This ensures a downhill slope for a sewer line extending from under the footing out to the stub. Also called a **septic tank stub**.

- **Sewer tap:** The physical connection point where the home's sewer line connects to the main municipal sewer line.

Sewerage: A system of sewers.

Shackle: Threaded rods to which elevator hoist cables are socketed and that bolt to the hitch plate and the counterweight.

Shade screen: A specially fabricated window screen of sheet material with small narrow louvers formed in place to intercept solar radiation striking a window; the louvers are so small that only extremely small insects can pass through. Also, an awning with fixed louvers of metal or wood construction. Also referred to as **sun screen**.

Shading: Slight differences in the color of materials that are used in construction. Shading occurs as a result of normal manufacturing operations, e.g., there may be slight differences in shine color from bundle to bundle.

Shading coefficient (SC): A measure of a window's ability to transmit solar heat, relative to that ability for 1/8-inch clear glass. The lower a unit's shading coefficient, the less solar heat it transmits, and the greater its shading ability. It is being phased out in favor of the **solar heat gain coefficient (SHGC)**. See **Solar heat gain coefficient (SHGC)**.

Shadow zone: The area of the earth from angular distances of 104–140° from a given earthquake that does not receive any direct P waves. The shadow zone results from S-waves being stopped entirely by the liquid core and P-waves being bent (refracted) by the liquid core.

Shaft: A vertical aligned series of penetrations through a buildings floor plates for elevators, ducts, etc.

Shaft liner: Gypsum board that has been fire rated and designed as an assembly for installation as a shaft wall.

Shag carpet: A carpet with a long pile. See **Wool carpet, Nylon carpet, Berber carpet, Indoor-outdoor carpet**, and **Sculptured carpet**.

Shake: A western red cedar (sometimes redwood) roofing and sidewall product. Shakes are produced by splitting a block of the wood along the grain line. Modern shakes are sometimes machine sawn on one side. See Shingle.

Cedar wood shake roof.

Also, there are **ring** and **wind shakes** (that are not shingles). Ring shakes are lengthwise grain separations between growth rings. Wind shakes break through the rings and are called radial shakes.

Ring and wind shakes are not usually associated with disease or insects but are caused by the wind. Nevertheless, there is usually a bacterial link and it is thought that some infection is usually present in the annual ring where the shake occurred. Ring shakes can be caused by high winds but are often initiated by other environmental factors such as fire and disease. Wind shakes are caused by the wind only; fire or bacteria play no role in their development.

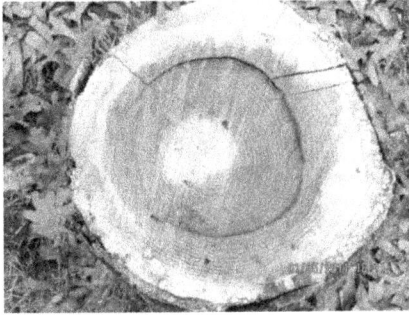

Ring and wind shakes.

Shake felt: Roofing felt that usually comes in rolls 18-inch wide.

Shake roof: A roof constructed from roofing material made from hand-split wood. Shakes come in three thicknesses: thin, medium, and heavy, and are usually made from cedar with relatively straight grain and free of knots. See also **Hand-split, Resawn shake**, and **Tapersawn shake**.

Shall: In contracts means mandatory. See **Should**.

Shallow well: A well with a pumping head of 25 feet or less, thus permitting use of a suction pump. See **Deep well**.

Shank: A device for locking inserted teeth in a circular saw.

Shape factor: The ratio of the plastic section modulus (Z) to the elastic section modulus (S) or the ratio of the plastic moment (Mp) to the yield moment (My).

Shape sawing: Sawing a log or cant by following the arc or curvature of the log or cant (A piece of wood produced by a canter [A machine that converts logs into a square, rectangular or two-sided cant for further processing.] that requires further breakdown.) Some systems follow an arc and other systems can follow a compound curve. Also referred to as **curve sawing** and **sweep sawing**.

Shaper: A machine with revolving cutters that is used to cut moldings and other irregular outlines.

A shaper uses linear relative motion between the workpiece and a single-point cutting tool to machine a linear toolpath. Its cut is analogous to that of a lathe, except that it is linear instead of helical. Axes of motion can be added to yield helical toolpaths.

A shaper is somewhat like a planer, but it is smaller and has the cutter riding a ram that moves above a stationary workpiece, rather than the entire workpiece moving beneath the cutter. The ram is moved back and forth typically by a crank inside the column.

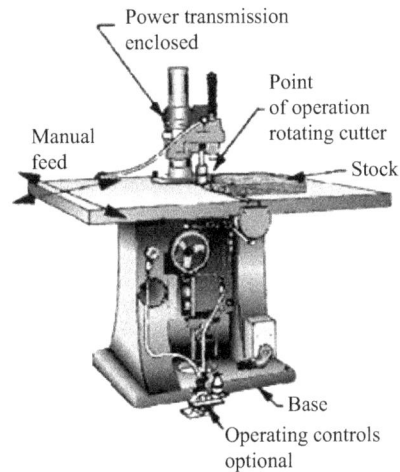

Shaper.

Shared savings: A contractor controlled contingency or buyout savings remaining after all approved contract cost-of-work has been accounted for, from which a percentage of the savings, as defined in the construction contract, is returned to the owner.

Shark fin: A curled corner or lap in a membrane.

Sharp sand: Sand that does not include fine silt or clay particles, making it more suitable for use in concrete and screed. See **Soft sand**.

Sharpening stones: Blocks of natural or artificial stones that have been dressed or smoothed. They are used with an oil or water lubricant to sharpen blades of woodworking tools such as chisels and planes. Most are rectangular in shape and come in many grades, from coarse to fine.

Shatter-proof glass: Two sheets of glass with a transparent plastic sheet sandwiched between to form a pane resistant to shattering. Also referred to as laminated glass.

Shaving: A thin strip of indefinite dimensions that falls off a surface during planning. This planning action produces a thin chip that is usually feathered and generally curled.

Shear: A system of internal forces whose resultant is a force acting perpendicular to the longitudinal axis of a structural member or assembly. See **Shear force**.

Other terms beginning with shear:

- **Shear block:** Plywood that is face nailed to short (2 × 4's or 2 × 6's) wall studs (above a door or window, for example). This is done to prevent the wall from sliding and collapsing.
- **Shear center:** The point in a cross section of a structural member to which a load may be applied and not induce any torsional stress in the cross section.
- **Shear diagram:** A diagram that represents graphically the shear at every point along the length of a member.
- **Shear distribution:** The distribution of lateral forces along the height or width of a building.
- **Shear force:** A force that is directed parallel to the surface element across which it acts. Therefore, the shear force at any section of a beam represents the tendency for the portion of the beam on one side of the section to slide (shear) laterally relative to the other portion.

  Shear force is also the pressure required to break the attachment between two members, causing them to slide across each other. For example, if the nails that attach two panels are severed by shear force, the members will slide.
- **Shear modulus:** The ratio of shear stress to shear strain of a material during simple shear.
- **Shear panel:** Usually a plywood or oriented strand board sheet that covers the wall from the top plate to the bottom plate. When nailed in place, this sheet resists shear forces applied to the wall that try to move it out of square.

- **Shear release:** A boundary condition that constrains a member end from axial displacement and rotation but allows movement in a direction perpendicular to the members longitudinal axis.
- **Shear strength:** Shear strength develops at maximum load or rupture in which the plane of fracture is centrally located along the longitudinal axis of the specimen. Shear strength describes the strength of a material or component against the type of yield or structural failure where the material or component fails in shear. A shear load tends to produce a sliding failure on a material along a plane that is parallel to the direction of the force. For example, when paper is cut with scissors, the paper fails in shear.

  In structural and mechanical engineering the shear strength of a component is important for designing the dimensions and materials to be used for, for example, beams, plates, or bolts. In a reinforced concrete beam, the main purpose of stirrups is to increase the shear strength of the beam.
- **Shear stress:** Stress acting parallel to an imaginary plane cut through an object.
- **Shear strain:** Strain measuring the intensity of racking in the material. Shear strain is measured as the change in angle of the corners of a small square of material.
- **Shear stud:** A steel dowel that is welded to a steel girder, beam, or joist, as part of a composite concrete assembly. The stud engages the concrete floor with the steel structure. Also referred to as a shear stud connector.
- **Shear wall:** A wall designed to resist lateral forces parallel to the wall. A shear wall is normally vertical, although not necessarily so.
- **Shear wave:** See **S wave**.

Sheathing: Sheets of rigid material (often plywood) attached to the framework of a building to strengthen it and to underlie siding or roofing. Also called sheeting.

Sheave: A wheel or roller with a groove along its edge for holding a belt, rope, or cable. When hung between two supports and equipped with a belt, rope or cable, one or more sheaves make up a pulley. The words sheave and pulley are sometimes used interchangeably.

A double sheave would be a pulley block with two grooved wheels.

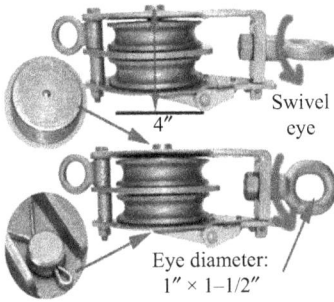

Double sheave.

Shed roof: A flat roof that slopes in one direction and may lean against another wall or building. Also known as **lean-to roof**. See **Single slope roof**.

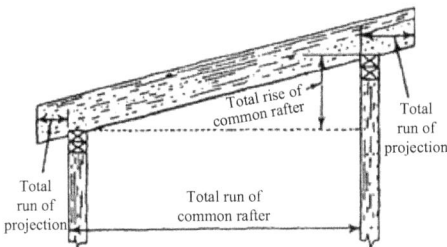

Shed roof.

Sheerleg: A lifting device using two poles fixed together at the top.

Sheerlegs.

Sheet: As used here a sheet is layer, stratum, covering, coating, coat, etc., for example a sheet of glass or a sheet of plywood.

Terms beginning with sheet:

• **Sheet glass:** A transparent, flat glass found in older windows, now largely replaced by float glass.

• **Sheet metal work:** All components of a house employing sheet metal, such as flashing, gutters, and downspouts.

  Sheet metal work also includes ducts that are used to distribute heated and air conditioned air throughout a building. This work is also referred to as sheet metal duct work. In a house, duct work is usually made of round or rectangular metal pipes and sheets.

• **Sheet rock:** A manufactured panel made out of gypsum plaster and encased in a thin cardboard. "Green board" type drywall has a greater resistance to moisture than regular (white) plasterboard and is used in bathrooms and other area that are frequently wet.

• **Sheet siding:** Exterior finish material that comes in sheets, usually 4 feet wide by 8, 9, or 10 feet long.

• **Sheet stock:** Materials such as plywood and medium density fiberboard that comes in sheets.

Sheeting: The members of a shoring system that retain the earth in position and in turn are supported by other members of the shoring system.

Shelf angle: A steel angle attached to the structure of a building frame to carry the load of a brick masonry facing. Sometimes referred to as a **relieving angle**.

Shelf life: The length of time between the manufacture of a material and when the material is no longer suitable for use.

Shellac: A coating made from purified lac dissolved in alcohol, often bleached white.

Shield/Shield system: A structure that is able to withstand the forces imposed on it by a cave-in and thereby protect employees within the structure. Shields can be permanent structures or can be designed to be portable and moved along as work progresses; they may be either pre-manufactured or job-built. Shields used in trenches are usually referred to as **trench boxes** or **rench shields.**

Shim: A small piece of scrap lumber or shingle, usually wedge shaped that, when forced behind a furring strip or framing member, forces it into position. Shims are also used when installing doors and placed between the door jamb legs and 2 × 4 door trimmers. Metal shims are wafer 1 1/2 × 2" sheet metal of various thickness' used to fill gaps in wood framing members, especially at bearing point locations.

Also, a split repaired in a piece of wood veneer, preferably from the same piece of veneer from which the face was made to ensure good color and grain match. The grain running in the same direction as the split to be inconspicuous to the naked eye, and free of any gaps where the shim joins the veneer. To be glued into the split and sanded after being made. The wood should be color matched.

Shiner: A term used to describe a nail that was not covered by the following course of roofing material.

Shingle: An individual unit of prepared roofing material designed for installation with similar units in overlapping rows or courses on inclines normally exceeding 3:12 slope (25%). Shingles may be made of asphalt, wood, tile, slate, or other material and are sold in stock lengths, widths, and thicknesses.

Also, to cover with shingles; to apply any roofing material in succeeding overlapping rows or courses similar to shingles. See **Shingling.**

Besides roofs, various kinds of shingles are used over sheathing as an exterior wall covering of structures in which case it is referred to as **siding.**

Also, a mass of small rounded pebbles, especially on a seashore are referred to as shingles.

Terms beginning with shingle:

- **Shingle fashion:** A term that refers to the way courses of like materials are overlapped in order to have multiple layer coverage.
- **Shingle molding:** Shingle molding, also known as. See **Molding.**
- **Shingle tack coat:** On composition shingles, the shingle mat after it has been saturated with asphalt or coal tar pitch but before granules, talc, or other materials have been embedded into the surface.

Shingling: The application of shingles; the procedure of applying shingles or laying parallel felts so that one longitudinal edge of each felt overlaps and the other longitudinal edge of the adjacent shingle or felts underlaps. Felts are normally shingled from a downslope portion of the roof to the upslope portion of the roof area so that runoff water flows over rather than against each felt lap. Felts are also applied in shingle fashion on relatively low slopes.

Shiplap siding: Horizontal siding that has been rabbeted on both long edges. A weathertight connection is formed when the rabbet joint on the upper piece overlaps the rabbet joint on the bottom piece.

Shipping dry: Having a moisture content (oven-dry basis) of 14–20%. Shipping dry reduces shipping weight and is less susceptibility to decay.

Shipping list: A list that gives each part or mark number, quantity, length of material, total weight, or other description of each piece of material to be shipped to a jobsite. See also **Bill of ladding.**

Shipping piece: Sometimes a single piece of steel but more typically an assembly of fabricated steel pieces that are transported to the field as a unit and that are erected into the structure as a single assembly.

Shoe molding: See **Molding.**

Shop drawings: Detailed drawings that are prepared by a manufacturer, fabricator, contractor, or installer for the purpose of insuring the proper fit of materials, assemblies, and equipment into the constructed project.

Shores/Shoring system: Supporting members that resists a compressive force imposed by a load, or the operation by which the supporting members are placed.

Also, a structure such as a metal hydraulic, mechanical, or timber shoring system that supports the sides of an excavation and which is designed to prevent cave-ins.

Also, temporary bracing and supports for structural reasons.

Short circuit: A situation that occurs when hot and neutral wires come in contact with each other. Fuses and circuit breakers protect against fire that could result from a short.

Short ton: 2,000 pounds or 0.9072 tonnes (Another term for **metric ton**.).

Shot blasting: Attacking the surface of a material with one of many types of shots usually to remove something on the surface such as scale, but sometimes to impart a particular surface to the surface being shot blasted. The shot can be sand, small steel balls of various diameters, granules of silicon carbide, etc. The device that throws the shot is either a large air gun or spinning paddles which hurl the shot off their blades.

Some applications of shot blasting are: concrete blasting that is used to prepare roadways, bridges, airport runways, etc.; removal of dirt, oil and old markings from concrete surfaces for overlays and markings; preparing metal and steel surfaces for painting; removing rust, non-skid coatings, paint and marine growth; and, removing tire rubber, residue and surface contamination from asphalt surfaces.

Shotcrete: A dense mixture of concrete designed to be blown from a high-pressure hose and nozzle via compressed air. Reinforcing filaments of fiber are either included or incorporated at the nozzle as part of the mix.

Should: In contracts means recommended. See **Shall**.

Shoulder: On a saw blade, the area of the tooth immediately in back of the cutting edge. The design of the shoulder interacts with the shape of the gullet to ensure efficient chip disposal, tooth rigidity, and quiet operation.

Shovel footing: A footing form that is typically made by thickening the concrete floor slab, usually formed by using a shovel to trench the area that is to be filled by the shovel.

Show through: On a door, a defect caused by the outline and/or surface irregularities, such as frame parts, core laps, voids, etc., that is visible through the face veneers. Also known as telegraphing.

Shower pan: Non-corrosive pan that covers the base of the shower and runs partway up the wall.

Shrinkage: The word is applied to many things. For example, in business it is an allowance made for reduction in the earnings due to wastage or theft. Wood may shrink due to water loss below the fiber saturation point; shrinkage of wood is expressed as a percentage of the green dimension. In concrete shrinkage cracks may occur because of a reduction in size that occurs during its hardening process, curing process, or both.

Shroud: A color-matched component under a wall-mount lavatory. The shroud covers the drain outlet for aesthetic purposes.

Shutter: Louvered decorative frames in the form of doors located on the sides of a window. Some shutters are made to close over the window for protection.

Shuttering: Formwork.

Side: As used here, an edge, border, boundary, margin, fringe(s), bank, perimeter, extremity, periphery, (outer) limit, limits, etc.

Terms beginning with and prefixed with side and sid:

- **Side clearance:** The distance that the side of the saw tooth projects beyond the body of the saw.

The term is also used when referring to a zoning set back requirements, e.g., a zoning ordinance might state, "Side yard setbacks of a minimum of six (6) feet are also required. Any encroachment into front, rear or side yard setbacks requires zoning reviews." See **Setback**.

- **Side dressing:** The adjusting of all saw teeth on a saw to project laterally the same distance from the plate.
- **Side gauge:** A measuring device that shows the lateral (sidewise) projection of saw teeth beyond the surface of the saw. Used to measure the tangential and radial side clearance of circular saw blades.

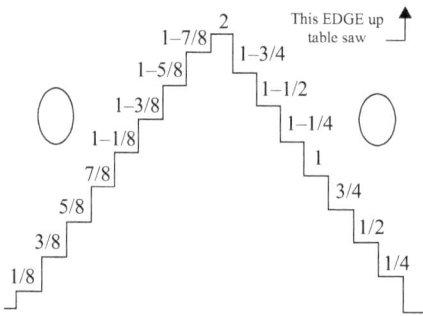

Side gauge.

- **Side jamb:** A specific name for the jamb located on each side of the inside of window and door openings.
- **Side lap:** The lap at the sides of a sheet of deck and is attached by side lap screws (A screw that is used to connect the sides of two adjacent sheets of deck together, #10 being the standard size.) or welds between supports.

Side lap screws.

- **Side light:** A fixed, often narrow, glass window next to a door opening (or window). Also referred to as margin light.
- **Side sewer:** The portion of the sanitary sewer that connects the interior waste water lines to the main sewer lines. The side sewer is usually buried in several feet of soil and runs from the house to the sewer line. It is usually "owned" by the sewer utility, but must be maintained by the owner. Usually it must be serviced by utility approved contractors. Sometimes called **sewer lateral**.
- **Side-view diagram:** A drawing or diagram that shows the outline of a structural member, e.g., a joist, with dimensions. Side-view drawings may also show the web system configuration and bridging rows. See **Profile drawing**.
- **Side wall:** An exterior wall that is parallel to the ridge of the building.
- **Sidelites:** Narrow fixed window units mulled or joined to operating door units to give a more open appearance.
- **Sidesway:** The lateral movement of a structure when subjected to lateral loads or unsymmetrical vertical loads.
- **Sidewalk lighting:** Lighting that is meant for pedestrians. It is located along paved or otherwise improved areas, is usually located within public street rights-of-way that also include roadways for vehicular traffic.
- **Sidewall pressure:** The force exerted on a cable as it is dragged around a bend. The longer the pull and the tighter the bend radius, the higher the sidewall pressure will become. High sidewall pressure damages cable.
- **Siding:** The exterior wall finish material that is applied to a light frame wood structure. There are many types of sidings. Only two are mentioned here:
  - **Batten siding:** It is made up of long, narrow strips of trim that are used to cover vertical joints on vertical exterior siding.

Batten siding.

- **Lap siding** is made up of wedge-shaped boards used as horizontal siding in a lapped pattern over the exterior sheathing. Varies in butt thickness from 1/2 to 3/4 inch and in widths up to 12".

Lap siding.

Silencer: Rubber inserts that are installed in a door stop to minimize noise when the door closes. The silencers also provide positive pressure against the door while in the latched position.

Silicate: A highly caustic compound (pH 11–13) containing metal salts.

Silicon: A chemical element (Si), atomic number 14, semi-metallic in nature, dark gray, that is an excellent semiconducting material and is the most common semiconducting material used in making photovoltaic devices.

Silicone: Not to be confused with the metalloid chemical element Silicon. Silicone is a resin that is used in the binders of coatings. Paints containing silicone are very slick and resist dirt, graffiti and bacterial growth, and are stable in high heat.

Silicone-based water repellants (any of the organopoly-siloxanes (silicone derivative)) are applied to masonry materials for damp-proofing or repelling water.

There are numerous silicone sealants available.

Sill: The main horizontal member forming the bottom of the frame of a window or door. A masonry sill or sub-sill can be below the sill of the window unit.

Also, the 2 × 4 or 2 × 6 wood plate framing member that lays flat against and bolted to the foundation wall (with anchor bolts) and upon which the floor joists are installed. Normally the sill plate is treated lumber.

Terms beginning with sill:

- **Sill cock:** An exterior water faucet. Also referred to as a **hose bib** or **hose bibb**.
- **Sill plate:** The bottom horizontal member of an exterior wall frame which rests on top a foundation, sometimes called mudsill. Also known as a mudsill, sole plate, and bottom member of an interior wall frame.
- **Sill flashing:** A flashing of the bottom horizontal framing member of an opening, such as below a window or door.
- **Sill cock:** Water faucet made for the threaded attachment of a hose; also called a hose bib.
- **Sill pan:** A product placed under a window or door during the installation process. It is intended for water drainage.
- **Sill seal:** Fiberglass or foam insulation installed between the foundation wall and sill (wood) plate. Designed to seal any cracks or gaps.
- **Sill-horn:** The extension of the lip of a window sill to the outside edge of the casing.

Silt barrier: Material, usually filter fabric, that is placed over the course aggregate of a drain system. It allows water to enter the drain system while preventing silt (i.e., dirt) from filtering down and clogging the system.

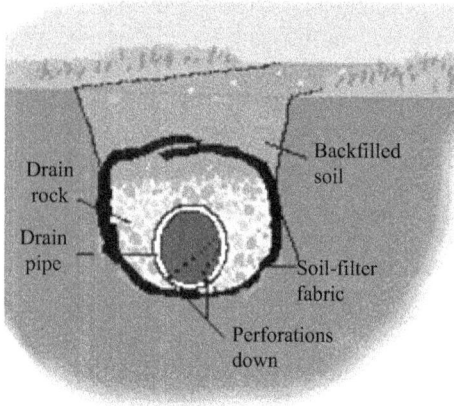

Showing a soil-filter/barrier fabric.

Silviculture: The process of growing and tending forests using both natural and enhanced methods. It is the practice of controlling the establishment, growth, composition, health, and quality of forests to meet diverse needs and values.

Simple harmonic motion: Oscillatory motion of a single frequency wave. It is essentially a vibratory displacement such as that described by a weight, that is attached to one end of a spring and allowed to vibrate freely.

Simple payback: The amount of time it will take to recover the initial investment through savings. The simple payback (in years) can be calculated by dividing first cost by annual savings. The simple payback method does not consider the time value (interest) of money. Therefore, its results should be considered rough, at best.

Simple span: A span with supports at each end, no intermediate support, that restrain only against vertical displacement with the ends of the member being free to rotate. Also referred to as **single span**.

Simplex basin: An ejector pump container usually 18 × 30" for single family homes.

Also, a single threshold type of shower base used in a three wall installation with the threshold being the side with the door.

Simplex communications system: A communications system in which data can only travel in one direction.

Simply supported beam: A beam that rests on a support at each end. A simply supported beam is not supported at more than two points, is not held rigidly by the supports, and does not form part of a larger framework.

Simpson connectors: A brand name. Simpson sells many types of connectors, but it is not uncommon to hear on a construction site people say simply "Simpsons" when referring to wood connectors.

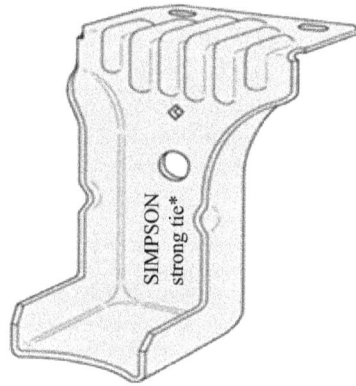

One of many Simpson wood connectors.

Simulated divided lites (SDLs): A type of grille or grid design that creates the appearance of a number of smaller panes of glass separated by muntins, but actually uses larger lites of glass with the muntins placed between and/or on the surfaces of the glass layers.

Single: Sole, lone, solitary, by itself/oneself, unaccompanied, alone. **Many terms begin with single:**

• **Single coverage:** Roofing material that provides one layer over the substrate to which it is applied. See **Single-ply membranes** and **Single-plyroofing**.

• **Single curvature:** When moments produce a deformed or bent shape of a structural member having a smooth continuous curve or arc.

• **Single cylinder door locks:** Single-cylinder door locks require a key to lock/unlock the door from the outside and they have a thumb-turn mechanism on the inside to lock/unlock the door. The single-cylinder function is an option for locks such as deadbolts and entry sets. The thumbturn on the inside allows for an easy exit in case of an emergency.

- **Single element transducer:** A transducer having one measuring element.
- **Single glazing:** Use of a single lite of glass in a window. Generally not as energy efficient as insulating glass or other forms of double glazing.
- **Single hole, single handle faucet:** A sink faucet with one handle only and needing one hole in the counter to install it. This style is well suited for kitchens and bathrooms because it is easy to clean and use.

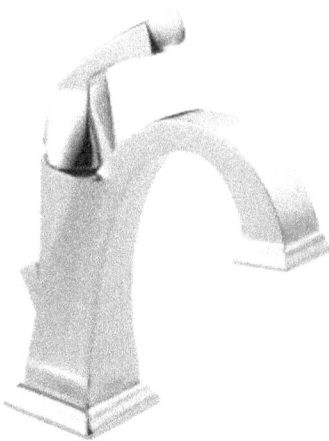

Single hole, single handle faucet.

- **Single hung window:** A window resembling a double-hung, or vertically sliding window, with a fixed, non-operating top sash.
- **Single membrane roof system:** A roof system with just one waterproof layer. The most common types of single membrane roofs are modified bitumen and elastomeric roof systems.
- **Single pane glass:** A window pane that has only one sheet of glass. See **Thermal pane** and **Triple pane**.
- **Single slope:** A sloping roof in one plane which slopes from one wall to the opposite wall. See Shed roof.

- **Single wall flue:** A flue consisting of a single metal pipe.
- **Single-lock standing seam:** A standing seam system with one overlapping inter-lock between two seam panels.

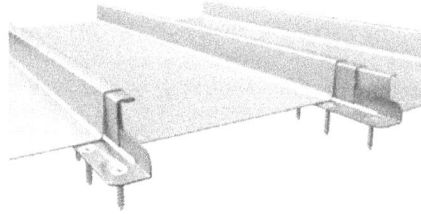

Single lock standing seam.

- **Single-phase:** Single-phase electric power refers to the distribution of electric power using a system in which the voltage is taken from one phase of a three-phase source. Single-phase distribution is used when loads are mostly lighting and heating, with few large elements.

  Single-phase implies a power supply or a load that uses only two wires for power. Some "grounded" single-phase devices also have a third wire that are used only for a safety ground, but not connected to the electrical supply or load in any other way except for safety ground.
- **Single-ply membranes:** Roofing membranes that are field applied using just one layer of membrane material (either homogeneous or composite) rather than multiple layers. See **Single membrane roof system**, **Single coverage** and **Single-ply roofing**.
- **Single-ply roofing:** A roofing system in which the principal roof covering is a single layer flexible membrane, often of thermoset, thermoplastic, or polymer modified bituminous compounds. See **Single membrane roof system, Single coverage**, and **Single-ply membranes**.

Generally, there are six types of **single-ply roofing systems**:
- **Fully-adhered**
- **Loose-laid**
- **Mechanically-fastened**
- **Partially-adhered**

- **Protected membrane roof**
- **Self-adhering.**
- **Single-strength glass:** Glass with thickness between 0.085 and 0.100 inch.

Sinkers: Teflon-coated common nails used to minimize the splitting of lumber because they are easier to hammer into the wood. Disapproved by some engineers because the Teflon coating that allows them to more easily slip into the wood may also allow them to more easily slip out of the wood.

Sinking fund: A fund designed to accumulate a designated amount of money over a specified period of time. The periodic amount of money deposited plus compound interest will accumulate to the designated amount of money over the specified period of time. For example, a hotel might establish a sinking fund for scheduled mattress replacement.

Sintered plate nickel cadmium battery: A type of rechargeable battery that uses nickel oxide hydroxide and metallic cadmium as electrodes. Commonly called a NiCd battery or NiCad battery; the abbreviation Ni-Cd is derived from the chemical symbols of nickel (Ni) and cadmium (Cd). The abbreviation NiCad is a registered trademark of SAFT Corporation, although this brand name is commonly used to describe all Ni–Cd batteries.

Siphon-jet water closet: A toilet having a siphonic trapway at the rear of the bowl and an integral flushing rim and jet. Also called a siphon wash closet.

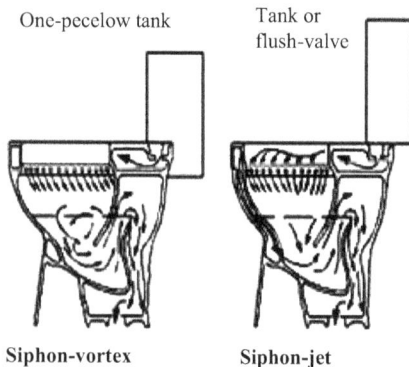

One-pecelow tank   Tank or flush-valve

Siphon-vortex   **Siphon-jet**

Siphon-vortex water closet: A toilet having a trapway at the rear of the bowl integral flushing rim and a water supply system with or without a jet that does not feed directly into the trap.

Sisson joint: A joint between pipes that is larger on one end than the other. Sisson joints are often prohibited for drainage system purposes.

Sistering: Reinforcing a structural member by nailing or affixing a strengthening piece to a weakened piece.

Site area: The total project area including all areas of property, constructed and non-constructed areas.

Site energy: The amount of heat and electricity consumed by a building, as reflected in utility bills.

Six points of estimation: A method for **estimating a loss** using six steps. Combining the first letter of each step spells the word points.

The six points of estimation are:

- **Perspective:** The estimator gains an understanding of the loss and decides where and how to proceed. Perspective includes such things as determining the type of structure, learning what caused the damage, noting subrogation issues, locating the damage point of origin, taking photos, and developing a theory of the total effect of the damaging event. Also, the estimator needs to determine if code changes have occurred since the original construction and whether or not the original work is grandfathered.
- **Organization:** The method that is used to document the loss. Organization often begins with an accurate and detailed diagram. Next, the loss is typically organized into interior and exterior areas. The interior is organized by levels, such as the basement, main floor, and attic. Each level is then broken down into rooms. The estimate should usually start in the room and level where the damage point of origin is found. Other rooms are estimated in a clockwise (or counter-clockwise) manner.

- **Identification:** The estimator decides what an item is. For example, identification occurs when an estimator decides that granite will be used on countertops to replace the loss.
- **Number:** The estimator calculates how many units of an item need to be replaced. For example, number calculations may include the cubic yards of concrete in a slab or the square feet of drywall on a ceiling.
- **Technique:** The estimator decides how a damaged item will be replaced. For example, the estimator may decide whether an item should be repaired, replaced, cleaned, or painted.
- **Supporting events:** The estimator includes work that must be done on undamaged items that are required in order to fix the damaged items. For example, if a countertop must be replaced, an undamaged sink must be detached and stored until the countertop has been replaced, and then it must be reset. Because it was not damaged, detaching, and resetting the sink is a supporting event to the countertop replacement.

Six Sigma: Simplistically Six Sigma means a measure of quality that strives for near perfection. Six Sigma is a disciplined, data-driven approach and methodology to eliminate defects.

The fundamental objective of the Six Sigma methodology is the implementation of a measurement-based strategy that focuses on process improvement and variation reduction through the application of Six Sigma improvement projects. This is accomplished through the use of two Six Sigma sub-methodologies: DMAIC and DMADV. The Six Sigma DMAIC process (define, measure, analyze, improve, control) is an improvement system for existing processes falling below specification and looking for incremental improvement. The Six Sigma DMADV process (define, measure, analyze, design, verify) is an improvement system that is used to develop new processes or products at Six Sigma quality levels. It can also be employed if a current process requires more than just incremental improvement.

Many consultancies have been established based on proprietary methodologies that are similar in nature to the change management philosophies and applications that are used for implementing Six Sigma quality.

Sizing: A compound that is placed on wood, plaster, or other porous surfaces to fill the pores; thus preparing the surface for additional finishes.

Sketch plan: A plan, generally not to exact scale although often drawn from measurements, where the features of a structure or landscape are shown in proper relation and proportion to one another. You might develop a sketch plan while conducting a reconnaissance.

Skew: The condition when two entities come together at an angle that is not 90°or perpendicular to each other.

Skin: A single piece of material that is used as a facing, e.g., the face of a door.

Skin effect: In an alternating current (ac) system, the tendency of the outer portion of a conductor to carry more of the current as the frequency of the ac increases.

Skirt board: Decorative trim that is usually made from a wood board. The skirt is installed on the wall adjacent to the stairs. The skirt trims the area around the ends of the treads and risers. Also called **skirt**. See **Stair bracket** and **Balustrade**.

Skirt board.

Skirting: See Molding.

Skylight: An opening or roof accessory in a roof or ceiling for admitting light. Also spelled sky light.

Slab: With regard to earthquakes, a slab is the oceanic crustal plate that underthrusts the continental plate in a subduction zone and is consumed by the earth's mantle.

Slab cabinet door: A cabinet door made from a single piece of material. A flush slab door is typically made from medium density fiberboard (MDF) or plywood and is either painted or covered with veneer. Sometimes decorative patterns are carved into its surface or decorative moldings are attached. See **Frame** and **Panel cabinet door**.

Slab on grade: A type of foundation with a concrete floor that is placed directly on the soil. The edge of the slab is usually thicker and acts as the footing for the walls. See **Concrete slab**.

Slag: A hard, air-cooled aggregate that is left as a residue from blast furnaces, that may be used as a surfacing material on certain (typically bituminous) roof membrane systems.

Slag cement: It has been used in concrete projects in the United States for over a century and earlier in Europe and elsewhere. When applied properly long-term durability is enhanced. Also known as ground granulated blast-furnace slag (GGBFS).

Also, a discontinuity in a welded joint resulting from a concentration of impurities in the base metal as it is fused and from deposit of electrode coating material in the weld; generally referred to as a **slag deposit**. This crust should be chipped away for inspection of the weld.

Slate: A fine-grain metamorphic rock that is easily split into thin slabs, making it ideal for flooring as well as roofing. Roofing slate comes in a variety of colors that are classified as unfading or weathering. Unfading colors stay very close to their original color throughout their life. Weathering colors change as they age.

Slating hook: A steep-slope roofing attachment device, shaped like a hook, that can be used for fastening roofing slate.

Sleeper: Usually, a wood member embedded in or on top of concrete, as in a floor, that serves to support and to fasten the sub-floor or flooring.

Sleepers.

Sleeper wall: A wall that supports a timber ground floor, and is often built in honeycomb brickwork to allow ventilation of the space under the floor.

Sleeper wall.

Sleeve(s): Pipe or box sections installed in concrete formwork or under concrete driveways and sidewalks, to create holes through the concrete for the later installation of utilities.

Also, preformed steel pipe or box sections placed in concrete formwork to create a hole through a concrete floor, wall, or slab.

Slenderness ratio: A ratio of the height of a column to its cross sectional properties. This ratio affords a means of classifying columns and is important for design considerations.

Sliced: The method by which most of the fine face veneers are cut. Three distinct types of equipment are used to produce veneer from hardwood logs: flat slicing, rotary cutting, and half round. After processing by one of these three methods, the veneer is further prepared for market.

Flat slicing is used to produce decorative face veneer. This veneer is produced by first cutting the log in half or sometimes into quarters or other proportions, each piece being called a **flitch**. The flitches are specially cut from the log to produce specific grain patterns. The flitches are heated in water vats to soften the wood, making it easier to slice or cut them. Heating or cooking schedules are dependent upon species and may vary by manufacturer.

With **rotary cutting**, the log or bolt is placed in a giant lathe and continuously turned against a knife. The log is "unrolled" much like a ribbon. The veneer is then clipped to width, objectionable defects are removed, and the veneer is then dried. Most construction-grade plywood is made in this manner from softwood species.

With the **half round method**, a half log or flitch is secured in place (dogged) and turned 360° against a stationary knife. As a result the log is cut somewhat, but not completely, along the circumference.

Slickensid: Polished striated rock surfaces caused by one rock mass moving across another on a fault.

Slide rule: The slide rule is a mechanical analog computer. They are now only on the desks of old engineers and in collections. The slide rule was used primarily for multiplication and division, and also for functions such as roots, logarithms and trigonometry, but they were not normally used for addition or subtraction.

Slide rules came in a diverse range of styles and generally appeared in a linear or circular form with a standardized set of markings (scales) essential to performing mathematical computations. Slide rules manufactured for specialized fields such as aviation or finance typically featured additional scales that aided in calculations common to that field.

William Oughtred and others developed the slide rule in the 17th century based on the emerging work on logarithms by John Napier. Before the advent of the pocket calculator, it was the most commonly used calculation tool in science and engineering. The use of slide rules continued to grow through the 1950s and 1960s even as digital computing devices were being gradually introduced. It was around 1974 when the electronic scientific calculator made it largely obsoleteand most suppliers left the business.

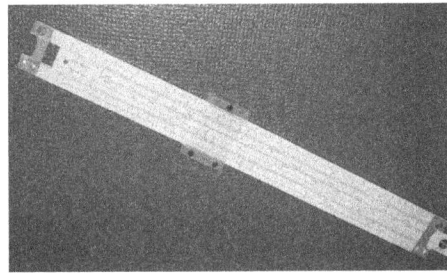

Slide rule.

Sliding French door: A sliding door utilizing French door style panels.

Sliding T bevel: A hand tool with an edge that can be adjusted and then locked into position to mark angles for specific layouts. It is helpful when marking odd angles and dovetail joints and when cutting a board to fit an existing angle. Also known as a **sliding bevel square**.

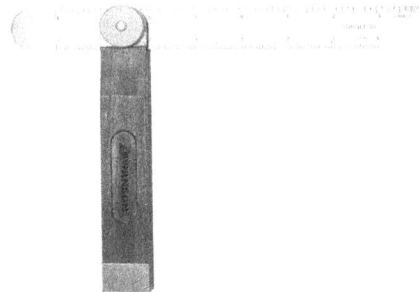

Sliding T bevel.

Sliding window: A window unit that opens by sliding one window sash past another horizontally.

Sling: The basic structural frame of an elevator car. The sling consists of two stiles, a crosshead and a bolster or safety plant that support the platform and cab of an elevator.

Sling psychrometer: A measuring instrument with two thermometers (dry-bulb and wet-bulb) used for determining the dewpoint and relative humidity of air. It ascertains the point at which moisture will condense on the inside surface of the glass.

Slip: The relative displacement of formerly adjacent points on opposite sides of a fault, measured on the fault surface.

Terms beginning with slip:

- **Slip gasket:** An elastomer (natural occurring elastic substance) gasket. It is used with a hub and spigot pipe because it has a tapered cross section.
- **Slip match:** See **Veneer**.
- **Slip matched veneer:** See **Veneer**.
- **Slip model:** A kinematic model that describes the amount, distribution, and timing of slip associated with an earthquake.
- **Slip rate:** The rate that the two sides of a fault are slipping relative to one another, as determined from geodetic measurements, from offset man-made structures, or from offset geologic features whose age can be estimated. It is measured parallel to the predominant slip direction or estimated from the vertical or horizontal offset of geologic markers.
- **Slip sheet:** Light roofing paper or thin fabric that allows the PVC roof membrane to easily slip over the foam insulation without rubbing and suffering damage.
- **Slip-critical joint:** A bolted joint in which the slip resistance of the connection is required.

Slit sample: A cut made in SPF roofing to measure coating thickness. The cut should be about 1.5" long by 3/4" deep by 1/2" wide.

Slope: The incline angle of a roof surface, given as a ratio of the rise (in inches) to the run (in feet). See **Pitch**.

Roof slope is often the primary factor in roof design. The slope of a roof has an effect on the interior volume of a building, drainage, style, and the choice of roofing materials.

The slope, or pitch, of the roof is determined by the vertical rise in inches for every horizontal twelve inch (12") length (called the "run"). A roof with x-rise/12 run slope means that for every 12 inches horizontally (run), it rises x inches. Some of the common roof slopes and the terms which classify them are:

- **Flat roof:** 2/12
- **Low slope:** 2/12–4/12
- **Conventional slope roof:** 4/12–9/12

Steep slope: 9/12 and higher.

Slope of grain: The angle between the direction of the grain and the axis of a piece of lumber, expressed as a ratio.

Sloping: A method of protecting employees from cave-ins by excavating to form sides of an excavation that are inclined away from the excavation so as to prevent cave-ins. The angle of incline required to prevent a cave-in varies with differences in such factors as the soil type, environmental conditions of exposure, and application of surcharge loads. See **Benching** and **benching system**. Also referred to as **sloping system**.

Slot weld: See Plug weld.

Sludge: The residual, semi-solid material left from industrial wastewater or sewage treatment processes. It also refers to the settled suspension obtained from conventional drinking water treatment and numerous other industrial processes.

The term is also used generally for solids separated from suspension in a liquid.

Slump: The "wetness" of concrete. A 3-inch slump is dryer and stiffer than a 5-inch slump. A **slump test** is performed on the jobsite by a concrete testing laboratory. The concrete is placed in a cone shaped form and rodded to distribute the mix. The cone is removed and the wet concrete slumps. The slump of the concrete is measured against the original cone form to establish its slump (usually 4" for most mixes).

Slump block: A masonry unit that is made by removing the forms before the concrete is completely dry. The concrete sags, or slumps, causing the block to have a rounded look. Slump block may be colored with a concrete dye admixture or by painting the surface of the block. Also called **slump stone**.

Smart window: A generic term, sometimes used for windows offering high-energy efficiency or windows featuring switchable glass to control solar gain.

Smoke chamber: The portion of a chimney flue located directly over the fireplace.

Smoke shelf: A ledge in the masonry flue that prevents downdrafts and moisture from entering the firebox.

Smooth surfaced roof: A roof membrane without mineral granule or aggregate surfacing.

Smoke purging: The mechanical system required to remove smoke from a building atrium in the event of a fire. The removed gases are replaced with fresh air as part of the smoke purging circuit.

Smooth plan: This tool is designed for general-purpose planning. It is often used after the jack plane for final polishing.

Smooth, tight cut: See **Veneer.**

Smoothwall texture: A drywall finish with no visible texture. To prevent flashing the entire surface is coated with a thin surface coat.

Smooth-surfaced roofing: Roll roofing that is covered with ground talc or mica instead of granules, i.e., coated.

Snap type standing seam: A standing seam roof in which the cap edge is snapped into place over the underlying edge to lock the edges in place and provide a water-tight seal.

A separate cap that snaps on over the vertical legs of some single standing or batten seam metal roof systems is called a snap-on cap.

Snap type standing seam.

Snaphook: A connector that is comprised of a hook-shaped member with a normally closed keeper, or similar arrangement that may be opened to permit the hook to receive an object and, when released, automatically closes to retain the object.

Snaphooks are generally one of two types: The locking type with a self-closing, self-locking keeper which remains closed and locked until unlocked and pressed open for connection or disconnection; or the non-locking type with a self-closing keeper which remains closed until pressed open for connection or disconnection.

The use of a non-locking snaphook as part of personal fall arrest systems and positioning device systems is prohibited.

Snatch block: A block (A single or multiple pulley.) that can be opened on one side to allow a cable or rope to be laid in the block, instead of threading it through from one end.

Snips: A tool that is used for cutting sheet metal, sheet brass, copper, plastic cloths, and many other materials. Snips are used by sheet metal workers, automotive mechanics, and in industrial plants. Also referred to a tinner's snips.

Snow: If you don't know what snow is, just don't worry about it.

Terms beginning with snow:

- **Snow avalanche:** A large mass of snow, falling, sliding, or flowing very rapidly under the force of gravity.
- **Snow drift:** The triangular accumulation of snow at high/low areas of structures expressed in PSF or PLF.

- **Snow guards:** Snow guards are used to break apart snow so it does not leave the roof surface in large pieces and harm people or property. An enhanced variation of these consists of snow fences. On standing seam roofs, these items need to be installed in a way that does not impede movement of the roofing system with thermal expansion and contraction.

- **Snow load:** A roof load resulting from snowfall. Snow load is a major structural consideration when roofs are designed in areas that receive heavy snow.

Snubber: An interlocking metal bracket attached at the center of the hinge side of a casement sash and frame. It pulls the sash tightly against the frame weather-strip to maximize performance. Snubbers prevent bowing of tall casement windows, or wide awning windows and have been found to be useful with double-hung windows in improving the seals of the bottom and top rails in the closed position.

Snubbers.

Soaker: A metal sheet bent at a right-angle, part of the waterproof flashing of the junction of a tiled or slated roof abutting a wall.

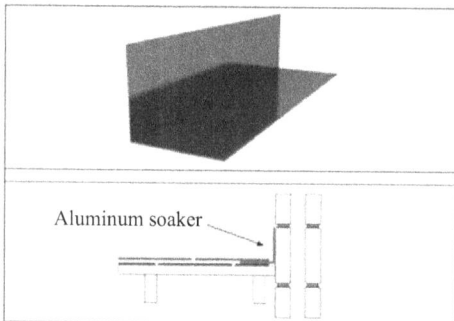

An aluminum soaker.

Soap: Slang for **cable pulling lubricant**.

Sodium rhodizonate: A chemical used to test a paint sample qualitatively for lead; a positve test is characterized by a pink or red discoloration of the paint film cross section.

Sodium sulfide: A chemical used to test a paint sample qualitatively for lead. A positive test is characterized by a gray or black discoloration of the paint film cross section.

Soffit: The area below the eaves and overhangs. The underside where the roof overhangs the walls. Usually the underside of an overhanging cornice.

A premanufactured or custom built air inlet source located at the downslope eave or in the soffit of a roof assembly is called a **soffit vent**, and the resulting ventilation is called **soffit ventilation**.

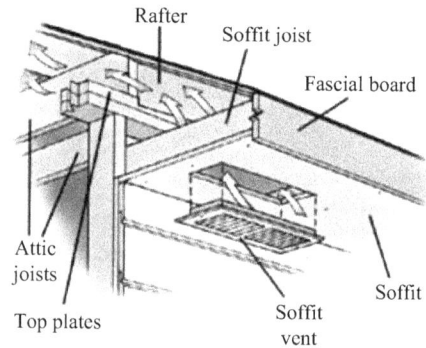

Soffit ventilation.

Soft: A very common word: pulpy, swampy, malleable, gentle, quiet, hazy, etc.

Terms beginning with soft:

- **Soft costs:** Cost items in addition to the direct construction cost. Soft costs include such items as architectural and engineering fees, legal, permits and fees, financing fees, construction interest and operating expenses, leasing and real estate commissions, advertising and promotion, and supervision.

- **Soft rot:** Decay that develops in the outer wood layers under very wet conditions. It is caused by micro-fungi that attack the secondary cell walls (and not the intercellular layer) and destroy its cellulose content.

- **Soft sand:** Sand that includes fine silt or clay particles making it more suitable for mortar or render than sharp sand. See **Sharp sand**.

- **Soft water:** Water with a low mineral (impurity) content.

- **Soft woods:** A term that is used to describe wood from trees that are known as gymnosperms (A plant that has seeds unprotected by an ovary or fruit.) Conifers are an example. These trees tend to be evergreen and have needlelike or scale-like leaves; notable exceptions being bald cypress and the larches. The term has no reference to the actual hardness of the wood.

- **Soft-coat glass:** A glass product that is coated in a secondary process known as sputter-coating, usually to offer low-emissivity or solar-control benefits. The term refers to the fact that these types of coatings generally require some additional care in handling and fabrication and must be used within an insulating glass unit. A hard-coat or pyrolytic glass is coated during the manufacturing process at the molten glass stage. This type of coating offers a surface that is generally as durable as an ordinary glass surface, and therefore requires no special handling and does not need to be used in an insulating glass unit.

Softening point: With regard to roofing, the temperature at which bitumen becomes soft enough to flow, as determined by a closely defined method: ASTM Standard test method D 36 or D 3461.

Softening point drift: A change in the softening point of bitumen. See also Fallback.

Soil: All excavated material exclusive of solid bedrock and organic matter that is divided into various classes and types.

Terms beginning with soil:

- **Soil mechanics:** The science of the strength of soil.

- **Soil pipe:** A pipe that conveys the discharge of water closets or fixtures having similar functions, with or without the discharges from other fixtures. Also called a **soil line**.

- **Soil pressure:** The load per unit area that a structure exerts through its foundation on the underlying soil.

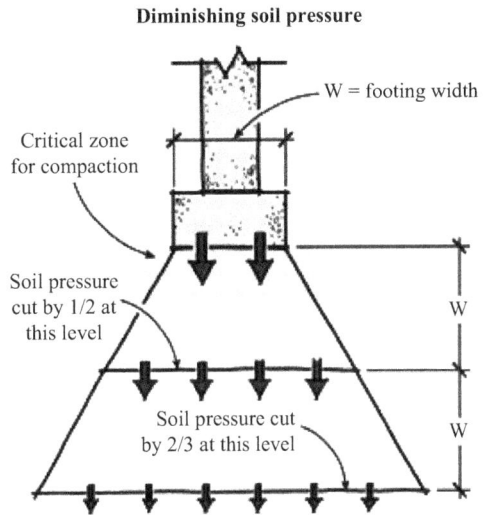

Diminishing soil pressure

Also, the lateral earth pressure that soil exerts against a structure in a sideways, mainly horizontal direction.

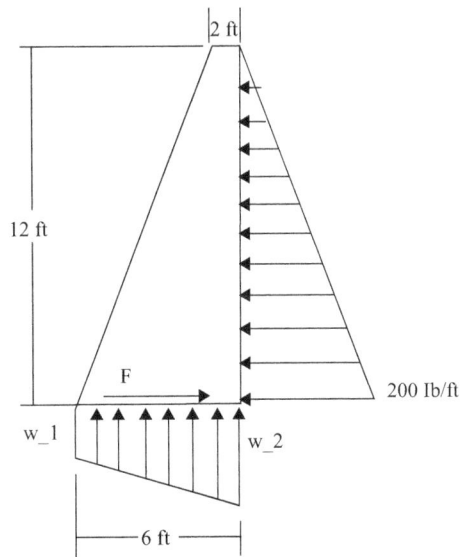

Soil pressure acting on a concrete dam.

- **Soil profile:** The vertical arrangement of soil horizons down to the parent material.

- **Soil stack:** A plumbing vent pipe that penetrates the roof.
- **Soil structure interaction:** The effects of the properties of both soil and structure upon response of the structure.
- **Soils engineer:** A licensed engineer who performs the necessary calculations to determine the types and sizes of footings, retaining walls and the like. Also called a **geotechnical engineer**.

Solar: Relating to the sun, coming from the sun or depending upon the sun.

Terms beginning with solar:

- **Solar cell:** The solar cells that you see on calculators, lighting systems and on satellites. Also called photovoltaic (PV) cells.
- **Solar energy:** Energy from the sun. The heat that builds up on surfaces exposed to the sun is an example.

  Governments at all levels are encouraging innovative uses of solar energy through tax incentives, research grants, and other "going green" initiatives. Therefore, the general term "solar energy" has a rapidly broader meaning with each passing day.
- **Solar heat gain coefficient (SHGC):** A rating that is now generally replacing shading coefficient. It measures a window's ability to transmit solar heat. It measures both the solar radiation that is directly transmitted, as well as the solar radiation absorbed by the glass and subsequently transmitted. The lower a unit's solar heat gain coefficient, the less solar heat it transmits, and the greater its shading ability. It is approximately equal to the shading coefficient divided by 1.15. It is expressed as a number without units between 0 and 1. See **Shading coefficient (SC)**.
- **Solar orientation:** A building placed on a lot so that the long dimension faces south and a majority of the windows are south-facing.
- **Solar reflectance:** The ratio of the reflected solar energy to the incoming solar energy over wavelengths of approximately 0.3–2.5 µm. A reflectance of 100% means that all of the energy striking a reflecting surface is reflected back into the atmosphere; none of the energy is absorbed by the surface. Also called albedo.

  The **solar reflectance index (SRI) measures a** material's ability to reject solar heat, as shown by a small temperature rise. Standard black (reflectance 0.05, emittance 0.90) is 0 and a standard white (reflectance 0.80, emittance 0.90) is 100. For example, a standard black surface has a temperature rise of 90°F (50°C) in full sun, and a standard white surface has a temperature rise of 14.6°F (8.1°C). Once the maximum temperature rise of a given material has been computed, the SRI can be computed by interpolating between the values for white and black.
- **Solar thermal systems:** Systems that collect or absorb sunlight via solar collectors to heat water that is then circulated to the building's hot water tank. The hot water can be used to warm swimming pools, provide domestic hot water, power pumps, cooking, and more.
- **Solar-control glass:** Glass produced with a coating or tint that absorbs or reflects solar energy, thereby reducing solar gain.

Solder: A lead/tin mixture that is melted and used to bond two pieces of some metals together.

Soldier: A vertical member in a retaining wall, especially in temporary works.

Soldiers.

Soldier course: A course of brick with each brick set vertically with the edge, the long-narrow side, of the brick exposed. See **Sailor course**.

A course of soldier bricks.

Sole plate: The bottom, horizontal framing member of a wall that's attached to the floor sheeting and vertical wall studs. Also called a **sill plate**. See **Sill plate**. Sometimes spelled **soleplate**.

Solid: As used here, hard, rigid, solidified, well-built, uninterrupted, etc.

Terms beginning with solid:

- **Solid bridging:** A solid member placed between adjacent floor joists near the center of the span to prevent joists or rafters from twisting.
- **Solid plastic countertop:** A class of countertops made from plastic resins. Includes cultured countertops and solid surface countertops. Solid plastic materials are also used to make tub and shower surrounds. Also referred to as a solid surface countertop.
- **Solid wood:** Wood as it is observed in a tree, log, or piece of lumber. Solid wood is free of manufactured voids and nonwood materials such as resins and other additives.
- **Solid wood flooring:** Flooring that is constructed from solid wood boards, rather than laminated or veneered boards. Normally three-quarters of an inch thick.
- **Solid-core door:** A flush door produced with a solid material placed within the door skins.

- **Solid-surface material:** Nonporous surface material that is resistant to germs and bacteria and is easy to clean. Because the color goes all the way through a solid-surface material, scrapes and cuts can be sanded away easily without harming the look of the surface. Used in countertops and shower walls. See **Solid plastic countertop**.

Solids: The part of the painted coating that remains on a surface after the vehicle has evaporated; the dried paint film. Also called **Nonvolatile**.

Solvent: A liquid, solid, or gas that dissolves another solid, liquid, or gaseous solute, resulting in a solution that is soluble in a certain volume of solvent at a specified temperature. Common uses for organic solvents are in dry cleaning (e.g., tetrachloro-ethylene), as a paint thinner (e.g., turpentine), as glue solvents (acetone, methyl acetate, ethyl acetate), in spot removers (e.g., hexane), in detergents (citrus terpenes), in perfumes (ethanol), nail polish and in chemical synthesis. The use of inorganic solvents (other than water) is typically limited to research chemistry and some technological processes.

In paints, the primary functions of solvents are to disperse/dissolve solids, help film coalescence, and control viscosity. Solvents are chosen for compatibility with the paint system used and for evaporation rate.

Solvent weld: To weld materials using a liquid solvent.

Solubility: In general terms, solubility is the property of a solid, liquid, or gaseous chemical substance called solute to dissolve in a solid, liquid, or gaseous solvent to form a homogeneous solution of the solute in the solvent. The extent of the solubility of a substance in a specific solvent is measured as the saturation concentration where adding more solute does not increase the concentration of the solution.

With regard to masonry, the loss in mass as a percent of original mass as determined by a test for solubility in sulfuric acid. This is a requirement for chemical-resistant masonry units.

Sone: A subjective unit of loudness equal to that experienced by a normal person hearing a 1 kHz tone at 40 dB.

Sonotube: A round, large cardboard tube designed to hold wet concrete in place until it hardens.

Soot mapping: A phenomenon that occurs when soot collects on a wall in a way that reveals or maps materials that are hidden in the wall finish such as drywall tape, the edges of drywall boards, and screws or nails.

Sound: As used here, sound refers to noise, din, racket, resonance, etc. and to well-built, solid, substantial, strong, etc., depending on how it is used.

Terms beginning with and prefixed with sound:

• **Sound attenuation:** Sound proofing a wall or sub-floor, generally with fiberglass insulation.

• **Sound transmission class (STC):** A rating that measures a window's acoustic properties or its ability to reduce sound transmission. An STC rating is determined by measuring the sound transmission over a selected range of sound frequencies. The higher the number, the less sound transmitted.

• **Soundness:** Freedom of a solid from cracks, flaws, fissures, or other variations from an accepted standard. For example, a **sound knot** is solid across its face, and remains intact. One may also be financially sound.

• **Sound-insulating glass:** Double glass fixed on resilient mountings and separated so as to reduce sound transmission. Also referred to as sound-resistive glass.

Source and use budget: A budget detailing the funds for all project-related costs and how they are to be expended.

Source reduction: A method of reducing the amount of unnecessary material brought into a building. Examples include purchasing products with less packaging and sustainable design.

Southern Yellow Pine (SYP): A species group, composed primarily of Loblolly, Longleaf, Shortleaf, and Slash Pines. Various subspecies also are included in the group. SYP grow well in the acidic red clay soil that is found in much of the Southeastern United States. It typically has a density value between 50 and 55 lbs/cubic foot when pressurized.

Space heat: Heat supplied to the living space, for example, to a room or the living area of a building.

Spacer: The linear object that separates and maintains the space between the glass surfaces of insulating glass.

Spacing: The distance between individual members or shingles in building construction.

Spaced sheathing: A sheathing material that is installed to allow air to flow between it and in and around wood shingles installed on it. Spaced sheathing is used because it helps wood shingles last longer by keeping them uniformly dry.

Spackle: To cover wallboard joints with plaster.

Spade bit: A bit that is used in electric and cordless drills and drill presses for fast drilling of holes in wood. Spade bits have a forged, flat paddle with a point and cutting edges.

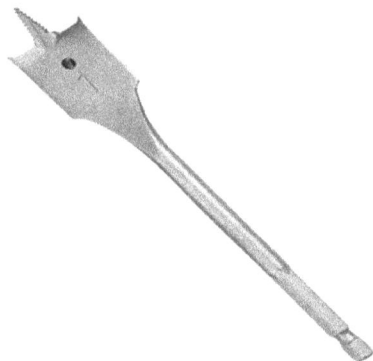

Spade bit.

Spalling: A condition where the surface of the concrete flakes off. It can be caused by premature trowling, overworking the concrete, exposure to high heat or chemicals, or water penetrating the surface and freezing. Shock waves or a single incident of varying internal temperature can cause passive spalling. Passive spalls may simply be repaired, but active spalls are harbingers of possible greater, and possibly dangerous problems.

Spalted: Wood that contains areas of natural decay, giving it distinctive markings.

Span: The clear distance that a framing member carries a load without support between structural supports. Also, the horizontal distance from eaves to eaves, and the distance between two poles of a transmission or distribution line.

Spandrel: The sometimes ornamented space between the right or left exterior curve of an arch and an enclosing right angle.

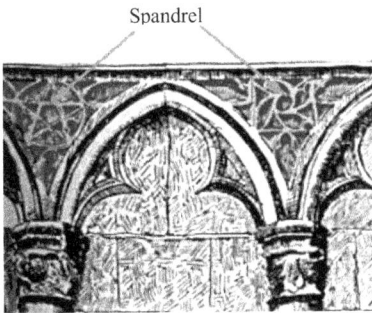

Spandrel

Also, the triangular space beneath the string of a stair.

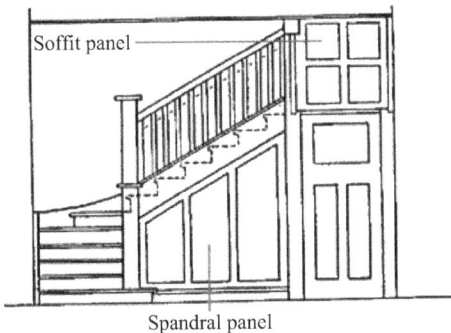

Soffit panel

Spandral panel

A spandrel panel.

Also, the wall between the head of a window and the sill of a window on the next floor. Thus, the spandrel spans between floors.

Spandrels that span between floors.

A **spandrel joist or beam** is a structural member at the outside wall of a building. It supports part of the floor or roof and possibly the wall above.

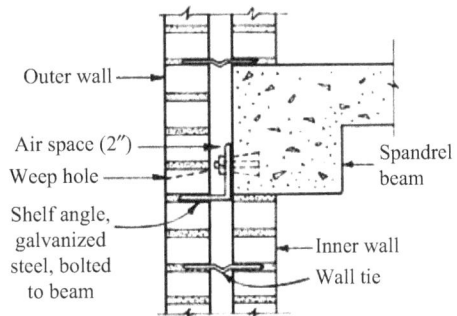

Outer wall

Air space (2″)
Weep hole

Shelf angle, galvanized steel, bolted to beam

Spandrel beam

Inner wall
Wall tie

Spandrel beam.

Spar: A tree, wood mast, or metal tower used to support rigging for one of the many cable yarding systems.

Also, the main longitudinal beam of an airplane wing.

Spark test: A high-voltage test that is performed on a certain types of conductor during manufacture to ensure the insulation is free from defects.

Special conditions: Amendments to the general conditions that change standard requirements to unique requirements, as appropriate for a specific project. Special conditions provide specific clauses setting forth conditions or requirements peculiar to the project under consideration, and that were not satisfactorily covered by the General Conditions. Special conditions are usually a separate section of the contract, other than the **general conditions** and **supplementary conditions**.

Special steep asphalt: A roofing asphalt conforming to the requirements of ASTM Specification D 312, Type IV. This asphalt can be used on roofs with slopes up to 6 in 12 (50%).

Specialty eaves flashing membrane: A self-adhering, waterproofing shingle underlayment designed to protect against water infiltration due to ice dams and wind-driven rain.

Species factor (ks): A coefficient that is used to adjust the evapotranspiration rate to reflect the biological features of a specific plant species.

The species factor (k_s) is one of three factors used to determine landscape water needs. A **microclimate factor (k_{mc})** and **planting density factor (k_d)** also need to be included in water needs estimates. The three factors are used in the following relationship to generate a **landscape coefficient (K_L)**:

Landscape coefficient = species factor × microclimate factor × density factor.

Specific heat: The amount of heat per unit mass required to raise the temperature by one degree Celsius. The relationship between heat and temperature change is usually expressed as

$$Q = cm\Delta T$$

$$\text{Heat added} =$$
$$\text{specific heat} \times \text{mass} \times (t_{final} - t_{initial})$$

where c is the specific heat.

The relationship does not apply if a phase change is encountered, because the heat added or removed during a phase change does not change the temperature.

Specification: A statement of requirements for a given job or project. Usually describes products, materials, and processes to be used. A specification may also contain terms of the contract.

Spectra: A plot indicating maximum earthquake response with respect to natural period or frequency of the structure or element. Response can show acceleration, velocity, displacement, shear or other properties of response.

Spectral acceleration (SA): SA is approximately what is experienced by a building, as modeled by a particle on a massless vertical rod having the same natural period of vibration as the building.

Spectrally selective glass: A coated or tinted glazing with optical properties that are transparent to some wavelengths of energy and reflective to others. Typically, spectrally selective coatings are designed to allow high levels of visible light or daylight into a building and reflect short-wave and long-wave infrared radiation.

Specular gloss: A mirror-like finish (usually 60° on a 60-degree meter).

Speed square: This versatile carpenter's layout tool combines the best features of a framing, try and miter square with the angle finding capability of a protractor. It is used to make basic measurements and mark lines on dimensional lumber, and as a saw guide for making short 45 and 90° cuts.

The speed square was invented in 1925 by Albert J. Swanson. While "Speed Square" is a registered trademark of the Swanson Tool Co., Inc., it has become a genericized trademark for similar products.

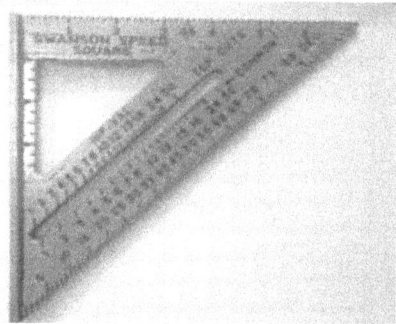

Swanson speed square.

Spill light: Unwanted light directed onto a neighboring property. Also referred to as **light trespass**.

Spine wall: In traditional domestic construction, a load bearing partition between the front and rear rooms of the house. It supports the upper floors and, usually, the roof.

Spiral-cut chain saw file: Featuring a special spiral-cut pattern, this file cuts faster than standard, round chain saw files.

Spiral-ratchet screwdriver: A spring-loaded shaft turns the driver bit, driving screws quickly and easily.

Spiral-ratchet screwdriver.

Splash block: A small masonry or polymeric block laid on the ground or lower roof below the opening of a downspout used to help prevent soil erosion and aggregate scour in front of the downspout. See **Splash guard**.

Splash guard: A fabricated metal pan or masonry block that is placed below a leader pipe or downspout and is used to help protect the roof membrane on a lower roof level or to prevent soil erosion when placed on the ground. See **Splash block**.

Splayed window: A window unit that is set at an angle in a wall.

Splice: A steelwork connection for joining, for example, two lengths of column to form a longer column. Beams can also be spliced, but the splice should not, if possible, be in the middle of the beam where the bending moment is greatest.

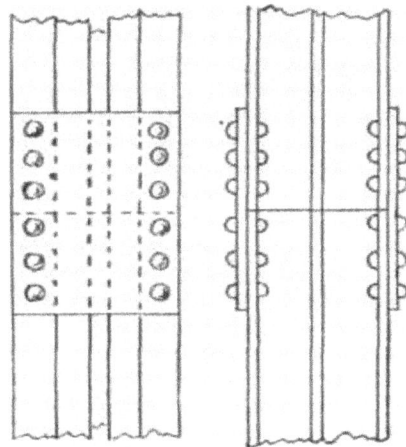

Splicing a steel column.

Also, bonding or joining of overlapping materials. See **Seam**.

Splicing re-bar.

Splice plate: A structural steel plate that joins two columns together at the flanges.

Splice-tape: Cured or uncured synthetic rubber tape used for splicing membrane materials.

Split: A rupture (generally linear) or tear in a material or membrane resulting from tensile forces.

Terms beginning with split:

• **Split finish:** When the exterior portion of door hardware has a different finish color than the interior portion.

• **Split heart:** A method of achieving an inverted "V" or cathedral type figure by joining two "flat-cut" face components of similar color and grain. Also called **manufactured cathedral**.

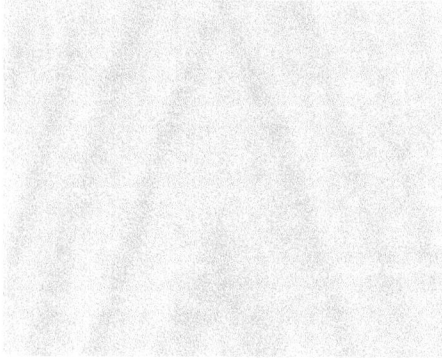

Split heart veneer.

- **Split phase:** A split-phase electric distribution system is a 3-wire single-phase distribution system, commonly used in North America for single-family residential and light commercial (up to about 100 kVA) applications.
- **Split sheet:** See **Nineteen-inch selvage**.
- **Split slab:** A term used to describe two separate concrete slabs. The first is placed as a slab-on-grade or suspended slab, and covered with waterproofing and a drainage system. The second slab, also referred to as a topping slab, is then placed over the underlying slab and waterproofing.

Spline: A strip of metal or fiber inserted in the kerfs of adjacent acoustical tile to form a concealed mechanical joint seal.

Splitting maul: A heavy long handled hammer used for splitting a piece of wood along its grain. One side is shaped like a sledge hammer and the other is a broad head axe shape.

Splitting maul.

Splitting wedge: A triangular shaped tool used to separate two objects.

Splits: Splits and cracks in wood are ruptures or separations in the grain of the wood that reduce the quality as measured by appearance, strength, and utility.

There are four categories or origins for splits and cracks in wood:

- **Resource based:** These are ruptures in the wood that occur in the standing tree or in the log.
- **Processing based:** The major splitting or rupturing of the wood that happens during machining is called loosened grain.
- **Changing moisture content based:** Splits and cracks in wood that occur due to changing moisture content.
- **Use based:** Splits and cracks that occur while the wood is in use.

Proper identification of these four categories is critical if the appropriate corrective action is to be taken to actually solve the problem.

Splitting tensile strength: An indirect method of applying tension to a concrete specimen to cause failure by splitting. Spliting tensile strength is used as a measure of diagonal tension capacity and of tensile strength.

Spoil: The dirt, rocks, and other materials removed from an excavation and either temporarily or permanently put aside.

Spool: A term loosely applied to almost any moving cylindrically device. Often used to wind flexible material such as rope, thread, film, etc. As part of a hydraulic component a spool moves to direct flow through the component.

Spray painting: A painting technique where a device sprays a coating (paint, ink, varnish, etc.) through the air onto a surface. The most common types of paint spray systems are: **air spray**, a paint and compressed air mixture that puts a fine spray on substrate (liquid); **airless spray**, that puts a high hydraulic pressure through a small orifice onto substrate (liquid); and, **electrostatic** that uses negatively charged paint sprayed onto positively charged objects (liquid and powder).

Paint sprayers are typically used for covering large surfaces with an even coating of liquid. Spray guns can be either automated or hand-held and have interchangeable heads to allow for different spray patterns.

Also available are single color aerosol paint cans are portable and easy to store.

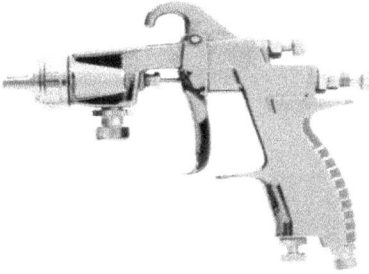

Low volume, low pressure spray gun.

Sprayed polyurethane foam (SPF): A foamed plastic material, formed by spraying two components: component A is polymeric methylene diphenylene diisocyanate (PMDI) and component B is a resin. Together they form a rigid, fully adhered, water-resistant, and insulating membrane.

SPF compounds are the isocyanate and resin components used to make polyurethane foam.

Spreader bar: A temporary metal channel at the sill of a door buck that keeps it square until installation.

Spring: As used here, to move or jump suddenly upward or forward.

Terms beginning with spring:

- **Spring bolt:** A fastener for holding a window sash in a fixed location by means of a spring-loaded bolt in the stile entering a hole in the jamb.
- **Spring buffer:** One type of buffer that cushions an elevator. A spring buffer is for an elevator with speeds less than 200 feet per minute. The buffer is located in the elevator pit.
- **Spring clamp:** This tool is used for holding mitered corners. Each jaw has tiny teeth that grip and hold angled work pieces, irregular moldings and tough-to-clamp joints. It is ideal for small work and light pressure. What a paper clip is to an office a spring clamp is to a shop.

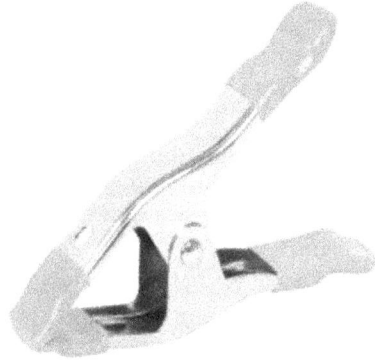

Spring clamp.

- **Spring set:** Alternately bending saw teeth to make the kerf wider than the blade.

Springing: The masonry supporting an arch.

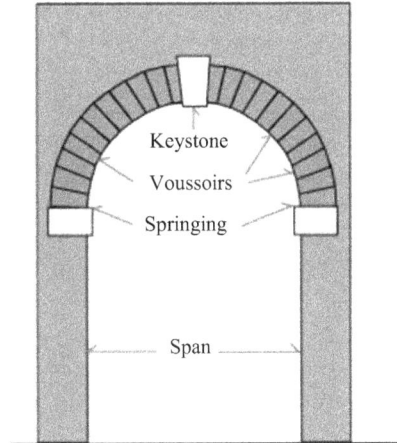

The parts of an arch; note springing.

Springwood: The portion of the annual growth ring that is formed during the early part of the growing season. It is usually less dense and weaker than latewood.

Sprinkler system: A system for fire protection usually consisting of overhead piping connected to a water supply to which automatic sprinklers are attached that discharges water in a specific pattern for extinguishment or control of a fire.

Sprinkler systems can be categorized as followes:

- **Wet pipe sprinkler systems:** The most common types of fire sprinkler systems. They are the most reliable with the only operating components being the automatic sprinklers and (commonly, but not always) the automatic alarm check valve. An automatic water supply provides water under pressure to the system piping.
- **Dry pipe systems:** These systems are installed in spaces where the ambient temperature may be cold enough to freeze the water in a wet pipe system. Dry pipe systems are most often used in unheated buildings, in parking garages, in outside canopies attached to heated buildings (in which a wet pipe system would be provided), or in refrigerated coolers. Dry pipe systems are the second most common sprinkler system type.

    The disadvantages of using dry pipe fire sprinkler systems include: increased complexity; higher installation and maintenance costs; lower design flexibility; increased fire response time; and, increased corrosion potential that often goes unnoticed until there is a failure.
- **Deluge systems:** All sprinklers connected to the water piping system are open, in that the heat sensing operating element is removed, or specifically designed as such. These systems are used for special hazards where rapid fire spread is a concern, as they provide a simultaneous application of water over the entire hazard.
- **Pre-action sprinkler systems:** Pre-action systems are used in locations where accidental activation cannot be tolerated, such as in museums with rare art works, manuscripts, or books, and data centers.
- **Foam water sprinkler systems:** These are special application systems that discharge a mixture of water and low expansion foam concentrate that results in a foam spray from the sprinkler. These systems are usually used with special hazards occupancies associated with high challenge fires,

such as flammable liquids, and airport hangars. Operation depends on the system type into which the foam is injected.

- **Water spray systems:** These systems are operationally identical to a deluge system, but the piping and discharge nozzle spray patterns are designed to protect a uniquely configured hazard, usually being three-dimensional components or equipment (i.e., as opposed to a deluge system that is designed to cover the horizontal floor area of a room). Water spray systems can also be used externally on the surfaces of tanks containing flammable liquids or gases.
- **Water mist systems:** Systems that are usually used where water damage may be a concern, or where water supplies are limited. They create a heat absorbent vapor.

Also, sprinkler systems other than fire protection systems are commonly used to irrigate gardens, agricultural fields, golf cources, etc.

Sprung saw: A twisted or bent saw. It will call attention to itself by chattering and over-heating.

Spud: To remove the top surfacing of a roof by scraping it with special tools called **spud bars** or **power spudders**.

A **spud bar** is a long-handle tool with a stiff flat blade on one end (usually 4- or 6-inches wide).

Spunbond: A type of nonwoven fabric formed from continuous fiber filaments that are laid down and bonded continuously, without an intermediate step.

Sputter-coating: A secondary manufacturing process in which a thin layer of materials, usually designed to offer low-emissivity or solar-control benefits, is applied to glass. Sputter-coatings are commonly referred to as soft-coats, as they generally require some additional care in handling and fabrication and must be used within an insulating glass unit. A hard-coat or pyrolytic glass is coated during the manufacturing process at the molten glass stage. This type of coating offers a surface that is generally as durable as an ordinary glass surface, and therefore requires no special handling and does not need to be used in an insulating glass unit.

Square: A situation that exists when two elements are at right angles to each other.

Also, a tool for checking this.

Also, a measure of the amount of material (e.g., shingles, decks, etc.) required to cover a surface area of 100 square feet when applied as recommended.

Terms beginning with square:

- **Square cut:** A cut to a structural member made at 90° to the length of the member.
- **Square foot:** A two-dimensional calculation, e.g., the area of a square.
- **Square file:** Used to make a round hole square; it is gradually tapered and cut on all four sides. The long, tapered profile makes it useful for cleaning up right-angle shapes such as slots and keyways.
- **Square hollow section:** A structural steel section in the shape of a square tube.
- **Square tooth:** A tooth style on a saw blade designed for heavy-duty cutting.
- **Square yard:** A two-dimensional calculation, e.g., the area of a square. There are 9 square feet in a square yard, i.e., 3 feet × 3 feet.
- **Square-recess   screw-driver:** This square-tipped tool drives square-recessed fasteners in items such as recreational vehicles, boats, mobile homes, hobby equipment, and furniture hardware.

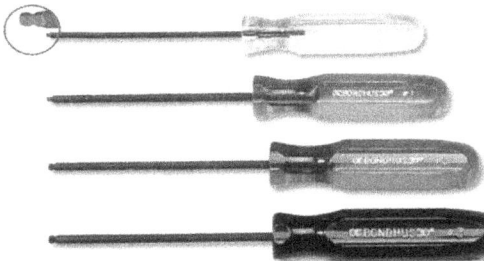

Square recess screwdrivers.

- **Square-tab shingles:** Shingles on which tabs are all the same size and exposure.

Square-tab shingles.

Squaring a wall: Pulling the corners of the wall so that the diagonal distance from corner to corner is equal. This means that the wall section forms a perfect rectangle. A wall is held in this shape by **let-in-bracing** or **shear panels**.

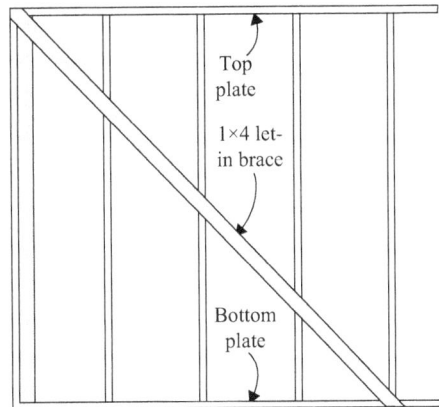

Top plate

1×4 let-in brace

Bottom plate

Showing let-in bracing.

Squeegie: Fine pea gravel that is used to grade a floor before concrete is placed.

Squeeze clamps: Light duty clamps that can be applied with one hand. They have an automatic advancing squeeze handle.

Squeeze clamp.

Squint: A special brick that is used on a corner that is not a right-angle.

Squint brick.

Stable rock: Natural solid mineral material that can be excavated with vertical sides and will remain intact while exposed. Unstable rock is considered to be stable when the rock material on the side or sides of the excavation is secured against caving-in or movement by rock bolts or by another protective system that has been designed by a registered professional engineer.

Stability: Resistance to displacement or overturning.

Also, stability is the opposite of the occurrence of large structural deformations that are not the result of material failure, i.e., instability.

Stabilization: The action that renders an unsafe, damaged, or deteriorated building, i.e., an unstable or potentially unstable building, protected from vandalism, weather, etc., while retaining its present form. Also referred to as mothballing.

Stabilizer plate: A steel plate at a column or wall inserted between the end of a bottom chord of a joist or joist girder to secure the bottom chord to or to restrain the bottom chord from lateral movement.

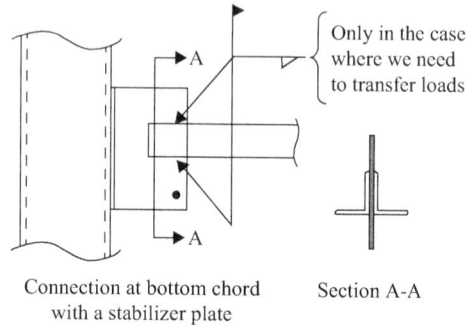

Only in the case where we need to transfer loads

Connection at bottom chord with a stabilizer plate

Section A-A

Stabilizer plate.

Wall plates: They serve the same function as stabilizer plates on frame construction.

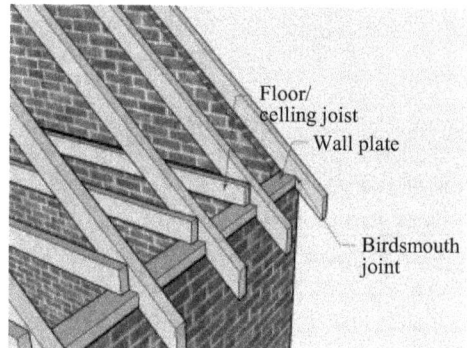

Floor/ceiling joist
Wall plate
Birdsmouth joint

Showing a wall plate.

Stack: A vertical column of gas; a vertical or pitched tube container for the removal of combustion gases in a building, piece of equipment, or vehicle. A chimney is a stack or smokestack.

With regard to trusses, to position trusses on the walls in their correct location. Similarly, almost anything can be stacked.

Stack bond: A course of brick in which each brick is directly over the brick in the course below it, making all the vertical joints form a line.

Stack effect: The occurrence where air escapes through opening in the upper part of a building and is replaced with outside air that enters through an opening lower down.

In roofing, the stack effect helps create proper air flow for attic or roofspace ventilation. The stack effect will be affected by atmospheric conditions such as temperature and wind. See **Soffit (soffit ventilation)**.

Stacked window units: A combined grouping of awning, hopper, casement, or non-operative windows to form a large glazed unit.

Stage: A subsection of a project, or a group of tasks that are performed together, and which have specified and scheduled outcomes.

Staging: A construction procedure for the storage and fabrication of material on a job site.

Stain-grade material: Trim material that has few flaws and is suitable for use in materials that will be stained, leaving the grain exposed.

Stainless steel: A highly corrosion resistant steel alloy containing either chromium, nickel, or copper.

Stained glass window: A window with a painted scene or pattern that has been fired into the glass. Windows with plain colored glass set in lead are also called stained glass. Also referred to as **lead lite (light)**.

Stairs: A series of steps or flight of steps for moving from one level to another. A stair is a single step of a stairway.

Terms beginning with stair:

- **Stair bracket:** Decorative trim that is attached to the wall or skirt below each stair tread. See **Skirt board** and **Balustrade**.

Stair bracket.

- **Stair carriage:** A supporting member for stair treads. Usually a 2 × 12-inch plank notched to receive the treads. Also called a **rough carriage, stringer**, **rough horse, cut jacks, jacks**, and **frame**.
- **Stair gauges:** Once the riser height and tread width are determined, the framer marks these dimensions on the framing square by attaching stair clamps to each leg of the square. The framing square with the stair gauges attached is used to lay-out the stringer (stair carriage). The framer places the framing square on the stringer board until the clamps touch the board and then traces the square.

Stair gauges.

Stair gauges.

- **Stair landing:** A platform between flights of stairs or at the termination of a flight of stairs. Often used when stairs change direction.

  Also, the area at the top and bottom of a staircase.
- **Stair rise:** The vertical distance from stair tread to stair tread.
- **Stair tower:** A tower comprised of scaffold components and that contains internal

stairway units and rest platforms. These towers are used to provide access to scaffold platforms and other elevated points such as floors and roofs. Also referred to as scaffold stairway/tower.

- **Stairway:** A staircase or stairwell.
- **Stairwell:** The framed opening in the floor that incorporates the stairs.
- **Stair-step pattern:** With regard to shingle roofs, the installation of several courses of shingles simultaneously with the lowest or bottom course extending further than the next course up and so forth. The result is a zig-zag or stair step outline.

Stanchion: A steel column.

Stand pipe: A system of piping for firefighting purposes. The system consists of connections to one or more sources of water supply, and serves one or more hose outlets. See **Fire stand pipe**. Also spelled **standpipe**.

Also, an open vertical pipe that receives water from a washing machine.

Standard: In scaffolding, a vertical tube.

Standards: A set of engineering calculations that define the procedure submitted by a fabricator for designing certain elements in the structure, e.g., a procedure for designing moment or truss connections. These procedures are usually approved by the engineer of record prior to the fabricator designing the connections.

Standard baseboard and casing:Standard baseboard is usually 2 1/4" or 3 1/4" high. Standard casing is usually 2 1/4" wide. See Molding.

Standard cabinet door hinge: A cabinet hinge that is attached to the door on one side and to the cabinet stile on the other side. Standard cabinet door hinges usually cannot be adjusted once they are installed. See European style cabinet door hinge.

Standard details: A drawing or illustration sufficiently complete and detailed for use on other projects with minimum or no changes; non-project-specific details.

Standard operating procedure(s) (SOP): Detailed, written instructions documenting a method to achieve uniformity of performance.

Standard practices of the trade(s): One of the more common basic and minimum construction standards. This is another way of saying that the work should be done in the way it is normally done by the average professional in the field.

Standing seam: Metal roof seam made by turning the long edges of the panels up and then over. The three common types of standing seams are rolled type, snap type, and batten type.

Start date: The date that an activity or project begins.

Starter: A device that initiates a flow of high voltage across the electrodes of a fluorescent lamp.

Also, one or something that starts.

Terms beginning with starter:

- **Starter course:** The first row of shingles laid at the eave line. The starter course for composition shingles usually consists of shingles that are installed wrong side down or is made from rolled starter strip material. The starter course for wood shingles or shakes is usually made by sawing two to three inches off the length and installing the wood shingles or shakes right side down. The first course of wood shingles or shakes completely overlaps the starter course. The starter course for tiles is also the first course.
- **Starter joist:** A joist that is spaced close, usually 6-inches, to a wall for deck support.
- **Starter sheets:** Felt, ply sheets, or membrane strips that are made or cut to widths narrower than the standard width of the roll. They are used to start the shingling pattern at an edge of the roof.

  Also, particular width sheets designed for perimeters in some mechanically attached and fully adhered single-ply systems.
- **Starter strip:** Asphalt roofing applied at the eaves that provides protection by filling

in the spaces under the cutouts and joints of the first course of shingles.

Starting tread: The first tread and riser at the bottom of a stair. Starting steps are usually rounded on the ends to accept volutes (a decorative way to start a stairway) or turn outs (a fitting used to start a rail system).

Statement of changes in financial position: The "statement of changes in financial position" (historically called a "funds statement") is designed to disclose the major financing and investing transactions undertaken by the business. Like the income statement, it covers a period of time. The statement show the business' sources of capital during the period and the major areas to which these funds were committed. In so doing, the statement discloses causes of changes in financial position during the period.

Static: Something that is stationary, motionless, immobile, unmoving, still, at a standstill, at rest, not moving, etc.

Also, of or relating to bodies at rest or forces that balance each other.

Also, random noise, such as crackling in a receiver or specks on a television screen, produced by atmospheric disturbance of the signal.

Terms beginning with static:

- **Static discharge head:** The vertical distance from the pump to the highest outlet in the water system. Also called Static lift.
- **Static equilibrium:** Equilibrium that does not include inertial forces.
- **Static head:** The pressure created in a pipe by a column of water.
- **Static load:** Any load, as on a structure, that does not change in magnitude or position with time.
- **Static pressure:** The pressure exerted by a liquid or gas, especially water or air, when the bodies on which the pressure is exerted are not in motion.
- **Static vent:** A vent that does not include a fan.
- **Static water level:** The undisturbed level of water in the well before pumping.

Statically determinate: A structure where there is only one distribution of internal forces and reactions that satisfies equilibrium. In a statically determinate structure, internal forces and reactions can be determined by considering nothing more than equations of equilibrium.

Statically equivalent: Two force systems that exert the same resultant force and resultant moment are said to be statically equivalent (they have the same effect on the motion of a rigid body). Physically, this means that the force systems tend to impart the same motion when applied to an object. However, the distribution of resulting internal forces in the object may be different.

Statically indeterminate: A statically indeterminate structure is one where there is more than one distribution of internal forces and/or reactions that satisfies equilibrium.

Stationary sash: A fixed sash. Also referred to as a picture, studio, vista, or view sash.

Statute of limitations: Any law that fixes the time within which parties must take judicial action to enforce rights or else be thereafter barred from enforcing them. Fixing the time when the period begins is often complex; legal advice should be obtained.

Staved lumber core (SLC): Made with any combination of blocks or strips of wood, not more than 2-1/2 inches (64 mm) wide, of one species of wood glued together (in butcher block fashion), with joints staggered in adjacent rows.

Steam contamination: In normal boiler operation steam passing the throttle valve is far from pure. It will contain water droplets carried along in the "steam wind." These droplets contain dissolved and suspended solids. Such particles build up in passages, ports, valve and piston heads/rings, gland packings and similarly in auxiliaries. Contaminated steam leads, amongst other things, to oil contamination. The contaminated oil can act as a fine grinding paste with obvious detriment to the equipment.

Steam fitters: Mechanics who are expert in the welding and installation of high-pressure water or steam pipe.

Steel: A generally hard, strong, durable, malleable alloy of iron and carbon, usually containing between 0.2 and 1.5% carbon, and often with other constituents such as manganese, chromium, nickel, molybdenum, copper, tungsten, cobalt, or silicon, depending on the desired alloy properties. Steel is widely used as a structural material.

Terms beginning with steel:

- **Steel angle:** A structural steel component that has an L-shaped cross section.
- **Steel beam:** Two common types are the wide flange steel beams that look like the letter "H" laid sideways and the I beams that look like the capital letter "I."
- **Steel decking:** A corrugated cold formed metal sheet that is used to span across joists, beams, and girders as part of a floor or roof assembly.
- **Steel fabricator:** One who fabricates structural steel to the specifications of the construction documents.
- **Steel inspection:** A municipal and/or engineers inspection of the concrete foundation wall, conducted before concrete is poured into the foundation panels. Done to insure that the rebar (reinforcing bar), rebar nets, void material, beam pocket plates, and basement window bucks are installed and wrapped with rebar and complies with the foundation plan.
- **Steel joist:** A horizontal supporting member that is normally used between beams or other structural members, suitable for the support of some roof decks. Also referred to as an **open web steel joist**.
- **Steel pan:** A preformed steel pan used as a concrete form for composite concrete stair treads. Steel pan stairs have concrete treads and platforms with steel risers, stringers, and railings.

Steep asphalt: Roofing asphalt having a high softening point. They are especially applied on roofs that are steep.

Steep-slope roof: A roof of suitable slope to accept the application of water shedding roofing materials. See **Slope**.

Steeple: A tower or spire, usually located on a church.

Stem: Part of the faucet that holds the handle on one end and the washer on the other. The stem assembly is comprised of the moving part of the valve that controls the amount and temperature of water released.

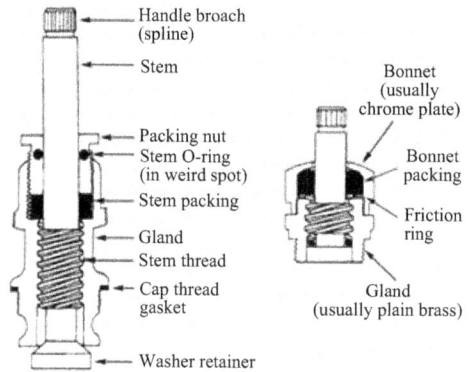

Stem.

Step: One unit of a stair. A step consists of a riser and a tread. A stair is a series of steps.

Step drill bit: A drill bit with a graduated design that permits drilling of variously sized holes without changing bits. It is designed for use with power drills and have self-starting tips that eliminate the need for center punching. Step bits can be used on all materials but are specially designed for use on metals.

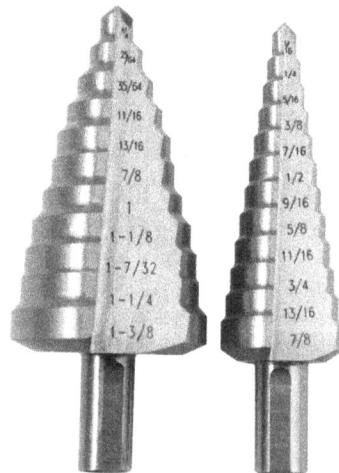

Step drill bits.

Step flashing: A flashing application method used where a vertical surface meets a sloping roof plane. 6 × 6" galvanized metal bent at a 90° angle, and installed beneath siding and over the top of shingles. Each piece overlaps the one beneath it the entire length of the sloping roof (step by step).

Step flashing.

Steradian: The SI unit of solid angle, equal to the angle at the center of a sphere subtended by a part of the surface equal in area to the square of the radius.

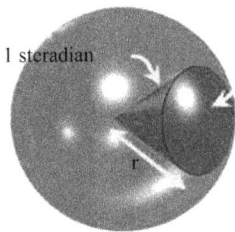

Steradian.

Stere: A metric measure of cordwood or pulpwood representing 1 × 1 × 1 meters. Approximately 0.27 cord. Also, the old metric name for a cubic meter. The word is seldom used, except on crossword puzzles.

Stick built: A house built without prefabricated parts. Also called conventional building.

Stick-slip: Refers to earthquakes. See **Fault**.

Sticker: A wooden strip laid between each layer of lumber as it is stacked for drying. Stickers permit transfer of heat into and removal of water from the lumber stacks.

Stiffener plate: A steel plate usually welded to the web of a beam or girder to support a concentrated load or to reinforce the web at a cutout section. Also referred to as a stiffener.

Stiffness: Rigidity, or resistance to deflection or drift. A measure of deflection or of staying in alignment within a certain stress.

Stile: An upright framing member in a panel door. Also, a vertical window sash member.

Stile-and-rail door: A traditional type of wood door constructed with vertical stiles and rails with openings filled with raised wood panels or glass.

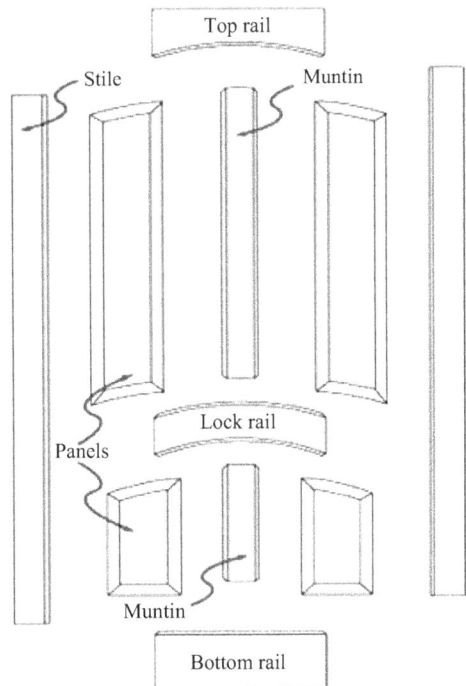

Components of a stile-and-rail-door.

Stillson wrench: A pipe wrench.

Stillson wrench.

Stilts: A pair of poles or similar supports with raised footrests, used to permit walking above the ground or working surface.

Working on stilts.

Stipulated sum agreement: A written agreement that sets a specific dollar amount as the total payment for performing the contract. See also **Lump sum contracts**.

Stirrup: Reinforcing steel in a U or box section used to resist tensile stresses diagonally transferred through a concrete beam, girder, or wall.

Preformed rebar shape stirrups.

Stochastic: A term applied to processes that have random characteristics.

Stone countertop: A countertop made from stone such as granite or marble. See **Plastic laminate countertop, Solid surface countertop, Solid plastic countertop, Wood block countertop, Cultured countertop, Cultured marble countertop**, and **Tile countertop**.

Stool: A member that forms the horizontal shelf at the bottom of the window.

Also, an interior trim piece sometimes used to extend a window sill and act as a narrow shelf.

Stop box: Normally a cast iron pipe with a lid that is placed vertically into the ground, situated near the water tap in the yard, and where a water cut-off valve to the home is located (underground). A long pole with a special end is inserted into the curb stop to turn off/on the water.

Stop: The molding on the inside of the window frame against which the window sash closes, or in the case of a double-hung window, the sash slides against the stop. Also referred to as bead, side stop, window stop, and parting stop.

Also, a device that is installed in a water supply line, usually near a fixture, that permits an individual to shut off the water supply to one fixture without interrupting service to the rest of the system. Also referred to as a **stop valve**.

Stop and drain fitting: A plug-type valve used to tap into a water main to control the flow to a branch line. The fitting has a side opening to shut off the water and allow it to drain out so the pipe won't freeze. Also referred to as a stop and waste valve.

Stop and drain fitting/stop and waste valve.

Stop order: A formal, written notification to a contractor to discontinue some or all work on a project for reasons such as safety violations, defective materials or workmanship, or cancellation of the contract.

Storm drain: A drain that receives and conveys rain water, surface water, and ground water away from buildings.

Storm sash: Additional window unit, complete with a window pane installed in a window sash, installed over the original window unit to provide an extra layer of glass insulation. Also called a storm window.

Storm sewer: A sewer system designed to collect storm water and is separated from the waste water system. In the past cellar drains were often connected to the storm sewer system. Nowadays these drains are usually required to tie into the sanitary sewer system.

Stormwater runoff: Water from precipitation that flows over surfaces into sewer systems or bodies of water. All precipitation that leaves project site boundaries on the surface is considered stormwater runoff.

Story: That part of a building between any floor or between the floor and roof.

Story drift: The difference in horizontal deflection at the top and bottom of a story.

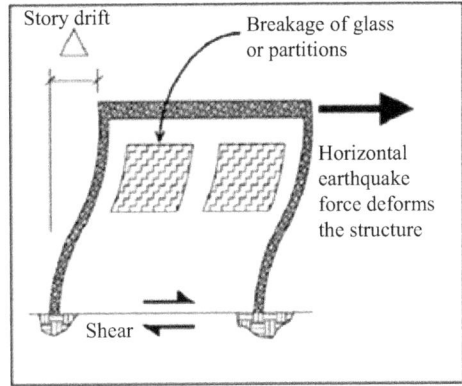

Story pole: A pole with lines on its surface to mark the height for each row of brick being laid.

Strain hardening: The condition when a material exhibits the capacity to resist additional load than that which caused initial yielding after undergoing deformation at or just above the yield point.

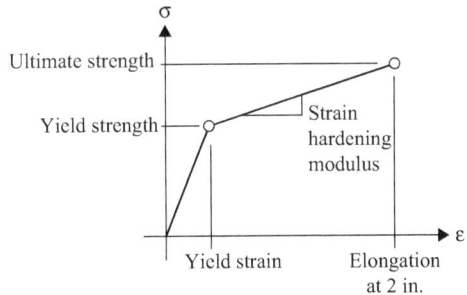

Straight seam: A carpet seam that is made by seaming two pieces of carpet that have been cut in a straight line. See **Serpentine seam**.

Straight wood floor installation: An installation where wood strips are installed in straight rows that are usually parallel to at least one of the walls. See Diagonal wood floor installation and Herringbone wood floor installation.

Strain: All materials distort under load. The intensity, or degree of distortion is known as strain. When the distortion disappears upon removal of the load it is known as elastic strain. If the metal remains distorted the strain is known as plastic strain.

Strain release: See Fault, strain release.

Strand: There is a difference between a strand and a cable, although a strand is often referred to as a cable.

Strand is two or more wires that are wound concentrically in a helix. They may or may not be wound around a center wire. Strand is normally referred to by the number of wires in the strand preceded by a 1. For example, a 1×7 is one group of seven wires. 1×19 is one group of 19 wires. See **Cable**.

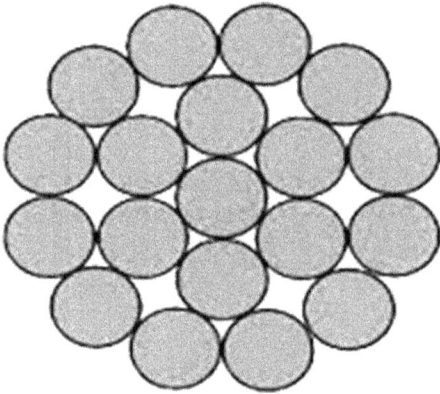

1×19 strand.

Strap: A component, usually steel, that is installed to ensure that structural elements are secure, e.g., a hurricane strap.

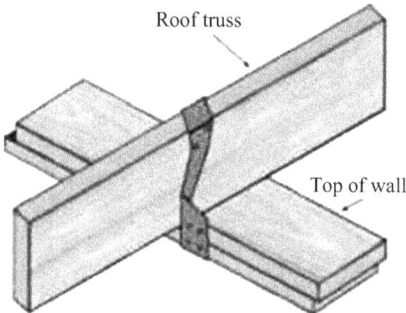

Roof truss

Top of wall

Hurricane strap.

Strap wrench: A tool with metal body and heavy cloth or metal mesh belt used for turning pipe.

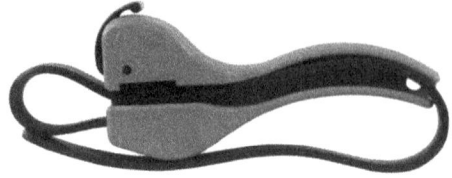

Strap wrench.

Strapping: A method of installing roofing rolls or sheet good materials parallel with the slope of the roof. Not a recommended installation method for slopes that are 1:12 or less.

Stratification: The segregation of over-wet or over-vibrated concrete into horizontal layers with increasingly lighter materials toward the top.

Also, with regard to geology, the formation of strata; deposition or occurrence in strata. Also a stratum.

Stratified random sampling: A method that groups members of a population into discrete subgroups, based on characteristics that may affect their responses to the survey. For example, a survey of building occupants' commuting behavior might separate people by income level and commuting distance. To yield representative results, the survey should sample the subgroups according to their proportions in the total population.

Straw nail: A long-shanked nail. Sometimes used for fastening over tile at hips and ridges.

Streamline flow: A condition where the fluid particles move in continuous parallel paths.

Streetscape: The distinguishing character of a particular street as created by its width, degree of curvature, paving materials, design of the street furniture and forms of surrounding buildings.

Strength: A measure of load bearing without exceeding a certain stress.

Strength is a very general term that may be applied to a material or a structure. In a material, strength refers to a level of stress at which there is a significant change in the state of the material, e.g., yielding or rupture. In a structure, strength refers to a level of loading that produces a significant change in the state of the structure, e.g., inelastic deformations, buckling, or collapse.

Strength of connections: A strength requirement of connections in welded or clipped deformed bar mats used in concrete reinforcement. Connectors are used to make the connections in precast structural concrete. Connections in cast-in-place structural concrete are integral with members. Therefore, there are no separate connections to be assessed.

Stress: The internal resistance, or counterforce, of a material to the distorting effects of an external force. These counterforces tend to return the atoms to their normal positions. The total resistance developed is equal to the external load and is known as stress. Stress can be equated to the load per unit area or the force applied per cross-sectional area perpendicular to the force. Stresses occur in all materials that are subjected to an applied force, or load.

Tensile stress: It occurs when two sections of a material on either side of a stress plane tend to pull apart, or elongate.

Compressive stress: It is the reverse of tensile stress. Adjacent parts of the material press against each other across a plane.

Sheer stress: It exists when two parts of a material tend to slide across each other in any plane of shear upon application of force parallel to that plane.

Stresses can be generally classified in one of six categories:

• **Residual stresses:** Residual stresses are due to the manufacturing process, e.g., welding, that leave stress in a material.
• **Structural stresses:** Structural stresses are stresses produced in structural members due to the weight they support.
• **Pressure stresses:** Pressure stresses are induced in vessels containing pressurized materials. This loading is provided by the same force producing the pressure.

• **Flow stresses:** Flow stresses occur when a mass of flowing fluid induces a dynamic pressure on a conduit wall. The force of the fluid striking the wall acts as the load. Flow stresses may be applied in an unsteady manner when flow rates vary. Water hammer is an example of transient flow stress.
• **Thermal stresses:** Thermal stresses exist where temperature gradients are present in a material. Different temperatures produce different expansions and subject materials to internal stress.
• **Fatique stresses:** Fatique stresses are due to due to cyclic applications of a stress, e.g., traffic on a bridge.

Terms beginning with stress:

• **Stress at 0.7% extension:** A mechanical requirement for wire strand and wire rope to establish the limiting load without fracturing of individual wires.
• **Stress concentration:** A localized stress which is considerably higher than average due to sudden changes in loading or sudden changes in geometry.
• **Stress drop:** See **Fault**.
• **Stress relaxation:** A significant problem when bolting at high temperatures. Creep occurs when a material is subjected to high temperature and a constant load. Stress relaxation occurs when a high stress is present that is relieved over time; the stress is relaxed with a subsequent reduction in the bolt's preload. The way to minimise the effects of stress relaxation is to use materials that have an adequate resistance to it.
• **Stress resultant:** A system of forces that is statically equivalent to a stress distribution over an area.
• **Stress-crack:** External or internal cracks within a material caused by long-term stress. Environmental factors, such as contact with corrosive material, usually accelerate stress-cracking.
• **Stress-strain diagram:** Graph of stress as a function of strain. It can be constructed from data obtained in any mechanical test where load is applied to a material, and continuous measurements of stress and

strain are made simultaneously. It is constructed for compression, tension and torsion tests.

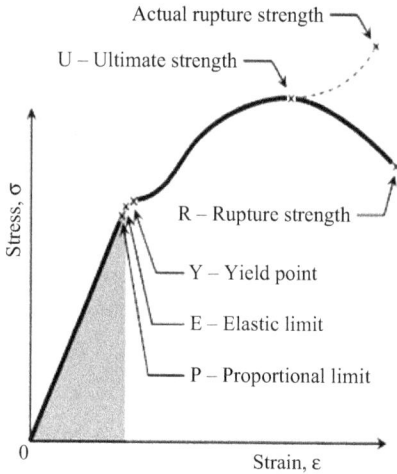

Typical stress-strain diagram.

Stretcher: A brick whose longest side is visible on the surface of the wall. Stretcher bond is brickwork consisting only of stretchers Stretcher bond is suitable for half-brick thick walls and cavity walls. See **Header.**

Stretcher bond.

Strike: The plate on a doorframe that engages a latch or dead bolt.

Also, with regard to earthquakes, the bearing relative to north of a line defined by the intersection of a planar geologic feature, such as a fault, and a horizontal surface.

Strike column: A column located inside an elevator car, and that extends the full height of the elevator door opening. This is the column against which the sliding door closes.

Strike plate: The plate that is located on a door buck jamb that receives the lock and latch when the door closes.

Strike-slip fault: See **Fault, strike-slip.**

Stringer: A timber or other support for cross members in floors or ceilings. Sometimes called a **string.**

Also, the supporting structural member of the treads and risers in a stair. See **Stair carriage.**

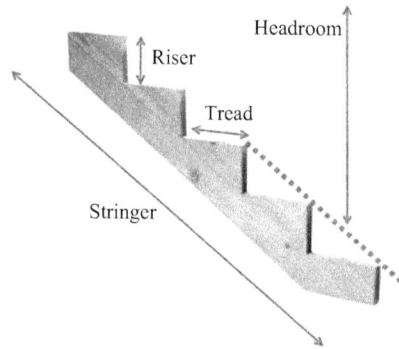

Stair risers, treads, and stringer.

Strip: As used here, piece, bit, band, belt, ribbon, slip, shred.

Terms beginning with strip:

- **Strip flashing:** Pieces of membrane material that are used to flash metal flashing flanges such as gravel stop. Also referred to as Stripping.
- **Strip flooring:** Wood flooring consisting of narrow, matched strips.
- **Strip mopping:** Hot bitumen applied in parallel bands. See **Mopping.**
- **Strip shingles:** Asphalt shingles that are manufactured in strips, approximately three times as long as they are wide.

Strippable wallpaper: Wallpaper with a face that easily strips from the backing. Also called **peelable wallpaper.**

Stripping: The procedure of removing formwork from concrete after the concrete has set.

Strike-through: A term used in the manufacture of roofing fabric-reinforced polymeric sheeting to indicate that two layers of polymer have made bonding contact through the scrim or reinforcement.

Stroke: The whole motion of a piston in either direction.

Also, to change the displacement of a variable displacement pump or motor.

Strong axis: The cross section that has the major principal axis.

Strong motion: Ground motion of sufficient amplitude to be of interest in evaluating the damage caused by earthquakes or nuclear explosions.

Struck and weathered pointing: Pointwork finished with a sloping surface, recessed slightly at the top and protruding slightly at the bottom of the joint. See **Pointing**.

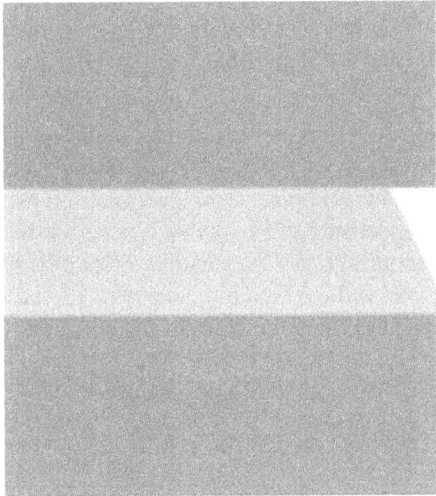

Struck and weathered pointing.

Structural: Something arranged in a specific way, or something that is strong enough for construction use. A **structure** is a mechanism that is designed and built or constructed of various parts jointed together in some definite manner to carry loads and resist forces. Examples are buildings of various kinds, monuments, dams, roads, railroad tracks, canals, millraces, bridges, tunnels, locomotives, nautical vessels, stockades, forts and associated earthworks, Indian mounds, ruins, fences, and outdoor sculpture.

Terms beginning with structural:

• **Structural bolt:** A heavy hexagon head bolt having a controlled thread length intended for use in structural connections and assembly of such structures as buildings and bridges. The controlled thread length is to enable the thread to stop before the joint ply interface to improve the fastener's direct shear performance. This term is used in civil and structural engineering, but is seldom used in mechanical engineering.

Structural bolt.

• **Structural damage:** Damage that affects the ability of a part or parts to hold and carry parts of the structure it was designed to hold and carry. Also see **Cosmetic damage**.
• **Structural engineer:** An engineer licensed to determine the material type, grade, size, and placement requirements for safe construction of the structural parts used in a building. See **Registered professional engineer**.
• **Structural features:** With regard to earthquakes, features in rocks that are produced by movement after the rocks were formed.
• **Structural floor:** A framed lumber floor that is installed as a basement floor instead of concrete. This is done on very expansive soils.
• **Structural glass:** Glass that is used where it supports more than just its own weight, e.g., glass balustrades, stairs, and floor panels.
• **Structural masonry brackets:** An installation bracket that is used with multiple high/wide window units or large doors for

added structural support. The brackets are also used to attach the unit in the rough opening in lieu of nailing through the casing, thus eliminating unsightly nail holes.

- **Structural model:** An idealization for analysis purposes of a real or conceived structure. A structural model includes boundaries limiting the scope of the analysis. Supports occur at these boundaries, representing things which hold the structure in place.

- **Structural panel:** A panel designed to be applied over open framing in which a structural deck is not required.

- **Structural part:** A part of a building that is essential in supporting a load or keeping the structure intact. A structural part cannot be removed without weakening the structure.

- **Structural performance:** Excessive cracking, bending or other indications of structural distress due to excessive applied loads, or deformations due to damage from fire or other external forces.

- **Structural ramp:** A ramp built of steel or wood, usually used for vehicle access. Ramps made of soil or rock are not considered structural ramps.

- **Structural shapes:** Standard steel shapes produced by steel mills, e.g., flanges, channels, pipes, angles, etc.

- **Structural steel:** The structural elements that make up the frame that supports the design loads: beams, columns, braces, trusses and fasteners. Structural steel does not include things such as cables, ladders, grating, or handrails.

- **Structural steels:** A large number of steels that are suitable for load-carrying members in a variety of structures because of their strength, economy, ductility, and other properties. Strength levels are obtained by varying the chemical composition and by heat treatment.

- **Structural systems:** The load bearing frame assembly of beams and columns on a foundation that is usually constructed of wood, cement block, reinforced concrete, or steel. Other systems such as non-load bearing walls, floors, ceilings and roofs are generally constructed within and on the structural system.

- **Structural timbers:** Relatively large size pieces of wood that are selected for use based upon their strength and stiffness. Structural timbers are used as trestle timbers, framing for buildings, Crossarms for poles, etc.

Struts: Member positioned between two other members to keep them a specific distance apart, giving them added strength.

Stub: To push through.

Stubbed out: A term that is used to describe leaving the end of a part exposed for easy connection later in the construction process. Rebar may be stubbed out of the footing for connection to the foundation concrete, or a short section of sewer line may be stubbed out from the septic tank or main sewer line for easy connection to the sewer lateral later in the construction process.

Stucco: Refers to an outside plaster finish made with Portland cement as its base.

Stucco sheathing: A wall covering on which synthetic stucco is installed. Common materials used are foam board and exterior grade gypsum board.

Stud: A vertical wood-framing member, also referred to as a wall stud, and attached to the horizontal sole plate below and the top plate above. Normally 2 × 4's or 2 × 6's, 8' long (sometimes 92 5/8"). One of a series of wood or metal vertical structural members placed as supporting elements in walls and partitions.

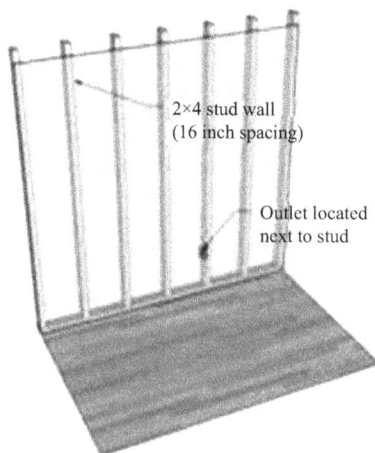

2 × 4 stud wall.

Also, with regard to steel, a vertical cylindrical bar of steel with a larger cylindrical cap fastened to metal decking used to form a mechanical connection between the metal decking and the poured-in-place concrete slab such that the two form a composite structural element. Studs are also used to produce composite beams.

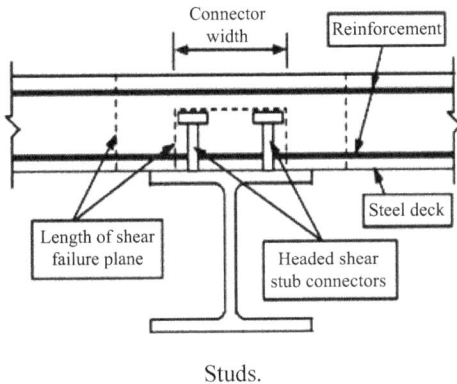

Studs.

Terms beginning with stud:

- **Stud finder:** This handheld device is used to determine the location of wood and metal framing studs in the walls of light-frame construction after the walls have veen closed-in.

  There are two main types of these devices: magnetic stud detectors that use a small magnet to detect the nails or screws that were placed into studs during construction and internal capacitor stud finders that use an internal capacitor plate to detect changes in the dielectric constant of the wall; a change in the dielectric constant indicates a dense object, hopefully a stud, behind the wall. Also called a **stud detector** or **stud sensor**.

- **Stud framing:** A building method that distributes structural loads to each of a series of relatively lightweight studs. Contrasts with post-and-beam.

- **Stud gun:** Tool that uses gunpowder contained in a cartridge to drive a nail into a hard surface like steel or concrete.

- **Stud shoe:** A metal, structural bracket that reinforces a vertical stud. Used on an outside bearing wall where holes are drilled to accommodate a plumbing waste line.

Styrene butadiene rubber (SBR): The most widely used elastomer for pipe and fitting gaskets worldwide.

High-molecular-weight polymers having rubber-like properties, formed by the random copolymerization of styrene and butadiene monomers.

Styrene butadiene styrene copolymer (SBS): A plasticiser used in the hot-mop type of modified bitumen roof systems.

High-molecular-weight polymers that have both thermoset and thermoplastic properties, formed by the block copolymerization of styrene and butadiene monomers. These polymers are used as the modifying compound in SBS polymer modified asphalt roofing membranes to impart rubber-like qualities to the asphalt.

Sub: A prefix that means under, below, beneath (subsoil), just outside of or near (subtropical), less than, or not quite (subpay), secondary or at a lower point in a hierarchy (subcontractor).

Terms beginning with the prefix sub (by no means all terms):

- **Subcontractor:** A contractor who specializes in performing a specific building trade such as drywall, masonry, or painting. A subcontractor will often enter into a subcontract with a general contractor to perform specific work in the construction for an agreed upon price.

- **Subcontractor bond:** A written document from a subcontractor given to the prime contractor by the subcontractor guaranteeing performance of his/her contract and payment of all labor, materials, equipment, and service bills associated with the subcontract agreement.

- **Subduction:** The sinking of a plate under an overriding plate in a convergence zone.

- **Subfloor:** The framing components of a floor to include the sill plate, floor joists, and deck sheeting over which a finish floor is to be laid.

- **Submetering:** A method of determining the proportion of energy used within a building attributable to specific mechanical end uses or subsystems (i.e., the heating subsystem of an HVAC system).
- **Submittals:** Required deliverables that a contractor submits to an owner. For example, submittals include shop drawings, manufacturers cut-sheets, and samples.
- **Subsidence:** A downwards movement, especially a movement of foundations.
- **Subsoil:** The layer or bed of earth beneath the topsoil.
- **Substrate:** A surface upon which a finish material is directly applied and that extends completely behind such finish material.
- **Substructure:** The portion of a building that is below grade.
- **Sub-sill:** The supplemental member of a frame that is used under most awning and casement units as an additional sill with the primary purpose being to hold multiple units together at the sill.

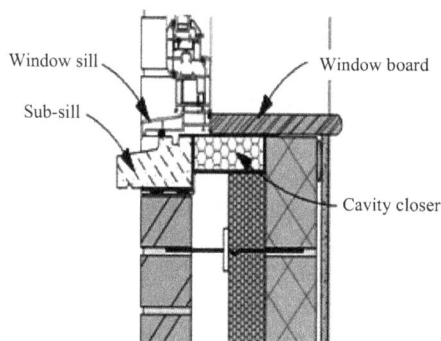

Showing a sub-sill.

- **Sub-surface investigation:** A term that is used to represent an examination of soil conditions below the ground. Sub-surface investigations may include geological investigations, soil borings and geotechnical laboratory tests for structural design purposes.

Suction: A partial vacuum due to wind loads on a building that cause a load in the outward direction.

Sulfate attack: With regard to concrete, a reaction between sulfate solutions and hardened cement paste. This reaction results from, for example, alkali magnesium, and sodium sulfates in groundwater and seawater. The products of the reaction lead to expansion and cracking of the concrete.

Summer beam: The largest beam spanning wall to wall, supporting the smaller floor joists timber-framed homes.

Summer beam.

Sump: A pit or large plastic bucket/barrel inside the home designed to collect ground water from a perimeter drain system. A submersible pump in a sump pit that pumps any excess ground water to the outside of the home is a sump pump.

Also, a depression around roof drains and scuppers to help promote roof drainage. A metal deck accessory that is used at drain locations to close the opening where holes are cut in the metal deck is a **sump pan**.

Also, simply a reservoir.

Sunburst light: See **Fan light**.

Sun screen: See **Shade screen**.

Super: Excellent, superb, superlative, first-class, outstanding, marvelous, magnificent, etc. Often used as a prefix.

Terms beginning with super:

- **Super window:** A generic term for a window with a very low U-value. Typically, it incorporates multiple glazings, low-E coatings, gas fills, and an insulating spacer.

- **Superimposed load:** Usually means a load that is in addition to the dead weight of the structure.
- **Superplasticizer:** An admixture for concrete that makes it more fluid without the addition of more water.
- **Supershear file:** This file is intended for use on aluminum, brass, bronze, copper, and other soft alloys, and to do heavy teardown work fast. The curved tooth pattern and unique double-purpose design has a shearing action that effectively lowers the cutting angle so the file smooths as it cuts.
- **Superstructure:** The portion of a building which is above grade.

Supplemental conditions: Supplemental or modifying conditions to the standard clauses of the general conditions to accommodate specific project requirements. Also referred to as supplementary conditions. See **Special conditions**.

Support: A support contributes to keeping a structure in place by restraining one or more degrees of freedom. In a structural model, supports represent boundary entities that are not included in the model itself, e.g., foundations, abutments, or the earth itself. For each restrained translation degree of freedom at a support, there is a corresponding reaction force; for each restrained rotation degree of freedom, there is a reaction moment.

Support system: A structure such as underpinning, bracing, or shoring, that provides support to an adjacent structure, underground installation, or the sides of an excavation.

Surety: A properly licensed firm or corporation that provides surety bonds that are payable to owners. The bonds secure the performance of contractors, either in whole or in part. Sometimes referred to as a company of surety.

Surface: As used here, outside, exterior; top, side; finish, veneer.

Terms beginning with surface:

- **Surface checks:** Checks that occur on the wood surface during seasoning. Checks may extend to varying depths into the wood.

Surface checks.

- **Surface drying:** A drying process where lumber is allowed to remain exposed to the air long enough to allow it to lose its excess moisture. Lumber dried in this way is marked with an S-Dry stamp.
- **Surface erosion:** The wearing away of a surface due to abrasion, dissolution, or weathering.
- **Surface force:** A force applied to the surface of an object.
- **Surface faulting:** See **Faulting**.
- **Surface-mount medicine cabinet:** Medicine cabinet which is attached to the surface of the wall. See **Recess mount medicine cabinet**.
- **Surface texture:** The geometric irregularities in the surface of a masonry unit. The measurement of surface texture does not include inherent structural irregularities, unless these are the characteristics being measured.
- **Surface waves:** Seismic waves that propagate along the surface of the earth, e.g., Love and Rayleigh waves.

Surfacing: The name of the process when rough cut lumber is planed down to make the surfaces smooth. Sharp knives are run over the surface of the lumber cutting away 1/4" of the board.

When lumber is surfaced on two sides it is called S-2-S and when it is surfaced on all four sides it is called S-4-S. Framing lumber is generally S-4-S. The resulting dimensions of the board are called nominal dimensions.

Also, the top layer or layers of a roof covering, specified or designed to protect the underlying roofing from direct exposure to the weather.

Surform tools: A surform tool resembles a food grater in that it has a perforated sheet metal surface. It consists of a steel strip with holes punched out and the rim of each hole is sharpened to form a cutting edge. The strip is mounted in a carriage or handle.

Surform planes have been described as a cross between a rasp and a plane. While looking like a food grater made of perforated sheet metal, surforms differ in having sharpened rims. They are typically used to shape material, not grate it.

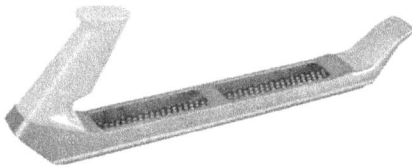

Surform plane.

Surge withstand: A measure of an electrical device's ability to withstand high-voltage or high-frequency transients of short duration without damage.

Surround: An attractive, protective trim that is secured to an energy panel by an adhesive or vinyl barb to give the glass panel a safe finished edge. Also the aluminum framework for most standard screens.

Suspended ceiling: A ceiling system supported by hanging it from the overhead structural framing.

Suspension bridge: A bridge in which the roadway deck is suspended from wire ropes that pass over two towers. The cables are anchored in housings at either end of the bridge.

Suspension bridge.

Suspension system: With regard to ceilings, a metal grid suspended from hanger rods or wires, consisting of main beams and cross tees, clips, splines, and other hardware which supports lay-in acoustical panels or tiles. The completed ceiling forms a barrier to sound, heat and fire. It also absorbs in-room sound and hides ductwork and wiring in the plenum.

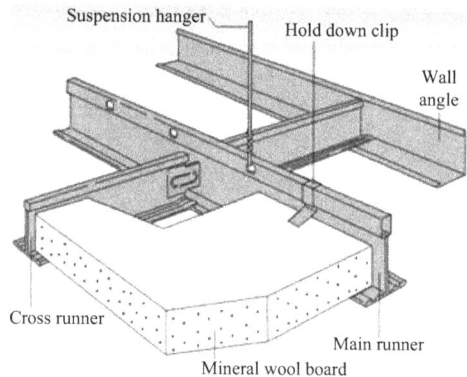

A suspension system for a ceiling.

Sustainable purchasing policy: A policy that gives preference to products that have little to no negative impacts on the environment and society throughout its life cycle, and also gives preference to those products that are supplied by companies whom also have little to no negative environmental and social impacts. The sustainable purchasing policy commits the organization to an overarching course of action, which empowers staff working at the operations level.

Sustainable purchasing program: The development, adoption, and implementation of a procurement strategy, which culminates in the purchase of products that have little to no negative impacts on the environment and society through its life cycle or that are supplied by companies whom also have little to no negative environmental and social impacts. The program is an operational working strategy aligned and in support of an organization's sustainable purchasing policy.

Swage: One of a pair of shaping tools that when moved toward each other produce a desired form in an object. In a swage tool the die is the movable part of the shaper, the anvil the fixed part. Also spelled swedge.

Also, a method of shaping a saw tooth to provide side clearance on both sides of each tooth.

Also, a groove, ridge, or other molding on an object.

Swan-neck mortise chisel: This tool is extremely helpful in smoothing out the bottom of mortises. Its curved blade is used like a lever to scrape the mortise bottom flat. It is also used for clearing deep, narrow, recesses when setting locks and other hardware into furniture. Also called a **lock mortise chisel**.

Swan neck chisel.

Swarm: With regard to earthquakes, a series of related earthquakes concentrated in location and time with no one earthquake of outstanding size.

Sway brace: Metal straps or wood blocks installed diagonally on the inside of a wall from bottom to top plate, to prevent the wall from twisting, racking, or falling over "domino" fashion.

Sweating: Soldering.

Also, the formation of condensation on the outside of pipes or toilet tanks.

Sweep: A gradual (but pronounced) bend in a log, pole, or piling. A sweep is considered a defect in the log. It is analogous to bow in a piece of lumber.

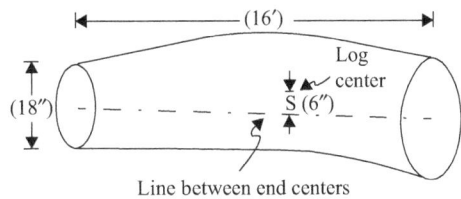

Sweep in a log.

Also, the curvature of a structural member in the perpendicular transverse direction of its vertical axis.

Also, a drain ell fitting with a long radius that allows for smooth passage of waste.

Sweep lock: A sash fastener located at the meeting rails of a double-hung window, which rotates and clamps the two rails closer together.

Swing: See **Door swing**.

Switch: A device that completes or disconnects an electrical circuit.

Terms beginning with switch:

- **Switch, bypass isolation:** A manually operated device used in conjunction with a transfer switch to provide a means of directly connecting load conductors to a power source and of disconnecting the transfer switch.
- **Switch gear:** A high-voltage (600 volts plus) switching mechanism and circuit breaker for distributing electric current throughout a building's primary circuitry.
- **Switch, general-use:** A switch intended for use in general distribution and branch circuits. It is rated in amperes, and it is capable of interrupting its rated current at its rated voltage.

- **Switch, general-use snap:** A form of general-use switch constructed so that it can be installed in device boxes or on box covers, or otherwise used in conjunction with wiring systems recognized by the NEC.
- **Switch, isolating:** A switch intended for isolating an electric circuit from the source of power. It has no interrupting rating, and it is intended to be operated only after the circuit has been opened by some other means.
- **Switch, motor-circuit:** A switch rated in horsepower that is capable of interrupting the maximum operating overload current of a motor of the same horsepower rating as the switch at the rated voltage.
- **Switch, transfer:** An automatic or non-automatic device for transferring one or more load conductor connections from one power source to another.

Switchboard: A large single panel, frame, or assembly of panels on which are mounted on the face, back, or both, switches, overcurrent and protection devices, buses, and usually instruments.

Switchboards are generally accessible from the rear as well as from the front and are not intended to be installed in cabinets.

Symbols: Found on plans, symbols are used to represent common objects such as doors and light switches.

Synthetic felt pad: A carpet pad made from man-made felt that is highly resistant to tearing. See **Waffle type sponge rubber pad, Rebond pad**, and **High-density urethane foam pad**.

Synthetic rubber: Any of several elastic substances resembling natural rubber. It is prepared by the polymerization of butadiene, isoprene, and other unsaturated hydrocarbons. Synthetic rubber is widely used in the fabrication of single-ply roofing membranes.

Synthetic stucco: Stucco that comes pre-mixed by the manufacturer. It is usually applied in two coats that are much thinner than common stucco. It is applied on stucco sheathing.

System of forces: One or more forces and/or moments acting simultaneously.

Système International [d'Unités] (SI): The international system of weights and measures (metric system).

Systems Narrative Introduction, LEED: A general description of each of the following types of base building systems installed in the project building: space heating, space cooling, ventilation, domestic water heating, humidification and/or dehumidification, and lighting. The narrative includes summaries of the central plant, distribution, and terminal units, as applicable. It also includes all the controls associated with these systems—central automatic, local automatic, or occupant control. It accounts for any differences in system types for different portions of the project building—for different floors, for interior vs. perimeter zones, etc. The systems narrative does not need to list each base building system individually (i.e., not each and every chiller) but does describe the distinct types of systems as listed above (i.e., all chillers having the same basic design and specifications). Other types of systems that may be included in the narrative if desired but are not required include process equipment, office equipment, plumbing systems, fire protection systems. As requirements change do check before making a submission.

Systematic sampling: A method of surveying every xth person in the population, using a constant skip interval. It relies on random sampling order or an order with no direct relationship to the variable under analysis (alphabetical order when sampling for commute behavior).

T lock shingle: Most common type of interlocking shingle. T lock shingles produces a basket-weave pattern by sliding the lower edge of the shingles into slots at the top of the downhill shingles.

T lock shingles.

T-handle screwdriver: The T-handle allows the user to apply extra high torque.

Tab: The exposed portion of strip shingles defined by cutouts.

Tabbed shingle: Common type of composition shingle. A tabbed shingle has from two to six tabs, but three is the most common number of tabs. Tabbed shingles may have an imprinted texture on their surface.

Tachometer: An instrument that measures the rotation speed of a shaft or disc, typically in revolutions per minute.

Tack hammer: This hammer is used for driving small brads and tacks. It has a magnetized head that holds the tack for one-hand starting. It is also called an upholsterer's hammer.

Table saw: A stationary or portable power tool with a circular saw blade mounted from under a table. Material to be cut is fed through the blade on top of the table.

Tag: Tags are used to identify electric equipment by class, group, and the temperature range for which it is approved.

Tagged end (T.E.): This is the end of a joist or joist girder where an identification or piece mark is shown by a metal tag. The member must be erected with this tagged end in the same position as the tagged end noted on the erection plan.

Tail beam: A relatively short beam or joist supported in a wall on one end and by a header at the other.

Talc: A soft mineral that is composed of magnesium silicate. It is heat-resistant and is used for stoves and firebricks. Also referred to as **soapstone** and **statite**.

Take off: An estimator's estimate of the materials necessary to complete a job.

Tamping: The process of pressing plastic material into a confined space using a bar or rod so that it compacts the material, removes air pockets, and causes it to mold completely to the shape of the space into which it is being pressed.

Concrete is also tamped so that the concrete flows around rebar and under and around window bucks while removing the air pockets that cause honeycombing.

Tangent: A straight line or plane that touches a curve or curved surface at a point, but if extended does not cross it at that point.

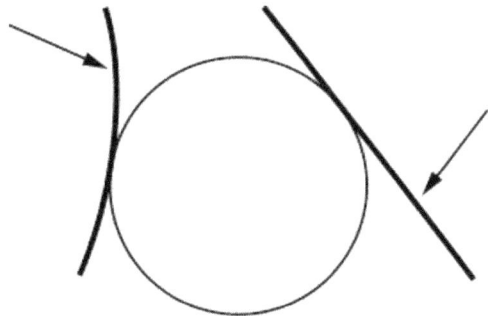

Tangent.

In mathematics the trigonometric function that is equal to the ratio of the sides (other than the hypotenuse) opposite and adjacent to an angle in a right triangle.

Tangent modulus: The slope of the stress-strain curve of a material in the inelastic range at any given stress level.

The tangent modulus can be taken at any point on the stress-strain curve. The figure shows that the tangent modulus is represented by the slope of the tangent to any point on the curve. The accuracy in calculating the tangent modulus is somewhat limited because the tangent is usually drawn by hand.

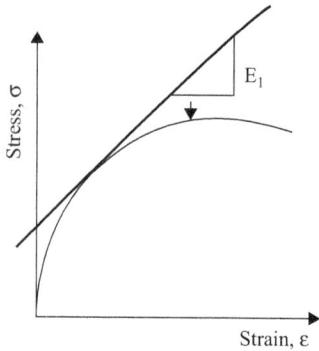

Tangent modulus.

Tank: A large receptacle or storage chamber, especially for liquid or gas; the container holding the fuel supply in a motor vehicle.

Also, a fixture reservoir for flush water. On a conventional toilet, the ballcock, flush valve, and trip lever are installed in the tank. A tank lid closes the top tank opening.

Tankless water heater: An instantaneous hot water heater.

Tap: A tool for cutting an internal thread in a drilled hole.

Tap adapter: A fitting with one plain end and one threaded female end.

Tap and reamer wrench: A hand tool adjusted by twisting one of the wrench handles to change the opening of the jaws.

Adjustable tap and reamer wrench.

Tape: Generally, a narrow strip of material, typically used to hold or fasten something for construction. There are many different tapes to deal with, among others, issues of moisture,

heat, color-coding, friction, venting, harsh weather, UV exposure.

Taper siding: Siding with one edge much wider than the other. The thicker edge may have a groove or rabbet cut out of it so that it fits snugly over the thin edge of the course of siding directly below it.

Tapersawn shake: Wood shake that is resawn on both faces.

Tapered edge strip: A tapered insulation strip used to elevate and slope the roof at the perimeter and at curbs, and to provide a gradual transition from one layer of insulation to another.

Taping: The process of covering drywall joints with paper tape and joint compound.

Tar: ASTM defines tar as "a brown or black bituminous material, liquid or semi-solid in consistency, in which the predominating constituents are bitumens obtained as condensates in the processing of coal, petroleum, oil-shale, wood, or other organic materials."

Tar boils: Bubbles of moisture vapor encased in a thin film of bitumen. Also known as **blackberries**.

Target housing: In lead inspecting it refers to any housing constructed prior to 1978, except housing for the elderly or persons with disabilities (unless any one or more children age 6 years or under resides or is expected to reside in such housing for the elderly or persons with disabilities) or any 0-bedroom dwelling.

Tax impact: The impact of taxes on investment income and rate of return. Tax impact should never be ignored when comparing alternatives.

Tax liability: Real estate taxable income multiplied by the tax rate.

Tax shelter: The ability of an investment to reduce an investor's tax liability through the use of cost recovery.

Taxable income: Adjusted gross income less allowable deductions and exemptions.

Taxation: How an investment is affected by tax laws and codes.

Teak: A deciduous tree native to South-East Asia. Teak is very hard and resistant to weathering. It is used for shipbuilding, furniture, fences, etc. Trees may reach 150 feet.

Tearout: The tendency to splinter the trailing edge of material when cutting across the grain.

Tear-off and reroof: The removal of all roof system components down to the structural deck, followed by installation of a completely new roof system. See **Re-cover**.

Tear resistance: The load required to tear a material, when the stress is concentrated on a small area of the material by the introduction of a prescribed flaw or notch. Expressed in psi (pounds force) per inch width or kN/m (kilo-newton per meter width).

Tear strength: The maximum force required to tear a specimen.

Technical feasibility: In the case of site selection, it is an evaluation of multiple sites to determine which sites should be considered further based upon their physical limitations, regulatory requirements, and environmental and legal considerations. Technical feasibility differs from highest and best use that refers to the determination of the possible uses of a particular site as based upon technical considerations.

Technical inspection: An inspection that matches the technical specification criteria with visual and mechanical tests to ascertain conformance.

Technical review: The critique of design solutions, or criteria used for design solutions, by a party other than the one providing the solutions or criteria, to determine adequacy and suitability of purpose.

Tectonic: Tectonic refers to rock-deforming processes and resulting structures that occur over large sections of the lithosphere.

Tectonic earthquakes: Earthquakes resulting from the release of strain by deformation of the earth. See **Fault**.

Tectonic plates: The large, thin, relatively rigid plates that move relative to one another on the outer surface of the Earth.

Tectonic province: A region characterized by uniform geologic structures.

Tectonics: A branch of geology dealing with structure and deformation of the earth's crust.

Teco: Metal straps that are nailed and secure the roof rafters and trusses to the top horizontal wall plate. Also called a hurricane clip. See **Hurricane clip**.

Tee: A hot rolled shape with symbol T and is shaped like a "T."

Tee-A: "T" shaped plumbing fitting.

Teflon tape: Tape made from Teflon that is wrapped around threads and helps to seal threaded pipe joints.

Tegular tile: Ceiling tiles with recessed edges that allow the tile to hang below the ceiling grid.

Telecommuting: Working by using telecommunications and computer technology from a location other than the usual or traditional place of business, for example, from home, a satellite office, or a telework center.

Telegraphing: A shingle distortion that may arise when a new roof is laid over and uneven surface.

Telegraphing underlying defects.

Asphalt pavement laid over a poorly prepared surface also telegraphs the underlying defects.

Teleseismic: Pertains to earthquakes at distances greater than 1,000 km from the measurement site.

Telltale: Lightweight cloth or plastic tape that has been tied to cable railings at the edge of a floor plate identifying the edge of a building (a warning device).

Tempered glass: Annealed glass that has been cut to size and heat-treated to produce a glass that has a strength of 15,000 psi.

Tempered glass will not shatter or create shards, but it will pelletize. Most codes require tempered glass in tub and shower enclosures, entry doors and sidelights, and in a windows when the window sill is less than 16" to the floor. Tempered glass is expensive.

Template: A shaped piece of metal, wood, card, plastic, or other material used to obtain dimensions of irregular shapes, cutting out, shaping, or drilling.

Temporary protection: Any barrier, warning device, railing, covering, etc., necessary and or required on a construction site for the protection of human life and any protection connected with the building process.

Tempory structure: Anything built that will not become part of the permanent structural system and will eventually be removed before or after the completion of the structure. Also referred to as **temporary works**.

Tenon: A projection formed on the end of a timber or the like for insertion into a mortise of the same dimensions.

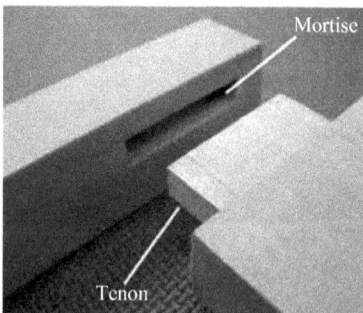

Tenon saw: This fine-toothed hand saw is used to cut tenons and other wood joints

accurately. Similar to a back saw, but shorter to offer greater control.

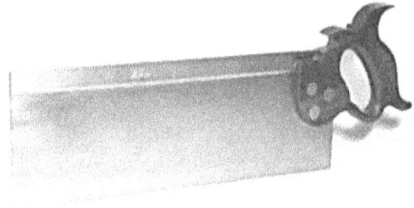

Tenon saw.

Tenoning head: Equipment that is used to form a cylindrical tenon on the end of round-wood posts or poles. Consists of a rotating cutterhead.

Tensile stress: Stress resulting from stretching a material.

Tensile strength: The maximum unit stress that a material can resist under axial tensile loading, based on cross-sectional area of the specimen before loading.

Also, regarding metal connectors, the stress calculated from the maximum load sustained by a metal connector during a tension test divided by the original cross-sectional area of the specimen. Tensile strength is a required physical property for bolts, rods, studs, wire strand, and wire rope.

Tension: A pulling or stretching force. Tension is the opposite of compression.

Tension ring: A support ring that resists the outward force pushing against the lower sides of a dome.

Tensional stress: The stress that tends to pull something apart. It is the stress component perpendicular to a given surface, such as a fault plane, that results from forces applied perpendicular to the surface or from remote forces transmitted through the surrounding rock.

Tensioning: A method of stretching the saw body in the inner area of either a circular saw or band saw to compensate for heating that expands the circular saw periphery or the band saw edges. The amount of tension required is affected by gauge, saw speed, number and kind of teeth, diameter of a circle saw, width of a band saw, wood species, horsepower, feed speed, etc.

As a general rule circular saws that are large, thin, fast, or heavily loaded require more tension than small, thick, slow, lightly loaded saws. Wide, thick band saws require more tension than narrow, thin band saws. A very thin, narrow band may need no tensioning.

People tend to guess at the proper tension when tensioning, as tension gauges are expensive. For large saws and serious sawing a tension gauge is recommended.

Tension wood: A type of wood that forms in angiosperms (The common name for this group is "flowering plants.") in response to environmental stresses. Tension wood helps the tree stay stable and upright. This type of wood is not useful as flooring, furniture, and other products, because it has an irregular texture; tension wood fibers hold together tenaciously, so that the sawed surfaces usually have projecting fibers, and planed surfaces often are torn or have raised grain. This wood also does not absorb paint, stain, and other treatments in the same way that regular wood does.

Termination: The treatment or method of anchoring and/or sealing the free edges of the membrane in a roofing or waterproofing system.

A **termination bar**, usually metal or vinyl, is used to seal and anchor the free edges of a roof membrane. Also referred to as a **term bar**.

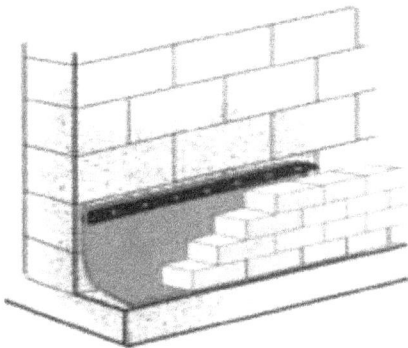

Termination bar.

Termites: Wood eating insects that superficially resemble ants in size and general appearance, and live in colonies.

Termite shield: A shield, usually of galvanized metal, placed in or on a foundation wall or around pipes to prevent the passage of termites.

Terne: Terne is used as a roofing material. Traditionally, terne was made of sheet iron or steel plated with an alloy of three or four parts of lead to one part of tin. Currently, lead has been replaced with the metal zinc and is used in the ratio of 50% tin and 50% zinc.

Terra cotta: A ceramic material molded into masonry units.

Terrazzo: Traditionally a type of stone flooring made from marble or other stone chips that are mixed in Portland cement, poured in place, allowed to dry, and then polished. In the 1970s, polymer-based terrazzo was introduced and is called thin-set terrazzo; thin-set terrazzo is a wet floor material made up of epoxy, marble dust, and marble aggregate that is ground smooth and polished after it has cured. Today, most of the terrazzo installed is **epoxy terrazzo.**

Test cylinders: Sample cylinders of concrete taken from each truck delivering concrete from a batch plant. The cylinders are cured and tested in a laboratory to verify the strength of the designed mix.

Test plug: A rubber plug that is used to seal off sections of pipe to allow testing for leakage.

Test tee: A tee with a removable plug for permitting access to a drainage line.

Thatch roof: The covering of a roof usually made of straw, reed, or natural foliage (palms) bound together to shed water.

Therm: In technical usage, the term is a convenient measure of the heating value of 100,000 Btu. One therm is roughly equivalent to the heating value of 100 cubic feet of natural (methane) gas.

Theodolite: An optical instrument used by land surveyors for surveying and by engineers and builders for setting-out lines and angles on the ground.

Theodolite circa 1913.

Modern laser theodolite.

Thermal: As used here, of, relating to, or caused by heat:

Terms beginning with thermal:

• **Thermal barrier:** A component that is a poor conductor of heat and is placed in an assembly containing highly conducting materials in order to reduce or prevent the flow of heat. For example, an insulating layer located between the inside and outside parts of an aluminum window frame to block the flow of heat through the window frame. Also referred to as thermal break and thermal block.

• **Thermal comfort:** A condition experienced by building occupants expressing satisfaction with the thermal environment.

• **Thermal conduction:** Heat transfer through a material by contact of one molecule to the next. Heat flows from a high-temperature area to one of lower temperature.

• **Thermal conductivity:** The heat transfer property of materials expressed in units of Btu per hour per inch of thickness per square foot of surface per 1°F temperature difference. Referred to by the letter "k."

• **Thermal conductance:** The same as thermal conductivity except thickness is 'as stated' rather than one inch. Referred to by the letter "C."

• **Thermal expansion:** The increase in dimension or volume of a material when subjected to heat. The coefficient of thermal expansion is the change in unit length or volume accompanying a unit change is temperature, at a specified temperature.

• **Thermal fatigue:** A form of fatigue failure that is due to thermal stresses that are produced by cyclic changes in temperature.

• **Thermal insulation:** A material that resists heat flow. Material having a high R-value.

• **Thermal movement:** Movement of a material resulting from temperature changes.

• **Thermal pane glass:** A window pane with two sheets of glass and a spacer between them. See **Single pane** and **Triple pane**.

• **Thermal protector:** As applied to motors, a protective device for assembly as an integral part of a motor or motor-compressor that, when properly applied, protects the motor against dangerous overheating due to overload and failure to start. The thermal protector may consist of one or more sensing elements.

• **Thermal resistance:** A property of a substance or construction that retards the flow of heat; one measure of this property is R-value. See **Heat transfer coefficient**.

- **Thermal shock:** The stress-producing phenomenon resulting from sudden temperature changes in a roof membrane when, for example, a cold rain shower follows brilliant hot sunshine, which may result in sudden cooling or rapid contraction of the membrane.
- **Thermal stress:** Stress introduced by uniform or non-uniform temperature change in a structure or material that is contained against expansion or contraction.

Thermally protected: As applied to motors, the words "thermally protected" appearing on the nameplate of a motor or motor-compressor indicate that the motor is provided with a thermal protector.

Thermocouple (TC): A sensor that measures temperature based on the voltage difference between two dissimilar metals. Thermocouples are on most gas- or oil-fired appliances that shut off the supply of fuel if the pilot light blows out, and are commonly found on furnaces and water heaters.

Also, an electronic device that converts thermal energy into electrical energy. Electrons flow between the hot junction and cold junction creating millivoltage. A thermopile is an apparatus that consists of a number of thermocouples combined so as to multiply the effect and is used for generating electrical current.

Thermoplastic: A plastic compound that will soften and melt with sufficient heat. Thermoplastic insulation compounds are used to manufacture certain types of electrical cables.

Thermoplastic olefin(TPO): Single-ply roofing membranes.

Thermoply™: An exterior laminated sheathing nailed to the exterior side of the exterior walls. Normally 1/4" thick, 4 × 8 or 4 × 10 sheets with an aluminumized surface.

Thermoset: A plastic compound that will not re-melt. Thermoset insulation compounds are used to manufacture certain types of cables.

Thermosetting glues and resins: Glues and resins that are cured with heat but do not soften when subsequently subjected to high temperatures.

Thermostat: A device that relegates the temperature of a room or building by switching heating or cooling equipment on or off.

Thermostatic valves: Valves that have a set and forget feature that allows a user to dial in a preferred water temperature for a shower or bath. Water temperature remains constant while showering, even with changes in water pressure or temperature such as when someone flushes the toilet. In addition to its convenience, the thermostatic valve is a safety feature that prevents scalding. Thermostatic valves are also used on radiators.

Thin kerf saw blade: A saw blade with a kerf, or cut width, between 0.065 and 0.070 inches.

Thinner: A liquid used to reduce the viscosity of coatings or mastic. Thinners evaporate during the curing process. Thinners may be used as solvents for clean-up of equipment.

Thinset tile: Tiles that are attached directly to a substrate such as drywall. See **Isolation membrane**.

Thixotropic: Describes a material having a viscosity that decreases when a stress is applied. Bentonite slurry acts as a thixotropic material. There are thixotropic paints that undergo a reduction in viscosity when shaken, stirred or otherwise mechanically disturbed and that readily recovers the original condition on standing.

Thread: Actually a screw thread, but usually shortened to and commonly called a thread, is a helical structure that is used to convert between rotational and linear movement or force. A screw thread is a ridge wrapped around a cylinder or cone in the form of a helix; the former is called a straight thread and the latter is called a tapered thread.

Terms beginning with thread:

- **Thread crest:** The top part of the thread. For external threads, the crest is the region of the thread which is on its outer surface, for internal threads it is the region which forms the inner diameter.

- **Thread flank:** The thread flanks join the thread roots to the crest.
- **Thread height:** The distance between the minor and major diameters of the thread measured radially.
- **Thread length:** Theportion of the fastener with threads.
- **Thread pitch gauge:** A measuring device used to determine the exact thread pitch needed for replacing screws and nuts.
- **Thread root:** The bottom of the thread, on external threads the roots are usually rounded so that fatigue performance is improved.
- **Thread runout:** The portion at the end of a threaded shank that is not cut or rolled to full depth, but that provides a transition between full depth threads and the fastener shank or head.

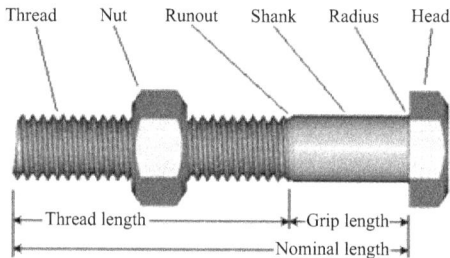

A bolt showing thread runout.

Three-dimensional calculations: A process by which the number of three-dimensional units (e.g., cubic yards) is determined for a given structural part. Three-dimensional units of measure include all those that are measured in three directions (e.g., length, width, and height). Three-dimensional units of measure deal with volume. Examples are cubic feet, cubic yards and board feet.

Three-dimensional shingles: Laminated shingles. Shingles that have added dimensionality because of extra layers or tabs, giving a shake-like appearance. May also be called "architectural shingles."

Three-way edging clamp: This tool is used to apply and repair moldings, decorative trim, and edging. It has a C-clamp design with a third screw that applies right angle pressure to the edge.

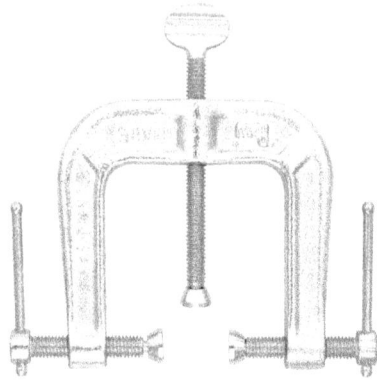

Three-way edging clamp.

Three-way switch: Electrical switches used to control the same fixture from two different locations, such as two ends of a hall.

Three-wide (3W): A term referring to any product or unit when three frames (i.e., separate jambs) are mulled together as a multiple unit.

Threshold: The bottom metal or wood plate of an exterior door frame.

Throat: The area at the top of the firebox between the face of the smoke shield and the top of the flue.

Also, the cutout of a shingle.

Throat plate: A plate atop a saw through which the saw blade goes.

Circular Saw Throat Plate.

Through-wall flashing: A water-resistant material that may be metal or membrane, extending through a wall and its cavities. Through-wall flashing is positioned to direct water entering the top of the wall or cavity to the exterior, usually through weep holes.

Thrust: The horizontal component of a reaction or an outward horizontal force.

Thrust block: A solid concrete block placed at the juncture (change in direction) of a water line that has been buried, preventing the line from breaking due to pressure surge caused by a water hammer.

Also, the blocking located at the bottom of a stair run that prevents stair stringers from sliding out of place.

Thrust fault: See **Fault, Thrust**.

Tidemark: A high-water or sometimes low-water mark left by tidal water or flood. Also a mark placed to indicate this point.

Tie: A rod, plate, or angle welded between a two angle web member or between a top or bottom chord panel to tie them together usually located at the middle of the member. See **Filler or Plug**.

Tie joist: A joist that is bolted at a column.

Tieback: A structural tie used to hold sheeting around an excavation by drilling the tie into the soil behind the sheeting and securing it through the sheeting as a tensile member.

Tie-in: The joining of two different roof systems.

Tie-off: A watertight seal used to terminate roof membranes at system adjuncts, terminations, flashings, or substrates. Can be temporary or permanent. See **Night seal**.

Tight side: In knife-cut veneer, that side of the sheet that was farthest from the knife as the sheet was being cut and containing no cutting checks (lathe checks).

Tile: A thin rectangular slab of baked clay, concrete, or other material, used in overlapping rows for covering roofs.

Also, a thin square slab of glazed ceramic, cork, linoleum, or other material for covering floors, walls, or other surfaces.

Tile base: A specialty tile trim piece installed on a wall that covers the corner of a floor and the wall. See **Cap mold piece, Cove piece**, and **Double bullnose**.

Tile countertop: A countertop made from tiles that are glued to a substrate. See **Plastic laminate countertop, Solid surface countertop, Solid plastic countertop, Wood block countertop, Cultured countertop, Cultured marble countertop,** and **Stone countertop**.

Tilt window: A single- or double-hung window whose operable sash can be tilted into a room to allow cleaning of the exterior surface on the inside.

Tilted joist: A joist that is supported in a manner such that the vertical axes of the joist is not perpendicular with respect to the ground.

Timber connector: Various kinds of steel fixings designed to make high-strength connections in timber construction.

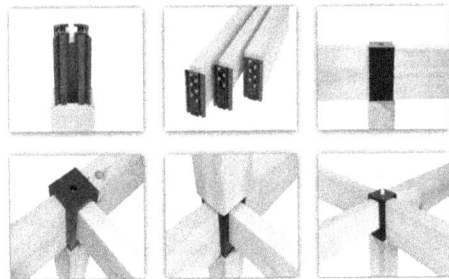

Timber connectors.

Timber framed: Construction in which the main load bearing elements are timber.

Timbers: Lumber that is nominally 5 inches or more in least dimension. Timbers may be used as beams, stringers, posts, caps, sills, girders, purlins, etc.

Timbers used in the original round form, such as poles, piling, posts, and mine timbers are round timbers.

Time and materials contract: A construction contract that specifies a price for different elements of the work such as cost per hour of labor, overhead, profit, etc. A contract that may not have a maximum price, or may state a "price not to exceed."

Time dependent response analysis: The study of the behavior of a structure as it responds to a specific ground motion.

Time is of the essence: A term that is used in contyracts that fixes time of performance as a vital term of the contract, the breach of which may operate as a discharge of the entire contract.

Time value of money (TVM): An economic principle recognizing that a dollar today has greater value than a dollar in the future because of its earning power.

Tin (Sn): A metal, tin has been used by man since prehistoric times. Today it is used primarily for tin-plating that prevents rust. Other uses include the manufacture of alloys such as brass, bronze and gun metal, tin foil, bearing metals, and household utensils.

Tin-can stud: Slang for a lighter-gauge metal stud.

Tinner: A slang name for the heating contractor. Another common name for these tradespeople is tinknockers.

Tinted glass: See **Heat-absorbing glass**.

Tip up: The downspout extension that directs water (from the home's gutter system) away from the home. They typically swing up when mowing the lawn, etc.

Titanium dioxide: White pigment that is in virtually all white paints. The prime hiding pigment in most paints.

Title: Evidence (usually in the form of a certificate or deed) of a person's legal right to ownership of a property.

Toe: The outside points of each leg of a structural angle.

Toe kick: Bottom portion of a lower cabinet unit that is recessed to reduce damage from shoes and hide marring that occurs as shoes hit against the finished material.

Toe nailing: Driving a nail in at a slant. A method used to secure floor joists to the plate. See **Face-nailing**.

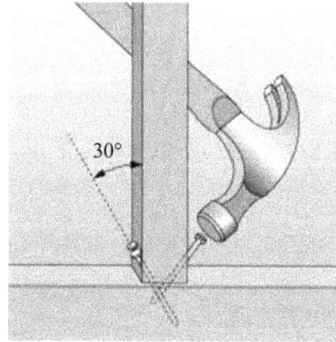

Toe nailing.

Toe of fillet: The end or termination edge of a fillet weld. Also, the end or termination edge of a rolled section fillet.

Toe of weld: The junction between the face of a weld and the base metal.

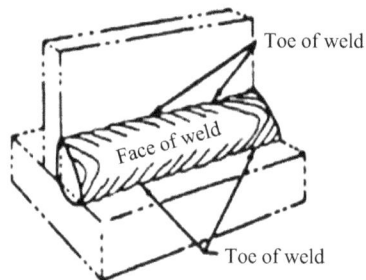

Toe of weld.

Toe nailing: Nailing at an angle or slant into one framing member and driving it through into a second member. See **Face-nail**.

Toeboard: A low protective barrier that will prevent the fall of materials and equipment to lower levels and provide protection from falls for personnel.

Toggle bolt: A bolt with a separate toggle end that can be flattened to fit through a pre-drilled hole and that springs outward to provide securement when the bolt is tightened.

Ton (T): US unit of weight equal to 2,000 lb; also called a Short ton. Also a British unit of weight equal to 2,240 lb (1,016 kg); also called a Long ton.

Ton: In cooling, it is the amount of cooling an air conditioning unit provides. One ton is equal to 12,000 BTUs.

Tongue and groove: A type of edge often found on materials to be used for sheathing and flooring. Each panel has one long edge with a tongue and the other long edge with a corresponding groove. The tongue of one sheet will fit into the groove of the next sheet to form a seam or joint.

Tongue and groove planks are one of the oldest types of dimensional structural wood used as roof decking.

Tonne: A unit of weight in the metric system equal to 1,000 kg or approximately 2,204 pounds. Also called a Metric ton.

Tool marks: Marks left in material by the knives used to create the shape such as those in a shaper or molder.

Tooled joint: A mortar joint between courses of brick in which the mason removes the excess mortar so that it is flush with the face of the brick. A tool is then used to shape the mortar.

Tooling: The process of removing unwanted material from a finish carpentry joint through the use of a chisel, rasp or other sharp instrument.

Tooth bite: The length of cut for each tooth on a saw as wood is fed to the saw. Tooth bite will vary based on arbor rpm, tooth pitch, saw diameter, and feed speed.

Tooth pitch: The distance between teeth on a saw.

Tooth pressure: The force each tooth exerts when cutting.

Tooth set: Bending alternate saw teeth to each side, causing their points to cut a kerf slightly wider than the plate thickness.

Terms beginning with top:

- **Top bevel:** An angle filed across the top of the tooth, usually staggered on alternate

teeth. If all teeth are angled in one direction, the saw will pull to one side.

- **Top chord:** The upper or top member of a truss.
- **Top chord bearing:** Flat trusses that are hung from their top chord.

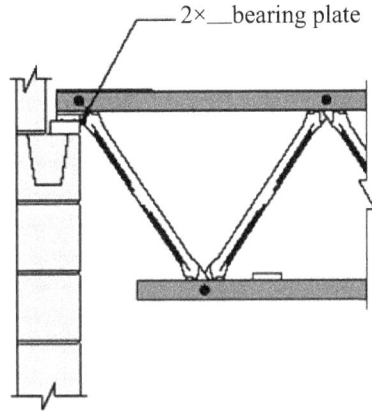

2×___bearing plate

Top chord bearing on a masonry wall.

- **Top chord extension (TCX):** The extended part of a joist top chord only. This type has only the two top chord angles extended past the joist seat. See Overhang.
- **Top clearance:** The angle of clearance on the top of a saw tooth. Also clearance angle and back clearance angle. See **Zero clearance plate**.
- **Top plate:** The top horizontal member of a frame wall. The top plate supports ceiling joists, rafters, or other members.
- **Top rail:** The top hand rail used on a balustrade. The tops of balusters are attached into the underside of the top rail. See **Balustrade** and **Bread loaf top rail**.

Topping: A thin finished layer of concrete, asphalt, or soil.

Topping mud: A type of drywall mud. Topping mud is used for final coats and contains less adhesive chemicals than does all purpose mud.

Topping out: The installation of the top most member in a buildings structural frame.

Torch down installation method: An Installation method in which roofing materials are heated with a torch until the material liquefies and forms a bond between layers, overlapping seams, or flashing. On modified bitumen roof systems, a method whereby APP type modified bitumen roofing is adhered to the base sheet.

Torque: A rotary thrust. The turning effort of a fluid motor usually expressed in inch pounds.

Torque converter: A device that transmits or multiplies torque generated by an engine.

Torque motor: A type of electromechanical transducer having rotary motion. These motors provide a servo actuator that can be attached directly to the driven load. A permanent magnet field and a wound armature act together to convert electrical power to torque.

Torque motors have relatively large diameter-to-length ratios, and short axial dimensions. The large diameter gives the motor a large lever arm to generate high levels of torque. Torque motors are "frameless" motors, i.e., they don't have housings, bearings, or feedback devices.

Torque motors are available in a wide range of sizes, with diameters from smaller than 100 mm to larger than 2 m. They are primarily designed for applications below 1,000 rpm.

Torque wrench: A tool for setting and adjusting the tightness of screws, nuts, and bolts to a desired value. A torque wrench allows the operator to measure the torque applied to the fastener so it can be matched to the specifications for a particular application.

Torpedo level: This back-pocket size (7–9 inches) level makes it ideal for work in restricted areas. Some have magnetic bases.

Torpedo level.

Torsion: Twisting around an axis. Simple torsion is produced by a couple or moment in a plane perpendicular to the axis.

Torsion loads cause a structural member to twist about its longitudinal axis.

With regard to earthquakes, torsion is when the center of the mass does not coincide with the center of the resultant force of the resisting building elements, thereby causing rotation or twisting action in plans and stress concentrations. Symmetry in general reduces torsion.

Total discharge head: The total pressure or head the pump must develop. It is the sum of the depth to pumping level, elevation, service pressure, and friction loss.

Total project budget: All costs, hard and soft, necessary to achieve completion of a project.

Total suspended solids (TSS): The total amount of suspended solids dispersed in a liquid.

TSS in mg/L can be calculated as:

(Dry weight of residue and filter − Dry weight of filter alone, in grams) ÷ (Ml of sample × 1,000,000).

Touchless faucet: A commercially installed faucet or urinal valve that functions using an infrared sensor system.

Toughness: The ability of a material to absorb large amounts of energy without being readily damaged.

Tower: The vertical structure in a suspension bridge or cable-stayed bridge from which cables are hung.

Also, used loosely as a synonym for the term skyscraper.

Tower crane: A crane with the jib mounted at the top of a tower. They may be static or tracked, with a rigid or "luffing" (vertically hinged) jib. They are usually electrically operated.

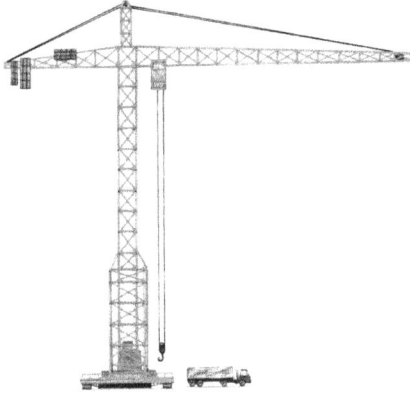

Tower crane.

Townhouse: A single-family dwelling unit constructed in a group of three or more attached units. Each unit extends from the foundation to the roof and each unit has open space on at least two sides.

Tracheids, longitudinal: The elongated cells or fibers that comprise the majority of the anatomical structure of softwoods. Also present in some hardwoods.

Traffic bearing roof: In waterproofing, a membrane formulated to withstand a predetermined amount of pedestrian or vehicular use with separate protection and a wear course.

Trammel points: Used to scribe a large diameter circle or arc and transfer measurements that are too great for dividers.

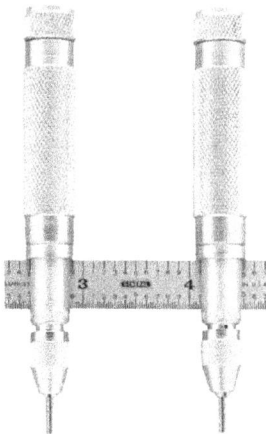

Trammel points

Transducer: A device that converts variations in a physical quantity, such as pressure or brightness, into an electrical signal, or vice versa.

Transfer column: A column that is supported by beams, girders, trusses or similar members and is reacting on two or more columns at a lower level.

Transform fault: See **Fault, Transform**.

Transformer: A device used to change the voltage of an alternating current in one circuit to a different voltage in a second circuit, or to partially isolate two circuits from each other.

Transitions: When a roof plane ties into another roof plane that has a different pitch or slope.

Translation: Motion of an object where the path of every point is a straight line.

Transmittal: A written document that is used to identify information being sent to a receiving party. The transmittal is usually the cover sheet for the information being sent and includes the name, telephone/FAX number, e-mail, and address of the sending and receiving parties. The sender may include a message or instructions in the transmittal. It is also important to include the names of other parties the information is being sent to on the transmittal form.

Transmitter: With regard to garage doors, the small, push button device that causes the garage door to open or close.

Transmissibility: The principle stating that a force has the same external effect on an object regardless of where it acts along its line of action.

Transom: A window used over the top of a door or window, primarily for additional light and aesthetic value.

Also, a horizontal member separating a door from a window panel above the door, or separating one window above another. Also referred to as a **transom bar**.

Transom light: The window sash located above a door. Also referred to as transom window.

Transit pipe: Pipe that was manufactured from asbestos and concrete and commonly used in water mains; asbestos is no longer permitted.

Transverse: Crossing from side to side or placed crosswise.

Trap: A plumbing fitting that holds water to prevent air, gas, and vermin from backing up into a fixture.

In plumbing, a trap is a U-, S-, or J-shaped pipe located below or within a plumbing fixture. The bend is used to prevent sewer gases from entering buildings. In refinery applications, it also prevents hydrocarbons and other dangerous gases from escaping outside through drains.

The most common of these traps in houses is referred to as a P-trap or sink trap. It is essentially a U-trap with a 90° fitting added on the outlet side giving it a P-like shape.

Traps retain a small amount of water after the fixture's use. This water creates a seal that prevents sewer gas from passing from the drain pipes back into the building. All plumbing fixtures such as sinks, bathtubs, and toilets must be equipped with either an internal or external trap.

Not infrequently sink traps capture (trap) objects such as rings that are inadvertently dropped into the sink. They also collect hair, sand, and other debris that ultimately stops the flow of water. For all of these reasons most traps can either be disassembled for cleaning or are provided with some sort of cleanout feature.

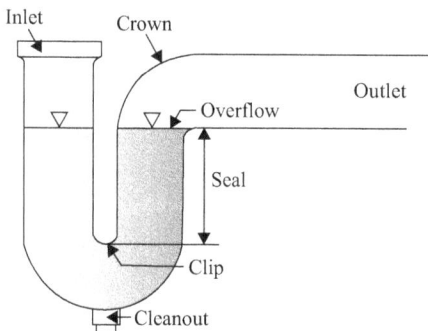

P-Trap.

Also, there are float and thermostatic traps that are used to separate condensate and air from the steam within the steam heating coil, and inverted bucket traps that operate on the difference in the density between steam and water.

Terms beginning with trap:

- **Trap arm:** The waste arm portion of a drainage trap.
- **Trap dip:** The U-bend portion of a drainage trap.
- **Trap primer:** A small feeder line that connects the cold water line directly to the drainage trap, which releases a small amount of water to the trap should it run dry to maintain the water seal.
- **Trap seal:** The height of water in a toilet bowl at rest. It provides the water seal that prevents sewer gases from entering the home. It is measured from the top of the dam down to the inlet of the trapway. Also referred to as deep seal.

Trapezoid: A 4-sided shape with only two parallel sides.

Travel time: Wages paid to workmen under certain union contracts and under certain job conditions for the time spent in traveling from their place of residence to and from the job.

Travel-time: A graph of time versus distance for the arrival of seismic waves at a series of stations. Also referred to as time-distance curve.

Travelers: Two leads that are connected between a three-way switch to allow power to a fixture to be switched on or off from either switch.

Traveling cable: A cable made up of electric conductors that provides an electrical connection between an elevator, dumbwaiter car, or material lift, and a fixed outlet in the hoistway or machine room.

Tread: The walking surface board in a stairway on which the foot is placed.

Treated lumber: A wood product which has been impregnated with chemical pesticides such as CCA (Chromated Copper Arsenate) to reduce damage from wood rot or insects. Often used for the portions of a structure that are likely to be in contact with soil and water. Wood may also be treated with a fire retardant.

Tree wire: A type of overhead distribution wire that is insulated for momentary contact with tree branches and used as a primary voltage conductor.

Treeing: Water treeing is a form of cable insulation degradation where micochannels, that often appear as a tree-like structure in the insulation, develop due to a complex interaction of water, electrical stress, impurities, and imperfections.

Treenail: A wooden peg made from dry compressed timber, made to swell when placed in its hold and moistened.

Treenails.

Tremie: A tube for the distribution of concrete in a controlled manner from the bottom of a form to the top. Tremie procedures are usually carried out through water or slurry displacing the liquid as concrete is placed.

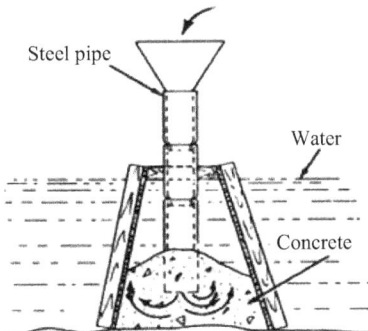

Tremie method.

Trench: With regard to earthquakes, a long and narrow deep trough in the sea floor. Trenches are interpreted as marking the line along which a plate bends down into a subduction zone.

Also, a trench is a narrow excavation (in relation to its length) that is made below the surface of the ground. In general, the depth is greater than the width, but the width of a trench (measured at the bottom) is not greater than 15 feet (4.6 m).

See **Shield, Trench box,** and **Trench shield**.

Trenchless: A system of installing or replacing underground infrastructure by pulling the lines through an existing pipe or a pilot hole, rather than laying pipe into long, pre-dug trenches.

Triangle: A three-sided shape. When one of the angles is a right angle, equal to 90°, it is called a right triangle.

Tributary width or area: The design area that contributes load to a structural member. It is one half the distance between members on either side of the member. Said differently, tributary width is the "width" of some space that contributes load (or something else) to a structural member.

Trim: The finish materials that are on the exterior a building, such as moldings applied around openings (window trim, door trim), siding, windows, exterior doors, attic vents, crawl space vents, shutters, etc., are, obviously, exterior trim.

The finish materials that are in a building, such as moldings applied around openings (window trim, door trim) or at the floor and ceiling of rooms (baseboard, cornice, and other moldings) are, again obviously, **interior trim**.

Trim, plumbing, heating, and electrical: The work that the HVAC, electrical and plumbing (subcontractors) contractors perform to finish their respective aspects of the work, and when the project is nearing completion and getting ready for occupancy.

Trimmer: The vertical stud that supports a header at a door, window, or other opening.

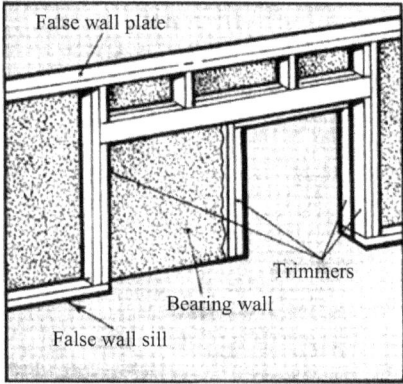

Trimmers.

Trimmer bit: A type of router bit that contains a roller that guides the blades along a straight edge.

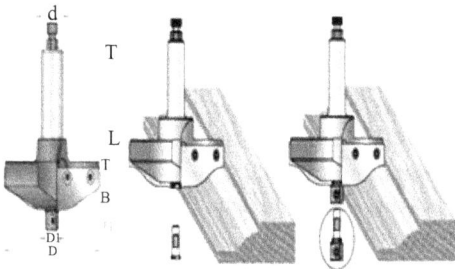

Trimmer bit.

Trimmer joist: One of the joists supporting a header. The header applies a concentrated load at that point on the trimmer joist.

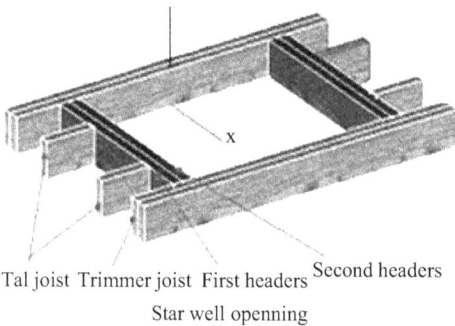

Stairwell opening showing trimmer joists.

Triple chip grind: A tooth style on a saw blade for cutting non-ferrous metal and plastic.

Triple pane glass: A window pane with three sheets of glass for extra insulation. Also referred to as triple glazing. See **Single pane** and **Thermal pane**. Also referred to as **triple glazing**.

Triple window: A term generally referring to any tripartite group of windows with square heads. These are frequently found on Colonial Revival houses; they suggest Palladian Windows but are less expensive to build.

Tri-polymer alloy (TPA): TPA is a thermoplastic tri-polymer alloy single ply roof membrane system.

Trombe wall: A masonry wall that is usually glazed on the exterior and is designed to absorb heat and release it into the interior of a building.

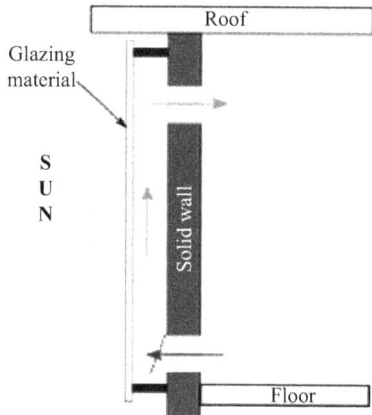

Trombe wall.

Trowel: A tool with a flat surface that is used to finish the concrete surface. A trowel usually the last tool used to smooth the concrete surface. Its use must be timed carefully: it should be used after the concrete has lost its weep moisture but before it loses all of its plasticity.

Trowel pattern finish: An exterior concrete finish that is created by skilled craftsmen when the concrete is ready to be troweled.

True divided lites (TDLs): Traditional window construction incorporating smaller panes of glass actually separated by muntins, rather than simulating such an appearance with larger lites of glass and a muntin grid or grille placed between or on the surfaces of the glass layers.

Trumeau: The stone mullion supporting the middle of a tympanum of a doorway.

Trumeau

Trumeau.

Truss: In general, a structural load-carrying member with an open web system that is designed as a simple span with each member designed to carry a tension or compression force. The entire structure acts as a beam.

A **truss bridge** is a bridge that is composed of connected elements (typically straight) that may be stressed from tension, compression, or sometimes both in response to dynamic loads. Truss bridges are one of the oldest types of modern bridges.

In house construction a truss is an engineered and manufactured roof support member that is essentially a small truss bridge. It is a structural part that is used to provide the primary support for the floor or roof sheathing. A roof truss system, including the trusses, sheathing, bracing, and fasteners and also provides support for the tops of the exterior bearing walls.

Trussed rafters: Wooden trusses, usually triangular in shape that span between the external walls to form a roof. They are relatively inexpensive and easy to use for new roofs. Trussed rafters generally do not require internal support from beams or partitions, but they do restrict the use of the attic space more than other types of construction.

Tsunami: A sea wave produced by large area displacements of the ocean bottom, the result of earthquakes or volcanic activity; a tidal wave caused by ground motion.

Tsunami magnitude (Mt): A number used to compare sizes of tsunamis generated by different earthquakes and calculated from the logarithm of the maximum amplitude of the tsunami wave measured by a tide gauge distant from the tsunami source.

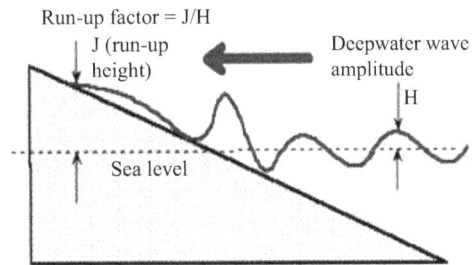

Tsunami magnitude (Mt).

Tsunamigenic: Refers to those earthquakes, commonly along major subduction-zone plate boundaries such as those bordering the Pacific Ocean, that can generate tsunamis.

Tub trap: Curved, "U" shaped section of a bath tub drain pipe that holds a water seal to prevent sewer gasses from entering the home through tubs water drain. See **Traps**.

Tube: A hollow structural steel member shaped like a square or rectangle used as a beam, column, or as bracing. Usually the nominal outside corner radius is equal to two times the wall thickness.

As pertains to plumbing, a hollow cylinder, especially one that conveys a fluid or functions as a passage. There is a difference between a tube and a pipe; the difference depends on how they are measured, and ultimately how they are used.

A pipe is a vessel while a tube is structural. A pipe is measured based on its inside diameter (ID) because it is a vessel, a tube is measured based on its outside diameter (OD) because it is structural.

Pipes have a consistent ID regardless of wall thickness. For example, a 1/2-inch high-pressure pipe may need a 2-inch thick wall, but the ID will still be 1/2-inch, even though the OD is 4.5 inches.

Common wall thicknesses of copper tubing in the USA are Type K, Type L, Type M, and Type DWV.

- **Type K** is the thickest wall section of the three types of pressure rated tubing and is commonly used for domestic water service and distribution, fire protection, etc.
- **Type L** is a thinner pipe wall section, and is used in residential and commercial water supply and pressure applications.
- **Type M** is an even thinner pipe wall section, and is used in residential and commercial water supply and pressure applications.
- **Type DWV** is the thinnest wall section, and is generally only suitable for unpressurized applications, such as drains, waste and vent (DWV) lines.

For comparison:

	Nominal Size	Actual OD	Wall Thickness
K	1"	1-1/8"	0.065"
L	1"	1-1/8"	0.05"
M	1"	1-1/8"	0.035"

Tube form: A cylindrical tube made from compressed and resin impregnated paper and used to hold wet concrete until it cures. Also known by trade names such as **sonotube, sleek Tube**, and **smooth Tube**.

Sonotube (a brand name).

Tuberculation: The process in which blister-like growths of metal oxides develop in pipes because of corrosion of the pipe metal. Iron oxide tubercles often develop over pits in iron or steel pipe, and can seriously restrict the flow of water.

Tuberculation: corrosion from the inside.

Tubular gauge glass: A tubular glass through which the boiler water shows the level.

Tubular lock or latch: Any door lock-set or latch-set that requires a bore hole and an edge-bore for installation.

A standard interior door handle requires one bore hole (usually 2-1/8" in diameter) and one edge-bore (usually 1" in diameter) whereas a tubular entry door handle set requires two such bores that are usually spaced either 3-5/8" or 5-1/2" apart from center-to-center.

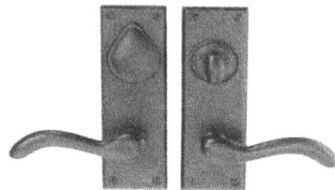
Tubular latchset front door handle.

Tubular welded-frame scaffold: See **Fabricated frame scaffold**.

Tuck carpet installation method: A method of installing carpet on a stair in which the installer wraps the tread with carpet, and as a separate step installs a piece of carpet on the riser, giving it a defined appearance. Think of it as upholstering the step with the carpet. See **Waterfall carpet installation method**. Sometimes called French tuck.

Tuck pointing: A difficult and expensive form of pointing. The joint is flush pointed with mortar colored to match the bricks, and a very thin false joint is cut into the mortar and pointed in lime putty of a contrasting colour. Tuck pointing is becoming a lost art. See **Pointing**.

Tuck pointing.

Tung oil: Tung oil is obtained from the seeds of trees of the genus Aleurites, family Euphorbiaceae, both of Chinese origin. The oil dries quickly and is therefore used in inks, paints and varnishes as a drying oil.

Tuned mass damper: A mechanical counterweight designed to reduce the effects of motion, such as the swaying of a skyscraper in the wind or in an earthquake.

A tuned mass wind damper.

Tuning: To modify the period of the building beyond the range of the site period to avoid resonance. Examples of tuning include lowering the height of a building; lowering the position of weight in a building; changing materials; changing fixity of base, etc. The longer the period, the less inertial forces can be expected. Short periods close to the fault and long periods far from the fault are usual.

Tunnel boring machine (TBM): A mechanical device that tunnels through the ground.

Tunnel shields: A cylinder pushed ahead of tunneling equipment to provide advance support for the tunnel roof. Tunnel shields are used when tunneling in soft or unstable ground.

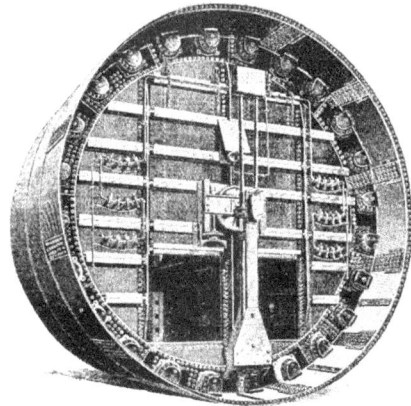

Tunnel shield.

Turbidites: Sea-bottom deposits formed by massive slope failures where rivers have deposited large deltas. These slopes fail in response to earthquake shaking or excessive sedimentation load. The temporal correlation of turbidite occurrence for some deltas of the Pacific Northwest suggests that these deposits have been formed by earthquakes.

Turbine: A rotary device that is actuated by the impact of a moving fluid against blades or vanes.

Turbine vent: A vent that creates a vacuum in the attic by turning as the warm air escapes thereby pulling more air out.

Turbo toilet: A specialty water closet that uses the water pressure from the plumbing lines to force water into the bowl. The turbo toilet uses less water than most other types of water closets. See **Low-profile water closet** and **Water closet**.

Turbulent flow: A condition where the fluid particles move in random paths rather than in continuous parallel paths. Also referred to as turbulence.

Turnbuckle: A rotating sleeve or link with internal screw threads at each end and used to tighten or connect the ends of a rod.

Galvanized turnbuckle.

Turnkey: A term used when the contractor or subcontractor provides all materials and labor for a job.

Turn-of-the-nut-method: A method for pre-tensioning high-strength bolts by the rotation of the wrench a predetermined amount after the nut has been tightened to a snug fit.

Turtle vent: A vent positioned several feet below the ridge. Turtle vents have no moving parts. As the air heats it becomes less dense and rises through the turtle vent.

Turpentine: A petroleum, volatile oil used as a thinner in paints and as a solvent in varnishes.

Tuscan order: The least ornate of the five classical orders of architecture, similar to the Doric but with a plain rather than fluted shaft.

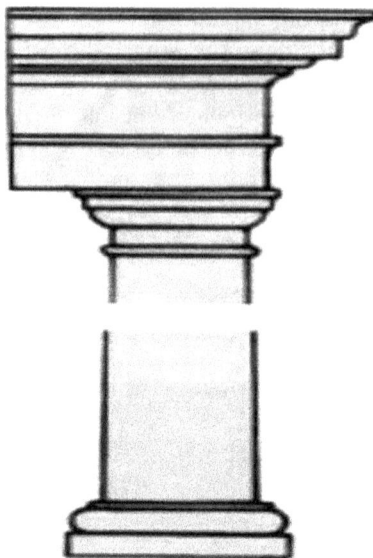

Tuscan order.

Tusk tenon joint: A traditional timber connection that is typically used to connect trimmers around a hearth; it is strong and ridged. The tenon extends through the main joist and is fitted with a wooden wedge to stop the joint from opening up. Also referred to as wedge tenon and keyed tenon. See **Mortise and tenon**.

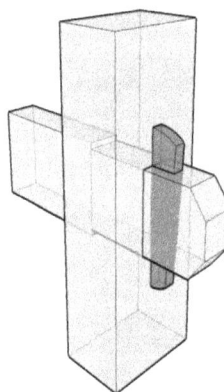

Tusk tenon.

Two-dimensional calculations: A process by which the number of two-dimensional units (e.g. square feet) is determined for a given structural part. Two-dimensional units of measure include all those that are measured in two directions. Determining the area of a surface that is measured in two-dimensional units such as square yards or square feet, is a typical two-dimensional calculation.

Two-step distributor: An industry term for a wholesale company that buys building products from the manufacturer and sells them to lumberyards and home centers that in turn sell to builders, contractors, and homeowners. A wholesaler that buys building products from a manufacturer and sells them to builders, contractors, and homeowners is referred to as a **one-step distributor**.

Two-way valve: A directional control valve with two flow paths.

Two-wide (2W): A term referring to any product or unit when two frames (i.e., separate jambs) are mulled together as a multiple unit.

Two-year, 24-hour design storm: The basis of planning stormwater management facilities that can accommodate the largest amount of rainfall expected over a 24-h period during a 2-year interval.

More generally, a design storm is a mathematical representation of a precipitation event that reflects conditions in a given area for design of infrastructure.

Twist: A spiral distortion along the length of a piece of timber so that the four corners of any face are no longer in the same plane.

Also, a deviation in which one or two corners of a door are out of plane with the other corners of the door.

Type X drywall: Drywall with a gypsum core that contains reinforcing fibers for added fire protection.

U-block: A block that looks the same as a standard block from the front or back, but whose cells are open on the top so that grout can flow outward to the other block on each side. The U-block provides for placement of horizontal reinforcing steel and grout to form a bond beam within the course.

U-block.

U-factor: The rate of heat flow-value through a building component, from room air to outside air. Also referred to as U-value. The lower the U-factor, the better the insulating value. U-factor, a rating more generally used in the window industry, is the reciprocal of R-value, a rating commonly used in the insulation industry.

Ufer ground: A type of electrical ground where the ground wire is connected to the rebar system inside a footing and foundation system. Named after Thomas Ufer, the first person to specify it.

UL-2218 impact resistance rating: A test criteria for measuring the impact resistance of roofing materials. UL-2218 rates roofing materials on a scale from I to IV with IV being products that best withstand the impact test. Insurance discounts area available to homeowners in some areas who choose Class IV roofs.

Ultimate load: The force necessary to cause rupture.

Ultimate strength: The maximum stress attained by a structural member prior to rupture which is the ultimate load divided by the original cross-sectional area of the member.

Ultimate strength design: Load and Resistance Factor Design.(LRFD) method. See **LRFD**.

Ultra square: A comprehensive builders square that reduces plate layout time by greater than 1/2 and makes rafter calculations automatically.

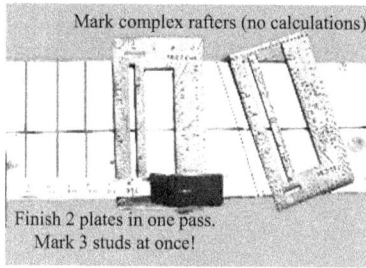

Ultra square.

Ultrasonic extensometer: An instrument that can measure the change in length of a fastener ultrasonically as the fastener is tightened or measure the length before and after it is tightened.

Ultrex: A pultruded composite material made of polyester resin and glass fibers. (Pultrusion is a molding process in which heated resin cures as it is pulled through a die.)

Unbraced frame: A frame that provides resistance to lateral load by the bending resistance of the frame members and their connections.

Unbraced length: The distance between points of bracing of a structural member, measured between the centers of gravity of the bracing members.

Unbraced top chord: The specific length where the top chord of a joist has no lateral bracing by deck, bridging, or any other means.

Undercoat: A coating that is applied prior to the finishing or top coats of a paint job. It may be the first of two or the second of three coats. Sometimes called the prime coat.

Undercounter sink: These sinks attach below the counter, creating a rimless, unobstructed juncture between counter and sink for easy cleaning and more counter room. Also referred to as an undermount sink. See **Drop-in sink**.

An undercounter sink.

Undercoursing shingles: Shingles that are made from low-quality wood with sapwood, flat grain, and possibly loose knots.

Undercoursing shingles.

Undercover parking: Undercover parking is underground, under a deck, under a roof, or under a building; its hardscape surfaces are shaded. See **Underground parking**.

Undercut: A notch or groove melted into the base metal next to the toe or root of a weld and left unfilled by weld metal.

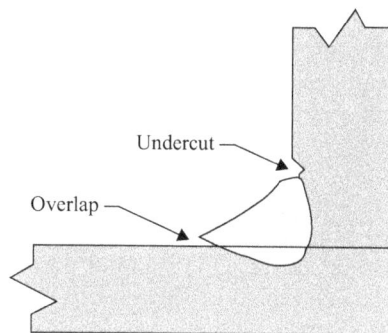

Weld undercut and overlap.

Underdriven: A term that is used to describe a fastener that is not fully driven flush to the shingles surface.

Underground parking: Underground parking is a tuck-under or stacked structure that reduces the exposed parking surface area. See **Undercover parking**.

Underground plumbing: The plumbing drains and waste lines that are installed beneath a basement floor.

Underlayment: A 1/4" material placed over the sub-floor plywood sheeting and under finish coverings, such as vinyl flooring, to provide a smooth, even surface.

Also, an asphalt-saturated felt or other sheet material (may be self-adhering) installed between the roof deck and the roof system, usually used in a steep-slope roof construction. Underlayment is primarily used to separate the roof covering from the roof deck, to shed water, and to provide secondary weather protection for the roof area of the building.

Synthetic roof underlayment.

Underpinning: The process of strengthening and stabilizing an existing building or other structure by extending its existing foundations downwards. This is usually done with mass concrete but other high- and low-tech methods are available, e.g., the use of micropiles and jet grouting.

Underpinning is accomplished by extending the foundation in depth or in breadth so it either rests on a more supportive soil or better distributes its load. are common methods in underpinning. An alternative to underpinning is the strengthening of the soil by the introduction of a grout.

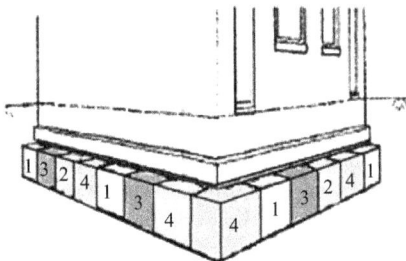

One approach to underpinning.

Underslung: A description of a joist that is suspended from upper support points where most of the mass of steel or wood is below the actual support points.

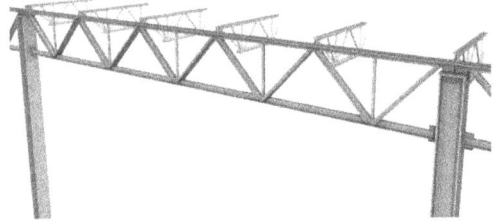

Underslung joist.

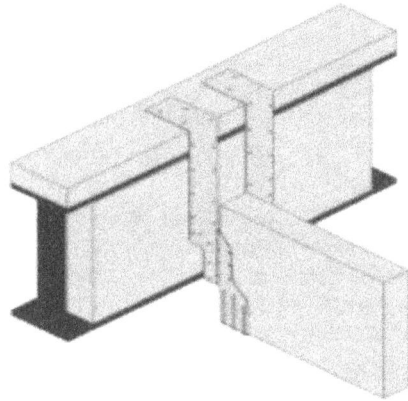

Another underslung joist.

Under-slab utilities: Heating, plumbing, electrical or other utilities that are placed under the floor slab. They are generally placed in trenches that are then covered with compactable fill before the concrete floor slab is poured.

Uniform Building Code (UBC): One of the family of codes and related publications published by the International Conference of Building Officials (ICBO) and other organizations, such as the International Association of Plumbing and Mechanical Officials (IAPMO) and the National Fire Protection Association (NFPA), that have similar goals as far as code publications are concerned. The Uniform Building Code is designed to be compatible with these other codes, as together they make up the enforcement tools of a jurisdiction.

Uniform load: A load that is distributed over a structural member, floor, or column without any concentrated loading condition. (pounds per square foot; pounds per square inch.)

A load or force that for practical purposes may be considered constant over the entire length or partial length of the member is a **uniformly distributed load**.

Uniformity of concrete: The degree or consistency of the properties of the concrete from one part of the structure to another.

The homogeneity of cement and aggregates that are dispersed throughout a concrete mix is termed the **uniformity of mix**.

Uninterruptible power supply (UPS): UPS provides conditioned power and battery backup for computers in the case of a power failure or brownout. A UPS differs from an auxiliary or emergency power system or standby generator in that it will provide near-instantaneous protection from input power interruptions. It does this by supplying energy stored in batteries or a flywheel. The on-battery runtime of most uninterruptible power sources is relatively short (only a few minutes) but sufficient to start a standby power source or properly shut down the protected equipment.

Uninterruptible power supply systemsare used to protect computers, data centers, telecommunication equipment, and other electrical equipment where a power disruption could cause injuries, fatalities, serious business disruption, or data loss.

Union: A plumbing fitting that joins pipes end-to-end so they can be dismantled.

A dielectric union that is joining dissimilar metals.

Unit dimension (UD): The actual dimension of a window, or door unit from the outside of the frame.

Unit masonry: A masonry module designed to be repetitive and modular with other masonry units.

Unit prices: A predetermined price for a measurement or quantity of work to be performed within a specific contract. The designated unit price would include all labor materials, equipment or services associated with the measurement or quantity established.

Universal arbor: The center portion of a saw blade, which is used to mount the blade onto a circular saw, consisting of a 5/8-inch round hole for mounting onto a standard circular saw, inside a diamond-shaped knockout which when removed will enable the blade to be mounted on a worm drive saw.

Universal design: Universal design accommodates the needs of users of varying ages, sizes, and abilities. By taking into account such things as door widths, counter heights, floor surfaces and handle shape, knowledgeable designers create homes and products that are comfortable and useful for a wide variety of users.

The seven Principles of Universal Design are:

- Low physical effort
- Tolerance for effort
- Simple, intuitive use
- Perceptible information (signage)
- Equitable use
- Flexibility
- Size and space for approach and use.

Universal notification: Notifying building occupants not less than 72 h before a pesticide is applied in a building or on surrounding grounds under normal conditions, and within 24 h after application of a pesticide in emergency conditions. Use of a least toxic pesticide or self-contained nonrodent bait does not require universal notification; all other pesticides applications do.

Universal Plumbing Code: A system of procedures designed to provide consumers with safe and sanitary plumbing systems. This code is used throughout the United States by local jurisdictions.

Unmatched veneer: See **Veneer**.

Unprotected sides and edges: Any side or edge (except at entrances to points of access) of a walking/working surface, e.g., floor, roof, ramp, or runway, where there is no wall or guardrail system at least 39 inches (1.0 m) high.

Unsound cement: The presence of compounds in the cement that results in appreciable expansion under conditions of restraint causing disruption of the hardened cement paste. This expansion may take place because of delayed or slow hydration of free lime, magnesia, and calcium sulphate.

Unsound concrete: Concrete that has undergone deterioration or disintegration during service exposure.

Unstable: Characteristic of a structure that collapses or deforms under a realistic load.

Unstable objects are those whose strength, configuration, or lack of stability may allow them to become dislocated and shift and therefore may not properly support the loads imposed on them. Unstable objects do not constitute a safe base support for scaffolds, platforms, or employees. Examples include, but are not limited to, barrels, boxes, loose bricks, and concrete blocks.

Untempered hardboard underlayment: An underlayment made from hardboard. Only untempered hardboard should be used as vinyl underlayment. See **Underlayment, Particleboard underlayment, Plywood underlayment, Lauan plywood underlayment, Cement board underlayment,** and **Gypsum-based underlayment**.

Upflow furnace: A furnace that forces air up and out the top of it.

Uplift: The wind load on a member that causes a load in the upward direction. The gross uplift is determined from various codes and is generally a horizontal wind pressure multiplied by a factor to establish the uplift pressure. The net uplift is the gross uplift minus the allowable portion of dead load including the weight of the joist and is the load that the specifying professional shall indicate to the joist manufacturer.

The bridging required by uplift design is **uplift bridging**. Usually always required at the first bottom chord panel point of a K-Series, LH- or DLH-Series joist and at other locations along the bottom chord as required by design.

Uprights: The vertical members of a trench shoring system placed in contact with the earth and usually positioned so that individual members do not contact each other.

Upper unit: Any cabinet unit that is designed to hang on the wall, usually above a lower unit or appliance. See **Lower unit**, **Vanity cabinet**, and **Full height cabinet**.

Upset: Any material that is raised above its primary surface yet integral with that surface.

Upside down roof: See **Protected membrane roof**.

Upstanding leg: The leg of a structural angle that is projecting up from you when viewing.

Upstream face: The side of a dam that is against the water.

Urethane: An important resin in the coatings industry. A true urethane coating is a two-component product that cures when an isocyanate (the catalyst) prompts a chemical reaction that unites the components.

Urinal: A plumbing fixture that receives only liquid body waste and conveys the waste through a trap seal into a gravity drainage system.

Usable storage: With regard to how water heaters, the percentage of hot water that can be drawn from a tank before the temperature drops to a point that it is no longer considered hot.

Usable wall space: Any section of wall along which a piece of furniture or an appliance may be placed. Hallways are generally not considered usable wall space.

USDA Organic: A certification for products that contain at least 95% organically produced ingredients (excluding water and salt). Any remaining product ingredients must consist of approved non-agricultural substances (as listed by USDA) or non-organically produced agricultural products that are not commercially available in organic form.

Utility easement: The area of the earth that has electric, gas, or telephone lines. These areas may be owned by the homeowner, but the utility company has the legal right to enter the area as necessary to repair or service the lines.

Utility pole auger bit: A specialty auger bit designed for drilling creosoted poles and heavy construction timber. Also called a **power pole auger bit**.

Utility sink: A deep fixed basin that is supplied with hot and cold water. Utility sinks are usually used for rinsing mops and disposing cleaning water.

V-clip: A clip shaped like the letter "V." It is used on a lavatory drain lift linkage assembly to easily adjust connection of the drain to the lift rod.

Vacuum breaker: An anti-siphon device that prevents the backflow of contaminated water into the water supply system.

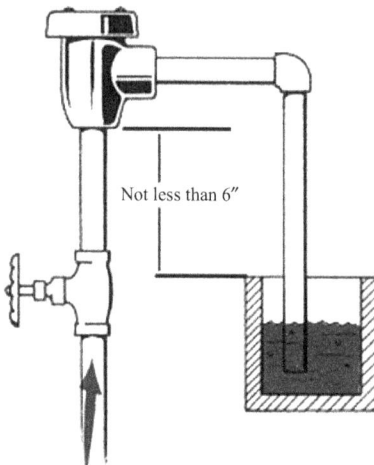

Atmospheric vacuum breaker typical installatio

Not less than 6″

Atmospheric vacuum breaker.

Valance: A short ornamental piece of drapery, wood, metal, etc., placed across the top of a window. Wooden shades usually have a valance to cover the brackets.

Also, a short curtain or piece of drapery hung from the edge of a canopy, the frame of a bed, etc.

Valley: The "V" shaped area of a roof where two sloping roofs meet. Water drains off the roof at the valleys.

Terms beginning with valley:

• **Valley flashing:** Sheet metal that lays in the "V" area of a roof valley.
• **Valley jack:** A type of jack rafter that runs from the valley rafter to the ridge board. See **Rafter, Jack**.
• **Valley rafter:** A rafter that runs along the valley, forming the valley line.

Valuation: An inspection carried out for the benefit of the mortgage lender to ascertain if a property is a good security for a loan.

Valuation fee: The fee paid by the prospective borrower for the lender's inspection of the property. Normally paid upon loan application.

Value Engineering (VE): A systems approach to saving money and, at the same time, provide equal or better value of goods and services; sometimes it is to improve on some other established value. VE began at the General Electric Co. during World War II when there were dire shortages of resources, and substitutes were needed. This approach has been institutionalized so that nowadays most large government contracts have a VE clause.

Valve: A device that controls fluid flow direction, pressure, or flow rate through a pipe or duct, usually allowing movement in one direction.

Valve seat wrench: A hexagonal end wrench inserted into the hexagonal opening in a valve seat for installing or removal.

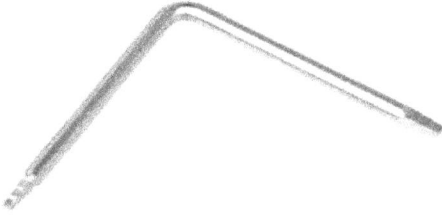

Faucet valve seat wrench.

Vanity cabinet: A type of lower unit cabinet that is designed to hold a bathroom sink. A standard vanity cabinet is slightly shorter than a standard lower unit. See **Lower unit**, **Upper unit**, and **Full height cabinet**.

Vapor barrier: A building product installed on exterior walls and ceilings under the drywall and on the warm side of the insulation. It is used to retard the movement of water vapor into walls and prevent condensation within them. Normally, polyethylene plastic sheeting is used.

Vapor retarder: Material installed to impede or restrict the passage of water vapor through a roof assembly.

Varnish: A coating that lacks pigment, offering a transparent finish for a wood surface.

Vault: An ancient form of construction consisting of masonry formed in an arched shape.

Vault construction.

Vector: A mathematical entity having a magnitude and a direction in space.

Variable interest rate: An interest rate that will vary over the term of the loan.

Varying distributed load: A load or force that for practical purposes may be considered varying over the surface of the member, for example a snow drift. See **Uniform load**.

Vegetation-containing artifices: Planters, gardens or other constructs intended to host flora.

Vegetative density factor (K_D): Vegetative density factor (K_D) in the equation for estimating K_L accounts for the fact that different collective areas of leaves lead to different losses of water.

The density factor (KD) is assigned a value between 0.5 and 1.3 within three groupings, as shown:

Density Factor			
Vegetation	**High**	**Average**	**Low**
Trees	1.3	1.0	0.5
Shrubs	1.1	1.0	0.5
Grd Cvr	1.1	1.0	0.5
Mixed	1.3	1.1	0.6
Turf Grass	1.0	1.0	1.0

Vehicle: The portion of a coating that includes all liquids and the binder. The vehicle and the pigment are the two basic components of paint.

Velocity: The speed of something in a given direction; the rate of change of distance traveled with time in a given direction.

Also, with regard to earthquakes, the time rate of displacement of a reference point in an earthquake or the speed with which a particular seismic wave propagates in a rock.

Velocity structure: A generalized model of the crust consisting of units of different seismic velocity.

Veneer: With regard to woodworking, veneer is thin slices of wood that are glued onto core panels of wood, particle board or medium-density fiberboard to produce flat panels. Veneers are used on furniture, doors, parquet floors, etc. They are also used in marquetry (inlaid work made from small pieces of variously colored wood or other materials).

The appearance of various veneers depends very much on how it was cut: either by peeling the trunk of a tree or by slicing large rectangular blocks of wood known as flitches.

Here is a sampling of matched veneer appearances (there are others):

• **Book match:** The most commonly used match. Every other piece of veneer is turned over so adjacent pieces are opened like adjacent pages in a book. The veneer joints match and create a mirrored image pattern at the joint line, yielding a maximum continuity of grain.

• **Barber pole effect in book match:** Because the "tight" and "loose" faces alternate in adjacent pieces of veneer, they might accept stain differently, and this might yield a noticeable color variation called barber poling. Proper sanding will help, but will not eliminate the effect.

Barber pole effect in book match.

• **Slip match:** Adjoining pieces of veneer are placed in sequence without turning over every other piece. The grain figure repeats, but joints won't show a mirrored effect.

• **Random match:** A random selection of individual pieces of veneer from one

or more logs. Random match produces a board-like appearance.

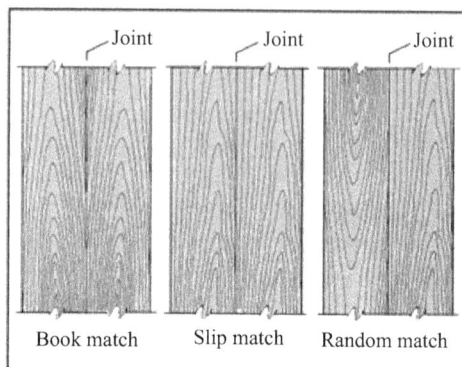

Book match   Slip match   Random match

• **Running match:** Running match gives a non-symmetrical appearance in any single door face. Veneer pieces are of unequal width and each face is assembled from as many veneer pieces as necessary.

• **Balance match:** Balance match gives a symmetrical appearance. Each face is assembled from pieces of uniform width before trimming.

• **Center match:** Center match also gives a symmetrical appearance. Each face has an even number of veneer pieces of uniform width before trimming. Therefore, there is a veneer joint in the center of the panel, producing symmetry.

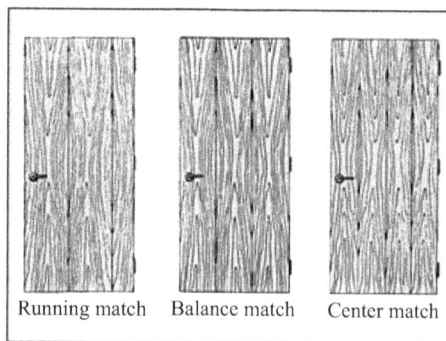

Running match   Balance match   Center match

• **Pair Match:** Doors may be specified as pair matched.

Pair matched doors.

- **Set Match:** Sets of doors may be specified as matching.

The term veneer also applies to other thin layers that are used to improve the appearance and durability of materials. For example, brick and stone can be used as veneers to cover the exterior of a structure.

Brick veneer wall.

Vent: A pipe or duct which allows the flow of air and gasses to the outside.

Also, another word for the moving glass part of a window sash, i.e., window vent.

Terms beginning with vent:

- **Vent header:** A vent pipe into which several vents connect. The vent pipe leads to the vent stack and out of the building.
- **Vent sleeve:** See **Vent collar**.
- **Vent stacks:** The pipes that are connected to soil lines. They allow sewer gasses to escape through to the roof or ground surface. The vent pipe also prevents plumbing traps from siphoning off. Also called a **soil stacks.** See **Vent header**.

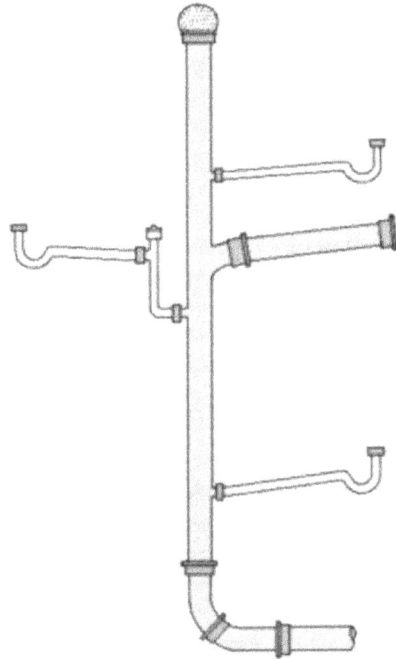

Vent stack.

Ventilation: The provision and removal of air to control air contaminant levels, humidity, or temperature within an indoor space. Ventilation is measured in air changes per hour—the quantity of infiltration air in cubic feet per minute (cfm) divided by the volume of the room.

Ventilation short circuit: With regard to a passive ventilation system (where the system is designed for air flow between intake and exhaust vents) a ventilation short circuit occurs when air is introduced into the ventilation system from an area higher than the intake vent thereby minimizing or defeating the effectiveness of the intake vent. One example can be a gable vent in a soffit-to-ridge ventilation system. Air intake from the gable vent can short circuit the stack-effect draw of air through the soffit vents, and interrupt the thorough venting of the roof cavity.

Ventilator: An accessory that is designed to allow for the passage of air.

Venting: A natural hairline crack in the stone.

Verge of popcorn texture: A rough surface texture of Sprayed Polyurethane Foam generally considered unsuitable to receive a base coating. Nodules on this surface are larger than the valleys and an additional 50% or more of coating material is necessary to properly cover and protect the surface.

Vermiculated: Stonework carved in a random pattern fancifully comparable with the appearance of worms.

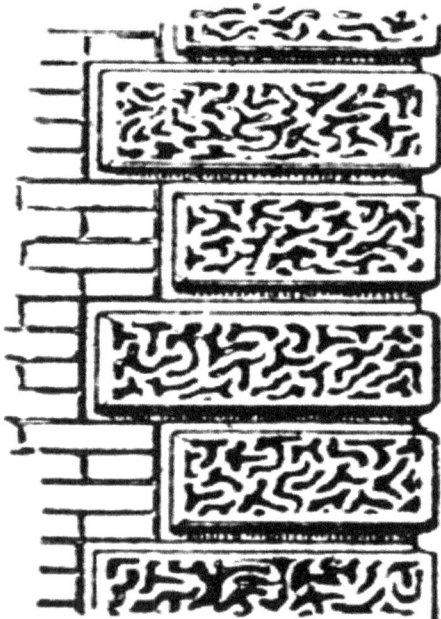

Vermiculated masonry.

Vermiculite: A mineral used as bulk insulation and also as aggregate in insulating and acoustical plaster and in insulating concrete floors.

Vernacular: A type of building method common to a specific region, often build with wood indigenous to the area.

Vertical grain (VG): Lumber that is sawn at approximately right angles to the annual growth rings so that the rings form an angle of 45° or more with the surface of the piece. Less than 25% of wood milled is vertical grain, making it considerable more expensive (up to twice the cost). However, vertical grain wood is more desirable than the cheaper and more plentiful flat grain.

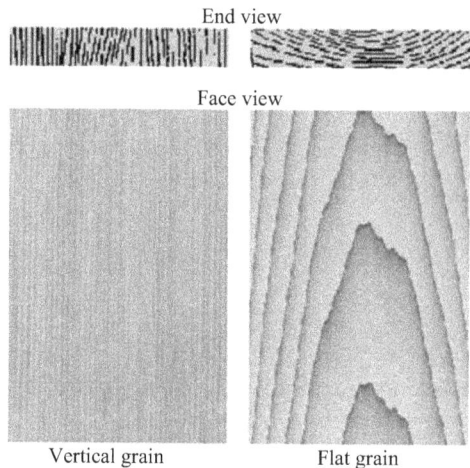

End view

Face view

Vertical grain   Flat grain

Vertical slip forms: Forms that are jacked vertically during the placement of concrete.

Vertically laminated timbers: Laminated timbers that are designed to resist bending loads applied parallel to the wide faces of the laminations.

VG-type joist girder: A type of joist girder where joists are located at panel points where vertical webs intersect the top chord only. This type of girder is used for ducts to pass through since the joists do not interfere with their passage.

Vibration: A periodic motion that repeats itself after a definite interval of time.

Also, a technique that involves the use of a mechanical device to shake concrete so that it settles tightly around the rebar and window bucks and removes the large air pockets which otherwise cause honeycombing.

Vibration causes the concrete to settle tightly and smoothly against the forms.

A mechanical or pnumatic piece of equipment that vibrates at a predetermined rate and is used to compact soil or distribute concrete in formwork is a **vibrator.**

A mounting apparatus, springs, or heavy resilient rubber, that isolates a piece of equipment from the building, thereby diminishing structure borne noise and vibration is a **vibration eliminator**.

Vierendeel girder: A type of truss consisting of vertical and horizontal members arranged like a ladder on its side.

Vierendeel girder/truss.

View: A specific way of looking at a building. There are five basic views on a plan: elevation, floor plan, plot plan, section, and detail.

Viga: A heavy rafter, most commonly a log, used for roof support in southwestern architecture.

Viga.

Vine streaks: Scars in the wood generally caused by the stems of clinging vines or by their hair-like roots, which cling to the tree trunk. Live vine streaks produce round scars. Dead vine streaks contain either dead residue of the vine, or the remaining pocket similar to bark pocket. Most vine streaks run across the grain, and therefore, all vine streaks are considered defects in accordance with restrictions described in veneer grading rules. Also referred to as vine marks.

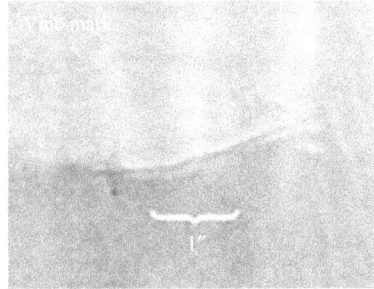

Vine Streak/Mark.

Vinyl: Polyvinyl chloride that can be extruded into solid sheets, pipes shapes etc. Vinyl has many uses, for example: Vinyl that wraps a short distance up the wall is a **vinyl cove**; A type of wall-covering with a vinyl face is **vinyl faced wall-covering;** and, a vinyl extrusion that is used on clad units that serves the same purpose as a wood glazing bead for wood units is a **vinyl glazing bead**.

Violin plane: A small plane that is used for precision finishing and decorative work. Also called a finger plane.

Violin plane.

Viscoelastic: A type of deformation in which a material behaves like an elastic solid when it is rapidly strained on time scales of seconds to hours, but deforms viscously by plastic flow over long periods of geologic time.

Viscosity: A measure of the internal friction or the resistance of a fluid to flow. The less viscous a fluid is, the greater its ease of movement.

Vise: A tool with movable jaws that are used to hold an object firmly in place while work is done on it. Vises are typically attached to a workbench.

Vise-grip locking pliers/clamps: Locking pliers. They come in various shapes to meet specific needs.

Visible light transmittance (Tvis): The ratio of total transmitted light to total incident light (i.e., the amount of visible spectrum, 380–780 nanometer light passing through a glazing surface divided by the amount of light striking the glazing surface). The higher the Tvis value, the more incident light is passing through the glazing.

Visible spectrum: That portion of the total radiation that is visible to the human eye and which lies between the ultra-violet and the infra-red portions of the electromagnetic spectrum. The colors associated with the visible spectrum range from violet, indigo, blue, green, yellow, orange, through red.

Vision glazing: Provides views of the outdoors to building occupants through vertical windows between 2'6" and 7'6" above the floor. Windows below 2'6" and windows above 7'6" (including daylight glazing, skylights, and roof monitors) do not count as vision glazing for this credit.

Visqueen: A 4 mil or 6 mil plastic sheeting.

Vitreous: A composition of materials that resemble glass. It is translucent and low on porosity.

Vitreous china: Ceramic materials that are fired at high temperature to form a non-porous body that has exposed surfaces coated with ceramic glaze fused to the body. This is used to form bathroom fixtures such as toilets, bidets, and lavs.

Void: An open space or break in consistency.

Volatile organic compounds (VOCs): Compounds that are volatile at typical room temperatures. Specifications may require you to use low or no VOCs as sealants and coatings. Check EPA Reference Test Method 24 (Determination of Volatile Matter Content, Water Content, Density Volume Solids, and Weight Solids, of Surface Coatings), Code of Federal Regulations Title 40, Part 60, Appendix A.

Volcanic earthquakes: An earthquake related to volcanic activity.

Volt: The unit by which electrical force or pressure is measured.

Voltage: A measure of electrical potential. Most homes are wired with 110 and 220-volt lines. The 110-volt power is used for lighting and most of the other circuits. The 220-volt power is usually used for the kitchen range, hot water heater and dryer.

An electrical device that is designed to eliminate any power surge or drop in current is a **voltage regulator**.

For **grounded circuits**, the voltage between the given conductor and that point or conductor of the circuit that is grounded; for ungrounded circuits, the greatest voltage between the given conductor and any other conductor of the circuit.

Volume solids: Solid ingredients as a percentage of total ingredients. The volume of pigment plus binder divided by the total volume, expressed as a percent. High-volume solids mean a thicker dry film with improved durability.

Volute: A decorative, circular handrail piece used at the bottom of a stair top rail that is installed over a newel post. See **Top rail**, **Gooseneck**, **One-quarter turn**, and **Balustrade**.

Volute: The scroll-like details on the capital of a Ionic column.

Voral tube: A flexible steel hose used for hand-held shower sprays.

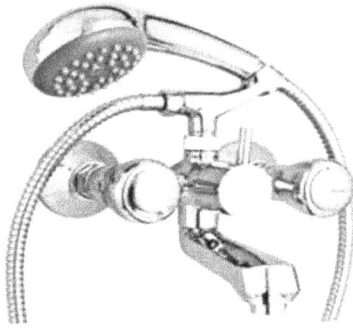

Voral tube.

Voussoir: One of the stones or bricks forming an arch. See **Springing**.

Vulcanization: Any of various processes by which natural or synthetic rubber or other polymeric materials may be cured or otherwise treated by, for example, exposure to chemicals, heat, or pressure, to render them non-thermoplastic, and that improves their elastic and physical properties.

W shapes: A hot rolled shape called a Wide Flange Shape with symbol W which has essentially parallel flange surfaces.

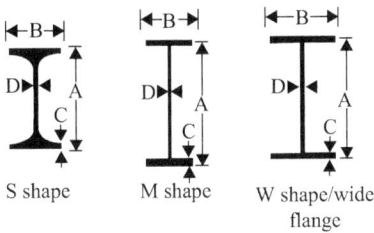

S-, M-, and W shape beams.

WT shape: A hot rolled structural tee shape with symbol WT that is cut or split from W Shapes.

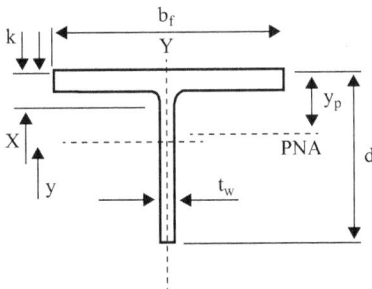

WT shape.

Wafer board: A manufactured wood panel made out of 1"–2" wood chips and glue. Often used as a substitute for plywood in the exterior wall and roof sheathing.

Waffle type sponge rubber pad: A sponge rubber pad that has been shaped to a pattern of alternating bumps and dimples. See Synthetic felt pad, Rebond pad, and High-density urethane foam pad.

Wainscot: A section of wall covering material that starts at the bottom of the wall, then proceeds upward until it is interrupted by a visible border such as a chair rail.

The wood panels or boards that cover the lower portion of a wall, often capped with molding is called **wainscoting**.

Wainscot cap: See **Molding**.

Wales: Horizontal members of a shoring system placed parallel to the excavation face whose sides bear against the vertical members of the shoring system or earth.

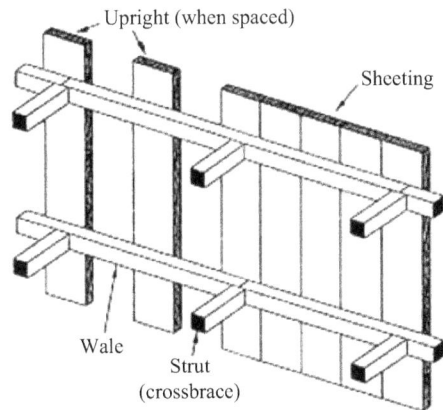

A shoring system.

Walers: Horizontal bracing, usually 2 × 4s, secured to concrete wall forms to stiffen them so they can be more easily straightened. After attaching the walers, the straightening is accomplished by placing a string parallel to the wall form, then moving the form into alignment with the string, and attaching bracing to hold the wall in position. The walers help hold the areas between the bracing in the straightened position.

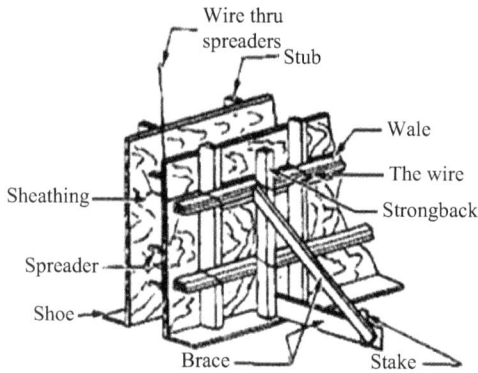

Formwork showing a wale.

Walk through inspection: A final inspection before acceptance of a project. Generally a walk through inspection looks for those last few punch list items that had not yet been accepted by the owner or owner's representative.

Walk-off mats: Mats placed inside building entrances to capture dirt, water, and other materials tracked inside by people and equipment. Walk-off mats are important items in climates that have snow and, thus, sand and salt tracked in during winter months.

Walking: The phenomena of having a joint filler, building assembly part, railing, piece of equipment, etc., move and or displace itself due to vibration, thermal loading, or gravity.

Also, any surface, whether horizontal or vertical, on which an employee walks or works, including, but not limited to, floors, roofs, ramps, bridges, runways, formwork, and concrete reinforcing steel, but not including ladders, vehicles, or trailers, on which employees must be located in order to perform their job duties. Also referred to as **working surface**.

Walkway: A portion of a scaffold platform used only for access and not as a work level.

Wall: A vertical or near vertical structure which encloses or separates spaces and may be used to resist horizontal or vertical forces or bending forces.

Wall anchor: A small piece of angle or other structural material that is usually bolted to a wall to which a starter joist or bridging angle is welded or bolted.

Wall covering: The exterior wall skin consisting of sheets or panels.

Wall out: When a painter spray paints the interior of a home.

Wall tie: Used at the intersection of two walls to provide a backing for the end stud of the connecting wall. The two types of wall ties are corner ties and wall channels.

Wallpaper: Wall covering that is available in a range of colors, patterns, textures, and materials for direct application to plaster or gypsum wallboard partitions.

Walnut: A deciduous tree that is widely cultivated, especially in France and Spain. The timber is used for fine furniture.

Wane: The absence of square wood on the edge of a board, usually because the timber section was cut too close to the edge of the trunk.

Warding file: Warding files are tapered to a point for narrow space filing. They have double-cut faces and single-cut edges. Warding files are used for lock repair or for filing ward notches in keys.

Warm-edge: A type of insulating glass construction using an airspacer offering lower thermal conductance than traditional aluminum spacer. Warm-edge IG units typically offer higher resistance to condensation and an incremental improvement in window energy performance.

Warm wall: The finished wall inside of a structure, used in roofing to determine how far up the deck to install waterproof underlayments at eaves.

Warning line system: A barrier erected on a roof to warn employees that they are approaching an unprotected roof side or edge, and which designates an area in which roofing work may take place without the use of guardrail, body belt, or safety net systems to protect employees in the area.

Warp: Any deviation from a true or plane surface in a board. Also, any distortion in a material.

Warranty: In construction there are two general types of warranties. The manufacturer of a product, such as roofing materials or any deviation from a true or plane surface in a board is an example of one type of warranty. The second warranty covers labor. For example, a roofing contract may include a 20-year material warranty and a 5-year labor warranty. Many new homebuilders provide a 1-year warranty. Any major issue found during the first year should be communicated to the builder immediately. Small items can be saved up and presented to the builder for correction periodically through the first year after closing.

Warranty phase: The time period during which a warranty is in force.

Warrington hammer: This lightweight hammer that is double headed. One end of the head of the hammer has the rounded appearance normally associated with hammers. The peen on the other end, however, is wedge-shaped or flattened. This allows people to start small nails and fasteners without having to worry about hitting their fingers. It is effectively used where a larger peen will not fit, such as tight corners.

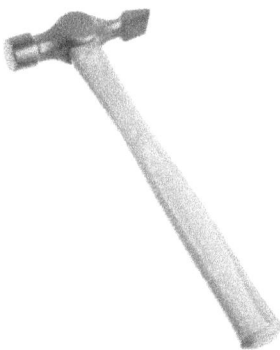

Warrington hammer.

Wash-down water closet: A water closet that has a siphon trapway at the front of the bowl, and an integral flushing rim.

Washer: A flat ring of metal with a hole in the middle used to give thickness to a joint or to distribute pressure under the head of a nut or bolt.

Washout: The process of removing any solids precipitated out of the boiler water as mud or scale. Normally either using cold pressurized water or hot pressurized water.

Waste: Material that must be purchased but cannot be used. Waste can result from trimming, rejection because of defect or other efforts to maintain acceptable quality of the structural part containing that material. A percent of waste should be included in virtually all material estimates.

Also, all materials that flow from the building to final disposal. Examples include paper, grass trimmings, food scraps, and plastics. For this credit, waste refers to all materials that are capable of being diverted from the building's waste stream of the building through waste reduction, including source reduction, recycling, and composting.

Terms beginning with waste:

- **Waste disposal:** Elimination of waste by means of burial in a landfill, combustion in an incinerator, dumping at sea, or any other way that is not recycling or reuse.
- **Waste diversion:** Disposal of waste other than through waste disposal as defined above. Examples are reuse and recycling.
- **Waste pipe and vent:** Piping that carries waste water to the municipal sewage system.
- **Waste reduction:** Includes source reduction and waste diversion through reuse or recycling.

  An organization's waste reduction program should: Describe the organization's commitment to minimizing waste disposal by using source reduction, reuse, and recycling.

  Assign responsibilities within the organization for implementation of the program.

  List the general actions that will be implemented to reduce waste.

Describe tracking and review procedures to monitor waste reduction and improve waste reduction performance.

• **Waste shoe:** A bathtub drain assembly.

Tub waste shoe.

• **Waste stream:** The overall flow of wastes from the building to the landfill, incinerator, recycling facility, or other disposal site.

Water: A chemical compound with the chemical formula H_{2O}; a water molecule contains one oxygen and two hydrogen atoms connected by covalent bonds. Water is a liquid at standard ambient temperature and pressure, but it often co-exists on Earth with its solid state, ice, and its gaseous state (water vapor or steam). Water also exists in a liquid crystal state near hydrophilic surfaces. Bottom line: we can't live without it.

Terms beginning with water and that have water as a prefix:

• **Water-based coatings:** Coatings in which the majority of the liquid content is water.
• **Water board:** Water resistant drywall that is used in tub and shower locations. Normally green or blue colored.
• **Water-cement ratio:** The ratio by mass of mixing water to cement. The water-cement ration is a significant indicator of concrete strength and durability.
• **Water closet:** Another name for toilet. See WC.
• **Water cure:** A method of curing a material, such as concrete, by applying a fine mist of water over the surface to control the rate of moisture evaporation from the material.

• **Water hammer:** A pressure surge or wave that occurs when a fluid in motion is forced to suddenly stop or change direction. Water hammer commonly occurs when a valve is closed suddenly at an end of a pipeline system and a pressure wave propagates in the pipe. (It may also be known as hydraulic shock.)

This pressure wave can cause major problems, from noise and vibration to pipe collapse. Water hammer is a potentially damaging condition in steam pipes where waves develop on the condensate surface within a pipe as steam rushes over it. If the peak of the wave becomes high enough to reach the roof of the pipe, the water in the wave is pushed violently to the far end of the pipe.

A device that is installed near a fixture to absorb the hydraulic shock caused by a sudden shutoff of water is called a **water hammer arrestor**.

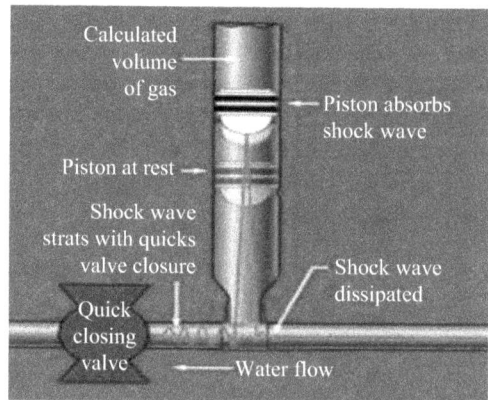

Water hammer arrestor.

• **Water meter:** A device used to measure the volume of water usage. Most commercial building water meters are designed to measure cold potable water.

The box, usually a cast iron bonnet with concrete rings, that contains the water meter is called a **water meter pit** or **vault**.

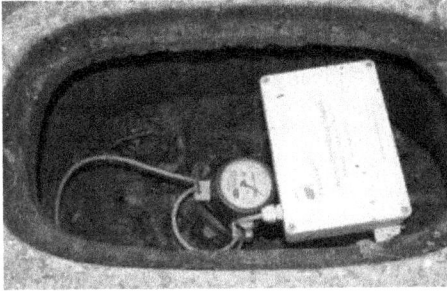

Water meter pit.

- **Water service pipe:** The pipe from the water main or other sources of potable water supply to the water-distributing system of the building served.
- **Water softener:** A device which removes minerals from water.
- **Water table:** The location of the underground water, and the vertical distance from the surface of the earth to this underground water.
- **Water tap:** The connection point where the home water line connects to the main municipal water system.
- **Water-repellent preservative:** A liquid applied to wood to give the wood water repellant properties.
- **Waterstop:** A synthetic rubber ribbon installed between concrete construction joints to prevent the passage of water. Bentonite clay is also used as a water stop in substructural joints.
- **Watertight:** Constructed so that moisture will not enter the enclosure under specific test conditions.
- **Waterproof:** Constructed or protected so that exposure to the weather will not interfere with successful operations.
- **Waterproof underlayments:** Modified bitumen based roofing underlayments. Designed to seal to wood decks and waterproof critical leak areas.
- **Waterfall carpet installation method:** A method of installing carpet on a stair in which the carpet flows over the tread and skims the riser to the next step all in one piece, giving it a waterfall appearance. See **Tuck carpet installation method**.

- **Waterproofing:** A process of coating the part of the foundation system that will be below the soil level with a material that can withstand long-term exposure to water. Not the same as damp proofing that can only withstandshort-term exposure to water.

Watt: The unit by which electric energy, or the ability of electricity to do work, is measured. A thousand watts, or one kilowatt, equals 1.34 horsepower.

Wattle and daub: Woven sticks smeared with clay to fill the spaces between the posts and beams of half-timbered homes.

Wattle (the sticks) and daub (the clay).

Wave and tidal power systems: A system that captures energy from waves and the diurnal flux of tidal power, respectively. The captured energy is commonly used for desalination, water pumping, and electricity generation.

Terms that are associated with waves:

- **Body waves:** Seismic waves within the earth.
- **Longitudinal waves:** Pure compressional waves with volume changes.
- **Love waves:** Surface waves that produce a sideways motion.
- **Rayleigh waves:** Forward and elliptical vertical seismic surface waves.
- **P-Waves:** The primary or fastest waves traveling away from a seismic event through the earth's crust, and consisting of a train of compressions and dilatations of the material (push and pull).

- **S-Waves:** Shear wave, produced essentially by the shearing or tearing motions of earthquakes at right angles to the direction of wave propagation.
- **Seismic surface waves:** Seismic waves that follow the earth's surface only, with a speed less than that of S-waves.

Wave length: The distance between successive similar points on two wave cycles. Also wavelength.

Waveform: A plot of the displacement produced by a seismic wave as a function of time.

Wavelet: A seismic pulse usually consisting of 1½ or 2 cycles.

Weak axis: The cross section that has the minor principal axis.

Weak well: A well is considered to be weak when the pump lowers the water level in the well faster than the well can replenish itself.

Wear layer: With regard to vinyl flooring, the top layer of surfacing that carries pedestrian or vehicular traffic. Sometimes referred to as wearing surface and wear course. See **Backing layer** and **Pattern layer**.

Weather: As used here, relating to meteorological conditions, climate, atmospheric pressure, temperature, elements.

Terms beginning with weather and that have weather as a prefix:

- **Weather head:** A device that prevents moisture from entering into the top of the conduit. Also known as **weatherhead, weathercap**, and **service entrance cap**.

Weather head.

- **Weather infiltration:** A condition where rain or snow penetrate the roof. This condition is typically wind-driven.
- **Weather strip:** A narrow sections of thin metal or other material installed to prevent the infiltration of air and moisture around windows and doors. Also spelled weatherstrip.
- **Weathering steel:** A type of high-strength steel that can be used in normal outdoor environments without being painted. However, wearthering steel should not be used in corrosive or marine environments.

Weathering steel.

- **Weatherization:** Work done to reduce energy consumption for heating or cooling. The term generally refers to exterior work that includes adding insulation, installing storm windows and doors, caulking cracks, and putting on weather-stripping. Also refered to as **winterization**.
- **Weathertight:** Sealed to prevent entry of air and precipitation into the structure.

Web: Diagonal supporting members that run between the top and bottom chords of a truss.

Also, the part of a structural steel wide flange section that holds the two flanges apart.

Web buckling: The buckling of a web plate.

Web configuration: The arrangement of the actual web system of a joist or joist girder that can be shown with a profile view of the member.

Web crippling: The local failure of a web plate in the region of a concentrated load or reaction.

Weep: An opening that permits the drainage of water or condensation from a wall, assembly, window, skylight, or piece of equipment.

Small openings left in the outer wall of concrete and masonry construction to permit water to escape from behind where it may do severe damage are called **weep holes**. Plastic weeps are used in modern masonry construction.

Weep holes are usually located near the base of structures, particularly brick buildings. Modern weep holes employ screens, constructed of flexible nylon or plastics to keep snakes and small animals from entering. Typically, drain tiles have weep holes, that allow water to enter the tile.

Also, small holes in storm window frames that allow moisture to escape are called weep holes.

Weep hole in masonry.

Excess water that is contained in the concrete mix that is not needed for hydration is called **weeping moisture**. It escapes through all the surfaces of the concrete as the mixture settles and forces it out.

Weeping mortar: Mortar between courses of brick that has not been trowled or otherwise smoothed after pressing the brick in place.

Weight: The force on an object resulting from gravity.

Weir: A long notch with a horizontal edge, as in the top of a vertical plate or plank, through which water flows. Weirs are used to measure the quantity of flowing water.

Weld: To fuse, bond, stick, join, attach, seal, splice, melt, solder, cement.

A weld.

Also, a term that is used for a type of corner construction. It is used with vinyl and other types of windows and doors, in which a small amount of material at the two pieces are melted or softened, then pushed together to form a single piece. This also is referred to commonly as a **fusion-weld**.

A fusion assembly.

Terms beginning with weld and that have weld as a prefix:

- **Weld shear strength:** A requirement for steel welded wire fabric used for concrete reinforcement to substantiate the bonding and anchorage value of the wires.
- **Weld washer:** A metal device with a hole through it to allow for plug welding of deck to structural steel.

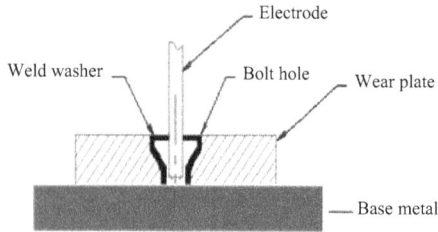

Weld washer.

- **Weldability:** The ability of a steel to be welded without its basic mechanical properties being changed.
- **Welded plate girder:** A large structural steel girder made up of plates that are welded together.
- **Welded splice:** A splice between two materials which has the joint made continuous by the process of welding.
- **Welded wire mesh (WWM):** A grid of heavy gauge wires welded together and used to reinforce concrete slabs. Also known as re-mesh or wire mesh.
- **Welding:** The process of joining materials together, usually by heating the materials to a suitable temperature.
- **Welding clamp:** A clamp that is specially treated to prevent welding spatter from adhering to, and eventually ruining, the clamp. It may be coated with a spatter-resistant copper or cadmium plating, and has special shields to protect the screw against damage.

Welding clamps.

Well casing: A steel or plastic pipe inserted into a drilled well to prevent dirt and debris from contaminating the water.

Wellpoints: Drilled wells at regular intervals for the purpose of removing ground water from an excavation.

Western red cedar: Thuja Plicata. This wood is soft, straight-grained, and extremely resistant to decay and insect damage. It is used extensively in roof coverings, exterior sidings, fences, decks, and other outdoor application.

Wet saw: A tool for sawing bricks, pavers, and tiles. It uses water to cool both the blade and brick cool and to decrease dust and flying debris.

Wet-set method: A roof tile installation method that is used on roofs with less than a 7/12 slope and usually over a mineral faced hot-mopped underlayment. Mortar is used to hold the tiles in place. The tile is wet before installation so that the mortar will better bond to it. Most commonly used in the Southeastern United States where high winds and high moisture combine. Also referred to as mortar-set method and mud-on method.

Whirlpool tub: A bathtub with circulation jets in various spots throughout the tub that provide therapeutic massaging action.

White lead: Lead carbonate.

White rot: A type of wood-destroying fungus that attacks both cellulose and lignin, producing a spongy and stringy mass that is usually whitish but which may assume various shades of yellow, tan, and light brown.

Whole house fan: A fan designed to move air through and out of a home and normally installed in the ceiling.

Whole piece: A piece of veneer that is large enough to cover an entire surface. see **Veneer**.

Wicking: The process of moisture movement by capillary action, as contrasted to movement of water vapor.

Wide flange section: A structural member that is hot rolled into the shape of an H section or I section with two flanges and one web. The structural section that replaced the I-beam American standard section.

Wildland/urban interface: An area where buildings are bounded by wild or natural areas in regions where wild fires are a concern. Some fire and code officials are looking at the establishment of fire-resistance requirements for exterior building products in these interface areas.

Wind: As used her, currents of air.

Terms beginning with wind and that have wind as a prefix:

- **Wind bracing:** Metal straps or wood blocks installed diagonally on the inside of a wall from bottom to top plate, to prevent the wall from twisting, racking, or falling over "domino" fashion.
- **Wind clip:** A clip that slips over the ends of tile, slate and other steep slope roofing materials in order to help prevent wind uplift damage.
- **Wind column:** A vertical member supporting a wall system designed to withstand horizontal wind loads. Usually between two main vertical load carrying columns.
- **Wind energy:** Electricity generated by wind turbines.
- **Wind load:** The cumulative load on a building due to wind pressure (positive or negative).
- **Wind locks:** Metal fastener inserted in the nail hole of a tile shingle and designed to overlap the lip of the next higher tile, providing additional means of holding it in place. Also called tile locks.
- **Wind pressure:** The pressure produced by stopping the wind velocity; the main cause of air infiltration.
- **Wind speed maps:** The 2012 International Building Code (IBC) has changed its wind speed maps. The IBC references ASCE 7-10 that now provides wind speeds for calculations of ultimate wind loading. This is a significant change from the previous ASCE 7-05 and changes the design pressure (psf) range from approximately 30% below to 20% above the previous values, although areas outside of hurricane prone regions remain for the most part unchanged.
- **Wind uplift:** The force caused by the deflection of wind at roof edges, roof peaks or obstructions, causing a drop in air pressure immediately above the roof surface. This force is then transmitted to the roof surface. Uplift may also occur because of the introduction of air pressure underneath the membrane and roof edges, where it can cause the membrane to balloon and pull away from the deck.
- **Windward:** The direction or side toward the wind. Opposite of leeward.

Window: A glazed opening in an external wall; an entire unit consisting of a frame, sash, and glazing, and any operable elements.

Terms beginning with window:

- **Window buck:** Square or rectangular box that is installed within a concrete foundation or block wall. A window will eventually be installed in this buck during the siding stage of construction.
- **Window fin:** Part of the window unit that serves as the flashing when siding is installed over it.
- **Window frame:** The stationary part of a window unit; window sash fits into the window frame.
- **Window hardware:** Various devices and mechanisms for the window including: catches, cords and chains, fasteners and locks, hinges and pivots, lifts and pulls, pulleys and sash. See **Weights, Sash balances**, and **Stays.**
- **Window pane:** The glass part of a window unit. Each light in a window unit has a windowpane.
- **Window sash:** The operating or movable part of a window; the sash is made of windowpanes and their border.
- **Window sill:** The bottom horizontal member of the window frame.
- **Window stop:** A horizontal or vertical piece that prevents the window from falling out of the window frame. The window stop also forms the groove that the window slides in across the surface of the jamb. See **Stop.**
- **Window unit:** A complete window with sash and frame.

Wire: As used here, a slender, stringlike piece or filament of relatively rigid or flexible metal, usually circular in section, that is manufactured in a great variety of diameters and metals depending on its application.

Terms beginning with wire:

• **Wire gauge:** A unit of measure used to indicate wire size. The thicker the wire, the smaller the gauge number. See **American standard gauge**.

• **Wire gauge drill bit:** A size range for a drill bit. Common sizes are 1-80. More commonly known as **number drill bits**.

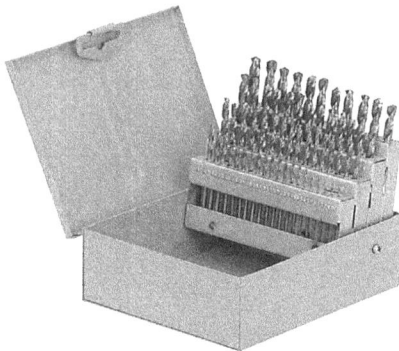

Wire gauge drill bit, aka number drill bits.

• **Wire glass:** A safety glass manufactured with a wire mesh imbedded in the glass. Also referred to **wired glass**.

• **Wire nut:** A plastic device used to connect bare wires together.

• **Wire reinforcement:** Metal reinforcing mesh placed inside the mortar joints along specified courses or rows of block to tie and reinforce the block. Like the bond beam, courses containing wire reinforcement within the mortar joint tie and reinforce the masonry wall horizontally. A structural engineer specifies which courses should contain the wire reinforcement.

• **Wire tie system:** A scheme of attachment for steep-slope roofing units (e.g., tile, slate, and stone) utilizing fasteners (nails and/or screws) in conjunction with wire to make up a concealed fastening system.

Withdrawal strength: The resistance to withdrawal of nails and screws in wood.

Also the resistance to withdrawal of bolts and other embedded items in concrete and other materials.

Under certain circumstances, such as strong gusts of wind, hurricanes or seismic activity, shingles or siding may be loosened or detached as nails are withdrawn. Alternatively, these lifting forces may pull roof or wall panels through nailheads. In either circumstance, the integrity of the structure may be severely impacted.

With: A wall dividing flues in a chimney.

Also, a single-width masonry wall in a building or home.

Also spelled **wythe**.

Wonderboard™: A panel made out of concrete and fiberglass usually used as a ceramic tile backing material. Commonly used on bathtub decks.

Wood: Wood is a hygroscopic (it has the ability to attract moisture from the air), anisotropic (its structure and properties vary in different directions) material of biological origin.

Terms beginning with wood:

• **Wood, biodeterioration:** The destruction and eventual reduction of wood to its component sugars and lignin elements through attack by organisms such as, fungi, and certain insects, for instance, termites.

• **Wood block countertop:** A countertop made from solid blocks of wood glued together. See **Plastic laminate countertop, Solid surface countertop, Solid plastic countertop, Cultured countertop, Cultured marble countertop, Stone countertop**, and **Tile countertop**.

• **Wood boring bit:** A drill bit designed specifically to bore holes into wood.

• **Wood destroying insects and organisms:** Wood destroying insects and organisms are a concern in any home with wooden structure or components. Failure to properly identify and deal with them can lead to untold damage and even complete destruction of a structure. Falling under this category are **termites, carpenter** bees, **carpenter ants**, **powder post beetles**, and **wood destroying organisms/ fungus**.

- **Wood-decay fungus:** A variety of fungus that digests moist wood, thereby causing it to rot. Some wood-decay fungi attack dead wood and some, are parasitic and colonize living trees. Fungi that not only grow on wood but actually cause it to decay, are called lignicolous fungi. Various lignicolous fungi consume wood differently. For example, some attack the carbohydrates in wood, and others decay lignin. Wood-decay fungi can be classified according to the type of decay that they cause.

The best-known types are brown rot, soft rot, and white rot.

- **Brown-rot:** Brown-rot fungi break down hemicellulose and cellulose. It diffuses rapidly through the wood, leading to a decay that is not confined to the direct surroundings of the fungal hyphae. Therefore, this type of decay shrinks the wood, causes a brown discoloration, and cracks into roughly cubical pieces.

Brown-rot.

- **Soft-rot:** Soft-rot fungi secrete cellulase, an enzyme, that breaks down cellulose in the wood. This leads to the formation of microscopic cavities inside the wood, and sometimes

to a discoloration and cracking pattern similar to brown rot. Soft-rot fungi need fixed nitrogen in order to synthesize enzymes, which they obtain either from the wood or from the environment.

Soft-rot.

- **White-rot:** White-rot fungi break down the lignin in wood, leaving the lighter-colored cellulose behind. Some white-rot break down both lignin and cellulose.

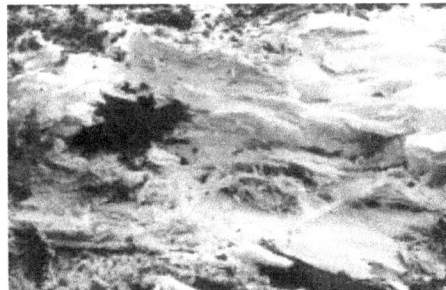

White rot.

- **Wood edge:** An edge on a countertop that is made by trimming the square front corner with a decorative strip of wood. See **Plastic laminate countertop**.
- **Wood filler:** An aggregate of resin and strands, shreds, or flour of wood that is used to fill openings in wood and provide a smooth, durable surface.
- **Wood flush door:** An assembly consisting of a core and one or more edge bands, with two plies of wood veneer with laminate, wood, or wood derivative on each

side. All parts are composed of wood, wood derivatives, fire resistant composites or decorative laminates.

- **Wood (lignostone) mallet:** This mallet is used with wood and plastic-handled chisels and gouges. It has a high resistance to cracking.

- **Wood shingle roof:** A roof constructed from wood shingles that are sawn out of logs and are about 3/8" thick. Grades of wood shingles are blue label, red label, black label, and undercoursing. No longer allowed in fire zones.

- **Wood strength:** Five basic factors influence the strength of wood:
  - **Wood specific gravity:** As a general rule the greater the specific gravity or density of a wood, the greater the strength.
  - **Slope of grain:** When the grain direction in a wood member is parallel to the two edges of the piece the wood is said to have straight grain. However, if the grain direction in a piece is not precisely parallel to the board edges, strength will be lower than if edges and fiber.
  - **Presence of knots or holes: Knots affect strength in two ways:** First the slope of grain around a knot causes a marked reduction in strength in the vicinity of the knot.
  - **Moisture content:** The moisture content of wood will change with changes in the conditions under which it is used. To give best service, the wood should be installed at a moisture content close to the midpoint between the high and low values it will usually attain in use.
  - **Time:** Wood is susceptible to creep that can lead to serviceability and strength reduction problems.

Woodscrew: A threaded fastener for use in wood.

Work order (WO): A written order (directive) that is signed by the owner or his representative, of a contractual status that requires performance by the contractor.

Working drawings: The complete set of architectural drawings prepared by a registered architect. Also known as contract drawings.

Working load: The actual load that is acting on the structure. Also referred to as service load.

Worm track: Marks that are caused by various types of wood attacking larvae. Worm tracks often appears as sound discolorations running with or across the grain in straight to wavy streaks. Sometimes referred to as pith flecks in certain species of maple, birch, and other hardwoods because of a resemblance to the color of pith. Also referred to as scar.

Woven valley: A method of valley construction in which shingles or roofing from both sides of the valley extend across the valley and are woven together by overlapping alternate courses as they are applied. This is done by alternately letting each row of shingles run past the valley. This requires that all the shingle rows line up correctly, which can be tricky. Also referred to as **laced valley**.

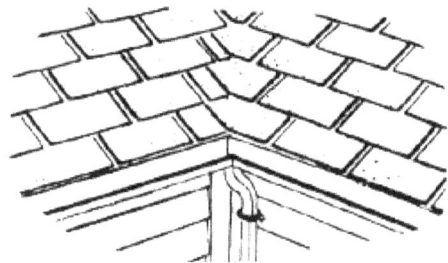

Woven valley.

Wrapped drywall: Areas that get complete drywall covering, as in the doorway openings of bi-fold and bi-pass closet doors. There is no visible hardware to distract from the appeal of the door.

Wrapped drywall shutters.

Wrought copper fittings: Plumbing fittings that are used to connect copper tubes made of mechanically worked and toughened copper as opposed to an as-cast brittle copper. See **Tubes**.

Wrought iron: An iron alloy with a very low carbon content in contrast to cast iron. Wrought and has fibrous inclusions, known as slag that give it a grain resembling wood. Wrought iron is no longer produced on a commercial scale.

Wye: A Y-shaped fitting with three openings used to create branch lines. It allows one pipe to be joined to another at a 45°angle.

Wythe: Single, vertical masonry wall one unit thick. A double wythe wall is two units thick.

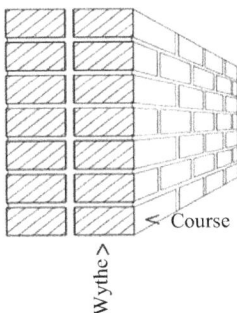

X-bracing: Cross bracing of a permanent or temporary nature to resist lateral loading.

Xeriscaping: A landscaping method designed for water conservation so that routine irrigation is not necessary. It includes using drought-adaptable and low-water plants, soil

amendments such as compost to conserve moisture, and mulches to reduce evaporation.

XO, XOO, XOOX: Designations of the arrangement of sliding and fixed panels in sliding glass doors and windows; X indicates a sliding panel; O indicates a fixed panel; see **OX, OXO, OXXO**.

Yard: Unit of length equal to 3 feet (91.5 cm).

Yard of concrete: One cubic yard of concrete is 3' × 3'× 3' in volume, or 27 cubic feet. One cubic yard of concrete will pour 80 square feet of 3 1/2" sidewalk or basement/garage floor.

Yellow tipping: A flame condition that is caused by too severe a reduction in primary air. The yellow color is caused by glowing carbon particles in the flame. It can be corrected by the injection of more primary air.

Yield: A measure of investment performance that gauges the percentage return on each dollar invested. Also known as rate of return.

Yield point (Fy): The unit stress at which the stress-strain curve exhibits a definite increase in strain without an increase in stress that is less than the maximum attainable stress.

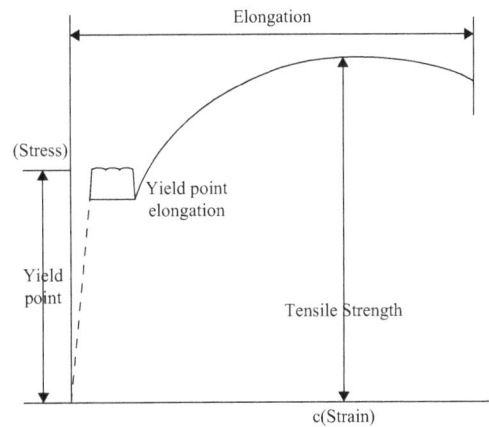

Yield point.

Yield strength: The stress at which significant increase in strain occurs without such increase in stress or as defined by the appropriate standard specification for the material.

Also referred to as yield point and yield stress. See Yield strength.

Yield stress: A material loaded beyond its yield stress no longer exhibits linear elastic behavior. Generally, metals, particularly mild steel have a very well defined yield stress compared to other materials. Yield stress is sometimes called yield strength.

Yoke: The location where a home's water meter is sometimes installed between two copper pipes, and located in the water meter pit in the yard.

Yoke.

Also, a, usually, brass casting that holds both the hot and cold valves and the mixing chamber for the water. Other metals that serve the same function, e.g., copper, are also used.

Yoke.

Yoke vent: A pipe connecting upward from a soil or waste stack to a vent stack for the purpose of preventing pressure changes in the stacks.

Young's modulus: Young's modulus, also known as the tensile modulus, is a measure of the stiffness of an elastic material and is a quantity used to characterize materials.

Young's modulus is the ratio of normal stress to corresponding strain for tensile or compressive stresses that are less than the proportional limit of the material. Since stress has units of pressure and strain is dimensionless, Young's modulus has units of pressure; in the United States it is commonly expressed as pounds (force) per square inch (psi) or thousands of psi (ksi), and the SI unit of modulus of elasticity (E, or less commonly Y) is the pascal. See **Modulus of elasticity**.

Z section: A structural section in the shape of a "Z" cold formed from a steel sheet.

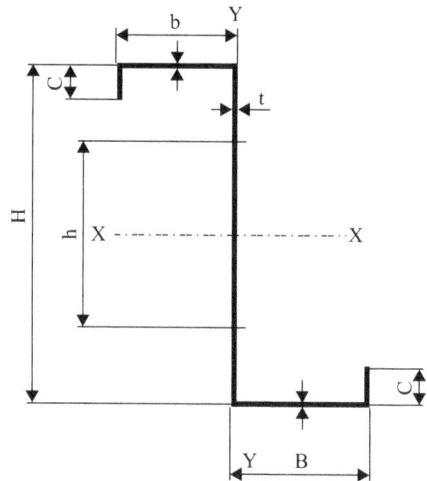

Steel Z section.

Z-bar flashing: Bent, galvanized metal flashing that is installed above a horizontal trim board of an exterior window, door, or brick run. It prevents water from getting behind the trim/brick and into the home.

Zeolite: Zeolites are microporous crystalline solids with well-defined structures. Generally they contain silicon, aluminum, and oxygen in their framework and cations, water and/or other molecules within their pores. Many occur naturally as minerals, and are extensively mined in many parts of the world. Others are synthetic, and are made commercially for specific uses, or produced by research scientists trying to understand more about their chemistry.

Because of their unique porous properties, zeolites are used in a variety of applications with a global market of several milliion tons per annum. They are often also referred to as molecular sieves. Synthetic zeolite is being used as an additive in the production process of warm mix asphalt concrete. It helps by decreasing the temperature level during manufacture and laying of asphalt concrete, resulting in lower consumption of fossil fuels, thus releasing less carbon dioxide, aerosols, and vapours. Other than that, the use of synthetic zeolite in hot mixed asphalt leads to easier compaction and, to a certain degree, allows cold weather paving and longer hauls.

When, added to Portland cement as a pozzolan, it can reduce chloride permeability and improve workability. It reduces weight and helps moderate water content while allowing for slower drying which improves break strength.

Zero clearance fireplace: Another term for factory built fireplace. The term is misleading. Zero clearance suggests that combustible materials can touch the assembly, when in reality no combustible materials should ever touch a factory built fireplace.

Zero clearance plate: A piece of material that is cut to the same size, shape and thickness of the standard throat plate. It fits into the throat on the tabs, just as a standard throat plate except that a zero clearance plate has no pre-cut slot in it. The plate is made from wood or plastic that is easily cut by the saw blade. Therefore, after installing the dado blade followed by the zero clearance insert, an operartor must merely turn on the saw and raise the blade. As the blade raises the slot is cut by the blade, leaving no gap around the blade. See **Throat plate**.

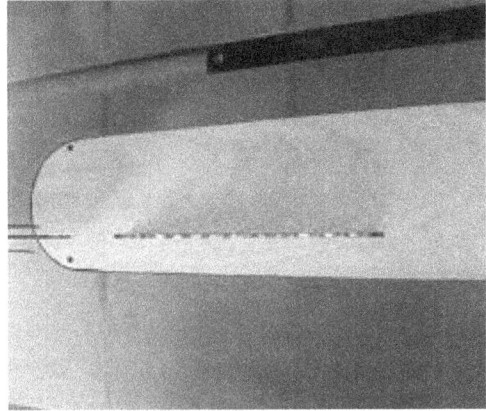

Zero clearance plate.

Zero force member: A member in a truss that has zero internal axial force when the truss is analyzed using the undeformed geometry of the truss. Zero force members occur when at a truss joint the total number of forces and members is three, and two are in-line, then the third must be zero to maintain equilibrium. Zero force members also occur when there are two members at a joint and the two are not in line; in that case, both members must have zero force to maintain equilibrium.

Zero lot line: Zoning that permits building to a property or lot line without a set back.

Zero soft: Water with a total hardness that is less than 1.0 grain per US gallon, as calcium carbonate.

Zigzag folding rule: A classic woodworking rule that is used for measuring longer runs.

Zinc: A metal that has application considerations including high expansion-contraction rates and low-temperature restrictions. The chief use of zinc is for galvanizing iron. Zinc oxide is used in making paint pigment, cosmetics, ointments, and dental cements.

Zone: Area, sector, section, region, territory, district, quarter, precinct, locality, neighborhood, province, etc.

With regard to a building, the section of a building that is served by one heating or cooling loop because it has noticeably distinct heating or cooling needs. Similarly, the section of property that will be watered from a lawn sprinkler system.

Terms beginning with zone:

- **Zone of aeration:** The layer in the ground above the water table where the available voids are filled with air. Water falling on the ground percolates through this zone on its way to the aquifer.
- **Zone of saturation:** The layer in the ground in which all of the available voids are filled with water.
- **Zone valve:** A device, usually placed near the heater or cooler, which controls the flow of water or steam to parts of the building, it is controlled by a zone thermostat.

Zoning: A governmental process and specification that limits the use of a property, e.g., single family use, high-rise residential use, industrial use, etc., Zoning laws may limit where you can locate a structure. Also see **Building codes**.

A local law specifying what is permitted to be constructed on a piece of land within that zoning district, county, municipality, township, village, etc., is a **zoning ordinance**.

Zoophoric column: A pillar supporting the figure of an animal.

ORGANIZATIONS THAT AFFECT CONSTRUCTION

Contractors are called upon to build many things, and not all are straightforward. Below are listed organizations that you may wish to contact when that unusual project comes your way or, perhaps, merely to broaden your knowledge base. You may also want to join one or more of them for purposes of networking, information, continuing education, and gaining the benefit of their advocacy.

Acoustical Society of America (ASA): The society provides information in the broad field of Acoustics.

Adhesive and Sealant Council (ASC): The council provides information, education, and representation to its members.

Air Barrier Association of America (ABAA): A resource for air barrier education and technical information.

Air Conditioning Contractors of America (ACCA): A trade association that promotes professional contracting, energy efficiency, and healthy, comfortable indoor environments.

Air Diffusion Council (ADC): The council represents manufacturers of flexible air-ducts and air connectors.

Air Infiltration and Ventilation Centre (AIVC): An international organization that provides industry and research organizations with technical support aimed at optimizing ventilation technology.

Air Movement and Control Association International (AMCA): An international association of air equipment manufacturers.

Air-conditioning, Heating and Refrigeration Institute (AHRI): Member companies produce residential and commercial air conditioning, heating, and water heating equipment. The institute provides members with a certification program, standards, advocacy, and other activities.

Allied Board of Trade, Inc. (ABT): ABT acts as a liaison between the designer and supplier to promote ethical standards of business conduct in the industry. Members have access to trade information, web site design, magazine program, source information, and business advice.

Alliance for Fire and Smoke Containment and Control (AFSCC): An alliance of building enforcement, construction, design, and manufacturing professionals. It promotes a balanced fire protection design in the built environment.

Aluminum Anodizers Council (AAC): Members are engaged in aluminum anodizing, are suppliers of products and services used in the anodizing of aluminum products, or purchasers of anodized finishes.

Aluminum Association (AA): The association promotes aluminum as a sustainable and recyclable automotive, packaging, and construction material. Member companies operate more than 200 plants in the United States, with many conducting business worldwide.

Aluminum Extruders Council (AEC): An international trade association that is dedicated to advancing the effective use of aluminum extrusion in North America.

American Architectural Manufacturers Association (AAMA): A national trade association that establishes voluntary standards for the window, door, and skylight industry.

American Arbitration Association (AAA): An organization that has a long history a long history and experience in the field of alternative dispute resolution. AAA provides services to individuals and organizations who wish to resolve conflicts out of court.

AAA goal is to move cases through arbitration or mediation in a fair and impartial manner until completion. Additional services include the design and development of alternative dispute resolution (ADR) systems for corporations, unions, government agencies, law firms, and the courts. The Association also provides elections services, education, training, and publications for people who want to be more deeply involved in alternative dispute resolution.

American Association of Automatic Door Manufacturers (AAADM): A trade association of power-operated automatic door manufacturers.

American Association of State Highway and Transportation Officials (AASHTO): A regulatory organization that governs the design and specifications of highway bridges.

American Backflow Prevention Association (ABPA): An organization dedicated to education and technical assistance that is focused on protecting drinking water from contamination through cross-connections.

American Boiler Manufacturers Association (ABMA): The national trade association of commercial, institutional, industrial, and electricity-generating boiler system manufacturing companies (>400,000 Btuh heat input). It is dedicated to the advancement and growth of the boiler and combustion equipment industry.

American Coal Ash Association (ACAA): A trade association that is dedicated to recycling the materials that are created when coal is burned to generate electricity.

American Coatings Association (ACA): An organization that is in support of the paint and coatings industry.

American Composites Manufacturers Association (ACMA): A trade group that represents the composites industry.

American Concrete Institute (ACI): The institute is dedicated to advancing concrete knowledge by conducting seminars, managing certification programs, and publishing technical documents.

American Concrete Pavement Association (ACPA): An organization comprised of concrete paving contractors, cement and material producers, equipment manufacturers and any company with an interest in concrete airports, highways, roads, streets, and industrial pavements.

American Concrete Pipe Association (ACPA): An advocate for the concrete pipe industry. Its members are committed to environmental improvement by producing quality concrete pipe for drainage and pollution abatement.

American Concrete Pressure Pipe Association (ACPPA): The association provides technical information and design assistance on the uses of concrete pressure pipe (CCP) to the water and wastewater industries.

American Fence Association (AFA): The association represents the fence, deck, and railing industry in the United States and parts of Canada. AFA offers educational, certification options, and networking to its members.

American Fire Sprinkler Association (AFSA): An international association that represents open shop fire sprinkler contractors. The associationpromotes the use of automatic fire sprinkler systems and offers educational advancement to its members.

American Floorcovering Alliance (AFA): The alliance promotes the industry's products and services, and educates its members and others through seminars, press releases, and trade shows.

American Galvanizers Association (AGA): A trade association that serves fabricators, architects, specifiers, and engineers, who use hot-dip galvanizing for corrosion control. The AGA maintains a large technical library, provides multimedia seminars, and offers toll-free technical support in North America.

The AGA is also a process technology research and information resource, liaison to environmental and safety regulatory authorities, and an active member of committees within specification, corrosion, and government transportation and infrastructure agencies.

American Hardboard Association (AHA): A trade organization of manufacturers of hardboard products for exterior siding, interior wall paneling, furniture, and industrial and commercial products.

American Hardware Manufacturers Association (AHMA): A trade organization with membership open to any company headquartered in the United States and is engaged in the manufacture of goods for hardware and home improvements, lawns and gardens, and painting and decorating.

American Institute of Architects (AIA): A professional organization that unites in fellowship the members of the architectural profession in the United States.

American Institute of Chemical Engineers (AIChE): A professional organization for chemical engineers, as distinct from chemists and mechanical engineers. There are student chapters at various universities around the world that tend to focus on providing networking opportunities in both academia and in industry.

American Institute of Steel Construction (AISC): A technical specifying and trade organization for the fabricated structural steel industry in the United States. It publishes the "Manual of Steel Construction."

American Institute of Timber Construction (AITC): A national technical trade association of the structural glued laminated (glulam) timber industry. AITC represents the glued laminated timber manufacturers in the United States and in addition has members who are installers, suppliers, sales representatives, engineers, architects, designers, and researchers.

American Insurance Association (AIA): A leading property-casualty represents over 300 insurers that write more than $110 billion in premiums each year. AIA is a resource for policy makers, the media and the public on property-casualty insurance issues.

American Iron and Steel Institute (AISI): The institute promotes the interests of the iron and steel industry.

American Lighting Association (ALA): A trade association that represents the lighting industry. Its membership includes lighting, fan and dimming control manufacturers, retail showrooms, sales representatives, and lighting designers.

American Lumber Standard Committee, Inc. (ALSC): ALSC is comprised of manufacturers, distributors, users, and consumers of lumber. ALSC serves as the standing committee for the American Softwood Lumber Standard (Voluntary Product Standard 20) and administers an accreditation program for grade-marking lumber produced under the system.

American National Standards Institute (ANSI): A clearinghouse organization for all types of standards and product specifications.

American Nursery and Landscape Association (ANLA): A national trade association for the nursery and landscape industry. Members grow, distribute, and retail plants of all types, and design and install landscapes for residential and commercial customers. ANLA provides education, research, public relations, and representation services to members.

American Rainwater Catchment Systems Association (ARCSA): ARCSA promotes rainwater catchment systems in the United States. Members include professionals working in city, state and federal government, academia, manufacturers and suppliers of rainwater harvesting equipment, consultants, and other interested individuals, membership is not limited to the United States.

American Shotcrete Association (ASA): An organization of contractors, suppliers, manufacturers, designers, engineers, owners and others with a common interest in promoting the use of shotcrete.

American Shutter Systems Association (ASSA): ASSA provides tested and approved hurricane shutter products for its licensed members who located throughout the coastal United States and the Caribbean.

American Society of Concrete Contractors (ASCC): Members include contracting firms, manufacturers, suppliers, architects, specifiers, and distributors of the concrete industry.

American Society of Civil Engineers (ASCE): The oldest national professional engineering society in the United States. It is dedicated to the advancement of the individual civil engineer and the civil engineering profession through education.

American Society of Furniture Designers (ASFD): The society is dedicated to advancing, improving, and supporting the profession of furniture design and its impact in the marketplace.

American Society of Heating, Refrigeration and Air Conditioning Engineers (ASHRAE): A national association that establishes standards for building energy performance.

American Society of Home Inspectors (ASHI)): A professional association for home inspectors. It establishes and advocates high standards of practice and has a strict code of ethics for its members.

American Society of Interior Designers (ASID): Comprised of designers, industry representatives, educators, and students, AISD provides education, knowledge sharing, advocacy, community building, and outreach to and for the interior design profession.

American Society of Irrigation Consultants (ASIC): The society strives to enhance the role of the independent professional irrigation consultant.

American Society of Landscape Architects (ASLA): ASLA provides education and participates in the stewardship, planning and design of cultural and natural environments.

American Society of Mechanical Engineers (ASME): The society promotes the art, science, and practice of multidisciplinary engineering and allied sciences worldwide.

American Society of Plumbing Engineers (ASPE): An international organization for professionals involved in the design, specification, and inspection of plumbing systems.

American Society of Professional Estimators (ASPE): The society serves construction estimators by providing education, fellowship, and opportunity for professional development.

American Society of Sanitary Engineers (ASSE): Members include all disciplines of the plumbing industry.

American Society of Theatre Consultants (ASTC): The society informs owners, users, and planners about the services that theatre consultants offer for projects large and small, whether for new construction or remodeling/renovation.

American Soil and Foundation Engineers (ASFE): ASFE aids geo-professionals to achieve business excellence and to manage risk through advocacy, education, and collaboration.

American Sports Builders Association (ASBA): ASBA provides information on tennis courts, running tracks, fields, and indoor sports facilities.

American Subcontractors Association, Inc. (ASA): A trade organization that is dedicated to improving the business environment in the construction industry.

American Supply Association (ASA): An Association that focuses primarily on plumbing (similar to AWWA).

American Walnut Manufacturers Association (AWMA): An international trade association that represents manufacturers of walnut lumber, dimension lumber, veneer, walnut squares, and gunstock blanks.

American Wire Producers Association (AWPA): A trade association for the ferrous wire and wire products industry in North America. Members include wire producers, manufacturers and distributors of wire rod, and suppliers of machinery, dies, and equipment to the wire industry.

American Wood Preservatives Association (AWPA): The association seeks to improve the performance and longevity of sustainable wood products. AWPA is a resource for knowledge on all aspects of wood protection.

American Welding Institute (AWI): An organization established to bridge the gap between the findings of basic welding research and the needs of the industry.

American Welding Society (AWS): An organization whose major goal is to advance the science, technology, and application of welding and related joining disciplines.

American Water Works Association (AWWA): An association for people associated with water supply (waterworks), including all types of materials.

American Wood Council (AWC): The council represents the North American traditional and engineered wood products industry. AWC develops engineering data, technology, and standards on structural wood products for use by design professionals, building officials, and wood products manufacturers. AWC also provides technical, legal, and economic information on wood design, green building, and manufacturing environmental regulations.

APA-The Engineered Wood Association: Founded in 1933 as the Douglas Fir Plywood Association and later recognized as the American Plywood Association, APA changed its name to APA–The Engineered Wood in 1994. The association represents engineered wood manufacturers. It provides quality testing, product research, market development, etc., for its members.

Appalachian Hardwood Manufacturers, Inc. (AHMI): AHMI promotes the use of logs, lumber, and products from the Appalachian Mountain region. It will assist users to make decisions relating to hardwoods.

Architectural Precast Association (APA): Members includes manufacturers and their suppliersof architectural precast concrete products. The association establishes and upholds quality assurance for member products.

Architectural Woodwork Institute (AWI): A trade association that represents architectural woodworkers, suppliers, design professionals, and students from around the world.

Architectural Woodwork Manufacturers Association of Canada (AWMAC): In many ways the Canadian version of AWI.

Art Glass Association (AGA): An international organization whose purpose is to create awareness, knowledge, and involvement in the art glass industry.

ASIS International: ASIS International was originally the American Society for Industrial Security (ASIS), the organization changed its name in 2002. ASIS develops educational programs and materials that address broad security interests.

Asphalt Emulsion Manufacturers Association (AEMA): An international trade association that represents the asphalt emulsion industry. AEMA's goal is to expand the use and applications of asphalt emulsions.

Asphalt Institute (AI): An international trade association of petroleum asphalt producers, manufacturers, and affiliated businesses. AI promotes the use of petroleum asphalt, through engineering, research, marketing, and educational activities, and through the resolution of issues affecting the industry.

Asphalt Pavers Alliance (APA): A coalition of the Asphalt Institute, the National Asphalt Pavement Association, and the State Asphalt Pavement Associations. Its mission is to establish asphalt pavement as the preferred pavement through research, technology transfer, engineering, education, and innovation.

Asphalt Roofing Manufacturers Association (ARMA): A trade association that represents North American asphalt roofing manufacturers and their raw material suppliers.

Asphalt Recycling and Reclaiming Association (ARRA): The association promotes the recycling of existing roadway materials through various construction methodologies.

Associated Air Balance Council (AABC): The council establishes industry standards for the field measurement and documentation of HVAC systems. It also provides education, technical training, and certification for its members.

Associated General Contractors of America (AGC): The association serves construction professionals through the education of its members and students (scholarship programs), research, recognition programs, and charities for needy.

Associated Locksmiths of America, Security Professionals Association, Inc. (ALOA): The association has recently changed its name from simply ALOA because it broadened its focus beyond "traditional" locksmithing. The association also encompasses electronic locksmiths, automotive locksmiths, and other security professions related to access control.

Associated Specialty Contractors (ASC): An umbrella organization of nine national associations of construction specialty contractors.

(International) Association of Foundation Drilling (ADSC): A trade association that advances the interests of people who are engaged in the design, construction, equipment manufacture and distribution of anchored earth retention, drilled shaft, micropiling, and related industries.

Association of Iron and Steel Technology (AIST): The association was formed in January 2004 from a merger of the Iron and Steel Society and the Association of Iron and Steel Engineers. It advances the technical development, production, processing, and application of iron and steel.

Association of Millwork Distributers (AMD): The association provides leadership, certification, education, promotion, networking, and advocacy to, and for, the millwork distribution industry.

Association of the Nonwoven Fabrics Industry (INDA): INDA focuses on networking events to help members increase sales and market share. It is a source for education, market leading data, global forecasts, testing standards, and trend reports.

Association of Physical Plant Administrators (APPA): Organized originally as the Association of Superintendents of Buildings and Grounds, the association later became the Association of Physical Plant Administrators of Universities and Colleges. In 1991, the name APPA: The Association of Higher Education Facilities Officers was adopted. In 2005, the association began to identify itself simply as APPA. It is a professional organization of college and university physical plant administrators, architects, and engineers.

Association of Pool and Spa Professionals (APSP): APSP is a trade organization that is dedicated to the growth and development of its members' businesses. It promotes the enjoyment and safety of pools and spas.

Association of Professional Landscape Designers (APLD): An international organization that advances the profession of landscape design and promotes the recognition of landscape designers as qualified and dedicated professionals.

Association of State Floodplain Managers (ASFPM): An organization of professionals who are involved in floodplain management, flood hazard mitigation, the National Flood Insurance Program, and flood preparedness, warning, and recovery.

Association of the Wall and Ceiling Industry (AWCI): A trade organization that provides services and undertakes activities that enhance its members' ability to operate a successful business.

Association of Zoos and Aquariums (AZA): AZA provides the standards and best practices needed for animal care, wildlife conservation and science, conservation education, the guest experience, and community engagement.

ASTM International: Formerly, the American Society for Testing and Materials (ASTM), ASTM International is an organization that establishes material standards.

Audio Engineering Society (AES): The society is devoted exclusively to audio technology. It is an international organization that unites audio engineers, creative artists, scientists, and students worldwide by promoting advances in audio and disseminating new knowledge and research.

Automatic Fire Alarm Association (AFAA): A trade organization that supports business advancement, code development, and development of training and educational programs.

Barre Granite Association, Inc.: It has been estimated that one-third of the public and private monuments and mausoleums in America are products of the Barre quarries and Barre's "international" community of sculptors, artisans, mechanics, and laborers.

Bath Enclosures Manufacturers Association (BEMA): The association represents industry manufacturers, suppliers, and dealers in the United States and Canada.

BC Wood: A trade association that supports British Colombia's wood products manufacturers.

Blow in Blanket Contractors Association (BIBCA): An industry support association. BIBCA requires members to abide by a strict code of ethics.

Brick Industry Association (BIA): BIA represents distributors and manufacturers of clay brick and suppliers of related products and services to regulators and legislators. BIA is comprised of regions that manage programs in the Midwest/Northeast, Southeast, and Southwest.

Builders Hardware Manufacturers Association (BHMA): A trade association for North American manufacturers of commercial builder's hardware. BHMA is involved in standards, code, and life safety regulations and other activities that specifically impact builder's hardware.

Building Environment and Thermal Envelope Council (BETEC): Part of the National Institute of Building Sciences, an organization representing government and industry, BETEC is involved in communicating government policy and influencing standards development within the industry.

Building Officials and Code Administrators International, Inc. (BOCA): One of the three model code groups in the United States that has now merged into the International Code Council. This code is a minimum model regulatory code for the protection of public health, safety, welfare, and property by regulating and controlling the design, construction, quality of materials, use, occupancy, location, and maintenance of all buildings and structures within a jurisdiction. The code is used primarily in the North Central and Northeast United States.

Building Owners and Managers Association International (BOMA): BOMA represents the owners and managers of all commercial property types. It advances the interests of the entire commercial real estate industry through advocacy, education, research, standards, and information.

Building Seismic Safety Council (BSSC): An independent, voluntary membership body that represents a wide variety of building community interests related to seismic safety. The BSSC was established in 1979 as a Council of the National Institute of Building Sciences.

The BSSC'sdeliberations and recommendations consider and assess social, technical, administrative, political, legal, and economic issues.

Building Stone Institute (BSI): The institute promotes and advances the use of natural stone. BSI provides its members with knowledge, information, products, and services for the design community and end user.

Business and Institutional Furniture Manufacturer's Association (BIFMA): A trade organization, it advocates, informs, and develops standards for the North American office and institutional furniture industry.

Cable Tray Institute (CTI): The institute supports the cable tray industry by engaging in research, development, education, and the dissemination of information designed to promote, enhance, and increase the visibility of the industry.

California Manufactured Housing Institute (CHMI): A professional and trade association that represents builders of factory constructed homes, retailers, financial services, developers, and community owners and their supplier companies.

California Redwood Association (CRA): An industry advocacy association that provides information on the use of redwood and its sustainability, "how-to" advice, etc.

Canadian Carpet Institute (CCI): An organization that represents Canada's carpet manufacturers and their suppliers.

Canadian Copper and Brass Development Association (CCBDA): A communications and advisory group for the copper industry. CCBDA represents and gives support to its members and users of copper and copper alloys, including educators and the general public.

Canadian Hardwood Plywood Veneer Association (CHPVA): A national association that represents the Canadian hardwood plywood and veneer industry with technical, regulatory, quality assurance, and product acceptance.

Canadian Institute of Steel Construction (CISC): The institute promotes good design, safety, and the efficient, economical, and sustainable use of structural steel.

Canadian Plywood Association (CAN-PLY): The association carries a complete line of products from the leading manufacturers. Certiwood™, a part of CANPLY, is an engineered wood products testing and certification agency.

Canadian Roofing Contractor's Association (CRCA): CRCA consists of companies that are actively engaged in Canada in the roofing and related sheet metal contracting business, along with companies engaged in manufacturing or supplying materials and services that are used in any branch of the roofing and sheet metal industry.

Canadian Society of Landscape Architects (CSLA): CSLA is dedicated to advancing the art, science, and business of landscape architecture.

Canadian Steel Producers Association (CSPA): The association is committed to a strong and internationally competitive Canadian steel sector.

Canadian Security Association (CANASA): An organization dedicated to the advancing the security industry and supporting security professionals in Canada.

Canadian Sheet Steel Building Institute (CSSBI): An industry association that is responsible for the development and dissemination of industry standards. CSSBI is a technical information and resources expert for both the general public and sheet steel manufacturers.

Canadian Standards Association (CSA): CSA is the Canadian equivalent of ASTM.

Canadian Wood Council (CWC): A trade organization that represents manufacturers of Canadian wood products that are used in construction.

Carpet Cushion Council (CCC): CCC educates carpet retailers, manufacturers, distributors, and cushion manufacturers about the benefits of carpet cushion.

Carpet and Rug Institute (CRI): The institute provides science-based facts about carpeting and rugs.

Cast Iron Soil Pipe Institute (CISPI): A trade organization that provides technical reports to advance interest in the manufacture, use and distribution of cast iron soil pipe and fittings. CISPI strives to improve the industry's products, achieve standardization, and provide a continuous program of product testing, evaluation, and development.

Cast Stone Institute (CSI): A self-governing association of producers and suppliers to the Cast Stone industry. CSI is a spokesperson for cast stone and provides counsel to the architectural and engineering communities.

Cedar Shake and Shingle Bureau (CSSB): An organization that promotes the use of Certi-label™ cedar roofing and sidewall products.

Ceiling and Interior Systems Construction Association (CISCA): The association serves the acoustical and specialty ceilings and interior finishes industry. CISCA provides networking, education, resources, and technical guidelines to its members.

Cellulose Insulation Manufacturers Association (CIMA): A trade association for the cellulose segment of the thermal/acoustical insulation industry.

Cement Association of Canada (CAC): The association represents the Canadian cement firms that have clinker and cement manufacturing facilities, granulators, grinding facilities, and cement terminals.

Ceramic Glazed Masonry Institute (CGMI): A trade organization of glazed ceramic product manufacturers and distributers.

Ceramic Tile Distributors Association (CTDA): An international association of distributors, manufacturers, and allied professionals of ceramic tile and related products. CTDA connects, educates, and strengthens tile and stone distributors.

Ceramic Tile Institute of America (CTIOA): CTIOA provides manufacturer's information but does not test products or validate manufacturer's claims. The institute has recently collaborated with the Los Angeles County Metropolitan Transportation Authority (Metro) and noted artists to implement tile artworks.

Certified Floorcovering Installers (CFI): An organization that identifies, trains, and certifies flooring installers according to skill and knowledge. It also provides the industry with educational programs.

Chain Link Fence Manufacturers Institute (CLFMI): An organization comprised of firms that manufacture chain link fence fabric, fittings, framework, accessories, and/ or the materials used to produce them. CLFMI provides advice to architects, engineers, contractors, etc., regarding chain link fencing.

Chicago Roofing Contractors Association (CRCA): Alocal trade association of roofing and waterproofing contractors in the greater Chicago area. Contractors, manufacturers, distributors, manufacturer's representatives, and consultants are members.

Chimney Safety Institute of America (CSIA): The institute fosters public awareness of issues relating to chimney and venting performance and safety. CSIA promotes technical training and certification opportunities.

Cold-Formed Steel Engineers Institute (CFSEI): The institute is comprised of structural engineers and other design professionals who provide designs for commercial and residential structures with cold-formed steel.

Commercial Food Equipment Service Association (CFESA): A trade association of professional service and parts distributors.

Composite Panel Association (CPA): CPA represents the North American composite panel and decorative surfacing industries. CPA sponsors all ANSI standards related to particleboard, MDF, hardboard, and engineered wood siding and trim, as well as CPA's Eco-Certified Composite™ sustainability standard and certification program.

Compressed Air and Gas Institute (CAGI): The institute serves the compressed air industry by providing technical, educational, and promotional support, and being involved in other matters that affect the industry.

Compressed Gas Association (CGA): CGA promotes the safe, secure, and environmentally responsible manufacture, transportation, storage, trans-filling, and disposal of industrial and medical gases and their containers.

Concrete Anchor Manufacturers' Association (CAMA): Members include manufacturers of concrete anchoring systems and associated technical representatives. CAMA works with code organizations to develop uniform codes and standards and advance the use of anchoring systems.

Concrete Countertop Institute (CCI): A trade organization that provides the concrete countertop industry with training, membership programs, and advocacy, consultation, and guidance.

Concrete Foundation Association (CFA): A trade association that provides promotional materials, educational seminars, networking opportunities, and technical and informative meetings for contractors who are in the residential concrete foundation industry.

Concrete Polishing Association of America (CPPA): The association builds and maintains standards and is an advocate for concrete that is processed to a polished finish.

Concrete Reinforcing Steel Institute (CRSI): A national trade association. It is a resource for information related to steel reinforced concrete construction. CRSI Industry members include manufacturers, fabricators, and placers of reinforcing bars and related products.

Concrete Sawing and Drilling Association (CSDA): The association promotes the use of professional specialty sawing and drilling contractors and their methods.

Confederation of International Contractors Association (CICA): A trade organization, CICA is dedicated to improving the business environment in the construction industry. The association promotes investment in engineering and building that enhances both our environment and the quality of life for all.

Construction Engineering Research Laboratory (CERL): Part of the U.S. Army Engineer Research and Development Center (USAERDC) that is the integrated Army Corps of Engineers' research and development organization. CERL conducts research and development in infrastructure andenvironmental sustainment.

Construction Management Association of America (CMAA): A North American organization that is dedicated exclusively to the interests of professional construction and program management.

Construction Materials Recycling Association (CMRA): CMRA promotes the safe and economically feasible recycling of recoverable construction and demolition (C&D). These materials include aggregates such as concrete, asphalt, asphalt shingles, gypsum wallboard, wood, and metals.

Construction Specifications Institute (CSI): The institute advances building information management and education of project teams to improve facility performance.

The CSI's Master Format is a system of numbers and titles for organizing construction information into a regular, standard order or sequence. By establishing a master list of titles and numbers Master Format promotes standardization and thereby facilitates the retrieval of information and improves construction communication. It provides a uniform system for organizing information in project manuals, for organizing project cost data, and for filing product information and other technical data.

Consumer Credit Counseling Service (CCCS): A nationwide, nonprofit organization that helps consumers get out of debt and improve their credit profile.

Consumer Electronics Association (CEA): The association unites companies within the consumer technology industry. Members receive market research, networking, educational programs and technical training, and advocacy.

Continental Automated Buildings Association (CABA): An international industry association that is dedicated to the advancement of intelligent home and intelligent building technologies. Membership companies are involved in the design, manufacture, installation, and retailing of products relating to home automation and building automation. Public organizations, including utilities and government, are also members.

Conveyor Equipment Manufacturers Association (CEMA): A trade association that serves North American manufacturers and designers of conveyor equipment. CEMA is focused on voluntary adherence to design standards, safety, manufacture, and applications to promote industry growth.

Cold Regions Research and Engineering Laboratory (CRREL): A U.S. Army Corps of Engineers Laboratory. It advances and applies science and engineering to complex environments, materials, and processes in all seasons and climates, with unique core competencies related to the Earth's cold regions.

Cool Metal Roofing Coalition: The coalition has as its mission to educate architects, building owners, designers, code and standards officials, and other stakeholders about the sustainable, energy-related benefits of cool metal roofing.

Cool Roof Rating Council (CRRC): An organization that maintains a third-party rating system for the radiating properties of roof surfacing materials. See **Cool Metal Roofing Coalition**.

Cooling Technology Institute (CTI): CTI advocates and promotes the use of environmentally responsible evaporative heat transfer systems (EHTS), cooling towers, and cooling technology through education, research, standards development and verification, government relations, and technical information exchange.

Copper Development Association (CDA): CDA supports the copper industry with market development, engineering, and information services.

Council of American Building Officials (CABO): CABO has joined with the **Southern Building Code Congress International (SBCCI)** and the **International Code Council (ICC)**, and they now write the internationally recognized One and Two Family Dwelling Code. These codes provide administrative and technical directions for all phases of residential construction.

Council of Forest Industries (COFI): The council works with governments, communities, organizations, and individuals to ensure that forest policies in British Columbia support the forest sector.

Council for Interior Design Accreditation (CIDA): The council ensures a high level of quality in interior design education through three primary activities: setting standards for postsecondary education, evaluating and accrediting colleges and universities, and facilitating outreach and collaboration.

Council of Landscape Architectural Registration Boards (CLARB): The council is dedicated to ensuring that all people who practice landscape architecture are fully qualified. Its members include the licensure boards in 48 states, two Canadian provinces and the territory of Puerto Rico.

Custom Electronic Design and Installation Association (CEDIA): An international trade association of companies that specialize in planning and installing electronic systems for the home.

Dade County: A Florida county, including Miami, that has set numerous standards and requirements for hurricane-resistant windows and doors.

Decorative Plumbing and Hardware Association (DPHA): An organization that represents independent retailers, manufacturers, and manufacturer's representatives. DPHA develops programs and publications to improve business practices, employee performance, and quality of service.

Deep Foundation Institute (DFI): The institute helps its members to improve in all aspects of planning, designing, and constructing deep foundations and deep excavations.

Door and Access Systems Manufacturers Association International (DASMA): A North American association of manufacturers of garage doors, rolling doors, high performance doors, garage door operators, vehicular gate operators, and access control products.

Door and Hardware Institute (DHI): DHI is dedicated to the architectural openings industry. It advances life safety and security within the built environment and is an advocate and resource for information, professional development, and certification.

Dry Stone Conservancy (DSC): The conservancy is dedicated to preserving dry-laid stone structures and promoting the ancient craft of dry stone masonry.

Ductile Iron Pipe Research Association (DIPRA): An association that is supported by ductile iron pressure pipe manufacturers in North America.

Earthquake Engineering Research Institute (EERI): A technical society of engineers, geoscientists, architects, planners, public officials, and social scientists. EERI members include researchers, practicing professionals, educators, government officials, and building coderegulators.

Electrical Generating Systems Association (EGSA): The association is exclusively dedicated to on-site power generation. The association is comprised of manufacturers, distributor/dealers, contractors/integrators, manufacturer's representatives, consulting and specifying engineers, service firms, end-users, and others that make, sell, distribute, and use on-site power generation technology and equipment, including generators, engines, switchgear, controls, voltage regulators, governors, and much more.

Electronic Security Association (ESA): A professional trade association that represents the electronic life safety, security, and integrated systems industry.

Elevator Escalator Safety Foundation (EESF): An organization that was created by the elevator/escalator industry to develop and disseminate safety materials to the public in order to eliminate preventable accidents on the industry's equipment.

Engineered Wood Association (EWA): The association tests and sets standards for all varieties of plywood used in the United States.

EnOcean Alliance: TheEnOcean Alliance develops and promotes self-powered wireless monitoring and control systems for sustainable buildings by formalizing the interoperable wireless standard.

(The United States) Environmental Protection Agency (EPA): An agency of the U.S. federal government that was created for the purpose of protecting human health and the environment by writing and enforcing regulations based on laws passed by Congress.

Erosion Control Technology Council (ECTC): The council consists of a broad range of professions and specialties, including site engineers, consultants, regulatory agencies, earthwork and seeding contractors, erosion control product suppliers, and manufacturers. ECTC has set as its mission to be the recognized industry authority in the development of standards, testing, and installation techniques for rolled erosion control products (RECPs), hydraulic erosion control products (HECPs) and sediment retention fiber rolls (SRFRs).

Ethylene Propylene Diene Monomer (EDPM) Roofing Association (ERA): A trade association in support of the EDPM industry.

Expanded Metal Manufacturers Association (EMMA): A division of the **National Association of Architectural Metal Manufacturers (NAAMM).**

Expanded Shale, Clay, and Slate Institute (ESCSI): An association for manufacturers of rotary kiln-produced expanded shale, expanded clay, and expanded slate lightweight aggregate.

Exterior Design Institute (EDI): EDI was founded to train and certify building envelope and EIFS (exterior insulation and finish systems) inspectors and moisture analysts. It promotes quality control within the construction industry.

Exterior Insulation and Finishing System (EIFS) Industry Members Association: A trade association that supports the industry by developing consensus technical, training, installation, and design standards for use by architects, designers, code bodies and officials, and other technical associations, by monitoring and positively influencing government actions, by working to assure the long-term availability of qualified contractors, and by providing other member services as appropriate.

EuroWindoor: A consortium of European window, door, and curtainwall industry associations that are involved in the development of common EU standards.

Federal Emergency Management Agency (FEMA): FEMA supports U.S. citizens and first responders to disasters. It also builds, sustains, and improves our nation's capability to prepare for, protect against, respond to, recover from, and mitigate all hazards.

Federal Housing Authority (FHA): The FHA sets construction standards throughout the United States.

Fenestration Canada: A trade organization that was formerly called the **Canadian Window and Door Association (CWDMA)**. Fenestration Canada represents and supports all aspects of the window and door manufacturing industry, including formulating and promoting standards of quality in manufacturing, design, marketing, distribution, sales, and application of all types of window and door products.

Fiberglass Tank and Pipe Institute: The Fiberglass Tank and Pipe Institute provides' a forum through which the fiberglass reinforced thermoset plastic (RTP) industry can advance the use of fiberglass products that are used in the underground tank and piping marketplace. Fiberglass Tank and Pipe Institute coordinates market studies, gathers statistics, and provides standard-setting organizations with technical data and it disseminates information to the government, industry, and the public.

Finishing Contractors Association (FCA): A trade organization that provides programs, products, and services and establishesrelationships with other relevant organizations.

Fire Equipment Manufacturers' Association (FEMA): Another FEMA. This FEMA is also committed to saving lives and protecting property. FEMA provides educational opportunities, advances best industry standards, and provides advocacy for the industry.

Fire Suppression Systems Association (FSSA): An international trade association whose members internationally. FSSA members are designer/installers, manufacturers, and suppliers.

Firestop Contractors International Association (FCIA): A trade organization of firestop contractors.

FLO-CERT GmbH: An independent International Certification company. FLO assists in the socio-economic Development of producers in the Global South. FLO-CERT allows people to identify products that meet agreed upon environmental, labor, and development standards.

Floor Covering Installation Contractors Association (FCICA): The association provides a network for problem solving, education, and support, to enhance its members' businesses and the flooring industry.

Floor Installation Association of North America (FIANA): An organization whose members are from Canada and the United States. Members must be manufacturers or distributors of floor installation products and/or flooring accessories.

Foodservice Consultants Society International (FCSI): Members are consultants with competencies that span the entire food service industry.

Forest Stewardship Council (FSC): An independent organization that protects forests for future generations. FSC sets standards under which forests and companies are certified. Its membership consists of three equally weighted chambers: environmental, economic, and social, to ensure these are balanced and the highest level of integrity. The members are in regular contact with their peers, customers, and suppliers.

Gas Technology Institute (GTI): A research, development and training organization that addresses energy and environmental issues.

GeoExchange® (GEO): A trade association, which promotes the manufacture, design, and installation of GeoExchange® systems. GEO supports its members' business objectives while promoting sustainable growth of the geothermal heat pump industry.

Geosynthetic Institute (GSI): A consortium of organizations that are interested in, and involved with, geosynthetics: geotextiles, geomembranes, geogrids, geonets, geocomposites, geosynthetic clay liners, geopipe, geocells, and geofoam.

Geothermal Energy Association (GEA): A trade association that supports the expanded use of geothermal energy and the development of geothermal resources for electrical power generation and direct-heat uses.

Geothermal Resources Council (GRC): An educational association that serves as a focal point for continuing professional development for its members through its outreach, information transfer, and education services.

Glass Association of North America (GANA): GANA places members in regular contact with their peers, customers, and suppliers. The association also provides members with educational programs, publications, networking opportunities, meetings, and conventions.

Green Roofs for Healthy Cities - North America Inc. (GRHC): GRHC promotes the industry throughout North America.

Gypsum Association (GA): A trade association that promotes the use of gypsum in the United States and Canada on behalf of its member companies.

Hardwood Plywood and Veneer Association (HPVA): A trade association that represents the interests of the hardwood plywood, hardwood veneer, and engineered hardwood flooring industries.

Hearth, Patio and Barbecue Association (HPBA): In 2002, the Hearth Products Association (HPA) merged with the Barbecue Industry Association (BIA) to form HPBA. It is an international trade association that includes manufacturers, retailers, distributors, manufacturers' representatives, service and installation firms, and other companies and individuals.

Heat Exchange Institute (HEI): A trade association that is committed to the technical advancement, promotion, and understanding of a broad range of utility and industrial-scale heat exchange and vacuum apparatus.

Heating, Air-conditioning and Refrigeration Distributers International (HARDI): The association's members market, distribute, and support heating, air-conditioning, and refrigeration equipment, parts, and supplies. HARDI Distributor members serve installation and service/replacement contractors in the residential, commercial, industrial, and institutional markets.

Heating, Refrigeration and Air Conditioning Institute of Canada (HRAI): A national association that represents heating, ventilation, air conditioning, and refrigeration (HVACR) manufacturers, wholesalers, and contractors, and provides information about HVACR to Canadians.

Hollow Metal Manufacturers Association (HMMA): The association promotes the advantages of hollow metal products. It is a division of the National Association of Architectural Metal Manufacturers (NAAMM).

Home Fire Sprinkler Coalition (HFSC): HFSC is a charitable organization. It provides independent, noncommercial information about residential fire sprinklers. HFSC offers educational material with details about installed home fire sprinkler systems, how they work, why they provide affordable protection, and answers to common myths and misconceptions about their operation.

Home Furnishings Independents Association (HFIA): A trade organization for member businesses.

Home Ventilating Institute (HVI): The institute certifies a wide range of home ventilating products that are manufactured by companies located throughout the world.

Hydraulics Institute (HI): An association of pump industry manufacturers, HI provides product standards and a forum for the exchange of industry information.

Illuminating Engineering Society (IES): The society is dedicated to promoting the art and science of quality lighting to its members, allied professional organizations, and the public.

INDA: See Association of the Nonwoven Fabrics Industry.

Independent Electrical Contractors (IEC): A national trade association for merit shop electrical and systems contractors. IEC develops and fostersa high level of quality and services within the industry.

Independent Office Products and Furniture Dealers Association (IOPFDA): A trade association for North American independent dealers of office products and office furniture. IOPFDA concentrates on providing independent dealers with information, tools, and knowledge needed to run their businesses.

Indiana Limestone Institute (ILI): A resource for architects, contractors, building owners, and others seeking information about the use of Indiana Limestone in construction.

Industrial Fabrics Association International (IFAI): A trade association comprised of member companies representing the global specialty fabrics marketplace.

Industrial Fasteners Institute (IFI): A globally recognized, North American focused, association that represents manufacturers of mechanical fasteners and formed parts, and suppliers to the industry.

Industrial Perforators Association (IPA): A highly specialized production resource for punching very large numbers of holes in a wide variety of materials. Hole sizes range from a few thousandths of an inch in diameter up to more than three inches, while the materials that can be perforated can be as thin as foil or as thick as 1' steel plate.

InfoComm International (InfoComm): A trade association that represents the professional audiovisual and information communications industries worldwide.

Innovative Pavement Research Foundation (IPRF): The foundation develops strategies and implements programs of research, technology advancement and transfer, and public education regarding concrete highways, streets, roads, and airports. IPRF sponsored by the American Concrete Pavement Association.

Institute of Electrical and Electronic Engineers (IEEE): IEEE fosters technological innovation and excellence for the benefit of humanity. IEEE is pronounced "Eye-triple-E."

Institute of Fire Engineers (IFE): The institute promotes, encourages, and improvesall aspects of the science and practice of fire engineering, fire prevention, and fire extinction.

Institute of Heating and Air Conditioning Industries, Inc. (IHACI): A trade association of contractors, manufacturers, distributors, utility firms, and related businesses actively engaged in the heating, ventilation, air conditioning, refrigeration, and sheet metal industries.

Institute of Inspection Cleaning and Restoration Certification (IICRC): IICRC identifies and promotes an international standard of care that establishes and maintains the health, safety, and welfare of the built environment. It is a certification and standard-setting organization for the inspection, cleaning, and restoration industries.

Institute of the Ironworking Industry (III): A labor-management trade association that protects, promotes, fosters, and advances the unionized erection industry.

Institute of Noise Control Engineering (INCE/USA): A professional organization whose primary purposeis to promote engineering solutions to environmental, product, machinery, industrial, and other noise problems. INCE/USA is a member society of the International Institute of Noise Control Engineering.

Institute of Transport Engineers (ITE): ITE is an international educational and scientific association of transportation professionals who are responsible for meeting mobility and safety needs. ITE facilitates research, planning, functional design, implementation, operation, policy development, and management for any mode of ground transportation.

Insulated Cable Engineers Association (ICEA): An association whose members are sponsored by many of North America's cable manufacturers. ICEA is dedicated to developing cable standards for the electric power, control, and telecommunications industries.

Insulating Glass Certification Council (IGCC): A trade organization for insulating glass unit manufacturers, consumers, specifiers, and others who are stakeholders this industry.

Insulating Glass Manufacturers Alliance (IGMA): A trade organization comprised of certified insulating glass manufacturers, their suppliers, and associates, window manufacturers, representatives from the architectural community, energy efficiency lobbies, code officials, and others interested in the design

and long-term performance of insulating glass units.

Insulation Contractors Association of America (ICAA): A trade organization that represents professional residential and commercial contractors.

Interior Design Educators Council®️ (IDEC): The council advances interior design education, scholarship, and service.

Interlocking Concrete Pavement Institute (ICPI): A trade association that represents the industry. Membership consists of interlocking paver manufacturers, design professionals, paver installation contractors, and suppliers of products and services related to the industry.

(United States) Internal Revenue Service (IRS): The organization that publishes all of the United States' tax laws and rules and forms that are associated with them.

International Association of Amusement Parks and Attractions (IAAPA): An international trade association for permanently situated amusement facilities worldwide.

International Association of Electrical Inspectors (IAEI): The association promotes safe products and safe installations. Members include electrical inspectors, testing agencies, standards organizations, manufacturers, distributors, installers, and contractors.

International Association of Lighting Designers (IALD): IALD promotes the advancement and recognition of independent, professional lighting designers.

InterNational Association of Lighting Management Companies (NALMCO®️): The association establishes and promotes professional standards for lighting management professionals through education, representation, the enhancement of professionalism, and distribution of information about the industry.

International Association of Plumbing and Mechanical Officials (IAPMO): IAPMO-provides code development assistance, education, plumbing and mechanical product testing and certification, building product evaluation, and a quality assurance program. The associationpublishes standards for mechanical products covering heating, ventilation, cooling, and refrigeration system products.Members contribute to the development of the Uniform Mechanical Code. The association also publishes standards covering products used in the recreational vehicle and manufactured housing industry called IAPMO Trailer Standards.

International Association of Professional Security Consultants (IAPSC): The association establishes and maintains standards for professionalism and ethical conduct in the industry.

The International Cast Polymer Alliance of the American Composites Manufacturers Association (ICPA): The alliance represents manufacturers, suppliers, fabricators, and installers of cast polymer composites, including cultured marble, cultured granite, cultured onyx, and solid surface kitchen and bath products.

International Code Council (ICC): ICC publishes the **International Building Code** that has been adopted throughout most of the United States;the **International Energy Conservation Code** (IECC) that sets forth compliance methods for energy-efficient construction of both residential and nonresidential construction; and the **International Residential Code** (IRC) that primarily covers low-rise residential construction.

International Concrete Repair Institute (ICRI): The institute serves to improve the quality of concrete restoration, repair, and protection, through education of, and communication among, the members and those who use their services.

International Cost Engineering Council (ICEC): ICEC promotes cooperation between national and multinational cost engineering, quantity surveying and project management organizations worldwide for their mutual well-being and that of their individual members.

International Council of Building Officials (ICBO): One of the three model code groups in the United States that has merged to form the International Code Council.

International Dark Sky Association (IDA): IDA is a recognized authority on light pollution. The association promotes "light what you need, when you need it." IDA works with manufacturers, planners, legislators, and citizens to provide energy efficient options that direct the light where you want it to go, not up into the sky.

International Door Association (IDA): The association provides programs and services to door and access systems dealers whose service products include residential and commercial doors and operators, and fire doors and gates.

InterNational Electrical Testing Association (NETA): The association serves the electrical testing industry by establishing standards, publishing specifications, accrediting independent, third-party testing companies, certifying test technicians, and promoting the professional services of its members. NETA also collects and disseminates information and data to the electrical industry and educates the public and end user about electrical acceptance and maintenance testing.

International Erosion Control Association (IECA): IECA is devoted to helping members solve the problems caused by erosion and its byproduct sediment.

International Facility Management Association (IFMA): IFMA is a widely recognized international association for facility management professionals. IFMA certifies facility managers, conducts research, and provides a wide range of educational courses and is a leading voice in the industry.

International Firestop Council (IFC): A trade association of manufacturers, distributors, and installers of passive fire protection materials and systems in North America. IFC promotes the technology of fire and smoke containment in modern building construction through research, education, and development of safety standards and code provisions.

International Furnishings and Design Association (IFDA): IFDA brings together professionals in the furnishing and design industries through networking, education, and professional development.

International Ground Source Heat Pump Association (IGSHPA): A not-for-profit organization that advances ground source heat pump (GSHP) technology on local, state, national, and international levels.

International Hurricane Protection Association (IHPA): IHPA is involved in all issues that affect the hurricane protection industry. IHPA brings together suppliers, manufacturers, contractors, engineers, architects, code writers, and government officials.

International Interior Design Association (IIDA): The association provides a forum to demonstrate design professionals' impact on the health, safety, well being, and virtual soul of the public. IIDA strives to balance good design and best business practices.

International Institute of Noise Control Engineering (I-INCE): I-INCE is a worldwide consortium of organizations concerned with noise control, acoustics, and vibration. The primary focus of the institute is on unwanted sounds and on vibrations producing such sounds when transduced.

International Masonry Institute (IMI): IMI offers training for craftworkers, professional education for masonry contractors, and free technical assistance to the design and construction communities. IMI is an alliance between the International Union of Bricklayers and Allied Craftworkers (BAC) and their signatory contractors.

International Organization for Standardization (ISO): The association certifies a company's ability to consistently manufacture quality products to ISO standards (ISO 9000, 9001, etc.).

International Parking Institute (IPI): The institute's members include professionals from cities, port authorities, civic centers, academic institutions, hospitals and health-care facilities, airports, corporate complexes, race tracks, transit and transportation agencies, retail, hospitality, and entertainment and sports centers, as well as architects, engineers, financial consultants, urban planners, and suppliers of equipment, products and services to the parking and transportation industries.

International Play Equipment Manufacturers Association (IPEMA): The association provides third-party product certification services for United States and Canadian public play equipment and public play surfacing materials in the U.S. IPEMA serves IPEMA-certified member companies, affiliated playground industry groups and anyone with an interest in playground equipment regulations.

International Sign Association (ISA): The association provides its members with information about current engineering research, EPA compliance issues, and other relevant matters.

International Society of Arboriculture (ISA): ISA promotes arboriculture fosters an awareness of the benefits of trees through research, technology, and education.

International Staple, Nail & Tool Association (ISANTA): An international organization of premier power fastening companies that are involved in the designing, manufacturing, and selling of pneumatic and cordless tools and the fasteners they drive.

International Safety Equipment Association (ISEA): An association dedicated to personal protective equipment and technologies. Its members design, manufacture, test, and use protective clothing and equipment.

International Slurry Surfacing Association (ISSA): An international trade association comprised of contractors, equipment manufacturers, public officials, research personnel, consulting engineers, and other industry professionals. ISSA promotes the concept of pavement preservation. ISSA provides members with information, technical assistance, and opportunities for networking and professional development.

International Surface Fabricators Association (ISFA): ISFA certifies member contractors who fabricate and install countertops.

International Tropical Timber Organization (ITTO): The organization's international membership is committed to achieving exports of tropical timber and timber products from sustainably managed sources. ITTO assists governments, industry, and communities to manage their forests and add value to their forest products, and to maintain and increase the transparency of the trade and access to international markets.

International Window Cleaning Association (IWCA): A trade organization that represents window cleaning companies to international, national, state, and local regulatory agencies and promotes the welfare of the industry through advocacy, education, training, and community involvement.

International Window Film Association (IWFA): A trade organization, IWFA partners with manufacturers and other members to increase consumer awareness and demand for all types of professionallyinstalled window film products.

International Wood Products Association (IWPA): The association advances international trade in wood products by providing education and leadership in business, environmental and public affairs.

International Zinc Association (IZA): An organization that is based in Brussels, Belgium. IZA is dedicated exclusively to the interests of zinc and its users by promoting such end uses as corrosion protection for steel and crop nutrition.

Intertek Testing Services—Warnock Hersey (ITS): Intertek tests to ensure products meet quality, health, environmental, safety, and social accountability standards.

Irrigation Association (IA): A trade organization for irrigation equipment and system

manufacturers, dealers, distributors, designers, consultants, contractors, and end users.

Kitchen Cabinet Manufacturers Association (KCMA): A voluntary trade association representing North American cabinet manufacturers and suppliers to the industry. KCMA promotes responsible environment practices in the industry.

Lighting Controls Association (LCA): The association is administered by the **National Electrical Manufacturers Association (NEMA)**. LCA is dedicated to educating the professional building design, construction, and management communities about the operation of automatic switching and dimming controls.

Lightning Protection Institute (LPI): The institute designs and develops information resources on complete lightning protection systems for consumers and designers. LPI also markets education products to members for use in the construction industry.

Lighting Research Center (LRC): A university-based research center (Rensselaer Polytechnic Institute) devoted to lighting. LRC offers graduate education in lighting, including one- and two-year master's programs and a Ph.D. program. LRC also provides training programs for government agencies, utilities, contractors, lighting designers, and other lighting professionals.

Lighting Safety Alliance (LSA): A not-for-profit corporation comprised of lightning protection manufacturers, distributors, and installers. LSA evaluates and responds to legislative, administrative, and regulatory issues facing the industry. Additionally, LSA serves as an informational clearinghouse for its membership.

Manufactured Housing Institute (MHI): A national trade organization that represents all segments of the factory-built housing industry. MHI provides industry research, promotion, education, and government relations programs.

Maple Flooring Manufacturers Association, Inc. (MFMA): A source of technical information about hard maple flooring.

MFMA publishes grade standards, guide specifications, floor care recommendations, and specifications for athletic flooring sealers and finishes.

Marble Institute of America (MIA): A source of information on standards of natural stone workmanship and practice and the suitable application of natural stone products. MIA promotes stone usage in the commercial and residential marketplaces. Membership includes natural stone producers, exporters/importers, distributors/wholesalers, fabricators, finishers, installers, and industry.

Mason Contractors Association of America (MCAA): A trade association that represents mason contractors. MCAA provides continuing education, promotes codes and standards, fosters a safe work environment, recruitsfuture tradespeople, and markets the benefits of masonry materials.

Masonry Advisory Council (MAC): MACis dedicated to providing the public with general and technical information about masonry design and detailing. MAC has an on-line technical library and industry directory.

Masonry Heater Association of North America (MHA): An association of builders, manufacturers, and retailers of masonry heaters. MHA promotes the industry, sponsors research and development, shapes regulations, standards, and codes, and provides information/educationto the public and its members.

Masonry Institute of America (MIA): A trade organization that is primarily supported by Southern California union signatory masonry contractors through a labor management contract between unions and contractors. MIA does not practice architecture or engineering or sell masonry building materials, but it is active in the development and distribution of seminars and publications on the use of masonry.

Masonry Veneer Manufacturers Association (MVMA): An incorporated trade association that represents the manufactured stone veneer industry's manufacturing companies and their suppliers. MVMA advances the growth of the manufactured masonry veneer industry through proactive technical, advocacy, and awareness efforts.

Master Painters Institute (MPI): The institute is dedicated to the establishment of quality standards and quality assurance in the painting and coating application industries.

Material Handling Industry of America (MHI): The institute was formed as the Material Handling Institute to advance the interests of material handling and logistics companies,systems and software manufacturers, consultants, systems integrators and simulators, and third-party logistics providers and publishers. It changed its name to Material Handling Industry of Americain the late 1980s; it continues to use both MHI as an abbreviation.

Materials Properties Council (MPC): The council was established in 1966 by the American Society of Mechanical Engineers, ASM International, ASTM and the Engineering Foundation. Industry, technical organizations, codes and standards developers, and government agencies support it.

Mechanical Contractors Association of America, Inc. (MCAA): An association of mechanical, plumbing, and service contractors. MCAA provides educational programs, a catalog of resources to help members manage and grow their businesses, periodicals, and other business services.

Medical Gas Professional Healthcare Organization, Inc. (MGPHO): The organization is made up of people and companies that are dedicated to advancing the safe design, manufacture, installation, maintenance and inspection/verification of medical gas and vacuum delivery systems through education.

Metal Building Contractors and Erectors Association (MBCEA): A trade organization that supports the advancement of metal building contractors, erectors, and the metal building industry.

Metal Building Institute (MBI): MBI was established to provide educational and training programs for metal building contractors, erectors, and students in construction. Members are manufacturers, contractors, and dealers in two distinct segments of the industry: permanent modular construction (PMC) and relocatable buildings (RB). Associate members are companies supplying building components, services, and financing.

Metal Building Manufacturers Association (MBMA): The association is instrumental in defining and promoting the interests of metal building systems manufacturers. MBMA sponsors research programs to improve the efficiency and quality of metal building systems, and to elevate the technology used to produce them.

Metal Construction Association (MCA): An organization of manufacturers and suppliers of metal products. MCA focuses on promoting the use of metal in the building envelope through marketing, education, and action on public policies that affect metal's use.

Metal Framing Manufacturers Association (MFMA): MFMA focuses is on the manufacture of ferrous and nonferrous metal framing (continuous slot metal channel systems) that consist of channels with in-turned lips and associated hardware for fastening to the channels at random points.

Metal Powder Industries Federation (MPIF): An association formed by the powder metallurgy (PM) industry to advance the interests of the metal powder producing and consuming industries.

Metal Roofing Alliance (MRA): MRA was founded to educate consumers about metal roofing. Membership includes paint companies, material suppliers, industry publications, and others.

Metals Service Center Institute (MSCI): A trade association that serves the industrial metals industry. MSCI provides data and education for operational efficiency, promotes industry advocacy, and creates a marketplace for efficient transactions, debate, discussion, and learning.

Mineral Insulation Manufacturers Association (MIMA): A source of information and advice on rock and glass mineral wool. MIMA promotes the benefits of mineral wool insulation and the contribution it makes to the energy efficiency of buildings and the comfort of their occupants.

Molding and Millwork Producers Association (MMPA): A trade association whose goals are to promote quality products, develop sources of supply, promote optimum use of raw materials, standardize products, and increase the domestic and foreign usage of molding and millwork products.

MSR Lumber Producers Council: A not-for profit corporation of the State of Washington. The council represents the interests of machine stress rated lumber producers in the manufacturing, marketing, promotion, utilization, and technical aspects of machine stress rated lumber.

National Air Duct Cleaners Association (NADCA): An association of companies engaged in the cleaning of HVAC systems. It promotes source removal as the only acceptable method of cleaning and establishes industry standards for the association. NADCA also refers to itself as the **HVAC Inspection, Maintenance, and Restoration Association**.

National Air Filtration Association (NAFA): A trade association whose members are from air filter and component manufacturers, sales, and service companies, and HVAC and indoor air quality professionals.

National Association of Architectural Metal Manufacturers (NAAMM): The association represents architectural metal products for building construction. NAAMM currently has six operating divisions: Architectural Metal Products (AMP), Detention Equipment Manufacturers Association (DEMA), Expanded Metal Lath Association (EMLA), Expanded Metal Manufacturers Association

(EMMA), Hollow Metal Manufacturers Association (HMMA), and Metal Bar Grating (MBG).

National Alarm Association of America (NAAA): A trade association that serves as a forum for alarm dealers and as a filter and provider of training programs and manuals for the education of installers, service personnel, and system designers.

National American Wholesale Lumber Association (NAWLA): NAWLA members include every aspect of the lumber industry from planting seedlings to selling building materials and wood. NAWLA is an advocate for wood's role in a green economy and a healthy planet.

National Association of Electrical Distributors (NAED): An organization that serves the electrical distribution channel. NAED provides its members with tools, information, and assistance to help them financially and improve the electrical distribution channel.

National Association of Elevator Contractors (NAEC): A trade association that serves the interests of independent elevator contractors and suppliers of products and services. NAEC promotes safe and reliable elevator, escalator, and short-range transportation and promotes in the management of member companies.

National Association of Home Builders (NAHB): A trade association that promotes housing as a national priority. NAHB's various groups analyze policy issues, work toward improving the housing finance system, analyze and forecast and consumer trends, and, generally, are involved in all aspects of the housing industry.

National Association of Pipe Coating Applicators (NAPCA): The association represents plant-applied pipe coating companies worldwide and promotes standardized protective coating practices. NAPCA includes as associate and international associate members firms that service or have a common industry interest in the pipe coating industry.

National Association of Reinforcing Steel Contractors (NARSC): NARSC furthers the interests of reinforcing steel and post-tensioning contractors throughout the United States and Canada. NARSC is in partnership with the **International Association of Bridge, Structural, Ornamental, and Reinforcing Iron Workers** and its local unions.

National Association for Surface Finishers (NASF): NASF represents the surface coatings industry. NASF advances an environmentally and economically sustainable future for the finishing industry and promote the role of surface technology in the global manufacturing value chain. The **American Electroplating and Surface Finishing Foundation (AESF)** is part of the NASF. It focuses exclusively on technical, educational, and research programs.

National Association of Sewer Service Companies (NASSCO): The association researches, evaluates, and develops new methods to train and educate its members about the importance of properly rehabilitated underground utilities.

National Association of State Fire Marshals (NASFM): NASFM is comprised of many senior fire officials in the United States. State Fire Marshals' responsibilities vary from state to state, but Marshals tend to be responsible for fire safety code adoption and enforcement, fire and arson investigation, fire incident data reporting and analysis, public education, and advising Governors and State Legislatures on fire protection. Some State Fire Marshals are responsible for fire fighter training, hazardous materials incident responses, wildland fires, and the regulation of natural gas and other pipelines.

National Association of Waterproofing and Structural Repair Contractors (NAWSRC): A professional trade association that serves the public and waterproofing, structure, and foundation repair industries.

National Bureau of Standards (NBS): An organization, founded in 1901, whose function is to establish and maintain standards for units of measurements.

National Clay Pipe Institute (NCPI): The institute does research and development of clay pipe technology. NCPI provides assistance in design, training, and evaluation of systems. Forensic analysis is performed when necessary.

National Coil Coating Association (NCCA): A trade organization that is dedicated to the growth of coil coated products. NCCA's member companies provide the coil coating service and are leading manufacturers and suppliers of metal, coatings, chemicals, and equipment.

National Collegiate Athletic Association (NCAA): NCAA is made up of three membership classifications: Divisions I, II, and III. Each division creates its own rules governing personnel, amateurism, recruiting, eligibility, benefits, financial aid, and playing and practice seasons. NCAA rules set the criteria for college/university athletic fields.

National Concrete Masonry Association (NCMA): A national trade association that represents the concrete masonry industry. NCMA is involved in technical, research, marketing, government relations, and communications activities. NCMA offers technical services and design aids through publications, computer programs, slide presentations, and technical training.

National Corrugated Steel Pipe Association (NCSPA): NCSPA promotes public policy relating to the use of corrugated steel drainage structures. The association collects and distributes technical information, assists in the formulation of specifications and designs, encourages greater knowledge of corrugated steel pipe's benefits and uses among college engineering students, and conducts seminars about the product and its application among designers.

National Council of Acoustical Consultants (NCAC): An international organization that supports the acoustical profession. NCAC is comprised of professional firms that specialize in acoustical consulting. To qualify for membership, the firm's principals who practice acoustical consulting must be full members of either the **Acoustical Society of America (ASA)** or **Institute of Noise Control Engineering (INCE).**

National Council of Examiners for Engineering and Surveying (NCEES): The council develops, administers, and scores the examinations used for engineering and surveying licensure in the United States.

National Council on Qualifications for the Lighting Professions (NCQLP): An organization that serves and protects the public through lighting practice. NCQLP establishes the education, experience, and examination requirements for baseline certification across the lighting professions.

National Council on Radiation Protection and Measurements (NCRP): NCRP was chartered by the US Congress as the National Council on Radiation Protection and Measurements. NCRP strives to prevent the occurrence of clinically significant radiation-induced deterministic effects of radiation and limit the risk of stochastic effects in exposed persons to an amount that is acceptable in relation to the benefits to the individual and to society.

The National Earthquake Hazards Reduction Program (NEHRP): Congress established NEHRP in 1977, directing that four federal agencies coordinate their complementary activities to implement and maintain the program. These agencies are **FEMA**, the **National Institute of Standards and Technology**, the **National Science Foundation**, and the **U.S. Geological Survey**. NEHRP leads the federal government's efforts to reduce the fatalities, injuries, and property losses caused by earthquakes.

National Electrical Contractors Association (NECA): The association represents electrical contractors from firms of all sizes performing a range of services. Most NECA contractors qualify as small businesses; however, many large, multinational companies are also members of the association.

National Electrical Manufacturers Association (NEMA): An association of electrical equipment manufacturers. Its member companies manufacture products such as power transmission and distribution equipment, lighting systems, factory automation and control systems, and medical diagnostic imaging systems.

National Elevator Industry, Inc. (NEII®): A national trade association of the building transportation industry. NEII promotes safety in building transportation, promotes laws and regulations that permit the introduction of safe, innovative technology, and endorses adoption of current model codes.

National Environmental Balancing Bureau (NEBB): Members perform testing, adjusting and balancing of heating, ventilating and air-conditioning systems, commission and retro-commission building systems, execute sound and vibration testing, building envelope testing, test and certify laboratory fume hoods, and electronic and biological cleanrooms.

National Fenestration Rating Council (NFRC): An organization that administers a uniform, independent rating and labeling system for the energy performance of windows, doors, skylights, and attachment products. NFRC is an **American National Standard Institute (ANSI)** accredited standards developer (ASD) that develops and administers comparative energy and related rating programs for fenestration products.

National Fire Protection Association (NFPA): An international organization that is dedicated to reducing the burden of fire on peoples' quality of life by proposing codes and standards, research, and education on fire-related issues.

National Fire Sprinkler Association (NFSA): NFSA strives to protect lives and property from fire by promoting the wide-spread acceptance of the fire sprinkler concept. NFSA also provides engineering and training to its members.

National Fireplace Institute® (NFI): A professional certification division of the **Hearth,**

Patio & Barbecue Education Foundation (HPBEF) that is an educational organization for the hearth industry.

National Floor Safety Institute (NFSI): A not-for-profit organization whose mission is to aid in the prevention of slips, trips-and-falls through education, research, and standards development.

National Frame Building Association (NFBA): A trade association that promotes the interests of the post-frame construction industry and its members professionals throughout the United States. NFBA's members are primarily post-frame builders, suppliers, manufacturers, building material dealers, code and design professionals, and structural engineers.

National Glass Association (NGA): A trade association that serves the architectural glass, automotive glass, and window and door industries. NGA provides education and training programs that pertain to technical skills, management practices, and quality workmanship.

National Guild of Professional Paperhangers, Inc. (NGPP): The guild is dedicated to superior craftsmanship in the hanging of every type of wallpaper, including the hanging of historic wallpapers, scenic murals, digital murals, silk papers, bamboo, grasscloth, English pulp papers, as well as traditional fabric-backed and paper-backed vinyl materials.

National Hardwood Lumber Association (NHLA): The association was the creator and is the keeper of the North American hardwood lumber grading rules. NHLA provides technical short courses to on-site company training by an NHLA National Inspector; NHLA conducts an Inspector Training School.

National Home Furnishings Association (NHFA): An organization devoted specifically to the needs and interests of home furnishings retailers.

National Institute of Standards and Technology (NIST): A non-regulatory federal agency within the US Department of Commerce. NIST's mission is to promote US innovation and industrial competitiveness by advancing measurement science, standards, and technology in ways that enhance economic security and improve our quality of life.

National Institute of Steel Detailing (NISD): An international association that advocates, promotes, and serves the interests of the steel detailing industry. NISD membership is offered to steel detailing firms and associated companies and individuals.

National Insulation Association (NIA): A trade association that represents both the merit (open shop) and union contractors, distributors, laminators, fabricators, and manufacturers that provide thermal insulation, insulation accessories, and components to the commercial, mechanical, and industrial markets throughout the nation.

National Kitchen and Bath Association (NKBA): Atrade association whose membership includes distributors, retailers, remodelers, manufacturers, fabricators, installers, designers, and other professionals. NKBA's certification program emphasizes continuing education and career development.

National Lighting Bureau (NLB): An organization founded to educate lighting decision-makers about the benefits of High-Benefit Lighting®. Professional societies, trade associations, manufacturers, utilities, and agencies of the federal government sponsor the NLB.

National Lightning Safety Institute (NLSI): The institute advocates a risk management lightning protection strategy. NLSI consults to identify vulnerabilities and to organize defenses and teaches and trains personnel.

National Lumber Grades Authority (NLGA): The authority is responsible for the establishment, issuance, publication, amendment, and interpretation of Canadian lumber grading rules and standards.

National Onsite Wastewater Recycling Association (NOWRA): NOWRA's principal purpose is to educate and serve its members and the public by promoting federal, state, and local policy, improving standards of practice, and advancing public recognition of areas that have no wastewater infrastructure.

National Ornamental and Miscellaneous Metals Association (NOMMA): A trade association of the ornamental and miscellaneous metalworking industry. NOMMA's members fabricate everything from railings and driveway gates to structural and industrial products. NOMMA provides continuing education to its members.

(U.S.) National Park Service (NPS): The National Park Service is a bureau of the U.S. Department of the Interior and is led by a director nominated by the President and confirmed by the U.S. Senate. The National Register of Historic Places, a comprehensive list of districts, sites, buildings, structures, and objects of national, regional, state, and local significance in American history, architecture, archeology, engineering, and culture is kept by the NPS under authority of the National Historic Preservation Act of 1966.

National Parking Association (NPA): NPA offers education, networking opportunities, advocacy, products, and services to its members.

National Pest Management Association (NPMA): A trade organization that is committed to the protection of public health, food, and property. It supports its members both technically and business wise.

National Precast Concrete Association (NPCA): An international trade association that provides technical and production information and education and networking opportunities to its members.

National Program for Playground Safety (NPPS): The program's mission is to help the public create safe and developmentally appropriate play environments for children. NPPS provides research, training, and development of play areas.

National Ready Mixed Concrete Association (NRMCA): The association represents and serves the ready mixed concrete industry through leadership, promotion, education, and partnering; it is an advocate for the industry.

The National Restaurant Association (NRA): The NRA is a foodservice trade association that represents and advocates for foodservice industry interests with state, local, and national policymakers. It has no connection to the National Rifle association.

National Roof Deck Contractors Association (NRDCA): A trade association of contractors, manufacturers, and associates who provide, install, or support the application of engineered composite roof deck substrates in the commercial roof top market. NRDCA develops guidelines, procedures, and educational programs.

National Roofing Contractors Association (NRCA): NRCA provides a forum for roofing contractors, manufacturers, and suppliers of programs and projects that contribute to the continual improvement of the roofing industry.

National Sanitation Foundation (NSF): A nationally recognized testing laboratory that certifies plumbing products to meet the standard to which they were created.

National Slate Association (NSA): NSA develops and disseminates technical information, standards, and educational resources on the materials and methods used in the manufacture, design, and construction of slate roofs and associated flashing systems. NSA's members include roofing contractors, slate quarries and distributors, architects, architectural conservators, roofing consultants, craftspeople, building owners, facilities managers, manufacturers, educators, and others concerned with the manufacture, design, construction, and care of natural slate roofs.

National Society of Professional Engineers (NSPE): In partnership with the State Societies, NSPE is an organization of licensed Professional Engineers (PEs) and Engineer Interns. NSPE provides education, licensure advocacy, leadership training, multidisciplinary networking, and outreach.

National Stone, Sand and Gravel Association (NSSGA): In 2001, the **National Stone Association** and **National Aggregates Association** merged to become the NSSGA. NSSGA represents the crushed stone, sand, and gravel—or construction aggregates—industries.

National Storm Shelter Association (NSSA): The association fosters quality by recognizing and distinguishing the shelter producers (and their products) who meet the association's standard of quality. Members include producers, installers, associate members, professionals, media partners and corporation, and individual sponsors.

National Sunroom Association (NSA): Aprofessional organization whose members include manufacturers, design professionals, and material suppliers and installers. It informs consumers, remodelers, and building officials about sunrooms, patio rooms, solariums, and conservatories, and insures that products are designed and manufactured to high standards, and are safety compliant, energy efficient, and environmentally friendly.

National Systems Contractors Association (NSCA): A trade association that represents the low-voltage industry, including systems contractors/integrators, product manufacturers, consultants, sales representatives, architects, specifying engineers, and other allied professionals.

National Terrazzo and Mosaic Association, Inc. (NTMA® Inc.): NTMA® Inc. is a trade association that establishes national standards for terrazzo floor and wall systems. NTMA also provides specifications, color plates, and other information to architects and designers. Membership is limited to terrazzo contractors.

National Tile Contractors Association (NTCA): A trade association that serves the tile and stone industry. NTCA includes manufacturers, distributors, contractors, architects, designers, and builders in its programs.

National Tile Roofing Manufacturers Association (NTRMA): NTRMA has changed its name to **Tile Roofing Institute.** TRI trains roofing installers, inspectors, and industry professionals on proper, code approved methods to installing concrete and clay tile roofs. See **Tile Roofing Institute**.

National Utility Contractors Association (NUCA): A trade association dedicated to the underground utility construction industry. NUCA represents contractors, suppliers, and manufacturers involved in water, sewer, gas, electric, telecommunications, site work, and other segments of the industry.

National Wood Flooring Association (NWFA): A trade association that represents the hardwood flooring industry: manufacturers, distributors, retailers, and installers. NWFA provides training and resources to wood flooring.

North American Association of Floor Covering Distributors (NAFCD): NAFCD promotes the wholesale distribution of floor coverings and provides members with resources for enhancing performance as industry suppliers. NAFCD represents its members through involvement and by providing education through its development programs and conferences.

North American Association of Food Equipment Manufacturers (NAFEM): A trade association of foodservice equipment and supplies manufacturers. NAFEM offers educational opportunities to its members.

North American Deck and Railing Association (NADRA): A trade association of the deck and railing building industry in North America. NADRA is made up of deck builders, manufacturers, dealers/distributors, wholesalers, retailers, and service providers to the industry.

North American Fiberboard Association (NAFA): A trade organization of manufacturers of cellulosic fiberboard products that are used for residential and commercial construction, commercial products, and packaging.

North American Insulation Manufacturers Association (NAIMA): Member companies manufacture fiber glass, rock wool, and slag wool insulations for residential, commercial, and industrial uses. NAIMA is a resource on energy-efficiency, sustainable performance, and the application and safety of fiberglass, rock wool, and slag wool insulation products.

North American Laminate Flooring Association (NALFA): A trade organization dedicated to the laminate flooring industry. NALFA is an accredited ANSI standards developing organization and publishes testing and performance criteria.

Northeastern Lumber Manufacturers Association (NELMA): NELMA is the rules writing agency for Eastern White Pine lumber and the grading authority for Eastern Spruce, Balsam Fir, Spruce-Pine-Fir (SPFs) grouping, and commercial eastern softwood lumber species. NELMA is an agency for export wood packaging certification. NELMA does marketing for the wood products industry in the Northeast.

Northwest Wall and Ceiling Bureau (NWCB): An international trade association for the wall and ceiling industry. NWCB's membership consists of subcontractors, manufacturers, suppliers, labor organizations, and other professionals in the wall and ceiling industry in the United States and Canada. NWCB works with architects, code bodies, designers, and construction professionals on the design and application of the industry's products.

Occupational Safety and Health Administration (OSHA): A federally funded agency in the Department of Labor that develops job safety and health standards. The states have parallel organizations, e.g., CAL/OSHA.

Painting and Decorating Contractors of America (PDCA): A trade association of painting and decorating contractors. PDCA offers contractor members education programs, attendance at local networking meetings, and use of PDCA industry standards.

Petroleum Equipment Institute (PEI): A trade association whose members manufacture, distribute, and service petroleum marketing and liquid-handling equipment. Members include manufacturers, sellers, and install-ers of equipment used in service stations, terminals, bulk plants, fuel, oil and gasoline delivery, and similar petroleum marketing operations.

Pile Driving Contractors Association (PDCA): The association advocates the use of driven piles for deep foundations and earth retention systems. PDCA promotes the use of driven pile solutions, supports educational programs for engineers and contractors, and encourages and supports research.

Pipe Fabrication Institute® (PFI): PFI promotes standards of excellence in the pipe fabrication industry. PFI initiates research and studies, proposes and maintains standards and technical bulletins, and organizes meet-ings and other technical exchanges within the industry.

Planet Professional Landcare Network (Planet): A national trade association that represents landscape industry professionals. PLANET was created when the Associated Landscape Contractors of America (ALCA) and the Professional Lawn Care Association of America (PLCAA) merged to create a sin-gle national trade association of lawn care and landscape professionals.

Plastic Lumber Trade Association (PLTA): A trade association that promotes standard-ized testing, standards of quality, recycling of plastics, and, generally, works in support of the plastic lumber industry.

Plastic Pipe and Fittings Association (PPFA): A trade association that promotes and defends plastic piping systems governed by construction codes. PPFA provides users with information to design, specify, and install plastic piping systems. PPFA promotes an understanding thermoplastic piping products as they pertain to the environment.

Plastic Pipe Institute (PPI): A trade associa-tion that represents all segments of the plas-tics piping industry. PPI promotes the use of plastics piping for water and gas distribution, sewer and wastewater, oil and gas production,

industrial and mining uses, power and com-munications, duct and irrigation.

Plumbing and Drainage Institute (PDI): PDI is an association of manufactur-ers of engineered plumbing products. PDI's members and licensees make products such as: floor drains, roof drains, sanitary floor drains, cleanouts, water hammer arresters, backwater valves, grease interceptors, fixture supports, and other drainage specialties.

Plumbing, Heating, Cooling Contractors Association-National Association (PHCC-National Association): A trade organization for plumbing, heating, and cooling profes-sionals. PHCC promotes advancement, edu-cation, and training.

Plumbing Manufacturers International (PMI): A trade association of manufacturers of plumbing industry products such as potable water supply system components, fixture fit-tings, waste fixture fittings, fixtures, flushing devices, sanitary drainage system compo-nents, and plumbing appliances.

Polyisocyanurate Insulation Manufacturers Association (PIMA): A national trade asso-ciation that represents polyiso insulation (a widely used insulation product) manufactur-ers and suppliers to the polyiso industry.

Porcelain Enamel Institute, Inc. (PEI): A trade organization that is dedicated to advanc-ing the porcelain enameling plants and sup-pliers of porcelain enameling materials and equipment.

Portland Cement Association (PCA): PCA-represents cement companies in the United States and Canada. The association conducts programs of market development, education, research, technical services, and government affairs on behalf of its members.

Post-Tensioning Institute (PTI): The insti-tute is dedicated to expanding post-tensioning applications through marketing, education, research, teamwork, and code development. PTI advances the quality, safety, efficiency, profitability, and use of post-tensioning sys-tems. Members include post-tensioning materials fabricators, manufacturers of pre-stressing materials, and companies supplying materials, services, and equipment used in post-tensioned construction.

Powder Actuated Tool Manufacturers' Institute, Inc. (PATMI): An association of manufacturers of powder actuated fastening systems. PATMI stresses training, certification, and safety awareness. (Powder actuated fasteners are used to bond various construction materials together, such as wood and concrete, or steel and concrete.)

Powder Coating Institute (PCI): A trade organization that represents the North American powder coating industry, promotes powder coating technology, and communicates the benefits of powder coating to manufacturers, consumers, and government.

Power and Communication Contractors Association (PCCA): A national trade association for companies that construct electric power facilities, including transmission and distribution lines, and substations and telephone, fiber optic, and cable television systems.Other areas of members' business activities include directional drilling, local area, and premises wiring, water and sewer utilities, gas and oil pipelines.

Precast/Prestressed Concrete Institute (PCI): PCI fosters understanding and use of precast and prestressed concrete.

Professional Awning Manufacturers Association of the Industrial Fabrics Association International (PAMA): An international trade association that is committed to supporting the awning industry. Membership is open to companies that are a member of **Industrial Fabrics Association International (IFAI)** and manufacture and/or supply material to the awning industry.

Professional Grounds Management Society (PGMS): Comprised of professional grounds managers and other people interested in the grounds management industry, PGMS promotes the dissemination of educational materials and information relevant to the execution of grounds management functions.

Professional Women in Construction (PWC): Actually, the National Association of Professional Women in Construction (PWC) is an organization that is committed to advancing professional, entrepreneurial, and managerial opportunities for women and other "non-traditional" populations in construction and related industries.

Quality Assurance Association (QAA): An organization of professionals from wholesalers, retailers, manufacturers, laboratories, government, suppliers, and others.

Rack Manufacturers Institute (RMI): An association that advances the standards, quality, and safety of industrial steel storage rack systems.

Radiant Professionals Alliance (RPA): The alliance promotes radiant heating on behalf of its members: contractors and dealers, manufacturers, distributers, designers, and others with an interest in the industry.

Reflective Insulation Manufacturers Association International (RIMA-I): A trade association that represents the reflective insulation, radiant barrier, and low-e reflective coatings industries.

Reflective Roof Coatings Institute (RRCI): A trade organization whose members are coatings manufacturers, raw materials suppliers, applicators, and industry consultants.

Research Council on Structural Connections (RCSC): A non-profit organization that is comprised of leading experts in the fields of structural steel connection design, engineering, fabrication, erection, and bolting.

Resilient Floor Covering Institute (RFCI): An industry trade association of resilient flooring manufacturers and suppliers of raw materials, additives, and sundry flooring products for the North American market.

Restoration Industry Association (RIA): A trade association for cleaning and restoration professionals. RIA provides leadership and promoting best practices through advocacy, standards and professional qualifications for the restoration industry.

Retail Contractors Association (RCA): A national organization of retail contractors who have united to provide a solid foundation of ethics, quality, and professionalism within the retail construction industry.

Roof Coatings Manufacturers Association (RCMA): An association of roof coating manufacturers and affiliates that advances, promotes, and expands the international market for roof coatings through education, technical advancement, and advocacy of industry issues.

Roof Consultants Institute, Inc. (RCI, Inc.): An international association of professional consultants, architects, and engineers who specialize in the specification and design of roofing, waterproofing, and exterior wall systems.

Roofing Industry Educational Association (RIEI): A roofing industry education resource. In 2000, RIEI merged with the **National Roofing Contractors Association (NRCA)**, and together they provide a variety of courses and training seminars.

Rubber Manufacturers Association (RMA): A national trade association for tire manufacturers that make tires in the United States.

Scaffolding, Shoring, and Forming Institute (SSFI): A trade association of manufacturers of scaffolding, suspended scaffolding, shoring, forming, planks, platforms, and related components. The institute focuses on the technical aspects and safe use of scope products.

Safety Glazing Certification Council (SGCC): A not-for-profit corporation comprised of manufacturers of safety glazing products, building code officials, and others concerned with public safety. SGCC maintains a program that provides for independent, third-party certification of safety glazing materials.

Scientific Certification Systems (SCS): A third-party provider of certification, auditing and testing services of forest management operations and wood product manufacturers. SCS is accredited by the Forest Stewardship Council (FSC). SCS evaluates forests according to the FSC Principles and Criteria for Forest Stewardship.

Scientific Equipment and Furniture Association (SEFA): A trade organization of lab designers and manufacturers of laboratory furniture. Members include companies whose work is principally in this industry.

Screen Manufacturers Association (SMA): A trade organization that represents the window and door screen industry. SMA participates in creating standards that meet various governmental and code entity requirements.

Sealant Waterproofing and Restoration Institute (SWR Institute): An international trade association that represents the commercial sealant, waterproofing, and restoration construction industry. SWR Institute's members include contractors, manufacturers, and design professionals in the industry.

Security Industry Association (SIA): A trade organization that advocates pro-industry policies and legislation, produces market research, creates open industry standards, provides education and training, and opens global market opportunities.

Sheet Metal and Air Conditioning Contractors National Association (SMACNA): An international trade association that promotes quality and excellence in sheet metal and air conditioning technology and construction.

Siding and Window Dealers Association of Canada (SAWDAC): Members are dealers and contractors of products and installations. Members must commit to SAWDAC's code of ethics and sign a 5-year workmanship guarantee statement.

Single Ply Roofing Industry (SPRI): SPRI represents sheet membrane and related component suppliers in the commercial roofing industry. SPRI serves as a commercial roofing components and system information resource for building owners, architects, engineers, designers, contractors, and maintenance personnel.

Slag Cement Association (SCA): A trade organization that represents companies that produce and ship slag cement (ground granulated blast furnace slag) in the United States.

Society of American Military Engineers (SAME): A professional military engineering association in the United States. SAME unites architecture, engineering, construction (A/E/C), facility management and environmental firms and individuals in the public and private sectors to prepare for and overcome natural and manmade disasters, and to improve security at home and abroad.

Society of Automotive Engineers International (SAE): An organization comprised of engineers and related technical experts in the aerospace, automotive, and commercial-vehicle industries. SAE International's core competencies are life-long learning and voluntary consensus standards development. The society provides standards, e.g., a thread size used on nuts and bolts but not pipe connections.

Society of Fire Protection Engineers (SFPE): A professional society that represents people who practice fire protection engineering. SFPE advances the science and practice of fire protection engineering and its allied fields, maintains a high ethical standard among its members, and fosters fire protection engineering education. SFPE supports the development of the annual Professional Engineer licensing exam in fire protection and the grading of those exams under the auspices of the National Council of Examiners for Engineering and Surveying.

Society of Glass and Ceramic Decorated Products (SGCDpro): The society is dedicated to the interests of the decorating, manufacturing, and marketing of glass and ceramics and associated businesses. SGCDpro provides information about business opportunities, and technical, educational, and regulatory information to its members. SGCDpro promotes the use of socially and environmentally responsible business practices.

Society of the Plastics Industry (SPI): SPI promotes business development, fosters the sustainable growth of plastics in the global marketplace, provides industry representation in the public policy arena and communicates the industry's contributions to society and the benefits of its products.

Society for Protective Coatings (SSPC): A professional technical society whose primary objective is to improve the technology and practice of prolonging the life of steel and concrete structures through the use of protective coatings. SSPC was originally **Steel Structures Painting Council**.

Society of Wood Science and Technology (SWST): The society develops and maintains knowledge that is specific to the science and technology of wood and other lignocellulosic materials, encourages the use of this knowledge, promotes policies and procedures that are aimed at wise and responsible use of wood and other lignocellulosic materials (any of several closely related substances constituting the essential part of woody cell walls of plants and consisting of cellulose intimately associated with lignin), assures high standards for members, fosters educational programs at all levels of wood science, other lignocellulosic materials and their technologies, and furthers the quality of such programs, and represents the wood science and technology profession in public policy development.

Soil and Water Conservation Society (SWCS): A scientific and educational organization that serves as an advocate for conservation professionals and for science-based conservation practice, programs, and policy. SWCS members include researchers, administrators, planners, policymakers, technical advisors, teachers, students, farmers, and ranchers.

Solar Energy Industries Association® (SEIA): A trade organization of the solar energy industry. SEIA's member companies research, manufacture, distribute, finance, and build solar projects domestically and abroad.

Solar Rating and Certification™ (SRCC™): An organization whose primary purpose is to provide authoritative performance ratings, certifications, and standards for solar thermal products.

Solar and Sustainable Energy Society of Canada Inc. (SESCI): The society advances the use and awareness of solar and sustainable energy in Canada.

Southern Building Code Congress: An active participant in the International Code Council; SBCCI helps to develop and maintain the ICC model building codes. It also continues to provide prints of its original codes that were published prior to merging with the ICC.

Stairway Manufacturers' Association (SMA): A trade association that serves stairway manufacturers. SMA provides code officials, design professionals, builders, and the stair industry and community schools with technical expertise through publications, seminars, and direct classroom experiences.

Southern Pine Inspection Bureau (SPIB®): SPIB® is comprised of family owned and publicly traded companies that place SPIB®'s logo on their products.

Specialty Steel Industry of North America (SSINA): A trade association that represents producers of specialty steel in North America. SSINA members produce products, including bar, rod, wire, angles, plate, sheet, and strip, in stainless steel and other specialty steels.

Spiral Duct Manufacturers Association (SPIDA): A trade organization that promotes the use of round duct, spiral duct (spiral pipe), and flat oval duct, supports testing and research of round duct and spiral pipe, and provides manufacturers with specialized information.

Spray Polyurethane Foam Alliance (SPFA): A trade organization that also serves as an educational and technical resource for the spray polyurethane foam industry.

Stairway Manufacturers' Association (SMA): A trade association that serves stairway manufacturers. SMA provides code officials, design professionals, builders, and the

stair industry and community schools with technical expertise through publications, seminars, and direct classroom experiences.

Steel Deck Institute (SDI): SDI keeps designers and constructors up to date on deck design and construction and to provide information about the SDI and the member companies.

Steel Erectors Association of America (SEAA): An organization that sets uniform standards among the many steel erectors and helps promote safety in the erection industry.

Steel Framing Alliance (SFI): SFI works to expand market share in the commercial and residential construction markets, with an emphasis on growth potential of structural cold formed steel (CFS) framing in the midrise sector.

Steel Joist Institute (SJI): An organization of active joist manufacturers that cooperates with business and government agencies to establish steel joist standards. The institute does continuing research of industry products to maintain the integrity of these products.

Steel Manufacturers Association (SMA): Most of SMA's members are electric arc furnace steel producers, or "minimills," that use a feedstock almost entirely composed of recycled steel scrap to make new steel. SMA's associate member companies provide equipment, supplies, and services to steel companies.

Steel Recycling Institute (SRI): An industry association that promotes and sustains the recycling of all steel products. SRI educates the solid waste industry, government, business, and ultimately the consumer about the benefits of steel's infinite recycling cycle.

Steel Stud Manufacturers Association (SSMA): A trade organization of the steel framing manufacturing industry. SSMA represents member firms engaged in the manufacture, marketing, and sale of cold-formed steel framing. Members include contractors, distributors, design professionals, code officials, and standards organizations.

Steel Tube Institute (STI): A trade organization that promotes and markets steel tubing. STI's active membership consists of producers of steel tube and pipe and its associate membership consists of companies that supply raw materials, equipment, and support services.

Steel Window Institute (SWI): A trade organization of manufacturers of windows made from either solid or formed sections of steel, and such related products as casings, trim, mechanical operators, screens, and moldings when manufactured and sold by members of the industry for use in conjunction with windows. SWI provides the public with general and technical information concerning the industry's products.

Structural Building Components Association (SBCA): An international trade association that represents manufacturers of structural building components. Membership also includes truss plate and original equipment manufacturers, computer engineering and other service companies, lumber mills, inspection bureaus, lumber brokers and distributors, builders, and professional individuals in the fields of engineering, marketing, and management.

Structural Insulated Panel Association (SIPA): A trade association that represents manufacturers, suppliers, dealer/distributors, design professionals, and builders committed to providing quality structural insulated panels for all segments of the construction industry.

Structural Stability Research Council (SSRC): SSRC offers guidance to specification writers and practicing engineers by developing both simplified and refined calculation procedures for the solution of stability problems and assessing the limitations of these procedures. SSRC is made up of representatives from government agencies, international organizations, private corporations, educational institutions, representatives of consulting firms, members-at-large selected from universities and design offices, and corresponding members from various countries.

Stucco Manufacturers Association (SMA): A trade association that is comprised of manufacturers of stucco in North America and their related suppliers.

Submersible Wastewater Pump Association (SWPA): A national trade association that represents and serves the manufacturers of submersible pumps for municipal and industrial wastewater applications. Regular members are manufacturers of submersible wastewater pumps for municipal and industrial. Component members are manufacturers of component parts and accessory products for submersible pumps and pumping systems. Associate members are non-manufacturers providing services related to industry products and who provide services to users of industry products.

Sustainable Forestry Initiative (SFI): An independent organization that promotes responsible forest management.

Telecommunications Industry Association (TIA): TIA is accredited by the **American National Standards Institute (ANSI)** to develop voluntary, consensus-based industry standards for a variety of information and communications technology (ICT) segments. TIA operates 12 engineering committees, that develop guidelines for private radio equipment, cellular towers, data terminals, satellites, telephone terminal equipment, accessibility, voice over internet protocol (VoIP) equipment, structured cabling, data centers, mobile device communications, multimedia multicast, vehicular telematics, healthcare ICT, smart device communications, smart utility mesh networks, and sustainable/environmental communications technologies.

Terrazzo Tile and Marble Association of Canada (TTMAC): A trade organization whose overall objective is to raise the profile of the industry within the marketplace, and the respective standards in order to achieve that goal.

Testing, Adjusting, and Balancing Bureau (TABB): The **Sheet Metal and Air Conditioning Contractors National Association (SMACNA)** endorses TABB's procedures' that includes a strict code of conduct for technicians performing hands on TAB work; TABB certification is encouraged for inclusion in project specifications.

The Masonry Society (TMS): Members are design engineers, architects, builders, researchers, educators, building officials, material suppliers, manufacturers, and other interested people. TMS gathers and disseminates technical information through its committees, publications, codes and standards, slide sets, videotapes, computer software, newsletter, refereed journal, educational programs, professors' workshop, scholarships, certification programs, disaster investigation team, and conferences.

Tile Contractors Association of America (TCAA): TCAA is a trade organization that promotes the tile industry. TCAA stresses professionalism, reliability, skilled craftsmanship, and technical performance in the industry. TCAA is in partnership with **International Masonry Institute (IMI)** to develop the "Trowel of Excellence."

Tile Council of North America, Inc. (TCA): A trade association that represents North American manufacturers of ceramic tile, tile installation materials, tile equipment, raw materials, and other tile-related products.

Tile Roofing Institute (TRI): Formerly named the **National Tile Roofing Manufacturers Association (NTRMA)**. Today, TRI trains roofing installers, inspectors, and industry professionals on proper, code approved methods to installing concrete and clay tile roofs. TRI is dedicated to growing the tile roofing market through technical expertise, training, and building awareness for the many benefits of tile.

Tilt-Up Concrete Association (TCA): A trade association that strives to expand and improve the use of Tilt-Up as a building system. TCA provides education and resources that enhance quality and performance.

Tree Care Industry Association (TCIA): A trade association of commercial tree care firms and affiliated companies. TCIA develops safety and education programs, standards of tree care practice, and management information for arboriculture firms around the world.

Tropical Forest Foundation (TFF): An international educational institution committed to advancing environmental stewardship, economic prosperity, and social responsibility through sustainable forest management (SFM).

Truss Plate Institute (TPI): A trade organization of the plate truss industry. TPI establishes methods of design and construction for trusses in accordance with the **American National Standards Institute's** accredited consensus procedures for coordination and development of American National Standards. TPI also provides a quality assurance inspection program and contributes its expertise in other technical areas

Turfgrass Producers International (TPI): A worldwide association committed to the advancement of the turfgrass sod industry. TPI provides education to members, product users, and various green industry and government entities. TPI encourages the use of turfgrass sod worldwide.

Underwriters Laboratories Inc., (UL): UL provides safety-related certification, validation, testing, inspection, auditing, advising, and training services to a wide range of clients, including manufacturers, retailers, policymakers, regulators, service companies, and consumers. A UL label is often displayed on packaging. For example, a UL label on the packaging of shingles indicates the level of fire and wind resistance of asphalt roofing.

Underwriters Laboratories of Canada (ULC): The Canadian equivalent of UL.

Uni-Bell PVC Pipe Association: An association of PVC pipe and fittings' manufacturers that was created to promote and provide technical support in the use of PVC pipe and fittings. Uni-bell writes installation guidelines that are used as a reference regarding installation failures.

United Lighting Protection Association (ULPA): A trade association of lighting protection manufacturers, engineers, contractors, and technicians.

United States Green Building Council (USGBC): USGBC is best known for its development of the Leadership in Energy and Environmental Design (LEED) green building rating systems and its annual Green build International Conference and Expo.

The council works toward its mission of market transformation through its LEED program, educational offerings, a nationwide network of chapters and affiliates, the annual Greenbuild International Conference & Expo, and advocacy in support of green buildings and communities.

United States Sign Council (USSC): An educational resource for the sign industry. USSC is open to any person or firm concerned with the advancement of the sign industry.

Valve Manufacturers Association of America (VMA): VMA represents the interests of North American manufacturers of valves and actuators.

Vibration Isolation and Seismic Control Manufacturers Association (VISCMA): A professional organization consisting of partnerships, companies, and corporations, that engage in the seismic restraint, vibration isolation, or noise isolation industry.

Vinyl Siding Institute, Inc., (VSI): A trade association for manufacturers of vinyl and other polymeric siding and suppliers to the industry. VSI is the sponsor of the VSI Product Certification Program and the VSI Certified Installer Program.

Warnock Hersey: A Canadian lab that is universally recognized in Canada. WH certifies products to CSA standards.

Wallcovering Association (WA): A trade association that represents wallcoverings manufacturers, distributors, and suppliers to the industry.

Walnut Council: The Walnut Council promotes sustainable forest management and utilization of American black walnut and other high-quality fine hardwoods. Its purpose is to assist in the technical transfer of forest research to field applications, help build and maintain better markets for wood products and nut crops, and to promote sustainable forest management, conservation, reforestation, and utilization of American black walnut (Juglansnigra) and other high-quality fine hardwoods.

Water Environmental Federation (WEF): WEF is similar to **AWWA**. Its members are people associated with water waste including all types of materials.

Water Quality Association (WQA): An international trade association that represents the residential, commercial, and industrial water treatment industry. WQA provides information to the industry, educators, and professionals. WQA has a laboratory for product testing, and is a communicator to the public.

Water and Sewer Distributors of America (WASDA): Comprised of distributors and manufacturers of waterworks and wastewater products, WASDA promotes the waterworks/wastewater products distribution industry.

West Coast Lumber Inspection Bureau (WCLIB): A service corporation for the benefit and protection of buyers, sellers, and consumers of softwood lumber. WCLIB's primary objective is the development and maintenance of uniform lumber standards.

Western Hardwood Association (WHA): A trade organization that promotes and markets western hardwoods. WHA provides education for stakeholders on sustainable and environmentally responsible resource management.

Western Red Cedar Lumber Association (WRCLA): A Vancouver-based association that represents producers of Western Red Cedar lumber products in Washington, Oregon, and British Columbia (Canada). WRCLA operates customer service programs throughout the United States and Canada to support its members' cedar products with information, education, and quality standards.

Western States Clay Products Association (WSCPA): A trade association of brick manufacturers in the Western United States. WSCPA develops technical information for enhancing quality use of clay brick construction with special attention directed to the seismic performance of clay brick.

Western States Roofing Contractors Association (WSRCA): Members are roofing contractors and vendors. WSRCA serves both members and consumers alike.

Western Wall and Ceiling Contractors Association (WWCCA): An organization that represents subcontractors and affiliates who have joined to promote the installation of quality construction. Members employ only union trained craftspeople.

Western Wood Preservers Institute (WWPI): A trade organization that represents the preserved wood products industry throughout western North America. WWPI's primary activity areas are regulatory and market outreach programs.

Western Wood Products Association (WWPA): A trade association that represents softwood lumber manufacturers in the western United States including Alaska. WWPA's mills produce lumber from Western softwood species, including Douglas Fir, Western Larch, Western Hemlock, True Firs, Engelmann Spruce, Ponderosa Pine, Lodgepole Pine, Sugar Pine, Idaho White Pine, Western/Inland Red Cedar, and Incense Cedar.

Window Coverings Association of America (WCAA): A national trade association whose members are from the retail window coverings industry and its dealers, decorators, designers, and workrooms. WCAA provides members educational opportunities, encourages a code of ethics for fair practices, and works for the betterment of the retail window coverings industry.

Window and Door Manufacturers Association (WDMA): Formerly the **National Wood Window and Door Association**, this trade organization has established many standards related to wood window and door products.

Wire Association International, Inc. (WAI, Inc.): A technical society for wire and cable industry professionals, WAI promotes, collects, and disseminates technical, manufacturing, and general business information to the ferrous, nonferrous, electrical, fiber optic, and fastener segments of the wire and cable industry.

Women Contractors Association (WCA): An organization composed of women owners and decision-making executives within the construction industry. WCA provide networking opportunities and information specific to the construction industry and the small business owner.

Wood Component Manufacturers Association (WCMA): The association represents manufacturers of dimension and wood component products that supply components for cabinetry, furniture, architectural millwork, closets, flooring, staircases, building materials, and decorative/specialty wood products made from hardwoods, softwoods, and a variety of engineered wood materials.

Wood Products Manufacturers Association (WPMA): The association provides members with information resources and services. WPMA acts as a clearinghouse for solving problems of mutual concern and assists members in controlling costs.

Woodwork Institute (WI): WI provides standards and quality control programs for the architectural millwork industry.

World Floor Covering Association (WFCA): A source of information on all types of flooring, including carpet, hardwood flooring, laminate floors, ceramic tile, area rugs, natural stone, cork, bamboo, and vinyl flooring.

World Waterpark Association (WWA): A trade organization that provides information on waterpark-business topics, industry trends, and other matters that are relevant to the industry.

Woven Wire Products Association (WWPA): A trade organization that represents manufactures of diamond woven wire mesh products for institutional, industrial, and architectural applications. Membership is open to persons, partnerships, or corporations involved in the manufacturing of diamond woven wire mesh products and those firms that supply materials and services to the manufactures of diamond woven wire mesh products.

THIS TITLE IS FROM OUR CIVIL ENGINEERING, CONSTRUCTION MANAGEMENT, FACILITIES MANAGEMENT, ENVIRONMENTAL ENGINEERING AND GENERAL REFERENCE COLLECTION. OTHERS TITLES YOU MIGHT BE INTERESTED IN...

Construction Crew Supervision: 50 Take Charge Leadership
Techniques & Light Construction Glossary
By Karl F. Schmid

Facilities Management: Managing Maintenance for Buildings and Facilities
By Joel D. Levitt

BACnet: The Global Standard for Building Automation and Control Networks
By Michael H. Newman

Construction Estimating: A Step-by-Step Guide to a Successful Estimate
By Karl F. Schmid

Professional Expression: To Organize, Write, and Manage
for Technical Communication,
By Morris

The Engineering Language: A Consolidation of the Words and Their Definitions
By Ronald Hanifan

Social Media for Engineers and Scientists
By Jon DiPietro

Building Inspection Manual: A Guide for Building Professionals for Maintenance,
Safety, and Assessment (coming in Dec 2014)
By Karl F. Schmid

Announcing Digital Content Crafted by Librarians

Momentum Press offers digital content as authoritative treatments of advanced engineering topics, by leaders in their fields. Hosted on ebrary, MP provides practitioners, researchers, faculty and students in engineering, science and industry with innovative electronic content in sensors and controls engineering, advanced energy engineering, manufacturing, and materials science. **Momentum Press offers library-friendly terms:**

- perpetual access for a one-time fee
- no subscriptions or access fees required
- unlimited concurrent usage permitted
- downloadable PDFs provided
- free MARC records included
- free trials

The **Momentum Press** digital library is very affordable, with no obligation to buy in future years.

For more information, please visit **www.momentumpress.net/library** or to set up a trial in the US, please contact **mpsales@globalepress.com**.

www.ingramcontent.com/pod-product-compliance
Lightning Source LLC
Chambersburg PA
CBHW062010190326
41458CB00009B/3029